OXFORD LOGIC GUIDES: 23

General Editors

ANGUS MACINTYRE
JOHN SHEPHERDSON
DANA SCOTT

OXFORD LOGIC GUIDES

1. Jane Bridge: *Beginning model theory: the completeness theorem and some consequences*
2. Michael Dummett: *Elements of intuitionism*
3. A. S. Troelstra: *Choise sequences: a chapter of intuitionistic mathematics*
4. J. L. Bell: *Boolean-valued models and independence proofs in set theory* (1st edition)
5. Krister Segerberg: *Classical propositional operators: and exercise in the foundations of logic*
6. G. C. Smith: *The Boole–De Morgan correspondence 1842–1864*
7. Alec Fisher: *Formal number theory and computability: a work book*
8. Anand Pillay: *An introduction to stability theory*
9. H. E. Rose: *Subrecursion: functions and hierarchies*
10. Michael Hallett: *Cantorian set theory and limitation of size*
11. R. Mansfield and G. Weitkamp: *Recursive aspects of descriptive set theory*
12. J. L. Bell: *Boolean-valued models and independence proofs in set theory* (2nd edition)
13. Melvin Fitting: *Computability theory: semantics and logic programming*
14. J. L. Bell: *Toposes and local set theories: an introduction*
15. Richard Kaye: *Models of Peano arithmetic*
16. J. Chapman and F. Rowbottom: *Relative category theory and geometric morphisms*
17. S. Shapiro: *Foundations without foundationalism*
18. J. P. Cleave: *A study of logics*
19. R. M. Smullyan: *Gödel's incompleteness theorems*
20. T. E. Forster: *Set theory with a universal set*
21. C. McLarty: *Elementary categories, elementary toposes*
22. R. M. Smullyan: *Recursion theory for metamathematics*
23. Peter Clote and Jan Krajíček: *Arithmetic, proof theory, and computational complexity*

Arithmetic, proof theory, and computational complexity

Edited by

PETER CLOTE
Department of Computer Science
Boston College, Massachusetts

and

JAN KRAJÍČEK
Mathematical Institute
Czechoslovak Academy of Sciences, Prague

CLARENDON PRESS • OXFORD
1993

Oxford University Press, Walton Street, Oxford OX2 6DP
Oxford New York Toronto
Delhi Bombay Calcutta Madras Karachi
Kuala Lumpur Singapore Hong Kong Tokyo
Nairobi Dar es Salaam Cape Town
Melbourne Auckland Madrid
and associated companies in
Berlin Ibadan

Oxford is a trade mark of Oxford University Press

Published in the United States
by Oxford University Press Inc., New York

A catalogue record for this book is available from the British Library

Library of Congress Cataloging in Publication Data
Arithmetic, proof theory, and computational complexity / edited by Peter Clote and Jan Krajíček.
(Oxford Logic Guides; 23)
1. Proof theory. 2. Computational complexity. I. Clote, Peter.
II. Krajíček, Jan. III. Series
QA9. 54.A75 1992 511.3—dc20 92–41659
ISBN 0–19–853690–9

Typeset by the authors using LaTeX
Printed in Great Britain by
Bookcraft (Bath) Ltd
Midsomer Norton, Avon

Preface

On 20 March 1956, K. Gödel wrote a letter to John von Neumann, while the latter was showing temporary signs of improvement under treatment in the hospital (von Neumann died of bone cancer on 8 February 1957). Gödel wrote:[1]

> "One can obviously easily construct a Turing machine, which for every formula F in first order predicate logic and every natural number n, decides if there is a proof of F of length n (length = number of symbols). Let $\psi(F, n)$ be the number of steps taken by the machine and let $\phi(n) = \max_F \psi(F, n)$. The question is how fast $\phi(n)$ grows for an optimal machine. One can show that $\phi(n) \geq k \cdot n$. If there were really a machine with $\phi(n) \sim k \cdot n$ (or even $\sim k \cdot n^2$), this would have consequences of the greatest importance."

Indeed, if there were such a machine as hoped by Gödel, it would follow that non-deterministic polynomial time is equal to deterministic polynomial time. The following reduction of the NP-complete set SAT of all satisfiable propositional logic formulas to Gödel's problem is due to S.A. Cook.[2] Suppose that A is a propositional logic formula with atoms $p_1, \ldots, p_n$. Let Q be a new monadic predicate symbol and $x_1, \ldots, x_n$ be variables of first order predicate logic. Define A' to be the result of replacing p_i by $Q(x_i)$ in A. Let the formula B be $(\exists x)(\exists y)(Q(x) \wedge \neg Q(y)) \rightarrow (\exists x_1) \ldots (\exists x_n) A'$. Then one can verify that A is satisfiable if and only if B is valid, and moreover, B admits a short proof in first order predicate logic.[3]

The theme of this book addresses problems related to Gödel's question: efficiency of propositional proof systems, length of proofs in propositional and predicate logic, theories of bounded arithmetic, and computational complexity theory. Research in this area has been supported from January 1990 until December 1991 by the National Science Foundation # INT-9014569 and by the Czechoslovak Academy of Sciences for the group consisting of S.R. Buss, P. Clote, P. Hájek, J. Krajíček, P. Pudlák and G. Takeuti. This funding allowed the organization of a workshop from June 1 – June 30, 1990 at the University of California at San Diego and from June 15 – July 15, 1991 at the Mathematical Insitute of the Czechoslovak Academy of Sciences and the Charles University in Prague. During each

---

[1]Transcription of the entire original German text and an English translation of Gödel's letter is at the end of the preface.

[2]Email correspondence 20 September 1991.

[3]Interested readers should consult the article of J. Hartmanis [34].

workshop, there was a 3 day international conference. The speakers at the San Diego conference from June 25-27, 1990 were: M. Ajtai, B. Allen, P. Clote, S. Feferman, F. Ferriera, P. Hájek, G. Mancivi, J. Krajíček, A. Nerode, P. Pudlák, T. Slaman, R. Sommer, G. Takeuti and A. Woods.[4] The speakers at the conference held in Prague from July 2- 5, 1991 were: Z. Adamowicz, A. Berarducci, S.R. Buss, P. Clote, S.A. Cook, C. Dimitracopoulos, J. Franco, A. Goerdt, V. Harnik, R. Kaye, J. Krajíček, A. Kučera, Z. Ratajczyk, J.-P. Ressayre, S. Riis, A. Sochor, G. Takeuti, R. Verbrugge, A. Wilkie, A. Woods and T. Yamakami. Many of the papers of this volume were presented at the Prague meeting. A number of informal talks were given by these and other authors during the month-long workshops in San Diego and Prague.

[4]Additional funding in July 1990 is gratefully acknowledged from UCSD for this conference.

Princeton, 20./III. 1956

Lieber Herr v. Neumann!
Ich habe mit grösstem Bedauern von Ihrer Erkrankung gehört. Die Nachricht kam mir ganz unerwartet. Morgenstern hatte mir zwar schon im Sommer von einem Schwächeanfall erzählt, den Sie einmal hatten, aber er meinte damals, dass dem keine grössere Bedeutung beizumessen sei. Wie ich höre, haben Sie sich in den letzten Monaten einer radikalen Behandlung unterzogen u. ich freue mich, dass diese den gewünschten Erfolg hatte u. es Ihnen jetzt besser geht. Ich hoffe u. wünsche Ihnen, dass Ihr Zustand sich bald noch weiter bessert u. dass die neuesten Errungenschaften der Medizin, wenn möglich, zu einer vollständigen Heilung führen mögen.
Da Sie sich, wie ich höre, jetzt kräftiger fühlen, möchte ich mir erlauben, Ihnen über ein mathematisches Problem zu schreiben, über das mich Ihre Ansicht sehr interessieren würde: Man kann offenbar leicht eine Turingmaschine konstruieren, welche von jeder Formel F des engeren Funktionenkalküls u. jeder natürl. Zahl n zu entscheiden gestattet, ob F einen Beweis der Länge n hat [Länge = Anzahl der Symbole]. Sei $\Psi(F, n)$ die Anzahl der Schritte, die die Maschine dazu benötigt u. sei $\phi(n) = \max_F \Psi(F, n)$. Die Frage ist, wie rasch $\phi(n)$ für eine optimale Maschine wächst. Man kann zeigen $\phi(n) \geq k \cdot n$. Wenn es wirklich eine Maschine mit $\phi(n) \sim k \cdot n$ (oder auch nur $\sim k \cdot n^2$) gäbe, hätte das Folgerungen von der grössten Tragweite. Es würde nämlich offenbar bedeuten, dass man trotz der Unlösbarkeit des Entscheidungsproblems die Denkarbeit des Mathematikers bei ja-oder-nein Fragen vollständig[5] durch Maschinen ersetzen könnte. Man müsste ja bloss das n so gross wählen, dass, wenn die Maschine kein Resultat liefert, es auch keinen Sinn hat über das Problem nachzudenken. Nun scheint es mir aber durchaus im Bereich der Möglichkeit zu liegen, dass $\phi(n)$ so langsam wächst. Denn 1.) scheint $\phi(n) \geq k \cdot n$ die einzige Abschätzung zu sein, die man durch eine Verallgemeinerung des Beweises für die Unlösbarkeit des Entscheidungsproblems erhalten kann; 2.) bedeutet ja $\phi(n) \sim k \cdot n$ (oder $\sim k \cdot n^2$) bloss, dass die Anzahl der Schritte gegenüber dem blossen Probieren von N auf $\log N$ (oder $(\log N)^2$) verringert werden kann. So starke Verringerungen kommen aber bei anderen finiten Problemen durchaus vor, z.B. bei der Berechnung eines quadratischen Restsymbols durch wiederholte Anwendung des Reziprozitätsgesetztes. Es wäre interessant zu wissen, wie es damit z.B. bei der Feststellung, ob eine Zahl Primzahl ist, steht u. wie stark im allgemeinen bei finiten kombinatorischen Problemen die Anzahl der Schritte gegenüber dem blossen Probieren verringert werden kann.
Ich weiss nicht, ob Sie gehört haben, dass "Post's problem" (ob es unter den Problemen $(\exists y)\phi(y, x)$ mit rekursivem ϕ Grade der Unlösbarkeit gibt) von einem ganz jungen Mann namens Richard Friedberg in positivem Sinn gelöst wurde. Die Lösung ist sehr elegant. Leider will Friedberg nicht Mathematik, sondern

[5] abgesehen von der Aufstellung der Axiome

Medizin studieren (scheinbar unter dem Einfluss seines Vaters).
Was halten Sie übrigens von den Bestrebungen, die Analysis auf die verzweigte Typentheorie zu begründen, die neuerdings wieder in Schwung gekommen sind? Es ist Ihnen wahrscheinlich bekannt, das Paul Lorenzen dabei bis zur Theorie des Lebesgueschen Masses vorgedrungen ist. Aber ich glaube, dass in wichtigen Teilen der Analysis nicht eliminierbare imprädikative Schlussweisen vorkommen.
Ich würde mich sehr freuen, von Ihnen persönlich etwas zu hören; u. bitte lassen Sie es mich wissen, wenn ich irgend etwas für Sie tun kann.
Mit besten Grüssen u. Wünschen, auch an Ihre Frau Gemahlin.

Ihr sehr ergebener

Kurt Gödel

Ich gratuliere Ihnen bestens zu der Auszeichnung, die Ihnen von der amerik. Regierung verliehen wurde.

The following is an approximate translation.

Princeton, 20 March 1956

Dear Mr. von Neumann:

With the greatest sorrow I have learned of your illness. The news came to me as quite unexpected. Morgenstern already last summer told me of a bout of weakness you once had, but at that time he thought that this was not of any greater significance. As I hear, in the last months you have undergone a radical treatment and I am happy that this treatment was successful as desired, and that you are now doing better. I hope and wish for you that your condition will soon improve even more and that the newest medical discoveries, if possible, will lead to a complete recovery.

Since you now, as I hear, are feeling stronger, I would like to allow myself to write you about a mathematical problem, of which your opinion would very much interest me: One can obviously easily construct a Turing machine, which for every formula F in first order predicate logic and every natural number n, allows one to decide if there is a proof of F of length n (length = number of symbols). Let $\psi(F, n)$ be the number of steps the machine requires for this and let $\phi(n) = \max_F \psi(F, n)$. The question is how fast $\phi(n)$ grows for an optimal machine. One can show that $\phi(n) \geq k \cdot n$. If there really were a machine with $\phi(n) \sim k \cdot n$ (or even $\sim k \cdot n^2$), this would have consequences of the greatest importance [Tragweite]. Namely, it would obviously mean that in spite of the undecidability of the Entscheidungsproblem,[6] the mental work of a mathematician concerning Yes-or-No questions could be completely[7] replaced by a machine. After all, one would simply have to choose the natural number n so large that when the machine does not deliver a result, it makes no sense to think more about the problem. Now

[6] decision problem for predicate calculus

[7] except for the setting up of axioms

it seems to me, however, to be completely within the realm of possibility that $\phi(n)$ grows that slowly. Since (1) it seems that $\phi(n) \geq k \cdot n$ is the only estimation which one can obtain by a generalization of the proof of the undecidability of the Entscheidungsproblem;[8] and (2) after all $\phi(n) \sim k \cdot n$ (or $\sim k \cdot n^2$) only means that the number of steps as opposed to trial and error can be reduced from N to $\log N$ (or $(\log N)^2$). However, such strong reductions appear in other finite problems, for example in the computation of the quadratic residue symbol using repeated application of the law of reciprocity. It would be interesting to know, for instance, the situation concerning the determination of primality of a number and how strongly in general the number of steps in finite combinatorial problems can be reduced with respect to simple exhaustive search.

I do not know if you have heard that "Post's problem", whether there are [intermediate] degrees of unsolvability among problems of the form $(\exists y)\phi(y, x)$, where ϕ is recursive, has been solved in the positive sense by a very young man by the name of Richard Friedberg. The solution is very elegant. Unfortunately, Friedberg does not intend to study mathematics, but rather medicine (apparently under the influence of his father).

By the way, what do you think of the attempts to build the foundations of analysis on ramified type theory, which have recently gained momentum? You are probably aware that Paul Lorenzen has pushed ahead with this approach to the theory of Lebesgue measure. However, I believe that in important parts of analysis non-eliminable impredicative proof methods do appear.

I would be very happy to hear something from you personally. Please let me know if there is something that I can do for you.

With my best greetings and wishes, as well to your wife,

Sincerely yours,

Kurt Gödel

P.S. I heartily congratulate you on the award that the American government has given to you.

Acknowledgements. We would like to thank the Institute for Advanced Study, as copyright holder, for kindly authorizing us to quote Gödel's letter in full. As well, we thank the Library of Congress for furnishing us with a copy of Gödel's letter. According to Professor John Dawson,[9] the existence of the letter among John von Neumann's papers at the Library of Congress was brought to the attention of the Gödel Editorial Committee by Gerhard Heise in a letter dated 27 May 1988.

We would like to thank the following persons for comments on the preface and on the transcription and translation of Gödel's letter: S.R. Buss,

[8] On 21 March 1992, S.R. Buss presented a proof that $\phi(n) \geq k \cdot n$ in the annual meeting of the Association for Symbolic Logic held at Duke University.

[9] Email correspondence 12 November 1991.

S.A. Cook, E. Dahlhaus, A. Goerdt, A. Hoene, M. Kusch, A. Urquhart. Despite the incorporation (or non-inclusion) of various suggestions of the persons just mentioned, the editors must take full responsibility for any and all remaining errors in the translation. Thanks to J. Dawson for remarks about the finding of Gödel's letter. Thanks as well to J. Hartmanis for kindly furnishing a copy of his article [34]. Finally, we would like to thank Boston College as well as the M.I.T. Laboratory for Computer Science, for providing workstation, email and telnet facilities which made the preparation and typesetting in LATEXof this book possible.

Peter Clote, Boston
Jan Krajíček, Prague
May 30, 1992

Contents

All papers in this volume have been refereed except G. Kreisel's postscript to the article of M. Baaz and P. Pudlák, and J.-P. Ressayre's extended abstract. We take this opportunity to thank all the referees for their work.

Dedication

We would like to dedicate this book to Professor Gaisi Takeuti on the event of his 65-th birthday. For more than two generations now, Professor Takeuti has been one of the most important leaders in the area of proof theory, in addition to his lasting contributions in other parts of mathematics and mathematical physics. Since 1984, Professor Takeuti has been working in the area of bounded arithmetic, efficiency of propositional proof systems and computational complexity theory.

Open problems

In the following sections, we distinguish between *fundamental problems* and *other problems*, the former which currently seem to be beyond existent techniques, but form the motivation for the latter which seem more accessible. This problem list was largely drawn up with the help of all participants during a problem session held during the conference held in Prague. We thank S. Buss and P. Pudlák for extensive comments on preliminary versions of this list. Attributions of problems are given to the best of our knowledge.

Peter Clote
Jan Krajíček

1 Bounded Arithmetic

The study of theories of bounded arithmetic was largely begun by work of R. Parikh [51] and the ground-breaking work of J. Paris and A. Wilkie [54, 73, 58, 57, 72, 25]. Especially with the work of S. Buss [10], proof theoretic problems of weak theories of bounded arithmetic were seen to be connected with problems of computational complexity theory. For instance, a recent result of Krajíček, P. Pudlák and G. Takeuti [47] shows that if the theory S_2 of [10] (or equivalently, the theory $I\Delta_0 + \Omega_1$ of [73]) is finitely axiomatizable then the polynomial time hierarchy collapses. See work of A. Wilkie [72] and A. Woods [75] for discussion of a result related to the question of collapse of the $\Delta_0^{\mathbf{N}}$ hierarchy, or equivalently of the linear time hierarchy.

1.1 Fundamental problems in bounded arithmetic

A. (Paris, Wilkie) Is $I\Delta_0$ or $I\Delta_0 + \Omega_1$ finitely axiomatizable? Buss [10] defined a finite conservative extension S_2 of $I\Delta_0 + \Omega_1$ and its subtheories $S_2^1 \subseteq S_2^2 \subseteq \ldots$ whose union is S_2. Each S_2^i is finitely axiomatizable and hence $I\Delta_0+\Omega_1$ is finitely axiomatizable iff $S_2^i = S_2$ for some i. By the theorem of [47] the answer is negative assuming that the polynomial hierarchy does not collapse.

B. Which theorems of finite combinatorics, number theory and complexity theory can be proved in $I\Delta_0$ and $I\Delta_0 + \Omega_1$? In particular, which counting techniques are available in these theories? Here are four specific questions.

(a) (Wilkie) Does $I\Delta_0$ prove the existence of infinitely many primes? Note that in [55], it is shown that $I\Delta_0 + \Omega_1$ (or equivalently S_2) proves the existence of infinitely many primes.

(b) (Wilkie, Wilmers) Does S_2^1 prove the little Fermat theorem ($a^{p-1} \equiv 1 \bmod p$, for p prime not dividing a) where the latter is formalized as a first order statement by using repeated squaring? A.J. Wilkie and G.M. Wilmers have observed[10] that an affirmative answer to this question implies the existence of a polynomial time integer factorization algorithm, which appears most unlikely. In [69], G. Takeuti introduces second order systems $U_2^{n,w}$ which are essentially equivalent to the S_2^n, and then shows that for all n, $U_2^{n,w}$ does not prove Fermat's little theorem, and $U_2^{n,w}$ does not prove Wilson's theorem $(p-1)! \equiv -1 \bmod p$, for p prime, where both are expressed using a (second order) relation symbol.

(c) (Macintyre) Does $I\Delta_0$ or $I\Delta_0 + \Omega_1$ prove the pigeonhole principle for bounded functions, i.e. that there is no Δ_0 - definable injection of $n+1$ into n? By Paris, Wilkie and Woods [55] $I\Delta_0 + \Omega_1$ proves the weak pigeonhole principle (with $2n$ in place of $n+1$). The answer is negative, if the language is extended by a binary predicate R (a graph of the function) and induction for all $\Delta_0(R)$-formulas is allowed, see [3, 8, 48, 60]. It is also open whether $I\Delta_0$ or $I\Delta_0(R)$ prove the weak pigeonhole principle.

(d) (Paris, Wilkie) Is the Matijasevič theorem for bounded formulas provable in $I\Delta_0$? That is: Is every bounded formula equivalent in $I\Delta_0$ to an existential formula? A.J. Wilkie observed in [72] that an affirmative answer to this question would imply that $NP = co-NP$.

1.2 Other problems in bounded arithmetic

The following problems seem to us to be not intractable and should give information about the above main problems.

1. (Paris,Wilkie) Show that the theory $I\Delta_0 + \Omega_1$ is not Π_1^0-conservative over $I\Delta_0$, possibly assuming some hypothesis from complexity theory (such as whether the linear time hierarchy is properly contained in the polynomial time hierarchy). Show the same result at least for the case where a new predicate symbol has been added to the language.

[10]Unpublished correspondence from spring 1984.

Krajíček and Wilkie (independent, unpublished proofs) showed that the Π_1^0-conservativity of $I\Delta_0 + \Omega_1$ over $I\Delta_0$ would imply that $I\Delta_0$ is not finitely axiomatizable.

2. Characterize the functions which are Σ_i^b-definable in the theory T_2^i. A more difficult problem is to determine those functions which are Σ_j^b-definable in T_2^i, for $j \leq i$. This seems to be related to question whether T_2^i is somehow conservative over S_2^i. Characterizations of functions Σ_{i+1}^b-definable in S_2^{i+1}, T_2^i and S_2^i are known by [10, 16, 39]. Also, Buss and Krajíček [13] showed that $T_2^i(\alpha)$ is not $\forall\Sigma_i^b(\alpha)$-conservative over $S_2^i(\alpha)$, and they characterized functions Σ_1^b-definable in T_2^1 in terms of polynomial local search problems.

3. (Krajíček, Pudlák, Takeuti) If S_2 proves that the polynomial time hierarchy (PH) collapses then S_2 is finitely axiomatizable, and by [47], if S_2 is finitely axiomatizable then PH collapses. In fact, $T_2^i = S_2^{i+1}$ implies that PH collapses. Buss found recently another proof of this theorem which does formalize in S_2. In particular, he showed that $T_2^i = S_2^{i+1}$ implies that T_2^i proves that PH collapses and $T_2^i = S_2$. Formalize in S_2 the original proof of [47]. In particular, does S_2 prove the "tournament principle", which states that given any tournament of n players (a directed graph $G = (V, E)$ with $|V| = n$, such that for all distinct i, j from V, $(i, j) \in E$ iff $(j, i) \notin E$) there exists a "dominating" set D of size $\log(n)$, where the dominating set D satisfies: For all $i \in V \setminus D$, there exists $j \in D$ such that $(j, i) \in E$. For the formalization one needs a hypergraph version of this principle. Note that Pudlák [63] proved in $I\Delta_0 + \Omega_1$ a related Ramsey theorem.

4. Does $S_2^i = T_2^i$, some i, imply that PH collapses? Krajíček [39] showed that the assumption implies that $L^{\Sigma_i^p} = \Delta_{i+1}^p$ but it is unknown whether the latter equality imply the collapse of PH. Also, does $S_2^i = T_2^i$ imply that $S_2^i = S_2$?

5. (Wilkie) Show that $I\Delta_0$ is not finitely axiomatized, provided that the linear time hierarchy does not collapse. By [47] this is true with the polynomial time hierarchy in place of the linear time hierarchy. Note that it is an open problem whether these two hierarchies coincide.

6. (Ajtai) Show with respect to $I\Delta_0$ and "counting" modulo q, one cannot "count" modulo p, for distinct primes p, q. Specifically, let $\Theta_p(R, S, X)$ say

   (i) $0 \in X$
   (ii) R is a partition of X into classes of size p
   (iii) S is a partition of $X \setminus \{0\}$ into classes of size p.

Then show that the theory

$$I\Delta_0(R,S,X)+\Theta_p(R,S,X)+\{\neg\Theta_q(\overline{R},\overline{S},\overline{X}) : \overline{R},\overline{S},\overline{X}\in\Delta_0(R,S,X)\}$$

is consistent, for p, q distinct primes. Here, $\Delta_0(R,S,X)$ denotes the collection of Δ_0-formulas with parameters R, S, X. Ajtai [4] showed that $I\Delta_0(R,S,X) + \Theta_2(R,S,X)$ is consistent with the pigeonhole principle for $\Delta_0(R,S,X)$ - definable functions.

7. (Wilkie) Let P^{top} be the theory of Peano arithmetic with a "top" element. Formally, P^{top} has the relational language $0, \max, \overline{S}, \overline{+}, \overline{\cdot}, \leq$, with constant symbols for zero and a maximum element (top), binary predicate symbol $\overline{S}$ for the graph of the successor function, ternary predicate symbols $\overline{+}, \overline{\cdot}$ for the graphs of addition and multiplication and the usual ordering relation. The axioms of P^{top} are the relational variants of the usual recursive axioms for the functions (here understood as partial functions), together with axioms stating that 0 is the least element, that max is the largest element, and that $\leq$ is a total ordering which satisfies $x \leq y \iff (\exists z)(\overline{+}(x,z,y))$, together with the scheme of induction (or equivalently of bounded induction). Is P^{top} finitely axiomatizable?

8. (Pudlák, Krajíček-Takeuti) Form the theory $I\Delta_0^{count}$ from $I\Delta_0$ by the iteration of the following step countably many times:

   For every predicate X, which is Δ_0-definable in previously introduced function symbols, add a new function symbol f_X and an axiom stating that f_X "counts" X; i.e. the following axioms

   $$\begin{aligned}&0 \in X \rightarrow f_X(0) = 1,\\ &0 \notin X \rightarrow f_X(0) = 0,\\ &(\forall n)\big((n \in X \rightarrow f_X(n) = 1 + f_X(n-1)) \wedge\\ &(n \notin X \rightarrow f_X(n) = f_X(n-1))\big).\end{aligned}$$

   and then add induction axioms for all bounded formulas in the language resulting from first step. Characterize the Π_1^0-consequences of the theory $I\Delta_0^{count}$. Note that all Π_1^0 consequences of $I\Delta_0^{count}$ are provable in the theory U_1^1 of [10], a second order extension of S_2, and also in $I\Delta_0 + Exp$, and that $I\Delta_0 + Exp$ is not Π_1^0-conservative over $I\Delta_0^{count}$ (see [41] for related results). It is unknown whether U_1^1 is Π_1^0-conservative over $I\Delta_0^{count}$. Note that by [49] U_2^1 is not Π_1^0-conservative over $I\Delta_0$.

9. (Takeuti) Let $S_2^{-\infty}$ be the equational theory involving equations $s = t$, where s, t are closed terms in the language of S_2, with natural rules based on recursive definitions of the functions symbols. Show that

$$S_2 \nvdash Con(S_2^{-\infty}).$$

Recall the fact that $I\Delta_0 \vdash Con(I\Delta_0^{-\infty})$, since in $I\Delta_0$ one can define values of closed terms in the language $\{0, 1, +, \cdot\}$. Values of terms of the language of S_2 are not definable in S_2 since e.g. the value of term $t = 2\#\ldots\#2$ is exponentially large in its Gödel number.

10. (Buss, Krajíček, Takeuti) Within the framework of Gentzen's sequent calculus, let the Δ_i^b-PIND rule be the inference: From the sequent

$$\rightarrow \phi(a) \equiv \neg\psi(a)$$

where ϕ, ψ are Σ_i^b and the sequent

$$\Gamma, \phi(\lfloor x/2 \rfloor) \rightarrow \phi(x), \Delta$$

infer the sequent

$$\Gamma, \phi(0) \rightarrow \phi(t), \Delta.$$

(Note that this is different from the Δ_i^b-PIND axiom, where the equivalence of ϕ and $\neg\psi$ is an antecedent of the formula.) In [14], it was shown by a model theoretic argument that S_2^i proves the Δ_{i+1}^b-PIND rule. Give a proof theoretic proof of this fact.

11. (Buss, Krajíček, Takeuti) Is R_2^{i+1} Σ_{i+1}^b-conservative over S_2^i? Note that in [14], it was shown that R_3^{i+1} is Σ_{i+1}^b-conservative over S_3^i.

12. (Buss, Ressayre) What is the strength of Σ_i^b-replacement relative to T_2^{i-1} and S_2^{i-1}? Note that R_2^i proves Σ_i^b-replacement, see [5] and that Σ_1^b-definable functions in R_2^1 are exactly those computable in the class NC of functions computable in polylogarithmic time with a polynomial number of processers on a parallel random access machine [5, 22].

13. (Buss, Krajíček, Takeuti) Assume that T_2^i proves $(\forall x)(\exists y)\phi(x, y)$, where ϕ is Σ_i^b. Does there exists ψ in Σ_i^b such that T_2^i proves $(\forall x)(\exists! y)$ $\psi(x, y)$ and T_2^i proves $(\forall x, y)(\psi(x, y) \rightarrow \phi(x, y))$? Same question for R_3^i.

14. (Verbrugge) Let ϕ be a Σ_1^0-sentence. Is it necessarily true that

$$S_2 \vdash \phi \rightarrow \mathrm{Pr}_{S_2}(\ulcorner \phi \urcorner).$$

Note that Parikh's theorem implies that if this holds for formulas instead of sentences then $NP = coNP$, see [71].

15. (Krajíček, Pudlák) Show that S_2^1, or a stronger system, does not prove superpolynomial lower bounds for extended Frege system proofs. It is known that the system $\forall\Sigma_1^b(S_2^1)$ does not, see [46].

2 Complexity of Proofs

As mentioned in the preface, in 1956 K. Gödel essentially raised the following k-symbol provability question, which is now known to be equivalent to $P = NP$: Does there exist a polynomial time algorithm to determine, given a first order formula F and integer k, whether there is a proof of F with at most k symbols? In [50], as reported in [62], Gödel as well raised the question of determining, given theories S, T, the length of shortest proof of "finitistic consistency" $Con_T(\underline{n})$ in S. Here $Con_T(\underline{n})$ is the first order statement that there is no proof with at most n symbols of falsehood $0 = 1$. See [62, 61] for some partial results. It is worth remarking that a superpolynomial lower bound to the length of proofs of $Con_T(\underline{n})$ in S implies that S does not prove $NP = coNP$.

From Cook's seminal result that SAT is NP-complete, it follows that $NP = co{-}NP$ iff there exists a polynomially bounded, or in terminology of [23] *super*, proof system for all propositional logic tautologies $TAUT$ [23]. In [45] this problem was shown to be equivalent to Gödel's question about finitistic consistency statements. In order to pinpoint the combinatorial difficulties arising in an attempt to prove $NP \neq co - NP$, to date there has been a focus on proving superpolynomial and exponential lower bounds for proof size (total number of symbols) of combinatorial families of propositional tautologies for *explicit* proof systems. Such superpolynomial bounds are now known for semantic tableaux (folklore), resolution [30, 15], Gentzen without cut [70] and constant-depth Frege systems [3, 4]. It is interesting to note that M. Ajtai's technique for proving a superpolynomial proof size in constant-depth Frege systems uses combinatorics similar to those employed in his lower bound [2] for boolean circuit depth. As well, unpublished work of Beame, Impagliazzo, Pitassi [60] and of Krajíček, Pudlák and Woods [48] (see [8]) uses Yao-Haståd style combinatorics in order to improve Ajtai's work to obtain an exponential lower bound for constant-depth Frege proofs of the pigeonhole principle. Exponential lower bounds for constant-depth systems and superpolynomial speed-up of depth $d + 1$ systems over the depth d systems were proved earlier in [42].

2.1 Fundamental problems in complexity of proofs

In this section, the most important and motivating open problems seem to be the following.

C. Prove superpolynomial (and exponential) lower bounds to the size of proofs of some combinatorial statements in any propositional proof system stronger than bounded-depth Frege systems, in particular for Frege and extended Frege systems. Not even superquadratic lower bounds are known (or superlinear when proof-steps are counted).

D. In predicate calculus, prove lower bounds for the length of proofs of naturally arising statements; in particular, improve lower bounds for the finitistic consistency statements. Specifically, for some S strong enough to meaningfully formalize syntactical notions and some $T \supseteq S$, both axiomatized schemes (e.g. $S = I\Delta_0 + \Omega_1, T = ZF$), show that there are no proofs of $Con_T(\underline{n})$ in S of size polynomial in n. Note that this implies that S does not prove $NP = coNP$.

E. Prove Kreisel's conjecture on "generalizing" proofs; i.e. if there exists an integer k for which $\phi(s^{(n)}(0))$ is provable in Peano arithmetic in k lines, for all $n \in \mathbf{N}$, then $\forall x \phi(x)$ is provable. This conjecture is sensitive to the language chosen for the theory and its form of axiomatization. (The conjecture is formulated for the axiomatization based on induction axioms and the language $\{0, 1, s, +, \cdot, =\}$.) It is true if addition and multiplication are treated as ternary relations [52], for any theory axiomatized by schemes with all function symbols at most unary [43, 26], for any finite theory [44], and for $L\exists_1$ (the least number principle for existential formulas) in the language $\{0, 1, s, +, \cdot, =\}$ [7]. See the survey [40].

2.2 Other problems in complexity of proofs

The following problems seem somewhat accessible.

16. (Razborov) Let $\Phi(\alpha, d, N)$ be a $\Sigma_1^{0,b}$-formula encoding the following statement:

> $d \leq |N|$, *α represents a Boolean circuit of depth at most d in* $|N|$ *variables (along with truth-tables of all Boolean functions appearing as intermediate results) and the top node of this circuit outputs* "SATISFIABILITY".

Let $\Sigma^{1,b} \rightleftharpoons \bigcup_{i \geq 0} \Sigma_i^{1,b}$. Does there exist a $\Sigma^{1,b}$-definable function $d(N)$ such that $\mathbf{N} \models \omega(\log |N|) \leq d(N) \leq |N|$ and $V_1^1 \vdash \neg\Phi(\alpha, d(N), N)$?

In another words, can V_1^1 prove superlogarithmic lower bounds on the depth of Boolean circuits (= superpolynomial lower bounds on the formula size) for SATISFIABILITY? Note that by the main result of [68, 67, 65] this is equivalent to the provability of an appropriate formalization of the same statement in S_2^1. Note also that V_1^1 (or S_2^1 if you prefer) is exactly the right theory to formalize proofs of lower bounds known in Boolean complexity so in particular it proves that no monotone circuit and no bounded-depth circuit over $\{\wedge, \vee, MOD_p\}$ of subexponential size can compute SATISFIABILITY.

A reasonable first step toward resolving this might be the following:

Show that $V_1^0 \not\vdash \neg\Phi(\alpha, d(N), N)$.

It is an easy corollary of the cut elimination theorem that V_1^0 is $\Sigma_1^{0,b}$-conservative over $I\Delta_0(\alpha)$, so this second question might be accessible with the techniques mentioned in the introduction to this section.

17. (W. Cook et al.) An extension of resolution called cutting planes proof system was investigated in [24]. Give a polynomial size family of propositional logic formulas for which every cutting plane proof family has superpolynomial length.

18. (Goerdt, Clote) A. Goerdt [28] has shown that Frege systems p-simulate cutting planes proofs, and P. Clote [20] has shown that a particular extension of cutting planes p-simulates constant depth Frege proofs. Does cutting planes with limited extension p-simulate constant depth Frege proofs?

19. (Clote) For p a positive integer, let CP_p denote the modification of the cutting planes proof systems, obtained by restricting the division rule to allow only division by p. For p, q relatively prime integers, is it the case that CP_p cannot polynomially simulate CP_q?

20. (Krajíček) Let PA denote the first order theory of Peano arithmetic in the language $\{0, 1, s, +, \cdot, =\}$. Is there a recursive function $f(k, \phi)$ such that

$$PA \vdash_{k \text{ steps}} \phi \Rightarrow PA \vdash_{f(k,\phi) \text{ symbols}} \phi?$$

By [44] this is known for finitely axiomatized theories, and for schematic theories whose language contains finitely many predicate symbols and at most unary function symbols [43, 27].

21. (Baaz) What happens with known results concerning the number of proof lines (Kreisel's conjecture) if Gentzen sequent calculus (or predicate logic) is augmented by the equality scheme, or when it is modified to allow the introduction of an entire block of like quantifiers in one step? See [6] for a partial result.

22. (Pudlák) Assume that $(\forall x)(\exists y)\phi(x, y)$ is provable in predicate logic. Introduce a new function symbol f and an axiom A_ϕ which states

$$(\forall x)\phi(x, f(x)).$$

Does there exist formula ϕ such that the extended system gives a superexponential speed-up over predicate calculus, with respect to number of symbols in proofs?

23. (Krajíček) For the theory RCF of real closed fields, is there a generalization result of the form: If there exists an integer k for which $\phi(1 + \ldots + 1)$ (with n occurrences of 1) is provable in k lines, for all $n \in \mathbf{N}$, then $\forall x \phi(x)$ is provable?

24. (Buss) Let LK_e denote Gentzen's sequent calculus with equality. Assume $LK_e \vdash_{k \text{ steps}} \phi$, where the formula ϕ has no occurrence of the symbol for equality. Does it follow that $LK \vdash_{k \text{ steps}} \phi$?

25. (Montagna) Letting PA denote Peano arithmetic, does $PA \vdash_{k \text{ steps}} \Box\phi \to \phi$ imply that $PA \vdash_{k \text{ steps}} \phi$, where $\Box\phi$ is the formalized statement that ϕ is provable?

26. (Buss) Let $f(n)$ be a polynomial (e.g. $f(n) = n^2$ or more generally, $f(n) = n^{O(1)}$). Find a sequence $\phi_0, \phi_1, \phi_2, \ldots$ of formulas and integer k such that

    (i) $LK \vdash \phi_i$, for all i,
    (ii) for each i,

    $$LK \vdash_{f(k) \text{ steps}} \phi_i \Rightarrow LK \vdash_{k \text{ steps}} \phi_i$$

    (iii) $\{i : LK \vdash_{k \text{ steps}} \phi_i\}$ is not recursive.

    This is known for $f(n) = n + 1$ by a construction from [12].

27. (Cook, Reckhow) Give a non-linear (resp. super-quadratic) lower bound to the number of steps (resp. of symbols) for Frege proofs. For instance find a family $\langle \phi_n : n \in \mathbf{N} \rangle$ of propositional tautologies, ϕ_n of length n, and positive rational ϵ, such that for every family

$\langle P_n : n \in \mathbf{N} \rangle$ of Frege proofs of these tautologies, it is the case that the number of steps (resp. of symbols) in P_n is at least $n^{1+\epsilon}$ (resp. $n^{2+\epsilon}$). In [23], the question was posed whether the pigeonhole principle, as a scheme in propositional logic, requires superpolynomial Frege proofs. In [11], S.R. Buss answered this in negative. With the results of [11, 28, 20, 19], it is clear that most combinatorial results proved by rudimentary counting have polynomial size Frege proofs. Apart from somewhat unnatural propositional versions of consistency statements of Peano arithmetic, ZF set theory, etc., there are currently few candidates of combinatorial tautologies requiring superpolynomial size proofs. In [63], P. Pudlák proved that a certain version of the finite Ramsey theorem has polynomial size constant depth Frege proofs. In [20] Clote suggested propositional formulations of the Paris-Harrington theorem [53] and related combinatorial independence results as candidates of tautology families requiring superpolynomial size proofs. For instance, work of A. Kanamori and K. McAloon [33] shows that (roughly) the Ackermann function $ack(m, m)$ is a lower and upper bound for the least N, such that given any integer coloring of increasing pairs i, j with $i < j$ satisfying the property that the color of (i, j) is at most i, there is a size m min-homogeneous set with the property that for all increasing pairs drawn from that set the color only depends on the first coordinate. This can be expressed in propositional logic as follows:

$$\bigwedge_{0 \leq i < j \leq N} \bigvee_{0 \leq k < i} p_{i,j,k}$$

$$\rightarrow$$

$$\bigvee_{0 \leq i_1 < \ldots < i_m \leq N} \; \bigvee_{k_1 \leq i_1, \ldots, k_m \leq i_m} \; \bigwedge_{1 \leq \alpha \leq m, \alpha < \beta \leq m} p_{i_\alpha, i_\beta, k_\alpha}.$$

Krajíček suggested the following sequence of tautologies based on theorem of Bondy as hard for Frege systems: for n let B_n be the formula with n^2 atoms $p_{ij} (i, j < n)$:

$$B_n = \left(\bigwedge_{i<j<n} \bigvee_{k<n} (p_{ik} \not\equiv p_{jk}) \right) \rightarrow \left(\bigvee_{\ell<n} \bigwedge_{i<j<n} \bigvee_{k<n, k \neq \ell} (p_{ik} \not\equiv p_{jk}) \right).$$

He showed that constant-depth Frege proofs of B_n require exponential size.

28. (Bonet) Assume that $\mathcal{F}$ is a Frege system. Does there exist a constant c (depending on $\mathcal{F}$) such that $\mathcal{F}, \phi \vdash_{k \text{ symbols}} \psi$ implies that

$\mathcal{F} \vdash_{c \cdot k \text{ symbols}} \phi \to \psi$? In other words, if $\mathcal{F}, \phi \vdash_{k \text{ symbols}} \psi$, then does it follow that $\mathcal{F} \vdash_{O(k) \text{ symbols}} \phi \to \psi$? This is known to be true with $O(k^2)$ in place of $O(k)$, or if there is a tree-like proof with k symbols of ψ from ϕ, see [9].

3 Fragments of Peano arithmetic

3.1 Fundamental problems in fragments

The main problems in the area of Peano arithmetic and its fragments which currently seem to be beyond existent techniques appear to concern a threshold for combinatorial independence results and the construction of model extensions. In [32], C.G. Jockusch asked whether there is a recursive partition of unordered pairs of integers into two classes, such that the r.e. complete set K of the halting problem is recursive in every infinite homogeneous set (this problem is related to the so-called $2-3$ problem in models of arithmetic). Recently, D. Seetapun (unpublished) gave a negative solution to Jockusch's problem, using a new forcing construction. See [37] for related work.

F. (Paris, Wilkie) Does every countable model M of $I\Delta_0 + B\Sigma_1$ have an end extension to a model of $I\Delta_0$ [59] ? This is the most popular problem in the area but appears to be difficult and several problems below are related to it (and hopefully more tractable).

3.2 Technical problems in fragments

The following technical problems are perhaps accessible and should give information about the above main problems.

29. (Paris, Wilkie) Does $I\Delta_0 + \neg exp \vdash B\Sigma_1$, where exp is the formula expressing $(\forall x)(\exists y)(2^x = y)$. A positive answer is known, if one assumes that the Matijasevič theorem is provable in $I\Delta_0$, cf. Problem B(d). See [59] for partial results and discussion. An affirmative answer to this problem implies a negative answer to the second fundamental problem above.

30. (Krajíček) Is there an interpretation of $I\Delta_0 + \neg B\Sigma_1$ in $I\Delta_0$? An affirmative answer to this problem implies a negative answer to the previous problem, since $I\Delta_0 + exp$ is not interpretable in $I\Delta_0$. (Here $\neg B\Sigma_1$ means the negation of any instance of the Σ_1-collection scheme.)

31. (Kaye) Is there a countable rigid model of $B\Sigma_1 + \neg I\Sigma_1$. Here, rigid means having no nontrivial automorphisms. Or even, is there a countable model of $B\Sigma_1 + \neg I\Sigma_1$ with at most countably many automorphisms, or even having no proper elementary substructure. It is known that countable rigid models of $I\Delta_0 + \neg I\Sigma_1$ exist. See [36, 38] for current results.

32. (Wilmers) Let IE_1 (resp. IE_1^-) denote the theory of arithmetic whose principal axiom scheme is induction for bounded existential formulas (resp. bounded existential formulas without parameters). Is IE_1^- a proper subtheory of IE_1? Does IE_1^- have the Tennenbaum property, meaning that it has no recursive countable non-standard models. See work of G. Wilmers [74], R. Kaye [35, 34], and Z. Adamowicz [1].

33. (Clote) For $n \geq 1$, if M is a countable model of $B\Sigma_{n+1}$ then does there exist an $n+1$-elementary end extension K of M such that $K \models B\Sigma_n$? See [18] for discussion and related results.

34. (Paris) For $n \geq 1$, is it the case that $I\Delta_n$ is equivalent to $B\Sigma_n$? It is known that $B\Sigma_n$ is equivalent to $L\Delta_n$ (Gandy, see [29]).

35. (Clote) Does $I\Delta_0 + exp$ prove $Con(B\Sigma_{n+1}) \equiv Con(I\Sigma_n)$? See [21] for a proof of $1 - Con(B\Sigma_{n+1}) \equiv 1 - Con(I\Sigma_n)$ over $I\Delta_0 + exp$.

36. (Hájek) J. Paris [56] and H. Friedman (unpublished) independently proved that $B\Sigma_{n+1}$ is Π_{n+2}-conservative over $I\Sigma_n$. A proof theoretic proof of this result was given by W. Sieg [66]. Though not explicitly mentioned in [66], for each fixed Π_{n+2} formula ϕ, Sieg's proof can be formalized to show that over $I\Delta_0 + exp$, if $B\Sigma_{n+1} \vdash \phi$ then $I\Sigma_n \vdash \phi$. A newer model theoretic proof due to P. Hájek, easily admits a similar formalization — see [21]. See also work of S. Buss [17] for a proof theoretic proof of this conservation result using the "witness" predicate. Give a sharp bound f such that for all Π_{n+2} formulas Θ, if there is a proof in $B\Sigma_{n+1}$ of Θ consisting of m symbols then there is a proof in $I\Sigma_n$ of Θ consisting of $f(m)$ symbols. Can f be polynomial?

37. (Quinsey) Let $N \rightarrow_* (k)^n_m$ mean that for every partition F of the unordered subsets of $\{0, \ldots, N-1\}$ of size n into m classes, there is a set $Y \subseteq \{0, \ldots, N-1\}$ of cardinality $\max(k, \min(Y))$, all of whose size n subsets are sent to the same class. With this notation, does Peano arithmetic prove

$$(\forall k)(\exists n) n \rightarrow_* (k+1)^k_2 \ ?$$

In [64], the notion of 'fulfillment' is used to sharpen the original Paris-Harrington result [53] to show that if $k+1$ is replaced by $k+2$, or the subscript 2 is replaced by 3, then the resulting combinatorial principle is unprovable in Peano arithmetic.

38. (Kanamori) In [33], the following elegant partition relation is introduced: $N \to (k)^n_{\text{reg}}$ means that for every regressive partition F of the size n unordered subsets of $\{0, \ldots, N-1\}$ there is a min-homogeneous set of size k. Here the function F is said to be regressive, if

$$F(\{a_1, \ldots, a_n\}) < a_1$$

holds for all $a_1 < \cdots < a_n < N$, and a subset $Y \subseteq \{0, \ldots, N-1\}$ is said to be min-homogeneous if

$$F(\{a_1, a_2, \ldots, a_n\}) = F(\{a_1, b_2, \ldots, b_n\})$$

for all $a_1 < a_2 < \ldots a_n$ and $a_1 < b_2 < \ldots < b_n$ drawn from Y. Does Peano arithmetic prove

$$(\forall k)(\exists n) n \to (k+2)^k_{\text{reg}} \ ?$$

In [33], it is shown that Peano arithmetic does not prove the combinatorial principle resulting from replacing $k+2$ by $2k$.

References

[1] Z. Adamowicz and G. Morales-Luna. A recursive model for arithmetic with weak induction. *Journal of Symbolic Logic*, 50:49–54, 1985.

[2] M. Ajtai. Σ_1^1-formulae on finite structures. *Annals of Pure and Applied Logic*, 24:1 – 48, 1983.

[3] M. Ajtai. The complexity of the pigeon hole principle. In *Proceedings of IEEE 29th Annual Symposium on Foundations of Computer Science*, 1988. pp. 346 – 355.

[4] M. Ajtai. Parity and the pigeonhole principle. In S.R. Buss and P.J. Scott, editors, *Feasible Mathematics*, pages 1–24. Birkhäuser, 1990.

[5] B. Allen. Arithmetizing uniform NC. *Annals of Pure and Applied Logic*, 53(1):1–50, 1991.

[6] M. Baaz. Note on the existence of most general semi-unifiers. In *this volume.*

[7] M. Baaz and P. Pudlák. Kreisel's conjecture for $L\exists_1$. In *this volume.*

[8] P. Beame, R. Impagliazzo, J. Krajíček, T. Pitassi, P. Pudlák, and A. Woods. Exponential lower bounds for the pigeonhole principle. In *Proceedings of the 24th Annual ACM Symposium on Theory of Computing, Victoria*, 1992.

[9] M.L. Bonet. Number of symbols in Frege proofs with and without the deduction rule. In *this volume.*

[10] S. Buss. *Bounded Arithmetic*, volume 3 of *Studies in Proof Theory.* Bibliopolis, 1986. 221 pages.

[11] S. Buss. The propositional pigeonhole principle has polynomial size Frege proofs. *Journal of Symbolic Logic*, 52:916 – 927, 1987.

[12] S. Buss. The undecidability of k-provability. *Annals of Pure and Applied Logic*, 53:72–102, 1991.

[13] S. Buss and J. Krajíček. An application of boolean complexity to separation problems in bounded arithmetic. Preprint, 1992.

[14] S. Buss, J. Krajíček, and G. Takeuti. Provably total functions in bounded arithmetic theories for R_3^i, U_2^i, and V_2^i. In *this volume.*

[15] S. Buss and G. Turán. Resolution proofs of generalized pigeonhole principles. *Theoretical Computer Science*, 62:311 – 317, 1988.

[16] S.R. Buss. Axiomatizations and conservation results for fragments of bounded arithmetic. In W. Sieg, editor, *Logic and Computation*, pages 57–84. American Mathematical Society, 1990. Contemporary Mathematics 106, Proceedings of a Workshop held at Carnegie Mellon University, June 30 - July 2, 1987.

[17] S.R. Buss. The witness function method and provably recursive fragments of peano arithmetic. In *Proceedings of Logic, Methodology and Philosophy of Science '91*. to appear.

[18] P. Clote. Partition relations in arithmetic. In C.A. DiPrisco, editor, *Methods in Mathematical Logic*, pages 32–68. Springer-Verlag, 1985. Lecture Notes in Mathematics 1130.

[19] P. Clote. ALOGTIME and a conjecture of S.A. Cook. In *Proceedings of IEEE Symposium on Logic in Computer Science*, 1990. Journal version in *Annals of Mathematics and Artificial Intelligence*, **6** (1992) 57–106. See related paper in *this volume.*

[20] P. Clote. Cutting planes and constant depth Frege proofs. In *Proceedings of IEEE Symposium on Logic in Computer Science*, 1992.

[21] P. Clote, P. Hájek, and J. Paris. On some formalized statements in arithmetic. *Archive for Mathematical Logic*, 30(4):201–221, 1991.

[22] P. Clote and G. Takeuti. Bounded arithmetics for NC, $ALOGTIME$, L and NL. *Annals of Pure and Applied Logic*, 56:73–117, 1992.

[23] S.A. Cook and R. Reckhow. On the relative efficiency of propositional proof systems. *Journal of Symbolic Logic*, 44:36 – 50, 1977.

[24] W. Cook, C.R. Coullard, and G. Turan. On the complexity of cutting plane proofs. *Discrete Applied Mathematics*, 18:25–38, 1987.

[25] C. Dimitracopoulos and J. Paris. Truth definitions for Δ_0 formulae. *Logic and Algorithmic. An International Symposium Held in Honor of Ernst Specker, Monogr. Enseign. Math. Univ. Geneve*, pages 317–329, 1982.

[26] W. Farmer. The Kreisel length-of-proof problem. Annals of Pure and Applied Logic, 1988.

[27] W. Farmer. A unification algorithm for second-order monadic terms. *Annals of Pure and Applied Logic*, 39:131–174, 1988.

[28] A. Goerdt. Cutting plane versus Frege proof systems. In E. Börger, editor, *Proceedings of Computer Science Logic 1990*, pages 174–194, 1992. Springer Lecture Notes in Computer Science **553**.

[29] P. Hájek and P. Pudlák. *Metamathematics of first order arithmetic.* Springer-Verlag, 1992.

[30] A. Haken. The intractability of resolution. *Theoretical Computer Science*, 39:297 – 305, 1985.

[31] J. Hartmanis. The Structural Complexity Column: Gödel, von Neumann and the $P = ?NP$ problem. *EATCS Bulletin*, 38:101–107, June 1989.

[32] C.G. Jockusch. Ramsey's theorem and recursion theory. *Journal of Symbolic Logic*, 37(2), 1972.

[33] A. Kanamori and K. McAloon. On Gödel incompleteness and finite combinatorics. *Annals of Pure and Applied Logic*, 33:23–41, 1987.

[34] R. Kaye. Parameter-free universal induction. *Zeitschrift für Mathematische Logik u. Grundlagenforschung*, 35(5):443–456, 1989.

[35] R. Kaye. Diophantine induction. *Annals of Pure and Applied Logic*, 46(1):1–40, 1990.

[36] R. Kaye. Model-theoretic properties characterizing Peano arithmetic. *Journal of Symbolic Logic*, 56(3):949–963, 1991.

[37] L. Kirby. Ultrafilters and types on models of arithmetic. *Annals of Pure and Applied Logic*, 27:215–252, 1984.

[38] R. Kossak. On extensions of models of strong fragments of arithmetic. *Proceedings of the American Mathematical Society*, 108:223–232, 1990.

[39] J. Krajíček. Fragments of bounded arithmetic and bounded query classes. *Transactions of the American Mathematical Society*, to appear.

[40] J. Krajíček. Generalization of proofs. In *Proc. 5-th Easter Model Theory Conference*, pages 82–99. Seminarberichte der Humboldt Universität, 1987.

[41] J. Krajíček. Exponentiation and second order bounded arithmetic. *Annals of Pure and Applied Logic*, 48:261–276, 1990.

[42] J. Krajíček. Lower bounds to the size of constant-depth propositional proofs. Submitted, 1991.

[43] J. Krajíček. On the number of steps in proofs. *Annals of Pure and Applied·Logic*, 52(1-2):143–154, 1991.

[44] J. Krajíček and P. Pudlák. The number of proof lines and the size of proofs in first order logic. *Archive for Math. Logic*, 27:69–84, 1988.

[45] J. Krajíček and P. Pudlák. Propositional proof systems, the consistency of first order theories and the complexity of computations. *Journal of Symbolic Logic*, 54(3):1063 – 1079, 1989.

[46] J. Krajíček and P. Pudlák. Propositional provability in models of weak arithmetic. In E. Börger et al., editor, *Computer Science Logic (Kaiserslautern, October 1989)*, pages 193–210. Springer-Verlag, 1989.

[47] J. Krajíček, P. Pudlák, and G. Takeuti. Bounded arithmetic and the polynomial hierarchy. *Annals of Pure and Applied Logic*, 52(1-2):143–153, 1991.

[48] J. Krajíček, P. Pudlák, and A. Woods. Exponential lower bound to the size of bounded depth Frege proofs of the pigeonhole principle. Submitted, 1991.

[49] J. Krajíček and G. Takeuti. On bounded Σ_1^1 polynomial induction. In S.R. Buss and P.J. Scott, editors, *Feasible Mathematics*, pages 259–280. Birkhäuser, 1990.

[50] G. Kreisel. Mathematical logic: What has it done for the philosophy of mathematics. In R. Schoenemann, editor, *Bertrand Russell: Philosopher of the Century*, pages 201–272. George Allen & Unwin, 1967.

[51] R. J. Parikh. Existence and feasibility in arithmetic. *Journal of Symbolic Logic*, 36:494 – 508, 1971.

[52] R. J. Parikh. Some results on the length of proofs. *Transactions of the American Mathematical Society*, 177:29–36, 1973.

[53] J. Paris and L. Harrington. A mathematical incompleteness in arithmetic. In J. Barwise, editor, *Handbook of Mathematical Logic*, pages 1133 – 1142. North Holland, 1977.

[54] J. B. Paris and A. J. Wilkie. Counting problems in bounded arithmetic. In C. A. di Prisco, editor, *Methods in Mathematical Logic*, pages 317 – 340. Springer Verlag Lecture Notes in Mathematics, 1985. Proceedings of Logic Conference held in Caracas, 1983.

[55] J. B. Paris, A. J. Wilkie, and A. R. Woods. Provability of the pigeonhole principle and the existence of infinitely many primes. *Journal of Symbolic Logic*, 53(4):1235 – 1244, 1988.

[56] J.B. Paris. Some conservation results for fragments of arithmetic. In C. Berline, K. McAloon, and J.-P. Ressayre, editors, *Model Theory and Arithmetic*, pages 251–262. Springer-Verlag, 1981. Lecture Notes in Mathematics 890.

[57] J.B. Paris and A.J. Wilkie. Δ_0 sets and induction. In W. Guzicki, W. Marek, A. Pelc, and C. Rauzer, editors, *Open Days in Model Theory and Set Theory*, pages 237–248. Leeds University Press, 1981. Proceedings of a conference held in September 1981 at Jadwisin, near Warsaw, Poland.

[58] J.B. Paris and A.J. Wilkie. Models of arithmetic and rudimentary sets. *Bull. de la Soc. Math. de Belgique*, 33(B):157–169, 1981.

[59] J.B. Paris and A.J. Wilkie. On the existence of end extensions of models of bounded induction. In J.E. Fenstad et al., editor, *Logic, Methodology and Philosophy of Science VIII*. North Holland, 1989.

[60] T. Pitassi, P. Beame, and R. Impagliazzo. Exponential lower bounds for the pigeonhole principle. Technical Report 257/91, University of Toronto, October 1991.

[61] P. Pudlák. On the lengths of proofs of finitistic consistency statements in first order theories. In J.B. Paris, A.J. Wilkie, and G.M. Wilmers, editors, *Logic Colloquium '84*, pages 165 – 196. North-Holland, 1986.

[62] P. Pudlák. Improved bounds to the length of proofs of finitistic consistency statements. In S.G. Simpson, editor, *Logic and Combinatorics*, volume 65. Contemporary Mathematics, *American Mathematical Society*, 1987.

[63] P. Pudlák. Ramsey's theorem in bounded arithmetic. In E. Börger, editor, *Proceedings of Computer Science Logic 1990*. 1992. Springer Lecture Notes in Computer Science **553**.

[64] J.E. Quinsey. *Some problems in logic: Applications of Kripke's notion of fulfillment.* PhD thesis, Oxford University, April 1980.

[65] A.A. Razborov. An equivalence between second order bounded domain bounded arithmetic and first order bounded arithmetic. In *this volume.*

[66] W. Sieg. Fragments of arithmetic. *Annals of Pure and Applied Logic*, 28:33–71, 1985.

[67] G. Takeuti. *RSUV* isomorphism. In *this volume.*

[68] G. Takeuti. S_3^i and $\overset{\circ}{V}{}_2^i(BD)$. *Archive for Math. Logic*, 29:149–169, 1990.

[69] G. Takeuti. A second order version of S_2^1 and U_2^1. *Journal of Symbolic Logic*, 56 (3):1038–1063, 1991.

[70] A. Urquhart. Hard examples for resolution. *Journal of the Association of Computing Machinery*, 34(1):209 – 219, 1987.

[71] R. Verbrugge. Feasible interpretability. In *this volume.*

[72] A.J. Wilkie. Applications of complexity theory to Σ_0-definability problems in arithmetic. In L. Pacholski, J. Wierzejewski, and A.J. Wilkie, editors, *Model Theory of Algebra and Arithmetic*, pages 363–369. Springer-Verlag, 1980.

[73] A.J. Wilkie and J.B. Paris. On the schema of induction for bounded arithmetic formulas. *Annals of Pure and Applied Logic*, 35:261 – 302, 1987.

[74] G. Wilmers. Bounded existential induction. *Journal of Symbolic Logic*, 50:72 – 90, 1985.

[75] A. Woods. Bounded arithmetic formulas and turing machines of constant alternation. In J.B. Paris, A.J. Wilkie, and G.M. Wilmers, editors, *Logic Colloquium 1984*. North Holland, 1986.

Note on the Existence of Most General Semi-Unifiers

Matthias Baaz

Abstract

This note proves the conditional existence of most general semi-unifiers, by giving a simple method of computing such a general solution from any one solution of a given semi-unification problem. Also a variant is discussed with applications in proof theory.

1 Introduction

Semi-unification problems arise in proof theory in connection with determining the set of proofs (by means of suitable rules) that satisfy given "structural" data, for example, a given skeleton, which is a structure labelled by (names for) inferences together with an end formula, or again a given size (in the sense of number of steps) and end formula, which generally determine a finite set of skeletons, and the like. The difficulty of the determination in question depends a good deal on the style of the (formal) rules considered, as follows:

The additional data to be determined are *terms*, which, in the case of first order rules, may be demanding in the completion of weak quantifier inferences. These take the general form $A(x/\cdot) \to (\exists x)A(x)$ in the given data for rules with cut, to be completed by suitable terms t: $A(t) \to (\exists x)A(x)$. Corresponding additional data in the case of propositional inferences are obtained easily (by ordinary unification).

With structural data for cut-free rules, the required terms are determined easily from the (given) end sequent. As a corollary, when cut elimination is available, the latter can provide (partial) information about the set of proofs determined by the structural data for rules with cut, too (cf. [KP88]). It may be remarked that the first ε-theorem can be used similarly.

When induction schemata are among the rules considered, cut elimination is not possible without some *evaluation* of terms. Then semi-unifiers (of suitable semi-unification problems) can be used to describe the sets above (cf. [BP92]). A most general semi-unifier is *minimal* with respect to term complexity. This translates into a proof of a generalization, say, $A(t^*)$, of the given end formula $A(t)$ with a new end formula, where t results from t^* by substitution (for a variable in t^*).

Finally, we go into a variant of semi-unification, which corresponds to rules of proof that permit substitution for only one variable at a time (and not for blocks). This turns out to be related to second order unification.

2 Most general semi-unifiers

Definition 1 A *substitution* (denoted by Δ, $\Sigma, \ldots$) is a list of replacements of variables by terms: $\Delta = \{t_1/z_1, \ldots, t_r/z_r\}$ means that $t_1, \ldots, t_r$ are simultaneously substituted for $z_1, \ldots, z_r$. The application of Δ to t is denoted by $t\Delta$. Application of substitutions is associative, so we omit parentheses. A substitution Δ is called a *unifier* of $(s_1, t_1), \ldots, (s_p, t_p)$, iff $s_1\Delta = t_1\Delta, \ldots, s_p\Delta = t_p\Delta$. The unification algorithm (cf. [CL73, §5.4]) decides the existence of a *most general unifier* and calculates it if it exists.

Definition 2 (cf. [Pud88], [KTU90]) A substitution Δ is called a *semi-unifier* of the semi-unification problem $(s_1, t_1), \ldots, (s_p, t_p)$ iff there exist $\Sigma_1, \ldots, \Sigma_p$ such that $s_1\Delta = t_1\Delta\Sigma_1, \ldots, s_p\Delta = t_p\Delta\Sigma_p$.

Example 1 $\Delta = \{f(x, f(x, x))/z\}$ is a semi-unifier of $\big(f(x, z), f(x, f(x, y))\big)$ because

$$f(x, z)\{f(x, f(x, x))/z\} = f(x, f(x, y))\{f(x, f(x, x))/z\}\{f(x, x)/y\}.$$

There is no semi-unifier of $\big(f(x, y), f(x, f(x, y))\big)$, since no simultaneous substitution will make the left side a substitution instance of the right side.

Definition 3 A substitution Δ is a *most general semi-unifier* of a semi-unification problem iff every semi-unifier is an extension of Δ. The most general semi-unifier is unique modulo renaming.

Remark. Not every extension of a most general semi-unifier of a semi-unification problem $(s_1, t_1), \ldots, (s_p, t_p)$ is again a semi-unifier, in contrast, for example, to the case of most general unifiers for ordinary unification problems. If one is interested in semi-unifiers $\Delta_x^t = \{t/x\}\Delta'$ for closed terms t as in proof theory when complex terms t in formulas $A(t)$ with short derivations are generalized, one considers the most general semi-unifier Δ of $\big(f(s_1, x), f(t_1, x)\big), \ldots, \big(f(s_p, x), f(t_p, x)\big)$. Δ represents Δ_x^t by $\Delta_x^t = \Delta\Sigma\Delta'$ with Σ the most general unifier of $x\Delta$ and t, and Δ has the property that $\Delta\Gamma[x\Delta]$ is a semi-unifier for arbitrary substitutions $\Gamma[x\Delta]$ which are the identity on variables not occuring in $x\Delta$ (cf. Theorem 1, Definition 4).

Example 2 In Example 1, the most general unifier is $\Delta' = \{f(x', y')/z\}$, not Δ.

Theorem 1 *There is an algorithm to find a most general semi-unifier of a semi-unification problem if any semi-unifier exists.*

Theorem 1 is a consequence of the following definitions and lemmas:

Definition 4 Let the semi-unification problem $(s_1, t_1), \ldots, (s_p, t_p)$ be given, and let $\alpha_1(n), \ldots, \alpha_p(n)$ be disjoint canonical renamings of all the variables occuring at level n (We omit n when the context is clear). The *approximations* Δ_n *to* Δ are defined as follows:

$$\Delta_0 = \text{id},$$

$$\Delta_{n+1} = \begin{cases} \Delta_n\Gamma_{n+1}, & \text{where } \Gamma_{n+1} \text{ is the most general unifier of} \\ & (s_1\Delta_n, t_1\Delta_n\alpha_1), \ldots, (s_p\Delta_n, t_p\Delta_n\alpha_p), \text{ and not} \\ & \text{just a renaming of variables on } \{s_1\Delta_n, \ldots, s_p\Delta_n\} \\ \Delta_n & \text{otherwise} \end{cases}$$

(Approximations $\Sigma_{1,n}, \ldots, \Sigma_{p,n}$ to the substitutions $\Sigma_1, \ldots, \Sigma_p$ in $s_1\Delta = t_1\Delta\Sigma_1, \ldots, s_p\Delta = t_p\Delta\Sigma_p$ are given by $\alpha_1(n)\Gamma_{n+1}, \ldots, \alpha_p(n)\Gamma_{n+1}$).

Lemma 1 *Let v be the variables in $\{s_1, t_1, \ldots, s_p, t_p\}$. If Ω is a semi-unifier of $\{(s_1, t_1), \ldots, (s_p, t_p)\}$, then there exist Ω_i such that for all m, $\Omega = \Delta_m\Omega_m$ on domain v and there is a most general unifier Γ_{m+1} of $(s_1\Delta_m, t_1\Delta_m\alpha_1), \ldots, (s_p\Delta_m, t_p\Delta_m\alpha_p)$.*

Proof. By induction on m:

(1) $m = 0$: Let $\Omega_0 = \Omega$. $\Omega = \Delta_0\Omega_0$ holds, since $\Delta_0 = \mathrm{id}$. Γ_1 exists because $s_i\Omega = t_i\Omega\Sigma_i$, by the assumption that $(s_1, t_1\alpha_1)$, ..., $(s_p, t_p\alpha_p)$ are unifiable (The most general unifier of $(s_1, t_1\alpha_1)$, ..., $(s_p, t_p\alpha_p)$ can be written as $\Omega\Sigma_i$, where Ω is a substitution of the variables in (s_1, t_1), ..., (s_p, t_p) and Σ_i is a substitution of the variables in $t_i\alpha_i$).

(2) $m > 0$: The induction hypothesis $\Omega = \Delta_m\Omega_m$ and the assumptions $s_i\Omega = t_i\Omega\Sigma_i$ imply that $s_i\Delta_m\Omega_m = t_i\Delta_m\Omega_m\Sigma_i$. Extend the left and right sides as follows:

$$s_i\Delta_m\big(\Omega_m(\bigcup_j \alpha_j^{-1}\Omega_m\Sigma_j)\big) = t_i\Delta_m\alpha_i\big(\Omega_m(\bigcup_j \alpha_j^{-1}\Omega_m\Sigma_j)\big).$$

(Note that the left-hand side is just $s_i\Delta_m\Omega_m$ while $\alpha_i\Omega_m\bigcup_j \alpha_j^{-1} = \mathrm{id}$ and so the right hand side is $t_i\Delta_m\Omega_m\Sigma_i$). This holds because

$$\mathrm{dom}\,\alpha_j^{-1} \cap \mathrm{range}(\Delta_m\Omega_m) = \emptyset =$$
$$= \mathrm{range}\,\alpha_i \cap \mathrm{dom}\,\Omega_m = \mathrm{range}\,\alpha_i \cap \mathrm{dom}\,\alpha_j^{-1}$$

for $i \neq j$. Therefore, $(s_1\Delta_m, t_1\Delta_m\alpha_1)$, ..., $(s_p\Delta_m, t_p\Delta_m\alpha_p)$ are unifiable. The unification theorem implies the existence of the most general unifier Γ_{m+1} and there is Ω_{m+1} such that

$$\Gamma_{m+1}\Omega_{m+1} = \Omega_m(\textstyle\bigcup_j \alpha_j^{-1}\Omega_m\Sigma_j).$$ □

Lemma 2 *If Ω is a semi-unifier of $\{(s_1, t_1), \ldots, (s_n, t_n)\}$, then there is an m such that $\Delta_m = \Delta_{m+1}$.*

Proof. $\Delta_m = \Gamma_0 \ldots \Gamma_m$ as long as the first case of the case distinctions of Definition 4 holds. The assumption and Lemma 1 imply that Γ_{j+1} always exists. Therefore the first case of the distinction holds iff Γ_{j+1} is not a renaming of variables. Such Γ_{j+1} substitute terms that are not variables or unify different variables. By assumption and Lemma 1 this can happen only finitely often. Therefore there is an m such that $\Delta_m = \Delta_{m+1}$. □

Lemma 3 *If $\Delta_{m_0} = \Delta_{m_0+1}$, where m_0 is minimal, and Γ_{m_0+1} exists, then Δ_{m_0} is a semi-unifier.*

Proof. From the assumptions it follows, that $s_i\Delta_{m_0}\Gamma_{m_0+1} = t_i\Delta_{m_0}\alpha_i\Gamma_{m_0+1}$ and that Γ_{m_0+1} is a renaming of variables on $\{s_i\Delta_{m_0}\}$. Use

$$\mathrm{INV}(\Gamma_{m_0+1}) = \begin{cases} \Gamma^{-1}_{m_0+1}, & \text{on the variables in } \{s_i\Delta_{m_0}\Gamma_{m_0+1}\} \\ \mathrm{id} & \text{otherwise} \end{cases}$$

for $s_i\Delta_{m_0} = s_i\Delta_{m_0}\Gamma_{m_0+1}\,\mathrm{INV}(\Gamma_{m_0+1}) = t_i\Delta_{m_0}(\alpha_i\Gamma_{m_0+1}\,\mathrm{INV}(\Gamma_{m_0+1})) = t_i\Delta_{m_0}\Sigma_i$ (Note that, $\alpha_i\Gamma_{m_0+1}\,\mathrm{INV}(\Gamma_{m_0+1}) = \Sigma_i$). □

Put together, the lemmas state that:

(1) If there is a semi-unifier, then Δ_{m_0} is a most general semi-unifier.

(2) There is a semi-unifier iff there is an m such that $\Gamma_1, \ldots, \Gamma_{m+1}$ exist and $\Delta_m = \Delta_{m+1}$.

Therefore there can be two reasons for unsolvability:

(1) There is an m such that Γ_{m+1} does *not* exist,

(2) For all m, Γ_m exists but $\Delta_m \neq \Delta_{m+1}$.

As the following examples demonstrate, (1) and (2) are both possible even for a singular semi-unification problem (s, t):

Example 3 $(f(u, v), f(f(x, y), u))$:

$$\begin{aligned} \Delta_1 &= \{f(x', y')/u\} \\ \Delta_2 &= \{f(x', y')/u, f(x'', y'')/v\} \\ \Delta_{n+2} &= \Delta_2 \end{aligned}$$

Δ_2 is the most general semi-unifier.

Example 4 $(f(u, g(v)), f(f(x, y), u))$:

$$\begin{aligned} \Delta_1 &= \{f(x', y')/u\} \\ \Delta_{n+1} &= \Delta_1 \end{aligned}$$

$(f(f(x', y'), g(v)), f(f(x'', y''), f(x''', y''')))$ is not unifiable, therefore there is no semi-unifier.

Example 5 $(x, f(x))$:

$$\begin{aligned}\Delta_1 &= \{f(x')/x\}\\ \Delta_2 &= \{f(f(x''))/x\}\\ \Delta_3 &= \{f(f(f(x''')))/x\} \quad \text{etc.}\end{aligned}$$

$\Delta_n \neq \Delta_m$ for all n, m, therefore, no semi-unification is possible.

Such situations as in Exmaple 4 and 5 do not arise if Definition 4 is reinterpreted, specifically, as a *total* algorithm with *inputs* consisting of pairs (a given semi-unification problem with a solution Δ) and most general semi-unifiers $\overline{\Delta}$ as *outputs*. This algorithm terminates in $\leq m \cdot 2^n$ steps, where n is the number of symbols in Δ and m is the number of variables in the semi-unification problem (of the input). It should be noted that this bound is independent of the number of equations in the semi-unification problem; in contrast to the significance of that number for decidability discovered in [Pud88]. To state a previously neglected property of the algorithm, we need the following definition:

3 Monadic semi-unification

Definition 5 A substitution Δ is *monadic*, iff $\left|\{x \mid x\Delta \neq x\}\right| \leq 1$. A semi-unifier Δ of (s_1, t_1),...,(s_p, t_p) is *monadic*, iff $s_1\Delta = t_1\Delta\Sigma_1$, ..., $s_p\Delta = t_p\Delta\Sigma_p$ with Σ_1, ..., Σ_p monadic.

By the parenthetical remark following Definition 4, Σ_1, ..., Σ_p need not be monadic for Δ, even if $\overline{\Sigma}_1$, ..., $\overline{\Sigma}_p$ are monadic for a solution $\overline{\Delta}$ extending Δ. This has the following consequence for the proof theoretic application outlined in the introduction: The proof π^* of the generalization $A(t^*)$ obtained by means of the algorithm generally has a skeleton different from that of the proof π of $A(t)$; in particular, in π^* substitutions for blocks of variables replace substitutions for a single variable in π.

It will now be shown, that this is not merely a property of the particular algorithm, but in general unavoidable. Familiarity with [KP88] is assumed, in particular, with the following definitions:

Definition 6 (cf. [Gol81, KP88]). Let L be a set of function symbols, and let x_1, ..., x_m be variables. Let $\mathbf{T} = (T, \mathrm{Sub}_1, \ldots, \mathrm{Sub}_m)$ be the algebra of terms, where T is the set of terms in L, x_1, ..., x_m (considered as constants), and for $i = 1, \ldots, m$:

$$\mathrm{Sub}_i(t, t') := t\{t'/x_i\}$$

are substitutions as binary operations on T. A *second-order unification problem* is a finite set of equations in the language $T \cup \{\mathrm{Sub}_1, \ldots, \mathrm{Sub}_m\}$ plus free variables for elements of T; a solution is called *second order unifier.* A second-order unification problem is called *strict,* if t' in $\mathrm{Sub}_i(t, t')$ is neither x_i nor a variable for elements of T, i.e., x_i must not be replaced by itself. A second order unifier is *most general,* iff every second order unifier is an extension of Δ.

The solutions of a second order unification problem

$$(u_1, \mathrm{Sub}_{i_1}(v_1, w_1)), \ldots, (u_s, \mathrm{Sub}_{i_s}(v_s, w_s)),$$

where x_1, ..., x_n are the variables in the definition of the operations $\mathrm{Sub}_i(t, t') := t\{t'/x_i\}$, can easily be represented (modulo renaming of variables) by the solutions of a finite set of strict second order unification problems. To see this, use case distinction for the w_i, which are variables for elements of T, according to whether w_i is replaced in a solution by a variable z or a term with outermost function symbol f. Substitute w_i by z (respectively by $f(v_1, \ldots, v_r)$ where v_1, ..., v_r are new variables). Finally, unify pairs (s, t) where $\big(s, \mathrm{Sub}_i(t, x_i)\big)$ occurs, and cancel this condition. If this unification process changes x_1, ..., x_n, cancel the subcase.

Definition 7 Π_1^x *is a map from recursively enumerable sets* $X \subseteq \omega$ *to strict second order unification problems* S, *such that* $S\{s^n(0)/x\}$ *has a solution iff* $n \in X$. *Let* $(\exists y_1) \ldots (\exists y_n) D_X(x, y_1, \ldots, y_n)$ *be any diophantine representation of* X, *where* D_X *is a conjunction of formulas of the form*

(a)	$y_i = u, u < w$	*(b)*	$y_i = y_j + y_l$
(c)	$y_i = y_j \cdot y_l$	*(d)*	$y_i = x.$

The argument in Theorem 4.1 of [KP88] completes the construction, when (b) and (c) are replaced by $s(y_i) = y_j + s(y_l)$ *and* $s(y_i') = y_i + s(y_l)$, $y_i' = s(y_j) \cdot y_l$, *using new variables, respectively.*

Proposition 1 *Let the recursively enumerable set* $X \subseteq \omega$ *and its complement be infinite. The second order unification problem* $X\Pi_1^x$ *has infinitely many solutions* Δ_n, *but no most general solution.*

Proof. There is a solution Δ_n of $X\Pi_1^x\{s^n(0)/x\}$ for every $n \in X$. If there were a solution Δ with $\Delta\Gamma_n = \Delta_n$ for all $n \in X$, $x\Delta$ would be of the form $s^k(z)$ (by the infinity of X) and $\Delta\{s^m(0)/z\}$ would be solvable for all m in contradiction to the infinity of the complement of X. □

Definition 8 Π_2^x *is a map from second order unification problems to semi-unification problems given by the following algorithm:*

We first rewrite the given second-order unification problem as

$$(u_1', \mathrm{Sub}_{i_1}(v_1', w_1)), \ldots, (u_s', \mathrm{Sub}_{i_s}(v_s', w_s)),$$

using new variables if necessary. Let $x_1, \ldots, x_n$ be the variables in the definition of the operations $\mathrm{Sub}_i(t, t') := t\{t'/x_i\}$. We obtain the semi-unification problem $\big(f(f(u_1', w_1), x), f(f(v_1', x_{i_1}), x)\big), \ldots, \big(f(f(u_s', w_s), x), f(f(v_s', x_{i_s}), x)\big)$ and add the pairs

$$\begin{aligned}
&\big(f(\cdots f(x_1, x_2) \ldots x_n)\{c/x_1\}, f(\cdots f(x_1, x_2) \ldots x_n)\big),\\
&\big(f(\cdots f(x_1, x_2) \ldots x_n)\{d/x_1\}, f(\cdots f(x_1, x_2) \ldots x_n)\big), \ldots\\
&\big(f(\cdots f(x_1, x_2) \ldots x_n)\{c/x_n\}, f(\cdots f(x_1, x_2) \ldots x_n)\big),\\
&\big(f(\cdots f(x_1, x_2) \ldots x_n)\{d/x_n\}, f(\cdots f(x_1, x_2) \ldots x_n)\big)
\end{aligned}$$

to avoid substitution into $x_1, \ldots, x_n$.

We obtain:

Proposition 2 *Let t be closed. Δ_x^t is a solution of the strict second order unification problem $S\{t/x\}$ iff Δ_x^t is a monadic solution of the semi-unification problem $S\Pi_2^x\{t/x\}$.*

Corollary 1 *For every infinite recursively enumerable set $X \subseteq \omega$, the semi-unification problem $X\Pi_1^x\Pi_2^x\{s^n(0)/x\}$ has a solution for all but finitely many n, but has a monadic solution iff $n \in X$.*

Proof. $X\Pi_1^x\Pi_2^x$ has a most general semi-unifier Δ because $X\Pi_1^x\Pi_2^x\{s^n(0)/x\}$ has a semi-unifier for $n \in X$. $x\Delta$ must be of the form $s^k(z)$ for a sufficiently large k as X is infinite. By definition of Π_2^x, $\Delta\{t/z\}$ is a solution for every closed term t. □

The two results correspond to the usual and a coarser meaning of "skeleton" (when "substitution" is applied to a single variable and to an arbitrary block, respectively).

For completeness, we mention a semi-unification problem dual to the original one: Given $(s_1, t_1), \ldots, (s_p, t_p)$, are there substitutions Δ and $\Sigma_1, \ldots, \Sigma_p$, such that $s_1\Delta = t_1\Sigma_1\Delta, \ldots, s_p\Delta = t_p\Sigma_p\Delta$? Γ_1 of the algorithm in Definition 4 provides a most general semi-unifier, because Δ, $\Sigma_1, \ldots, \Sigma_p$ can be rewritten as

$$\Delta' = \begin{cases} \Delta\Sigma_i & \text{on range}(\alpha_i) \\ \Delta & \text{elsewhere} \end{cases} \qquad\qquad \Sigma_i' = \alpha_i.$$

Δ' is therefore an extension of Γ_1. This semi-unification problem is obviously decidable in contrast to the original one (cf. [KTU90] for undecidability).

4 Addendum to [BP92]

In Section 6 (following Claim 2) of [BP92] a switch to substitution for blocks of variables is made, but not made explicit. In fact, the input is a (Herbrand) proof with a usual skeleton which corresponds to a suitable second order unification problem S. The skeleton of the generalized proof has to be modified such that blocks of variables are substituted where only one variable is replaced in the original skeleton. The generalization is given by the most general semi-unifier of $S\Pi_2^x$.

References

[BP92] M. Baaz and P. Pudlák. Kreisel's conjecture for $L\exists_1$. in this volume.

[CL73] C.-L. Chang and R. C.-T. Lee. *Symbolic Logic and Mechanical Theorem Proving*. Academic Press, New York and London, 1973.

[Gol81] W. D. Goldfarb. The undecidability of the second-order unification problem. *Theor. Comp. Sci.*, 13:225–230, 1981.

[KP88] J. Krajíček and P. Pudlák. The number of proof lines and the size of proofs in first order logic. *Arch. Math. Logic*, 27:69–84, 1988.

[KTU90] A. J. Kfoury, J. Tiuryn, and P. Urzyczyn. The undecidability of the semi-unification problem. *Proc. 22nd ACM STOC*, 468–476, 1990. journal version to appear in *Inf. Comp.*

[Pud88] P. Pudlák. On a unification problem related to Kreisel's conjecture. *Comm. Math. Univ. Carol.*, 29(3):551–556, 1988.

MATTHIAS BAAZ
Institut für Algebra und Diskrete Mathematik
Technische Universität Wien
Wiedner Hauptstraße 8–10
A-1040 Wien

KREISEL'S CONJECTURE FOR $L\exists_1$

Matthias Baaz Pavel Pudlák
(Including a Postscript by Georg Kreisel)

Introduction

The famous conjecture of Kreisel states:

> *If there exists k such that the formal system $\mathbf{T}$ proves $\varphi(S^n(0))$ in $\leq k$ steps for every n, then $\mathbf{T}$ proves $\forall x \varphi(x)$.*

Here $S^n(\cdots)$ denotes n-times iterated successor function. The conjecture was proposed for $\mathbf{T}$ = Peano Arithmetic in the formalization of Kleene (cf. [Parikh], [Friedman] problem N.34, [Kreisel1]), but it is interesting for any $\mathbf{T}$ which contains some fragment of number theory. We shall prove Kreisel's conjecture for a formal system with finitely many (basic) axioms and the schema of the least number principle for existential formulas $\mathbf{L\exists_1}$.

It should be stressed that the answer to the conjecture depends not so much on the sentences provable in $\mathbf{T}$ as it does on the presentation of $\mathbf{T}$ and the logical calculus used. For example, it is well known that $\mathbf{L\exists_1}$ is equivalent to $\mathbf{I\Sigma_1}$ [Kaye] and that $\mathbf{I\Sigma_1}$ is finitely axiomatizable. But our result does not imply anything about $\mathbf{I\Sigma_1}$ and it is not implied by the fact that for every sufficiently strong *finite* axiomatic system Kreisel's conjecture holds. Thus by writing $\mathbf{L\exists_1}$, $\mathbf{I\Sigma_1}$ etc. we shall always mean a formal system, not just a deductively closed set of sentences. The reason is that the fact that $\mathbf{S}$ and $\mathbf{T}$ have the same logical consequences does not imply that we have the same bounds to the number of steps in the proofs. For instance one can prove the following statements:

Proposition 0.1 *(a) Let $\mathbf{T}$ be a consistent number theory axiomatized by the quantifier-free induction scheme or identity scheme $t = s \rightarrow \psi(t) =$*

$\psi(s)$ ($\psi(x)$ arbitrary term), and let $\mathbf{T}_0$ be a finite axiomatization whose theorems include those of $\mathbf{T}$. *Then there is no k such that all axioms of* $\mathbf{T}$ *can be proved from $\mathbf{T}_0$ within k steps.*

(b) For each recursively enumerable consistent number theory $\mathbf{T}_0$ there is a number theory $\mathbf{T}$ *with exactly the same derivable formulas such that Kreisel's conjecture does not hold for* $\mathbf{T}$.

Proof. *(a)* By Theorem 6.1 of [Krajíček/Pudlák] and cut-elimination, for each formula $\varphi(x)$ and k there exists a sequence of terms $\tau_1, \ldots, \tau_n$ such that

(*) $\mathbf{T}_0 \vdash \varphi(\tau_i)$, $1 \leq i \leq n$, and $\mathbf{T}_0 \vdash^k \varphi(\tau)$ implies that τ is substitution instance of some τ_i. (For the definition of $\vdash^k$ c.f. Section 1.)

Let $d_0 := 0$, $d_{r+1} := d_r + 0$. Then

(**) there is a k such that for all r $\mathbf{T} \vdash^k d_r = 0$.

This was shown in [Yukami] p. 33 using $\forall x(x = 0 \rightarrow \psi(x) = \psi(0))$ for an adequate ψ (Derive $d_r + \ldots + (d_2 + (d_1 + d_0)) = d_{r-1} + \ldots + (d_1 + (d_0 + d_0)) = d_{r-1} + \ldots + (d_1 + d_0)$ from $0 + 0 = 0$ and $d_r = 0$ from $d_r + t = t$). The principle $\forall x(x = 0 \rightarrow \psi(x) = \psi(0))$ is derivable using the same skeleton for all ψ by quantifier-free induction on $x = 0 \rightarrow \psi(x) = \psi(0)$. (*) contradicts (**) if there is a k such that all axioms of $\mathbf{T}$ are derivable from $\mathbf{T}_0$ within k steps.

(b) Consider a formula $\varphi(x)$ such that $\mathbf{T}_0 \vdash \varphi(S^n(0))$ for all n, but $\mathbf{T}_0 \nvdash \forall x\varphi(x)$. Let $\mathbf{T} := \mathbf{T}_0 \cup \{\varphi(S^n(0)) | n \in \mathsf{N}\}$. □

Clearly, it is also important that we talk about the number of steps (proof lines, sequents, formulas in the proof) but not about the total size of the proof (e.g., the length of a string encoding the proof). For the latter concept the conjecture does not have sense. The point is that there can be infinitely many proofs with $\leq k$ steps (for k sufficiently large) even if we consider the proofs of a fixed formula. The essential reason is that there can be arbitrarily large terms in short proofs.

Kreisel's conjecture does not have any immediate consequences relevant to other topics in logic and number theory. However it is very stimulating if we want to learn something about the role of terms in first order logic. This may have eventually quite practical consequences, since manipulation with terms is very important in computer science. In fact, our main tool — semiunification of terms — appeared in this area and in computer science almost at the same time (cf. [Baaz2], [Kfoury et al.]). The idea of using semiunification for Kreisel's conjecture is due to the first author, who used Hilbert's ε–calculus. The second author adapted it so that Herbrand's theorem is used instead.

1 Preliminaries

We shall consider a standard Hilbert style system of first order logic, say as in [Barwise]. *A proof of* φ *in* **T** *with* k *steps* is a sequence of formulas $< \varphi_1, \ldots, \varphi_k >$ such that each φ is either an axiom (logical or in **T**) or follows from preceding ones by a logical rule; $\mathbf{T} \vdash^k \varphi$ means that φ has a proof in **T** with $\leq k$ steps. The language of **T** should contain the constant 0 and the successor function $S(x)$ and a finite number of function and predicate symbols including $\leq$. For the general result from which we derive Kreisel's conjecture we even do not need 0 and S.

In the proof we shall need arrays of variables and terms. A d-dimensional array will always be marked with the left superscript d, e.g., ${}^d a$. By writing, say ${}^2\mathrm{a}$, we mean *the string* $a_{11}, a_{12}, \ldots, a_{1n}$, $a_{21}, a_{22}, \ldots, a_{2n}$, $a_{m1}, a_{m2}, \ldots, a_{mn}$; ${}^2 a_i$ is the string $a_{i1}, \ldots, a_{in}$; ${}^2 a_{ij}$ is just a_{ij} (thus we can omit the superscript here). A similar convention will be used for $d > 2$.

The schema $\mathbf{L}\exists_\mathbf{1}$ are all formulas

$$\exists y_1 \forall z_1 (\exists z_2 \ldots \exists z_n \gamma({}^1 p, {}^1 z) \rightarrow \exists y_2 \ldots \exists y_n \gamma({}^1 p, {}^1 y) \wedge y_1 \leq z_1), \tag{1}$$

where γ is an open formula. This schema is equivalent to

$$\exists y_1 \forall^1 z \exists y_2 \ldots \exists y_n (\gamma({}^1 p, {}^1 z) \rightarrow \gamma({}^1 p, {}^1 y) \wedge {}^1 y_1 \leq {}^1 z_1), \tag{2}$$

which will be denoted by $\overline{\mathbf{L}\exists_\mathbf{1}}$. The number of steps of the proof of this equivalence depends only on n. We shall see below that for a fixed k, we can bound n in proofs with k steps which use $\mathbf{L}\exists_\mathbf{1}$. Thus for every k and r, there exists K such that if a formula of rank $\leq r$ (see below for the definition) can be derived from $\mathbf{L}\exists_\mathbf{1}$ in k steps, then the formula can be derived from $\overline{\mathbf{L}\exists_\mathbf{1}}$ in K steps.

Because of the deduction theorem, if Kreisel's conjecture holds for **T**, then it holds for **T** augmented by any finite set of axioms. Thus we may assume that **T** consists only of the scheme $\overline{\mathbf{L}\exists_\mathbf{1}}$.

We should point out that we use logic without equality which means that we allow the equality sign, but we treat it as a nonlogical predicate. As we assume a finite language, we need only finitely many equality axioms. It is an open problem what happens if we add infinitely many equality axioms, for instance the *schema of equality*

$$t = s \rightarrow \psi(t) \equiv \psi(s),$$

where $\psi(x)$, t, s are arbitrary terms. Kreisel's conjecture may still hold true, but our tool — Theorem 4.2 — fails (cf. Proposition 0.1).

2 Bounding the rank of formulas

Definition 2.1 *The* rank *of a formula* φ *is inductively defined as follows:*

(i) $rank(\varphi)=0$ *if* φ *is atomic;*

(ii) $rank(\varphi \wedge \psi) = 1 + \max(rank(\varphi), rank(\psi))$,
$rank(\neg\varphi) = 1 + rank(\varphi)$,
$rank(\exists x\varphi) = 1 + rank(\varphi)$,
etc. for other connectives and $\forall$.

The rank of a proof d is the maximal rank of a formula of d.

Our aim is to show that the rank of a proof with $\leq k$ steps can be bounded in the following sense.

Lemma 2.1 *For every k and r there exists R such that for every formula* φ, *if* $rank(\varphi) \leq r$ *and* $\mathbf{L}\exists_\mathbf{1} \vdash^k \varphi$, *then there exists a proof d such that* d: $\mathbf{L}\exists_\mathbf{1} \vdash^k \varphi$ *and* $rank(d) \leq R$.

Proof. The idea of the proof goes back to [Parikh] (see also [Krajíček/Pudlák] and [Krajíček]). But since we use a special scheme, we have to check that the same idea works also here.

Let $\varphi_1, \ldots, \varphi_k$ be a proof of φ (i.e., $\varphi = \varphi_k$) of rank $\leq r$. We construct a unification problem. We consider the language consisting of a constant c for atomic formulas, distinct variables v_β for each subformula β of the proof and function symbols for each connective and each quantifier: $f_\rightarrow$, $f_\neg$, $f_\exists$ etc.

The equations of the unification problem will be the following:

(1) For each application of a propositional rule in the proof, say

$$\frac{\alpha, \quad \alpha \rightarrow \beta}{\beta}$$

we add the equation

$$v_{\alpha\rightarrow\beta} = f_\rightarrow(v_\alpha, v_\beta).$$

(2) For each application of a quantifier rule, say

$$\frac{\alpha \to \beta}{\exists x \alpha \to \beta}$$

we add the equations

$$v_{\alpha\to\beta} = f_{\to}(v_\alpha, v_\beta).$$

$$v_{\exists\alpha\to\beta} = f_{\to}(f_{\exists}(v_\alpha), v_\beta).$$

If we have a quantifier axiom, say

$$\alpha(t) \to \exists x \alpha(x), \quad t \text{ a term},$$

we add the equation

$$v_\beta = f_{\to}(v_\gamma, f_{\exists}(v_\gamma)).$$

where β, γ are respectively

- the instance of the axiom above,
- $\alpha(x)$.

(3) For each instance of $\mathbf{L\exists_1}$ which occurs in the proof, say

$$\exists y_1 \forall z_1 (\exists z_2 \ldots \exists z_n \gamma({}^1p, {}^1z) \to \exists y_2 \ldots \exists y_n \gamma({}^1p, {}^1y) \wedge y_1 \leq z_1),$$

we add an equation

$$v_\alpha = f_{\exists} f_{\forall} f_{\to}(v_\beta, f_{\wedge}(v_\beta, c_\delta)),$$

where α, β, δ are respectively

- the instance of $\mathbf{L\exists_1}$ above,
- $\exists z_2 \ldots \exists z_n \gamma({}^1p, {}^1z)$,

(4) Finally we add

$$v_\varphi = \tau,$$

where τ is the term constructed from the constant corresponding to atomic formulas of φ and function symbols corresponding to the connectives and quantifiers in φ and which is isomorphic to φ.

Claim 1. *The size of the unification problem depends only on k and r.*

To prove the claim we just note that the size of each equation in (1) – (3) is fixed, and there are at most $2k$ such equations; the size of the equation (4) is bounded in terms of $rank(\varphi)$ which is $\leq r$.

Observe that the proof $\varphi_1, \ldots, \varphi_k$ determines a solution to the unification problem, if we do not distinguish different atomic formulas and bound variables. By the claim the most general unifier has size bounded in terms of k and r. Using such a unifier we construct a proof whose rank is bounded in terms of k and r. As the original proof is obtained by substitution from the most general unifier, we can think of it as a sequence $\varphi'_1, \ldots, \varphi'_k$ obtained from $\varphi_1, \ldots, \varphi_k$ by

(i) replacing some atomic formulas by the constant c,

(ii) replacing some subformulas by variables,

(iii) omitting the bound variables from quantifiers.

Pick a suitable formula ξ so that it has small size (a constant independent of k and r), it is open, and does not contain logical variables occurring in the proof; say $0 = 0$. Construct $\varphi''_1, \ldots, \varphi''_k$ as follows. If c or a variable of φ'_i corresponds to some atomic formula α in φ_i, replace it by α. If a variable of φ'_i corresponds to a non-atomic formula, replace it by ξ. Finally, add the bound variables to quantifiers according to $\varphi_1, \ldots, \varphi_k$.

Claim 2. $\varphi''_1, \ldots, \varphi''_k$ *is an* $\mathbf{L\exists_1}$ *proof of* φ.

We shall check that for every $i \leq k$, if φ_i is an instance of $\mathbf{L\exists_1}$ or follows from some previous formulas by a rule, then φ''_i is an instance of $\mathbf{L\exists_1}$ or etc. For propositional rules this is almost trivial. For quantifier rules we have to check that the conditions on occurrences of variables are satisfied. By our choice of ξ no clashes of variables can be introduced.

Now suppose φ_i is $\alpha(t) \to \exists x \alpha(x)$. Since φ_i is a substitution instance of φ'_i and by the construction of φ''_i from φ'_i, φ''_i must have the following form: it is φ_i with some pairs of subformulas $\psi(t)$ of $\alpha(t)$ and $\psi(x)$ of $\alpha(x)$ replaced both by ξ. Thus φ''_i is of the form $\alpha'(t) \to \exists x \alpha'(x)$ for some α. Exactly the same argument works for instances of $\mathbf{L\exists_1}$.

It remains to observe that $\varphi''_k = \varphi$ because of the equation (4). □

Corollary 2.2 *For every k and r there exist K and R' such that for every formula φ, if $rank(\varphi) \leq r$ and $\mathbf{L\exists_1} \vdash^k \varphi$, then there exists a proof d' such* *that* $d' : \overline{\mathbf{L}\exists_1} \vdash^K \varphi$ *and* $rank(d') \leq R'$.

Proof. Let d be the proof from Lemma 2.1. To obtain d' we insert a proof of the particular instance of $\mathbf{L}\exists_1$ from $\overline{\mathbf{L}\exists_1}$. The number of steps and the rank of the inserted proofs can be bounded using the ranks of the instances of $\mathbf{L}\exists_1$ which in turn are bounded in terms of k and r. □

Lemma 2.3 *For every* k *and* r *there exist* K *and* R *such that for every formula* φ, *if* $rank(\varphi) \leq r$ *and* $\mathbf{L}\exists_1 \vdash^k \varphi$, *then there exist a finite set* A_0 *of universal closures of instances of* $\overline{\mathbf{L}\exists_1}$ *and a proof* d *in pure predicate calculus of* $\bigwedge\!\!\bigwedge A_0 \rightarrow \varphi$ *such that* d *has* $\leq K$ *steps and* $rank(d) \leq R$.

Proof. Apply the deduction theorem. Clearly the increase of the number of steps and rank is bounded in the original parameters. □

Now we can go on in the same manner and show that we can transform $\bigwedge\!\!\bigwedge A_0 \rightarrow \varphi$ into a prenex normal form while keeping the number of steps and the rank bounded. In fact we can choose any of the possible prenex normal forms.

3 Herbrand's theorem

Let A be a formula in prenex normal form

$$\exists^2 x_1 \forall^2 y_1 \exists^2 x_2 \forall^2 y_2 \ldots \alpha({}^1p, {}^2x_1, {}^2y_1, {}^2x_2, {}^2y_2, \ldots).$$

We define $He(A)$, the Herbrand form of A, to be

$$\exists^2 x_1 \exists^2 x_2 \ldots \alpha({}^1p, {}^2x_1, {}^2f_1({}^2x_1), {}^2x_2, {}^2f_2({}^2x_1, {}^2x_2), \ldots).$$

where ${}^2f$ is an array of new function symbols.

Lemma 3.1 A *is provable in* $\mathbf{T}$ *iff* $He(A)$ *is provable in* $\mathbf{T}$; $\mathbf{T}$ *is any theory not containing the new function symbols of* $He(A)$. *Moreover the number of steps in the proof of* $He(A)$ *can be bounded using the number of steps in the proof of* A *and* $rank(A)$ *and vice versa.*

Proof. It is a well-known fact. □

Let us write $He(A)$ in the form

$$\exists^1 u\beta(^1p, {}^1u).$$

Then we call a term instance of $He(A)$ any formula of the form

$$\beta(^1p, {}^1t).$$

where ^1t is an array of terms.

Lemma 3.2 (Herbrand's Theorem) *(i)* *For every k and r, there exists a K such that for every formula in prenex normal form A, if A is provable in pure predicate calculus by a proof with $\leq k$ steps and rank $\leq r$, then there exist $\vartheta_1, \ldots, \vartheta_l$, $l \leq K$, term instances of $He(A)$ such that*

$$\vartheta_1 \vee \ldots \vee \vartheta_l$$

is a propositional tautology.

(ii) *If $\vartheta_1 \vee \ldots \vee \vartheta_l$ is a propositional tautology, where $\vartheta_1, \ldots, \vartheta_l$ are term instances of $He(A)$, A in a prenex normal form, then A is provable.*

Proof. *(i)* Apply any of the standard constructive proofs of Herbrand's theorem. For instance let us consider cut-elimination in a Gentzen style calculus. There the increase w.r.t. the original size can be bounded by the rank of cut formulas and their number.

(ii) This is an obvious fact. □

We shall consider Herbrand proofs of only a special class of formulas, namely $\bigwedge\!\!\!\bigwedge A_0 \to \varphi$ where A_0 are universal closures of instances of $\overline{\mathbf{L}\exists_1}$ and φ is existential. They can be presented in a special form which we shall describe now.

Let φ be $\exists^1 x\psi(^1x)$, ψ quantifier–free, and let the elements of A_0 be

$$\forall^1 p\exists^2 y_i\forall^2 z_i(\gamma_i(^1p, {}^2z_i) \to \gamma_i(^1p, {}^2y_i) \wedge {}^2y_{i1} \leq {}^2z_{i1}).$$

Then we take the following prenex normal form of $\bigwedge\!\!\!\bigwedge A_0 \to \varphi$:

$$\exists^1 p\forall^2 y\exists^2 z\exists^1 x\bigvee\!\!\!\bigvee_i[(\gamma_i(^1p, {}^2z_i) \to \gamma_i(^1p, {}^2y_i) \wedge {}^2y_{i1} \leq {}^2z_{i1}) \to \psi(^1x)]. \tag{1}$$

We shall abbreviate the part in square brackets by

$$\Gamma_i({}^1p, {}^2z_i, {}^2y_i, {}^2y_{i1}, {}^2z_{i1}) \to \psi({}^1x).$$

Hence an Herbrand proof of the formula has the form

$$\bigvee_{j.i}[\Gamma_i({}^2s_j, {}^3q_{ji}, {}^2f_i({}^2s_j), {}^2f_{i1}({}^2s_j), q_{ji1}) \to \psi({}^2u_j)],$$

where ${}^2s$, ${}^3q$, ${}^2u$ are arrays of terms and ${}^2f$ is the array of new function symbols.

It is possible that for some $j \neq j'$ the strings ${}^2s_j$ and ${}^2s_{j'}$ are equal, while some $q_{jim} \neq q_{j'im}$ or $u_{jm} \neq u_{j'm}$. Therefore we group the disjuncts according to the strings ${}^2s_j$. After introducing new indices and renaming the arrays we get thus

$$\bigvee_{j.i,l}[\Gamma_i({}^2s_j, {}^4q_{jil}, {}^2f_i({}^2s_j), f_{i1}({}^2s_j), q_{jil1}) \to \psi({}^3u_{jl})].$$

In order to be able to prove that there exists a most general Herbrand proof with a given skeleton (see below for definition), we shall allow looser structure than the above one. Namely we allow Γ_i to depend also on j, hence also ${}^2f_i$ will depend on j. Thus we get

$$\bigvee_{j.i,l}[\Gamma_{ji}({}^2s_j, {}^4q_{jil}, {}^3f_{ji}({}^2s_j), f_{ji1}({}^2s_j), q_{jil1}) \to \psi({}^3u_{jl})].$$

This transformation is quite formal, we only added another index to Γ_i's and f_i's. However there may be propositional tautologies of this form where $\Gamma_{ji} \neq \Gamma_{j'i}$ and they can be interpreted as Herbrand proofs of φ from $\overline{\mathbf{L}\exists_1}$. Thus for the purpose of proving Kreisel's conjecture for $\mathbf{L}\exists_1$ we shall use the above shape of a formula as our notion of a *Herbrand proof.*

We shall analyze this formula further. It is possible (and actually happens, if we derive the formula as above) that $f_{jim} = f_{j'im}$ for $j \neq j'$. But note that, since ${}^2s_j \neq {}^2s_j'$ for $j \neq j'$, the terms

$$f_{jim}({}^2s_j) \quad \text{and} \quad f_{j'i'm'}({}^2s_{j'})$$

are always different if $i \neq i'$ or $j \neq j'$ or $m \neq m'$. Furthermore we can order the disjunction in such a way that

$$j < j' \Rightarrow \max_m |s_{jm}| \leq \max_m |s_{j'm}|,$$

where $|\ldots|$ denotes the size of a term, i.e., the number of symbols in the term. We shall include these two conditions into our definition of a Herbrand proof.

Definition 3.1 *A* Herbrand proof *of* $\exists^1 x\psi(^1x)$, ψ *quantifier-free, in* $\mathbf{L}\exists_1$ *is a formula (1) such that* γ_{ij} *are quantifier-free and such that for every* $j < j'$

$$\max_m |s_{jm}| \le \max_m |s_{j'm}|$$

and $s_{jm} \neq s_{j'm}$ *for some* m.

Lemma 3.3 *(i) For every* k *and* r*, there exists* K *such that if* $\mathbf{L}\exists_1 \vdash^k \exists^1 x\psi(^1x)$ *by a proof of rank* $\le r$ *then, for some quantifier-free formulas* γ_{ji} *and terms* ^2s, ^4q, ^3u

$$\bigvee_{j.i,l}[\gamma_{ji}(^2s_j, {}^4q_{jil}) \to \gamma_{ji}(^2s_j, {}^3f_{ji}(^2s_j)) \wedge {}^3f_{ji1}(^2s_j) \le {}^4q_{jil1}) \to \psi(^3u_{jl})] \tag{2}$$

is an Herbrand proof where the size of the arrays and the rank of each γ_{ji} *is boundeded by* K.

(ii) If (2) is an Herbrand proof, then $\mathbf{L}\exists_1 \vdash \exists^1 x\psi(^1x)$.

Proof. The first part has been proved above. The second part is proved in the usual way. Namely, we derive *(1)* from *(2)*, where we will have γ_{ji}'s instead of γ_i's. First apply the rule of introducing $\exists$ to terms ^4q and ^3u. Thus, using the abbreviation above, we get

$$\bigvee_{j.i} \exists^3 z \exists^2 x(\Gamma_{ji}(^2s_j, {}^3z_{ji}, {}^3f_{ji}(^2s_j), f_{ji1}(^2s_j), {}^3z_{ji1}) \to \psi(^2x_j)). \tag{3}$$

Clearly, we can replace different strings x_j by a single one; formally it means to replace ^2x_j by ^1x. Let the maximal j be J. We consider the disjuncts with $j = J$. By our definition of Herbrand proof

(i) all $^3f_{Jim}(^2s_J)$ are different,

(ii) they do not occur elsewhere in *(3)*, since they contain s_{jm} of maximal size.

Replace all $^3f_{Jim}(^2s_J)$ by different new variables ^3y_J. By conditions (i), (ii) we can apply the generalization rule to ^3y_J. Then apply the $\exists$-rule to ^2s_j. Thus we get

$$[\exists^1 p \forall^3 y_J \exists^3 z \exists^1 x \bigvee_i (\Gamma_{Ji}(^1p, {}^3z_{Ji}, {}^3y_{Ji}, {}^3y_{Ji1}, {}^3z_{Ji1}) \to \psi(^1x))] \vee \Phi,$$

where Φ is (3) with the restriction $j < J$. Now we do the same with Φ (note that $^3f_{jim}(^2s_j)$ do not occur in the first disjunct) and so on. Eventually we obtain

$$\exists^1 p \forall^3 y \exists^3 z \exists^1 x \bigvee\!\!\!\bigvee_{j.i} (\Gamma_{ji}({}^1p, {}^3z_{ji}, {}^3y_{ji}, {}^3y_{ji1}, {}^3z_{ji1}) \rightarrow \psi({}^1x)),$$

which is essentially (1) except γ_{ij}'s are indexed by two indices j and i. □

A formula will be called *term-free* if it does not contain occurrences of function and constant symbols (variables are allowed). Let $\beta_1, \ldots, \beta_l$ be term-free formulas and ${}^2v$ an array of variables. A sequence $\beta_1({}^2v_1) \ldots, \beta_l({}^2v_l)$ will be called a *Herbrand skeleton* if

$$\beta_1({}^2v_1) \vee \ldots \vee \beta_l({}^2v_l) \tag{4}$$

is a propositional tautology. Here we do not assume that all the variables in ${}^2v$ are distinct.

If $\beta_1({}^2t_1), \vee \ldots \vee \beta_l({}^2t_l)$ is an Herbrand proof and

$$t_{ij} = t_{i'j'} \Rightarrow v_{ij} = v_{i'j'},$$

then we say that *(4)* is the skeleton of the Herbrand proof.

We shall use the following two easy facts about Herbrand skeletons:

(i) For a given r and l there are only finitely many Herbrand skeletons with formulas of rank $\leq r$ and with l disjuncts.

(ii) Any substitution instance of an Herbrand skeleton is also a propositional tautology.

Note that because of *(ii)*, when constructing an Herbrand proof from an Herbrand skeleton, we need only to substitute terms with a suitable shape.

4 Main theorem

Theorem 4.1 (Main Theorem) *Let A_0 be a finite set of the universal closures of instances of $\overline{\mathbf{L}\exists_1}$, let $\varphi(t)$ be an existential formula which may contain function symbols not present in $\overline{\mathbf{L}\exists_1}$. Suppose the prenex normal form of $\bigwedge\!\!\!\bigwedge A_0 \rightarrow \varphi(\tau)$ has an Herbrand proof with skeleton S. Then there exists a term τ_0 such that*

(i) $\overline{\mathbf{L}\exists_1} \rightarrow \varphi(\tau_0)$ (in fact we have an Herbrand proof of $\bigwedge\!\!\!\bigwedge A_1 \rightarrow \varphi(\tau_0)$ with the skeleton S for some finite set A_1 of universal closures of instances of $\overline{\mathbf{L}\exists_1}$);

(ii) if τ' *is a term such that* $\bigwedge\!\!\!\bigwedge A_2 \to \varphi(\tau')$ *has an Herbrand proof with skeleton* S *for some finite set* A_2 *of universal closures of instances of* $\overline{\mathbf{L}\exists_1}$*, then* τ' *is a substitution instance of* τ_0 *(in fact the whole Herbrand proof of* $\bigwedge\!\!\!\bigwedge A_2 \to \varphi(\tau')$ *is a substitution instance of the Herbrand proof of* $\bigwedge\!\!\!\bigwedge A_1 \to \varphi(\tau_0)$*).*

Before proving Kreisel's conjecture for $\mathbf{L}\exists_1$ we shall derive another theorem from Theorem 4.1 from which the conjecture follows easily (cf. similar theorem for finitely axiomatized theories [Krajíček/Pudlák]). The proof of Theorem 4.1 is postponed to Section 6.

Theorem 4.2 *Let* $\varphi(x)$ *be an arbitrary formula, let* k *be a positive integer. Suppose that*

$$\mathbf{L}\exists_1 \vdash^k \varphi(\tau).$$

Then there exist terms $\tau_0, \ldots, \tau_n$ *such that*

(i) $\mathbf{L}\exists_1 \vdash \varphi(\tau_0), \ldots, \varphi(\tau_n)$*;*

(ii) if $\mathbf{L}\exists_1 \vdash^k \varphi(\tau')$*, then* τ' *is a substitution instance of* τ_i *for some* i*.*

The proof of Theorem 4.2 is based on Theorem 4.1 and the following lemma. Let us denote by $\forall\overline{\mathbf{L}\exists_1}$ the universal closures of formulas of $\overline{\mathbf{L}\exists_1}$.

Lemma 4.3 *For every* k *and* r *there exist skeletons* $S_0, \ldots, S_n$ *such that for every formula* φ*, if* $\mathbf{L}\exists_1 \vdash^k \varphi$ *and rank*$(\varphi) \leq r$*, then for some* $A_0 \subseteq \forall\overline{\mathbf{L}\exists_1}$*, every prenex form of* $\bigwedge\!\!\!\bigwedge A_0 \to \varphi$ *has an Herbrand proof with a skeleton* S_i *for some* i*.*

Proof. By Lemma 2.3, if the number of steps in a proof of $\mathbf{L}\exists_1 \vdash \varphi$ and the rank of φ are bounded, we can find $A_0 \subseteq \forall\overline{\mathbf{L}\exists_1}$ such that $\bigwedge\!\!\!\bigwedge A_0 \to \varphi$ has a proof in pure predicate calculus whose rank is bounded. Moreover, we can take any prenex normal form of $\bigwedge\!\!\!\bigwedge A_0 \to \varphi$. By Lemma 3.2 we have an Herbrand proof of $\vartheta_1 \vee \ldots \vee \vartheta_l$ of the prenex form of $\bigwedge\!\!\!\bigwedge A_0 \to \varphi$, where l is bounded in terms of k and r. The rank of each ϑ_i is also bounded, since it is equal to the rank of (the prenex normal form of) $\bigwedge\!\!\!\bigwedge A_0 \to \varphi$. Let $\beta_1({}^2v_1) \vee \ldots \vee \beta_l({}^2v_l)$ be the skeleton of the Herbrand proof. Since

$\beta_1, \ldots, \beta_l$ are term-free and of bounded rank, there are only finitely many such skeletons, up to renaming of variables ${}^2v$. Thus let $S_0, \ldots, S_n$ be these skeletons. □

Proof of Theorem 4.2: By Lemma 3.1 it suffices to consider only $\varphi(x)$ existential. Let k be given and suppose $\mathbf{L}\exists_\mathbf{1} \vdash^k \varphi(\tau)$ for some term τ. Let $r = rank(\varphi(x))$; note that $rank(\varphi(\tau)) = r$ for every term τ. Hence by Lemma 4.3, $\mathbf{L}\exists_\mathbf{1} \vdash^k \varphi(\tau)$ implies that the prenex form of $\bigwedge\!\!\!\bigwedge A_0 \to \varphi(\tau)$, for some $A_0 \subseteq \forall \overline{\mathbf{L}\exists_\mathbf{1}}$, has an Herbrand proof whose skeleton is one of $S_0, \ldots, S_n$ (n depends only on k and r). By Theorem 4.1 we have "a most general" term τ_i for each S_i. □

Corollary 4.4 (Kreisel's conjecture for $\mathbf{L}\exists_\mathbf{1}$) *Let* $\mathbf{T}$ *be the theory axiomatized by* $\mathbf{Q}$ *(Robinson Arithmetic) and* $\mathbf{L}\exists_\mathbf{1}$. *Suppose* k *is an integer and* φ *is a formula such that for every* m:

$$\mathbf{T} \vdash^k \varphi(S^m(0)).$$

Then

$$\mathbf{T} \vdash \forall x \varphi(x)).$$

Proof. Apply Theorem 4.2 to $\bigwedge\!\!\!\bigwedge \mathbf{Q} \to \varphi(S^m(0))$ instead of φ. Let $\tau_0, \ldots, \tau_n$ be terms given by (ii). Take m bigger than the depth of all $\tau_0, \ldots, \tau_n$. Let τ_i be such that $S^m(0)$ is a substitution instance of τ_i. Then, clearly, τ_i is of the form $S^l(x)$ for some variable x.

By (i)

$$\mathbf{L}\exists_\mathbf{1} \vdash \bigwedge\!\!\!\bigwedge \mathbf{Q} \to \varphi(S^l(x)),$$

whence

$$\mathbf{L}\exists_\mathbf{1} + \mathbf{Q} \vdash \forall x \geq S^l(0) \quad \varphi(x).$$

Since

$$\mathbf{Q} \vdash x = 0 \lor x = S(0) \lor \ldots \lor x = S^{l-1}(0) \lor x \geq S^l(0),$$

and since all numeric instances of $\varphi(x)$ are provable, we have

$$\mathbf{L}\exists_\mathbf{1} + \mathbf{Q} \vdash \forall x \varphi(x). \quad \square$$

Note that the answer to Kreisel's conjecture depends heavily on the representation of the n-th number by $S^n(0)$: Let $\mathbf{T}$ be a sufficiently strong number theory and let $\mathbf{T} \vdash \varphi(S^n(0))$ for all n. There is a sequence of terms $\tau_0, \tau_1, \ldots$ consisting of $0, S, +, \cdot, -$ such that $\mathbf{T} \vdash \tau_n = S^n(0)$ for all n, and there is a k such that $\mathbf{T} \vdash^k \varphi(\tau_n)$ for all n ([Baaz1]). If $\mathbf{T}$ is r.e., we can find $\varphi(x)$ so that $\mathbf{T} \nvdash \forall x \varphi(x)$ moreover.

On the other hand results similar to Kreisel's conjecture may be proved for other sequences of terms $0, F(0), F^2(0), \ldots$. For example take $F(0) := x + 0$. Then we get

Corollary 4.5 *Let* $\mathbf{T}$ *be finite and* $\mathbf{T} + \mathbf{L\exists_1} \vdash \forall x(x + 0 = x)$. *Let* $d_0 := 0$, $d_{r+1} := d_r + 0$. *Then for every* k *and* φ, *there exists an* n_0 *such that for all* $n \geq n_0$

$$\mathbf{T} + \mathbf{L\exists_1} \vdash^k \varphi(d_n) \Rightarrow \mathbf{T} + \mathbf{L\exists_1} \vdash \forall x \varphi(x).$$

Proof. Derive $\mathbf{T}+\mathbf{L\exists_1} \vdash^k \varphi((\ldots((x+0)+0)+\ldots)+0)$ from $\mathbf{T}+\mathbf{L\exists_1} \vdash^k \varphi(d_n)$ for sufficiently big n using the Main Theorem. □

To summarize, the validity of Kreisel's conjecture depends heavily on the *structure* of the terms, but not on their *values*. This corresponds to the fact that we speak about *formal systems* and not about *theories* (i.e., sets of formulas closed under derivations): Theories identify terms with their values.

Theorem 4.2 proves also that there will be no bounds on the derivations of $s^n(0) + s^m(0) = s^{n+m}(0)$ or $s^n(0) \cdot s^m(0) = s^{n \cdot m}(0)$ for arbitrary n, m: If there were bounds, $\exists y(x = y + y)$ respective $\exists y(x = y \cdot y)$ could be derived for all even respectively square numbers x with fixed length, contradicting theorem 4.2. (Successor induction implies the existence of such bounds for addition, cf. [Richardson], §1.)

Independently of theorem 4.2 one can show that the validity of Kreisel's conjecture for a consistent formal system $\mathbf{T}$ and notations $t_0, t_1, \ldots$ with $\mathbf{T} \vdash t_i = s^i(0)$ implies that there cannot be uniform bounds on both addition, $t_n + t_m = t_{n+m}$, and multiplication, $t_n \cdot t_m = t_{n \cdot m}$: Take polynomials p, q in $+$ and $\cdot$ such that $A(x)$ defined by $\exists^1 y p(x, {}^1y) = q(x, {}^1y)$ is true for all x, but $\forall x A(x)$ is not derivable in $\mathbf{T}$. If there were uniform bounds for addition and multiplication, $A(t_n)$ could be proved for all n within a fixed number of steps using notations for suitable values for ${}^1y$. (The number of steps would depend only on the number of $+$ and $\cdot$ in the polynomials.) Uniform bounds for addition are provided by uniform bounds for multiplication and successor $s(t_n) = t_{s(n)}$ by the relation $i + j = k$ iff $s(s(i) \cdot s(s(k))) \cdot s(s(j) \cdot s(s(k))) = s(s(s(i) \cdot s(j)) \cdot (s(s(k)) \cdot s(s(k))))$. In particular, for t_n beeing $s^n(0)$ we have trivially a bound for $s(t_n) = t_{s(n)}$, thus Kreisel's conjecture implies that there is no bound for multiplication.

5 Semiunification

Our main tool to prove Theorem 4.1 will be semiunification.

Definition 5.1 *A semiunification problem is given by a set of pairs of terms* $(s_1, t_1), \ldots, (s_n, t_n)$. *A solution to the semiunification problem is a substitution* δ *such that there exist substitutions* $\sigma_1, \ldots, \sigma_n$ *such that*

$$s_1\delta = t_1\delta\sigma_1, \ldots, s_n\delta = t_n\delta\sigma_n;$$

a solution will be also called a semiunifier. A most general semiunifier *is a semiunifier* δ_0 *such that for every semiunifier* δ *there exists a substitution* δ' *such that* $\delta = \delta_0\delta'$ *(i.e.,* $\delta(x) = \delta'(\delta_0(x))$*).*

Theorem 5.1 ([Baaz2]). *If a semiunification problem has a semiunifier then it has a most general one.*

We shall need a somewhat more general situation than the one of the definition.

Lemma 5.2 *Theorem 5.1 holds also if we impose additional conditions of the following kind:*

(i) for some pairs i, j: $\sigma_i = \sigma_j$;

(ii) for some variables v: $v\delta = v$;

(iii) for some pairs i, j: *if* ξ *is a variable such that* $\xi\sigma_i \neq \xi$, *then* ξ *does not occur in* $s_j\delta$;

(iv) for some pairs, a variable v *and a number* j: v *does not occur in* $s_j\delta$ *provided that* $v\delta = v$.

Proof. *(i)* Replace equations (s_i, t_i) and (s_j, t_j) by a single one $(f(s_i, s_j), f(t_i, t_j))$, f a new binary function symbol.

(ii) Let $v_1, \ldots, v_l$ be the variables for which we require $v\delta = v$. Choose $l+1$ different constants $c_0, \ldots, c_l$ which do not occur elsewhere, and add equations $(v_1, c_0), \ldots, (v_l, c_0)$ and $(f(\ldots f(v_1, v_2), \ldots, v_l), f(\ldots f(c_1, c_2) \ldots, c_l))$. Let δ_0 be a most general semiunifier for the extended problem. The extra equations ensure that $u_1 = v_1\delta, \ldots, u_l = v_l\delta$ are distinct variables. Let ρ be the partial mapping defined by

$$\rho(u_1) = v_1, \ldots, \rho(u_l) = v_l.$$

Since ρ is one-to-one, it can be extended to $\bar{\rho}$ a total one-to-one mapping (permutation) defined on all variables. Then, clearly $\delta_0\bar{\rho}$ is a most general semiunifier satisfying

$$v_1\delta_0\bar{\rho} = v_1, \ldots, v_l\delta_0\bar{\rho} = v_l.$$

(iii) For each such pair i, j add the equation (using *(i)* above)

$$s_j\delta = s_j\delta\sigma_i.$$

(iv) First add an equation (c, v) with the corresponding substitution σ, where c is a constant. Then, if

$$v = v\delta \quad \text{and} \quad c\delta = v\delta\sigma$$

then $v\sigma \neq v$. Now use *(iii)* to ensure that v does not occur in $s_j\delta$. □

Let us also note that we may use also equations $s_i\delta = t_i\delta$ (i.e., without σ_i), because we can first unify these (using the existence of a most general unifier) and then use Theorem 5.1 for the rest.

6 Proof of the main theorem

In this section we shall prove Theorem 4.1.

Let $A_0 \subseteq \forall\overline{\mathbf{L}\exists_1}$ and $\varphi(\tau)$ be given. By definition a Herbrand proof of $\bigwedge A_0 \rightarrow \varphi(\tau)$ has the following form:

$$(0) \qquad \bigvee_{j,i,l} \Gamma_{ji}({}^2s_j, {}^4q_{jil}, {}^3f_{ji}({}^2s_j), f_{ji1}({}^2s_j), q_{jil1}) \rightarrow \psi({}^3u_{jl})).$$

where ${}^2s, {}^4q, {}^3u$ are arrays of terms and ${}^3f$ are new function symbols, $\Gamma_{ji}({}^1p, {}^2z, {}^2y, {}^2y_1, {}^2z_1)$ is an abbreviation of $\gamma_{ji}({}^1p, {}^2z) \rightarrow \gamma_{ji}({}^1p, {}^2y) \wedge {}^2y_1 \leq {}^2z_1$, and we have the following conditions:

$$j < j' \Rightarrow \max|s_{jm}| \leq |s_{j'm}| \quad \text{and} \tag{1}$$

$$^2s_j \neq {}^2s_{j'} \quad \text{(i.e., } s_{jm} \neq s_{j'm} \quad \text{for some } m),$$

$$\text{all} \quad {}^3f_{ji}({}^2s_j) \quad \text{are distinct, i.e.} \tag{2}$$

$$f_{jir}({}^2s_j) = f_{j'i'r'}({}^2s_{j'}) \quad \text{iff} \quad i = i', r = r', j = j',$$

Because of condition (1) we have

$$^3f_{ji}({}^2s_j) \quad \text{do not occur in} \quad {}^2s_{j'}, \quad \text{for} \quad j' \leq j, \tag{3}$$

(because a term of equal size or bigger cannot be a proper subterm of a given term).

Now replace all maximal subterms which have the outermost symbol some f_{jir} by distinct variables. Thus we obtain

$$\bigvee_{j.i,l} \Gamma_{ji}({}^2s_j^*, {}^4q_{jil}^*, {}^3v_{ji}, v_{ji1}, q_{jil1}^*) \rightarrow \psi({}^3u_{jl}^*)). \tag{4}$$

where ^3v is an array of variables and $^2s^*$, $^4q^*$, $^3u^*$ are arrays of terms. The conditions (2) and (3) are now transformed to

$$\text{all variables in } {}^3v \text{ are different, and} \tag{5}$$

$$\text{the variables of } {}^3v_{ji} \text{ do not occur in } {}^2s_{j'}^* \text{ for } j' \leq j. \tag{6}$$

Claim 1. *If a formula of the form* (4) *satisfies* (5) *and* (6), *one can substitute terms in it in such a way that a formula of the form* (0) *is obtained.*

Proof. For $j = 1, 2, \ldots$, gradually replace v_{jir} by f_{jir} applied to the array of terms obtained from $^2s_j^*$ at the previous stage. Thus first we replace v_{1ir} by $f_{1ir}({}^2s_1^*)$; in this way $^2s_2^*$ may be changed to some $^2s_2^{**}$, hence in the second stage we replace v_{2ir} by $f_{2ir}({}^2s_2^{**})$ and so on. Note that $^2s_1^*$ is never changed; after the first stage $^2s_2^{**}$ is never changed, etc. Thus the formula gradually acquires the form (0).

Next we shall rewrite (4) using terms and term-free formulas. We get

$$\bigvee_{j.i,l} (\delta_{ji}({}^2t_i({}^2s_j^*, {}^4q_{jil}^*)) \rightarrow \delta_{ji}({}^2t_i({}^2s_j^*, {}^3v_{ji})) \wedge v_{ji1} \leq q_{jil1}^* \rightarrow$$

$$\beta({}^1w(\tau, {}^3u_{jl}^*))), \tag{7}$$

where δ_i and β are term-free. Let us abbreviate this formula by

$$\bigvee_{j.i,l} \Delta_{ji}({}^2t_i({}^2s_j^*, {}^4q_{jil}^*), {}^2t_i({}^2s_j^*, {}^3v_{ji}), v_{ji1}, q_{jil1}^*, {}^1w(\tau, {}^3u_{jl}^*)). \quad (8)$$

Put

$${}^3t_{ji}^*({}^2V_j, {}^3v_{ji}) :={}^2 t_i({}^2s_j^*, {}^3v_{ji}),$$

where ${}^2V_j$ are the variables of ${}^2s_j^*$. Hence the condition (6) can be expressed by

$${}^3v_{ji} \text{ do not occur in } {}^2V_{j'} \text{ for } j' \leq j, \quad (9)$$

and (8) becomes

$$\bigvee_{j.i,l} \Delta_{ji}({}^3t_{ji}^*({}^2V_j, {}^4q_{jil}^*), {}^3t_{ji}^*({}^2V_j, {}^3v_{ji}), v_{ji1}, q_{jil1}^*, {}^1w(\tau, {}^3u_{jl}^*)). \quad (10)$$

Claim 2. *If a formula of the form* (10) *satisfies* (5) *and* (9), *then it can be put into the form* (0) *by a suitable substitution.*

Proof. Just note that (10) is a special case of (7), where we have variables instead of terms ${}^2s_j^*$. As (7) is only a more explicit form of (4) we get Claim 2 from Claim 1. □

The skeleton of the above proof is

$$\bigvee_{j.i,l} \Delta_{ji}({}^4a_{jil}, {}^3b_{ji}, c_{ji}, {}^3d_{jil}, {}^3e_{jl}) \quad (11)$$

where ${}^4a, {}^3b, {}^2c, {}^3d, {}^3e$ are arrays of variables which are not supposed to be mutually distinct.

We have shown that any Herbrand proof with the above skeleton determines a formula of the form (10) with conditions (5) and (9) and vice versa, such a formula determines an Herbrand proof of some $\bigwedge A_0 \to \varphi(\tau)$. Now we shall show that the structure of terms in (10) with conditions (5) and (9) can be determined by a semiunification problem. Hence a most general semiunifier will determine a most general formula of the form (10) (plus (5),(9)), thus also in a sense a most general Herbrand proof with the above skeleton. In particular we obtain a most general term τ_0 in the place of τ.

The semiunification problem is defined as follows:

variables: ${}^4a, {}^3b, {}^2c, {}^3d, {}^3e, h, {}^3x$,
substitutions: δ, σ_{jil},
equations:

(i) ${}^1w(h, {}^3x_{jl})\delta = {}^3e_{jl}\delta$,

(ii) ${}^4a_{jil}\delta = {}^3b_{ji}\delta\sigma_{jil}$,

(iii) $d_{jil}\delta = c_{ji}\delta\sigma_{jil}$,

restrictions:

(iv) $c_{ji}\delta = c_{ji}$,

(v) if $\sigma_{jil}(\xi) \neq \xi$, then ξ does not occur among the variables of ${}^3b_{j'i'}\delta$ for $j' < j$ or ($j' = j$ and $i' \neq i$),

(vi) if $\sigma_{jil}(\xi) \neq \xi$, then ξ is different from every $c_{j'i'}$ for $j' \neq j$ or $i' \neq i$ (or both),

(vii) c_{ji} does not occur among the variables of $b_{j'i'}\delta$ for $j' < j$ or ($j' = j$ and $i' \neq i$).

Furthermore, we identify c_{ji} with v_{ji1}.

We must stress that the variables ${}^4a, {}^3b, {}^2c, {}^3d, {}^3e$ are not distinct. They come from the Herbrand skeleton. (Otherwise the equations (i)–(iii) had a trivial solution.)

First we shall show that (10) with (5) and (9) determines a solution to the semiunification problem. Put

$$\begin{aligned}
{}^4a_{jil}\delta &= {}^3t^*_{ji}({}^2V_j, {}^3q^*_{jil}),\\
{}^3b_{ji}\delta &= {}^3t^*_{ji}({}^2V_j, {}^3v_{ji}),\\
c_{ji}\delta &= v_{ji1},\\
d_{jil}\delta &= q^*_{jil1},\\
{}^3e_{jl}\delta &= {}^1w(\tau, {}^3u^*_{jl}),\\
h\delta &= \tau,\\
{}^3x_{jl}\delta &= u^*_{jl},\\
\sigma_{jil}({}^3v_{ji}) &= {}^4q^*_{ji},\\
\sigma_{jil}({}^2V_j) &= {}^2V_j
\end{aligned}$$

This definition is possible, since we assume that (10) has the Herbrand skeleton (11). It remains to check the restrictions. (iv) is satisfied by definition; if $\sigma_{ji}(\xi) \neq \xi$, then ξ must be some v_{jim}, hence we get (v) from (5) and (9); to see (vi), recall that $c_{ji} = v_{ji1}$, and that (5) says that the variables in ${}^3v$ are distinct; finally (vii) means that v_{ji1} does not occur among ${}^2V_{j'}$ for $j' \leq j$ and among ${}^3v_{ji}$ for $j' < j$ or ($j' = j$ and $i' \neq i$), which follows from (9) and (5) respectively in the same way as above.

The last thing to do is to show that any solution to the semiunification problem determines a formula of the form (10) satisfying conditions (5) and (9). Trivially the terms $v_{ji1}(= c_{ji})$, $q^*_{jil}(= d_{jil}\delta)$ and ${}^1w(\tau, {}^3u^*_{jl})(= e_{jl}\delta)$ have the required form. We get some terms ${}^4a_{jil}\delta$ and ${}^3b_{ji}\delta$ and we must write them in the form $^3t^*_{ji}(^2V_j, ^3\overset{*}{q}_{jil})$ and $^3t^*_{ji}(^2V_j, ^3v_{ji})$, respectively. This is done as follows: ^2V_j are the variables that occur in $^3b_{ji}\delta$ for some i and such that σ_{jil} are identity on them for *all* i and l, i.e., $\sigma_{jil}(^2V_j) = {^2V_j}$. The variables $^3v_{ji}$ are the remaining variables of $^3b_{ji}\delta$ and the variables $c_{ji}(= v_{ji1})$. We shall prove the conditions (5) and (9).

Proof of (5): Consider v_{jim}, $m \neq 1$. This means that some $\sigma_{ji'l}$ is not identity on v_{jim} (i.e., $\sigma_{ji'l}(v_{jim}) \neq v_{jim}$). By (v) it is only possible that $i' = i$. Also by (v) v_{jim} must be different from all $v_{j'i'm'}$ with $j' < j$ or ($j' = j$ and $i' \neq i$), and $m' \neq 1$ while (vi) gives this also for $m' = 1$. For $m = 1$ we get the same from (vii). Thus we have for all v_{jim}: v_{jim} is different from $v_{j'i'm}$ with $j' < j$ or ($j' = j$ and $i' \neq i$), but this simply means that they are all different.

Proof of (9): Let v_{jim} be given. It cannot occur among ^2V_j by definition. For $j' < j$, v_{jim} cannot occur among $^2V_{j'}$ by (v) (for $m \neq 1$) and (vii) (for $m = 1$).

At last we should comment on the fact that $\varphi(x)$ may use some function symbols which are not present in $\overline{\mathbf{L}\exists_1}$. It is easily seen that these function symbols cannot occur in the terms t^*_{ji} obtained from a most general semiunifier, since the original t_{ji}'s do not contain them and they are substitution instances of t_{ji}'s. □

POSTSCRIPT: A BROADER, NOT ONLY PROFESSIONAL VIEW

Georg Kreisel

The material below may be of use to some of those readers who have gone through the paper above, and have its mathematical ideas at their fingertips; in whatever sense this phrase is appropriate, to temperament and education of a particular reader. Beyond this, as a corollary to a simple, but important home truth repeated since Pindar, the readers should not — even feel to — be so wise that the facts are enough for them. The material is concerned with an interpretation, in particular, by means of relations to broad proof-theoretic experience. Since it is not only possible, but common, to assume that scholarly traditions are suitable for such an enterprise, a general disclaimer is necessary here; to be bolstered by specifics when occasions arise.

7.1 The risk of contamination by detail

This phrase comes from the literature on chaotic elements in the sense of contemporary (population) dynamics. The details meant here are those concerning a local view, not, for example basins of attraction (and their fractional dimension). In terms of this metaphor, not meant as some kind of 'theory', it cannot be excluded that some of the *rewarding* elements in the evolution of proof-theoretic experience have the character above; with themes, as they are called in the literary tradition, corresponding to basins of attraction (which do not even make sense for a local view). Depending on the situation, (generally short-term) causal elements and, of course, randomness will have more or less weight than the chaotic variety. The latter are emphasized here, because they are systematically neglected in the scholary tradition; in effect, of course, not on purpose since it is not even aware of such a possibility.

7.2 A brief look back

By the end of the 50s there was, by and large, consent that proof theory was largely wasted; tacitly as long as it confined itself to results of the kind most prominent at the time. These referred only to the set of theorems of the (formal) rules of proof considered. Naturally, this consent was expressed in different ways (again, according to temperament and education). For example, Tarski focussed attention on the matter by the then-new terminology 'theory', and, quaintly, proposed to 'identify' rules of proof with the theory they determine. In any case, he pursued one — and surely the most immediately compelling — practical conclusion from that consent: to see where you get when rules of proof are — of course used, but — not made objects of study, in short, model theory.

As with any other (sensible) philosophy exceptions present themselves (if one does not shut one's eyes to them). For example, some aspects of some theories are particularly easily studied by use of particular descriptions, for example, cut-free rules; just as some geometric problems are best solved by use of particular coordinates, and some arithmetic problems by use of some particularly suitable notation for the natural numbers. Again, the balance shifts when theories determined by non-classical rules are looked at, and new kinds of models are concocted. It is a matter of temperament to what extent the *rarity value* of successful exceptions balances the relatively broad, if generally slight, scientific success of the model-theoretic option.

7.3 Loose ends

Dubious doubts (to be ignored or looked at, for better or for worse), here about the option above:

What about the metamathematical methods used?

(in model theory compared to the proof theory of the 50s). A wide-spread impression was that — tacitly, by the order of priority implicit in the logical tradition — the proof-theoretic methods were more economical, and so, again by that tradition, at a premium. The impression was right when applied to what most people in the field knew of metamathematics. So a first loose end to tie up comes from general experience in mathematics: A subject has to be understood pretty well before there is a chance of guessing correctly what can and what cannot be eliminated; even in principle, let alone with advantage.

By the end of the 60s, those who had thought about the topics in the last paragraph, knew the broad answer (again of course with occasional exceptions). The key reminder (for them) is: conservation results for so-called higher order principles convenient for the metamathematics of model theory; over those (often explicitly) stated in texts on proof theory. This was one theme in the broad area under discussion.[1]

Another theme concerns the doubts associated with restrictions on - or, simply, attention to — metamathematical methods. Those doubts come from the foundational literature under such headings as 'finitist' or 'predicative' or other informal concepts of proof of the logical kind. 'Logical' inasmuch as the parallel in mathematics — attention to abstractions, for example, from theorems about the integers to, say, rings with or without unit — does not go far. But the delicacy of determining rewarding selections, which is notorious in mathematics, is enough as a warning against regarding those restrictions as sacrosanct. In simplest terms: a *principal* purpose of studying those restrictions is to examine the corresponding foundational doubts, *whether or not* they are dubious. This is not all:

Proof-theoretic experience concerns (formal) rules that — are designed to — represent logical aspects of those phenomena that have made 'proof' a household word (where 'logical' has the sense(s) established in mathematical logic). Examination of those dubious doubts is an — albeit not the only — introduction to the question:

> Where, if anywhere, do logical aspects contribute to understanding the phenomena mentioned?

as opposed to distracting from more rewarding aspects. Only the most innocent would assume that mere — necessarily informal — rigour concerning those informal concepts is enough to provide an even remotely adequate idea of a balanced answer, though some degree of precision is often necessary.

Evidently, also for progress in *mathematical* logic, not only for a broader view, it can be fruitful to include the examination of (dubious) doubts, not only to pursue traditional ideals relentlessly. The examination exploits relations to other areas of knowledge (at the price of having such knowledge).

7.4 Broad (realistic) options: samples

Even those with broad experience (in the household of mathematics) *and* sound intellectual reflexes often have trouble saying what they know about the question above if they are not competent in logic. Few logicians benefit from knowledge that is not said at all or said incompetently. Yet such knowledge constitutes a *threshold for informed discussion* of that question.

Those of us who do not have it have different, generally more modest, options, for example, as a corollary to the common consent that logical

[1] This theme is mentioned here for perspective. In contrast, the next theme is quite central to the interpretation of results on the length of proofs.

aspects contribute *rarely*. In this case they may be expected not to strike the (educated) mind's naked eye, and thus to be missed when they could contribute, especially (from personal experience) when they are related to other aspects that also contribute rarely. Thus, even within the area of computation, the (mere) rate of growth of functions is usually marginal compared to the choice of parameters, their relative weight and many more things besides. But when growth *is* a central concern then socalled logical complexity may be, too; and — to the educated mind — big differences in logical complexity, say, of (induction) axioms or (cut) rules will appear small; fittingly, inasmuch as *generally* they are of no consequence.

Another option is familiar from centuries of socalled applied mathematics, in particular, of idealizations in theoretical physics. Time and again (mathematical) ideas, which may be almost embarrassingly inept for a traditional (physical) interpretation, were found to have *discovered* interest, sometimes for other physical interpretations, sometimes primarily for pure mathematics. 'Embarrassingly', of course, only *after* one knows enough both of the mathematics and the physics. — So much for broad background.

7.5 A shift of emphasis

Now, there is a perfectly obvious alternative to — what was called above — Tarski's option. It is a shift of emphasis *from* provability *to* proofs. After such a, so to speak strategic, step a selection of suitable parameters (= aspects of proofs) and of some rewarding properties (relating the parameters to other things) becomes central. Naturally, a main topic of this postscript is *length* as in the present paper, and in the sense explained there.

One general and one specific rule of thumb are decisive below. First, as with all shifts of emphasis, it is necessary to keep one's distance from the former emphasis, and so properties (of formal rules) are at a premium for which the former emphasis is *obviously* inadequate; most simply, when they differ for (some) different formal rules with the same theory; cf. 0.1(b) w.r.t. the conjecture in the title.[2]

Remark. Experience shows that a warning is needed here. 'Conjecture' is a synonym for 'guess', not meant in its technical sense appropriate

[2] Readers familiar with [Parikh] will find there a much more substantial, but less easily interpreted example of the kind of property required. Some of the *lemmas*, patently, apply only to rules, in fact, to socalled languages, that exclude function symbols with more than one argument. This is in contrast to — what has been known since the stone age (of logic) about — provability, which is not sensitive to the passage between functions and (corresponding functional) relations. However, [Parikh] — in contrast to 0.1(b) — leaves open, whether the — property expressed by the — conjecture is sensitive in the way required by the rule of thumb above.

(and used) in mature branches of mathematics. Elsewhere that sense is premature, and thus scientifically immature.

More specifically, by 7.4 on the limitations of logical aspects w.r.t. the phenomena around the household word 'proof', (cl)aims about length concerning, say, complexity (of proofs) are suspect. Thus forewarned (and forearmed), it is easy to be specific; cf. pp. 252-253 of [Kreisel2] for some recent reminders. At this stage the second option of 7.4 becomes relevant: the possibility of discovering properties of length that are significant (besides the conjecture); specifically, (ii) on p. 253 of [Kreisel2] mentions examples with the same general computational flavour as in the corollary in 7.4.

7.6 Another shift of emphasis

From the conjecture in the title of this paper, to the theme of generalizing theorems with proofs that are short compared to the syntactic depth of terms in the theorem considered. The style of this section is autobiographical, and hence needs due regard to the risk of contamination by detail. Some corollaries for a broader view follow in 7.8.

(a) In 1984, one of the authors (M.Baaz) told me a proof of the main result in [Parikh] by use of *unification.* (This prepared me for the dissertation of W.M. Farmer.) When the new proof had sunk in I was struck by another aspect of the (same) argument, reported in footnote 3 on p. 400 of [Kreisel1] under the heading (different from the theme above):

generalizing short proofs of theorems

(with terms that are deep compared to the length of the proof considered). The difference concerns the logical form, aka skeleton, of the proof(s) of the generalized theorem. In the case of [Parikh], for a given proof π of $A[x/s^n(0)]$ with suitably large n and suitable $d \in \omega$, and for all $m \in \omega$, there are proofs π_m of the *same logical form* as π s.t.[3]

$$\pi_m \quad \text{proves} \quad A[x/s^{n+d\cdot m}(0)].$$

The exponents (in the iterations) of s above are, in familiar terms, elements of the arithmetic progression starting at n with the common difference d. So A is not only proved (by π) for $x = n$, but by π_m for each x

[3]Relations to footnote 3 of p. 400 [Kreisel1]. (i) In l. 10, m and n should be interchanged. Then this m corresponds to d above. (ii) In l. 9–10 there is some sort of attempt — in terms of some 'similar' logical form — to state the generalisation above (with metavariable m) by a theorem of the system itself. The attempt is not only formally clumsy, but probably premature.

on that progression. In the rest of this section, the slogan 'generalization to arithmetic progressions' will be used; naturally, with the understanding that it does not have to do primarily with arithmetic properties of s, let alone of $+$ (which is very prominent in [Parikh]), as stressed explicitly in [Kreisel1] and in 4.2 of this paper. But for a broader (mathematical) view there is more to be said, cf. the next section.

(b) The emphasis on (generalizations to) arithmetic progressions was consolidated — so as to become a basin of attraction — when I learnt (again from M. Baaz) a fact buried in [Yukami]: all equations $s^n(0) + s^m(0) = s^{n+m}(0)$ have proofs of bounded length, and so, in particular, $A[x/s^{2n}(0)]$ when A is $\exists y(x = y + y)$.

Thus, in general, theorems $A[x/s^{2n}(0)]$ with short proofs *cannot* be generalized brutally, for example to $\forall x \geq n\ A$.

(c) As noted in the footnote loc.cit. the original result follows easily from the new formulation. This is written out on pp. 81-82 of [Krajíček/Pudlák]. But this is only a formal, as it were legalistic, case for the shift of emphasis. More compellingly, the new formulation implies that, in contrast to the case of $+$ in (b), there is no uniform bound on proofs of $s^n(0) \cdot s^m(0) = s^{n \cdot m}(0)$. Otherwise, if A is $\exists y(x = y \cdot y)$ and n is square, there would be a short proof of $\exists y(s^n(0) = y \cdot y)$. But the squares do not form (nor even lie on) an arithmetic progression.

This may be compared to the observation of M. Baaz (last paragraph of Section 4) that, for arbitrary formal systems for $+$ and $\cdot$, if the conjecture holds then there is no uniform bound for proofs of all numerical identities *both* for $+$ *and* $\cdot$.

(d) The reference to logical form in (a) remained no more than a kind of memo (to myself) until relatively recently (in fact, after pp. 252-255 of [Kreisel2] were written, where the matter is not mentioned). It became a theme when I had the good fortune to encounter patently absurd claims that deciding k-provability was relevant to the conjecture (for which preservation of logical form is *not* relevant). 'Patently absurd' when made by a reader of [Krajíček/Pudlák] who also knows that k-provability is not decidable for the rules on pp. 82–83 for which the conjecture is proved.

(e) More specifically in connection with the present paper, several years ago M. Baaz drew my attention to the description of the first–order theory of arithmetic provided by the (rules known as) *least number* schema, here meant w.r.t. the order relation; w.r.t. the successor relation, it is the familiar schema of *induction*. (The former is familiar in connection with transfinite ordinals.) What he told me about — his impressions of — a proof was quite enough to make it a suitable candidate according to the first strategic requirement of Section 7.5. Specifically the *forms* of generalization differ for them, cf.4.2 with generalizations to *all* sufficiently large

n, and not only arithmetic progressions.

7.7 A touch of (mathematical) breeding

By the warning at the end of 7.6(a), the loose use of 'arithmetic progression' requires a little thought; here under the heading: the risk of contamination by premature abstraction (and corresponding terminology).

A well known parallel comes from the common practice of labelling one of van der Waerden's theorems by reference to arithmetic progressions. Here too it is evident that the result in question extends to (subdivisions of) plenty of other semigroups (into finitely many classes). But as matters stand around this high spot of recreational mathematics it is wide open for which semigroups and for which kinds of subdivisions it contributes to effective knowledge. For the insider, to whom the choice of 'arithmetic' is *obviously* haphazard, common practice provides a safeguard against premature views (= illusions) on what is essential about that theorem.

Returning to 7.6(a) and generalizations of $A[x/s^n(0)]$, as a minimum one would look first at other deep terms (besides $s^n(0)$ with large n) and at formulae A with more than one variable. These points dominate p. 254 of [Kreisel2], but the broad background above was not spelt out.

In this connection it should be emphasized that the word 'unification' was used in 7.6(a) also only because of its familiarity; as it were despite the reservations in l. -11 to l. -8 of [Kreisel2]. Roughly, 'unification' means solution for a particular kind of word equation; cf. the references in footnote 3 of [Kreisel1] to Makanin's work. Now, by p. 261 of [Kreisel2] the time may be ripe to stop pussy footing around unification, and not only those other branches of theoretical computer science mentioned there. On the one hand this 'science' teems with new words, which obscure (totally obvious) relations to already existing intellectual resources; both elementary, as in l. -11 to l. -8, and others. For example, in the geometry of numbers, applied to linear diophantine equations, 'geometric' arguments often hold for plenty of (suitable semi-)groups. But so far no rewarding occasion has been encountered in number theory to abstract from these arguments: *practical* computational problems may provide one. On the other hand, the 'science' is full of remarks and observations, of which some are obviously sensible, while most turn into gibberish when called 'theorems'; cf. the remark on 'conjecture' in 7.5 above. (Incidentally, this 'science' shows that not being a science is not enough for being Art.)

7.8 Out of professional bounds

At the end of the introduction to the present paper the topic of consequences (of the conjecture) is raised. For a professional view, the observation of M. Baaz in 7.6(c) is a correct answer. In contrast, for the broader view adopted in this postscript it is always an open question where — not only some logical result, but simply — logical experience is particularly rewarding; without *assuming* that it must correspond to first impressions! Viewed this way, the story in 7.6 serves very well to underline the reservations, at the beginning of the postscript, about the scholarly tradition, for example, punctilious chronologies.

(a) Perhaps, it is fitting that the conjecture is, haphazardly, attributed to me though (no doubt) it must have crossed many other minds, too: 30 years ago it was an haphazard guess. But much intervened before — something like — it became a theme.

(b) Are relations of cause and effect, aka influences in the scholarly tradition, at all appropriate in connection with 7.6(a)? As so often, one perfectly legitimate option is to ignore the matter (of influences, and with it this question). But it happens to enter the broad view adopted in this postscript, and so one sees what, if anything, is to be said about it (on this side of some point of diminishing returns).

At one extreme would be the relative contribution, of oral instruction by Baaz and of reading Farmer's dissertation, for drawing (my) attention to unification.[4] Probably not even the scholarly tradition would assume the evolution of (my) knowledge to be free enough of chaotic elements to expect a sensible answer (though the Freudian idea(l) of overdeterminism certainly does).

At another extreme there is the fact that [Parikh] certainly contains *instances* of unification. What was missing (objectively and/or in my understanding)? Above all — and when forgotten, such trivial facts can have profound consequences —, [Parikh] contains instances of infinitely many other general ideas, too! So one 'missing piece of the puzzle' is *emphasis* on a relation to the particular idea of unification (in an aside, joke, (genuine) theorem or whatever literary form may suit a particular temperament). A consequence, which is certainly central for the shift in 7.6, is this:

The detour (via reflection for Presburger arithmetic) in [Parikh] is so far from a, let us say, implicit focus on unification that it spoils the potential use of the latter for generalizing proofs (so as to preserve logical form; recall π and π_m in 7.6(a) above).

(c) Certainly, consciously I was put off by Farmer's emphasis on - what

[4]The points under discussion are so general, and hence so elementary that they are not affected by the (more) demanding section 7.7.

is called in the trade — a new result, which extends [Parikh] by adding an odd special case of the identity schema. For the record, today I see the identity schema as a quite promising item in the present context. Specifically, together with its obvious propositional counterparts it is a, if not the, principal *logical* schema. And experience, for example, of decision problems, shows that logical methods are often easier to apply to logical objects (even if later combinations with more specific subject matter are more rewarding).

(d) When speaking of 'education' — and not only in the particular cases in (b) of this section — what degree of 'accuracy' is warranted by the data, when it comes to the extent to which one understands what the other means or (and this is often not the same) says? All this applies with knobs on to understanding what one has read. Thus this section has not been understood (as meant by me) unless one has spontaneously concluded, for example: (a) Where the horizon of predictability, familiar from chaotic dynamics, ends the idea(l) of responsibility begins to be a *contradictio in adjecto*, or: (b) If that horizon is close in one direction, there may be another that is scientifically more rewarding; in other words, an opportunity for a (rewarding) shift of attention. The latter is called 'shift of emphasis' in this postscript, and illustrated throughout.

7.9 A glaring, but innocuous asymmetry

There has been a good deal about my education by one of the authors (M. Baaz), and nothing about the other. Since I have never met the latter, there are simply no conversations on which I could muse (or report). Besides, his (principal) part in the paper is very clear.

He did not merely reformulate ideas from the ε-calculus, thus making the material accessible to some of those who do not know it (but do know Herbrand's theorem, which incidentally was first proved properly, in Hilbert-Bernays, by use of the first ε-theorem.[5]) He actually wrote up the paper; without footnotes, asides, allusions (or pomposity). The future may show that at least for some sectors in the commerce of ideas this is decisive.

[5] Incidentally, closer attention to that proof, especially as applied to Herbrand's original formulation, is recommended (also by M. Baaz) to those who wish to extend the methods of this paper to $\mathbf{L}\exists_\infty$, too. For the record I find that enough is as good as a feast, and should be more interested in (new) aspects presented by the ε-calculus, especially when identity is included as in the second ε-theorem.

For the present view it must be stressed that here (and so far in all rewarding contexts) Herbrand's theorem is applied to proofs of *prenex* formulae. It is wide open whether elaborations are on this side of a point of diminishing returns. Accordingly, the proof of the theorem in the *first* edition of Hilbert–Bernays is meant. (I hold no brief for their formulation of the second ε–theorem.)

The last parenthesis (on pomposity) has to do with [Baaz1]. At least as I see it, my education by M. Baaz has been pretty satisfactory, much less the reverse. Thus neither the last sentence of footnote 3 in [Kreisel1] nor many similar remarks in conversation since then (in stronger terms like l. −2 and l. −1 on p. 254 of [Kreisel2]) have prevented the — to me, breathtaking — pomposity of [Baaz1], of which, admittedly, the title gives fair warning; specifically of two unfounded assumptions: First, what is true for all generalizations of proofs is likely to be rewarding in many particular cases, and secondly, something like the kind of generalization considered in this paper, that is, of theorems with syntactically complex terms and short proofs, is often encountered in mathematical practice; cf. 2 and 7.4 on rarity value. (There are plenty of sensible remarks and observations in [Baaz1].) For the record, 'gifted computer-minded mathematicians' in [Kreisel2] was meant to include Euler, in particular, the computation that is often interpreted as containing a proof of Wedderburn's theorem (on the commutativity of finite fields).

References

[Baaz1] M. BAAZ: *Über den allgemeinen Gehalt von Beweisen.* Contributions to General Algebra 6, Holder-Pichler-Tempsky, B.G. Teubner, (1988), pp. 21-29.

[Baaz2] M. BAAZ: *Note on the Existence of Most General Semi-Unifiers.* In this volume.

[Barwise] J. BARWISE: *An Introduction to first-order Logic* in J. BARWISE: *Handbook of Mathematical Logic.* 5th edition, (1989), North-Holland.

[Friedman] H. FRIEDMAN *One hundred and two problems in mathematical logic*, JSL 40/2 (1975), pp. 113-129.

[Kaye] R. KAYE: *Diophantine Induction.* Ann. Pure Appl. Logic 46, No. 1 (1990),pp. 1-40.

[Kfoury et al.] A.J. KFOURY, J. TIURYN, AND P. URZYCZYN. *The undecidability of the semi-unification problem.* Proceedings of the 22nd Annual ACM Symposium on Theory of Computing, pp. 468-476 (1990), journal version to appear in *Information and Computation.*

[Krajíček/Pudlák] J. KRAJÍČEK, AND P. PUDLÁK: *The Number of Proof Lines and the Size of Proofs in First Order Logic.* Arch. Math. Logic 27 (1988), pp. 69-84.

[Krajíček] J. KRAJÍČEK: *On the number of steps in proofs.* Ann. Pure and Appl. Logic 52,1–2 (1991), pp. 143-154.

[Kreisel1] G. KREISEL: *Some personal recollections*, an appendix in G. TAKEUTI: *Proof Theory*, 2nd edition, (1986), pp. 395-445.

[Kreisel2] G. KREISEL: *Logical Aspects of Computation: Contributions and Distractions* in P. ODIFREDDI: *Logic and Computer Science.* London, (1990), pp. 205-278.

[Parikh] R. PARIKH: *Some Results on the Length of Proofs.* TAMS 177 (1973), pp. 29-36.

[Pudlák] P. PUDLÁK: *On a Unification Problem Related to Kreisel's Conjecture.* Comm. Math. Univ. Carol. 29,3 (1988), pp. 551-556.

[Richardson] D. RICHARDSON: *Sets of Theorems with Short Proofs*, JSL 39/2 (1974), pp. 235-242.

[Yukami] T. YUKAMI: *Some Results on Speed-Up.* Ann. Jap. Assoc. Philos. Sci. 6 (1984), pp. 195-205.

Matthias Baaz
Institut für Algebra
und Diskrete Mathematik
Technische Universität Wien
Wiedner Hauptstraße 8-10
A–1040 Wien
Austria
email: baaz@csdec1.tuwien.ac.at

Pavel Pudlák
Mathematical Institute
ČSAV
Žitná 25
11567 Praha 1
Czechoslovakia
email: pudlak@csearn.bitnet

Number of Symbols in Frege Proofs with and without the Deduction Rule

Maria Luisa Bonet*

Abstract

Frege systems with the deduction rule produce at most quadratic speedup over Frege systems using as a measure of length the number of symbols in the proof. We study whether that speedup is in reality smaller. We show that the speedup is linear when the Frege proofs are tree-like. Also, two groups of formulas, permutation formulas and transitive closure formulas, that seemed most likely to produce an almost quadratic speedup when using the deduction rule, are shown to produce only $\log n$ and $\log^2 n$ factors respectively.

1 Introduction

A Frege proof system is an inference system for propositional logic in which the only rule of inference is Modus Ponens.

Definition 1 *A Frege Proof System consists of:*

1. *A finite complete set of propositional connectives. We use the following set:* $\neg, \vee, \wedge, \supset$.
2. *A finite set of axiom schemata.*
3. *A proof in this system is a sequence of lines* $A_1, \ldots, A_n$ *where each* A_i *is either a substitution instance of an axiom scheme, or is inferred by Modus Ponens(MP) from some formulas* A_j *and* A_k *where* $j, k < i$.

*Supported in part by NSF Grant DMS-8902480.

4. The proof system must be consistent and complete.

By Modus Ponens we mean the rule:

$$\frac{A \quad A \supset B}{B}$$

We write $A_1, \ldots, A_n \vdash_{\overline{k}} B$ to mean that there is a sequence of $\leq k$ formulas each of which is one of the A_i's, is an axiom, or is inferred by modus ponens from earlier formulas such that B is the final formula of the proof.

Although it suffices to have modus ponens as the single inference rule to obtain a complete proof system, it is well-known that other modes of inference are also sound. The most notable example of this is the deduction rule which states that if a formula B has a proof from an additional, extra-logical hypothesis A (in symbols, $A \vdash B$) then there is a proof of $A \supset B$.

$$\frac{A \vdash B}{A \supset B} \qquad \text{Deduction Rule}$$

This paper establishes upper bounds on the proof speedups obtained by this rule on certain types of proofs and certain types of formulas. By a "speedup" of a proof, we mean the amount that proofs can be shortened with additional inference rules. In this paper, the *length* of a proof is the number of symbols in the proof.

If S and T are proof systems we say that S can linearly (respectively, quadratically) *simulate* T if, for any T-proof of k symbols, there is an S-proof of the same formula of $O(k)$ symbols (respectively, of $O(k^2)$ symbols). We say that T provides at most linear (respectively, quadratic) *speedup* over S if S can linearly (respectively, quadraticly) simulate T.

Counting length in terms of number of symbols, any two Frege systems simulate each other with at most polynomial increase in the length of the proofs if we change the set of connectives. See Reckhow [8] for a proof of this fact. If we only change the set of axiom schemata, again any two Frege systems simulate each other with only linear increase in the length (number of symbols) of the proofs (see [1]). Therefore in this paper we fix the set of connectives, but not the set of axiom schemata. For an example of a complete set of axiom schemata see [7].

In section 2 we analyze the proof of the deduction theorem to show that Frege systems with the deduction rule provide at most quadratic speedup over the Frege systems. The second theorem proves that Frege systems linearly simulate tree-like Frege systems with the deduction rule.

The open question is whether Frege systems can do better than a quadratic simulation of the deduction Frege systems. The last two sections of this paper will consist of the analysis of some examples that seem to be the most likely to get an almost quadratic speedup when using the deduction rule. But in all our examples, we prove that the speed up is much less than quadratic.

Also there is an intermediate section (section 3) about ways to associate parenthesis on formulas, and how that leads into different ways of balancing trees.

Work on lengths of propositional proofs counting the number of symbols in a proof was done also by Cook-Reckhow [6, 8] and Statman [10]. Also work on strengthenings of the deduction theorem but counting number of lines as a measure of length was done by Bonet-Buss [1, 4, 2].

2 The Deduction Theorem Counting the Number of Symbols

Let us make the following important comment before we prove any theorems in this section:

Note Say a fixed tautology B has $p_1, \ldots, p_n$ as propositional variables, and a proof of B has m symbols. Then if we substitute $p_1, \ldots, p_n$ by $A_1, \ldots, A_n$ respectively, then the resulting formula has a proof with at most $m(|A_1| + \ldots + |A_n|)$ symbols.

Theorem 1 *If $A \vdash B$ in n symbols, then $\vdash A \supset B$ in $O(n^2)$ symbols.*

Proof By the hypothesis, there is a proof of B from the assumption A that looks like:

$$\left[\begin{array}{l} C_1 = A \\ C_2 \\ \vdots \\ C_m = B \end{array}\right.$$

To find a proof P of $A \supset B$ in the Frege system, we will transform the former proof into the sequence P':

$$A \supset C_1$$
$$A \supset C_2$$
$$\vdots$$
$$A \supset C_m$$

This is not really a Frege proof, but we can easily get a Frege proof by filling in the gaps the following way:

The first line is $A \supset A$. We need to introduce a proof of it at the beginning of P to justify that first line. This means an increase in length of $O(|A|)$ symbols. This is because if $p \supset p$ has a proof of a constant k number of symbols, then by the note $A \supset A$ has a proof in at most $k|A|$.

We need to fill in the rest of the lines to make up a proof. Let's justify $A \supset C_i$ for any i such that $2 \leq i \leq n$. We have two cases:

- C_i is an axiom. The proof of $A \supset C_i$ will consist of: axiom C_i, a proof of $C_i \supset (A \supset C_i)$ and a use of MP to get $A \supset C_i$. All this will have length $O(|A| + |C_i|)$.

- C_i was obtained by MP on C_j and C_k. Say $C_k = C_j \supset C_i$. Then in some former lines of P' we have $A \supset C_j$ and $A \supset C_k$. From these two former lines, we can justify $A \supset C_i$, and the number of symbols to do such justification is $O(|A|+|C_i|+|C_j|)$, or alternatively $O(|A| + |C_k|)$.

This way we get a proof P of $A \supset B$. In general each line $A \supset C_i$ can be justified with

$$d(|A| + |C_i| + |C_k|)$$

symbols, where d is a constant. Each formula in the proof of $A \vdash B$ will play the role of C_k only once, since each formula will be proven only once.

If we count the number of symbols of P, we get that it is at most

$$d(m \cdot |A| + \sum_{i=1}^{m} |C_i| + \sum_{i=1}^{m} |C_k|) \qquad (1)$$

Also, all we know about A and m is that $|A| < n$ and $m < n$. So P has less than

$$d(n^2 + n + n) \qquad (2)$$

symbols. Therefore $A \supset B$ has a proof in $O(n^2)$ symbols. □

If we could bound m and $|A|$ so that $m \cdot |A| = O(n)$, then the above argument shows that we can obtain a proof of $A \supset B$ with $O(n)$ symbols. But it is possible to have a proof where both $|A|$ and m are $O(n)$. In fact the examples we are going to be working with have big $|A|$'s and m's. Before we look into these examples, let us show that Frege systems linearly simulate tree-like Frege systems with the deduction rule.

Definition 2 *We say a proof P is tree-like if each line in P is used only once in the proof. By used in a proof we mean to be a hypothesis of a MP inference. Also, if $A \vdash B$, then A can occur more than once in the proof.*

Definition 3 *Given $A \vdash B$, we say A is used in a linear way if all the lines $A_1, \ldots, A_s$ in the proof that depend on A are obtained the following way: there are lines $\varphi_1, \ldots, \varphi_s$ in the proof that don't depend on A (i.e. they can be proved without any hypothesis) such that $\forall i, 1 \leq i \leq s$, A_i is obtained by MP on φ_i and A_{i-1}, A_1 is obtained by MP on φ_1 and A, and $A_s = B$. Pictorically, $A \vdash B$ looks like*

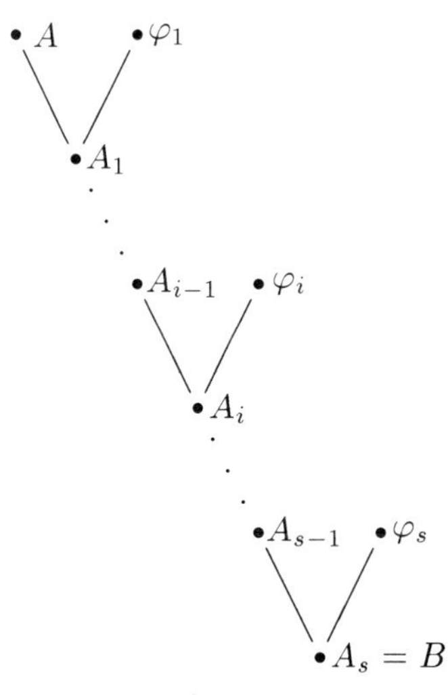

Figure 1

where the φ_i's have their proofs possibly non tree-like.

Lemma 2 *If $A \vdash B$ is tree-like and A occurs only once in the proof, then also A is used in a linear way.*

However, the opposite is not true, since the formulas $\varphi_1, \ldots, \varphi_s$ could have been obtained in a non tree-like manner. The next theorem is due to S. Buss.

Theorem 3 *Suppose that $A \vdash B$ in n symbols, with A used in a linear way. Then $\vdash A \supset B$ in $O(n)$ symbols.*

Proof Say P is the proof of B from the hypothesis A in n symbols. Say $A_1, \ldots, A_s$ are the lines in the proof that depend on the hypothesis A. Also there are lines $\varphi_1, \ldots, \varphi_s$ such that each A_i is obtained by MP from φ_i and A_{i-1}. Note that $A_s = B$, since B should depend on A, otherwise B would be provable without assuming A.

Let us create the following sequence P':

$$
\begin{array}{l}
\varphi_1 \supset (A \supset A_1) \\
\varphi_2 \supset (A_1 \supset A_2) \\
\vdots \\
\varphi_s \supset (A_{s-1} \supset A_s) \\
\\
A \supset A_1 \\
A_1 \supset A_2 \\
\vdots \\
A_{s/2-1} \supset A_{s/2} \\
A_{s/2} \supset A_{(s/2)+1} \\
\vdots \\
A_{s-1} \supset A_s \\
\\
A_{(s/2)-1} \supset A_{(s/2)+1} \\
A_{(s/2)-2} \supset A_{(s/2)+2} \\
A_{(s/2)-3} \supset A_{(s/2)+3} \\
\vdots \\
A_1 \supset A_{s-1} \\
A \supset A_s
\end{array}
$$

The first s lines of the sequence above have less than $O(n)$ symbols because each formula A_i appears only twice, and $|\varphi_i|$ is $O(|A_{i-1}| + |A_i|)$. The second set of s lines has again less than $2 \cdot n$ symbols because each formula of the proof P appears at most twice. The last $s/2$ lines of the sequence above have less than n symbols because each formula appears only once. Therefore, the sequence above has at most $5 \cdot n$ symbols.

From the sequence P' we want to create a proof P_1 of $A \supset B$ in $O(n)$ symbols. To do that we need to justify each line, i.e., to fill in the blanks in P'.

In the first group of s lines each line $\varphi_i \supset (A_{i-1} \supset A_i)$ is justified by adding to the proof $O(|\varphi_i| + |A_{i-1}| + |A_i|)$ symbols. Since each formula A_i shows in the first lines of P' at most twice, and $|\varphi_i|$ is $O(|A_{i-1}| + |A_i|)$, the first s lines can be justified in $O(n)$ symbols. The reason we worry about $|\varphi_i|$ is because some φ_i's could be identical, and they wouldn't be repeated in the original proof as they would in P'.

The second set of s lines can be justified very easily. P contains a subproof that doesn't depend on A and proves all the φ_i's. Insert this subproof between the first and second set of s lines. This subproof has $< n$ symbols since it is just part of P. Now with s uses of MP (using the first s lines of P'), we can get the second set of s lines. The total number of symbols added is $O(n)$.

Now let us explain how to justify the last $s/2$ lines. We get each $A_{(s/2)-i} \supset A_{(s/2)+i}$ from the former lines, assuming we have justified lines $A_{(s/2)-i} \supset A_{(s/2)-(i-1)}$, $A_{(s/2)-(i-1)} \supset A_{(s/2)+(i-1)}$, and $A_{(s/2)+(i-1)} \supset A_{(s/2)+i}$. To obtain $A_{(s/2)-i} \supset A_{(s/2)+i}$ we add $O(|A_{(s/2)-i}| + |A_{(s/2)-i+1}| + |A_{(s/2)+i-1}| + |A_{(s/2)+i}|)$ symbols. Note that $A_{(s/2)-(i-1)}$ and $A_{(s/2)+(i-1)}$ were used in the line before, but they won't be used anymore in the sequence. Also $A_{(s/2)-i}$ and $A_{(s/2)+i}$ were not used before, but they will be used to justify the next line and then not used anymore. So each line A_i from P is used a constant number of times to justify the last $s/2$ lines. Therefore the number of symbols added to justify the last $s/2$ lines is $O(n)$ also.

Therefore the proof P_1 obtained by filling in the blanks of P' as explained above has $O(n)$ symbols. $\square$

If we analyze the former proof, and compare it with what we did in theorem 1, we will see that the reason why we don't get the quadratic increase is because we don't have to put $A \supset$ in front of each line. In fact, for all A_i, we put $A_i \supset$ in front of a line only a constant number of times in the sequence.

Corollary 4 *Suppose that $A \vdash B$ in n symbols, A occurs only once and the proof is tree-like. Then $\vdash A \supset B$ in $O(n)$ symbols.*

Theorem 5 *Suppose that $A \vdash B$ in n symbols, and the proof is tree-like. Then $\vdash A \supset B$ in $O(n)$ symbols.*

Proof We prove that $\vdash A \supset B$ in $c(3n - |A| - |B|)$ symbols for some constant c, by induction on the number of occurences of A in the tree-like proof. Suppose the theorem is true for all proofs where A occurs less than l times. We are going to prove it for $l \geq 2$ (if $l = 1$ it's true by the former corollary).

Say $A \vdash B$ has l occurences of A, and it looks as figure 1, where now the φ_i's have tree-like proofs, and some of them depend on A. Say A_i is such that A_{i-1} and φ_i depend on A, but for all $j > i$ φ_j doesn't depend on A (this always happens since $l \geq 2$). Also say the proof of $A \vdash A_{i-1}$ has m_1 symbols, the proof of $A \vdash \varphi_i$ has m_2 symbols, and the proof of $A_i \vdash B$ (the rest) has m_3 symbols. So $m_1 + m_2 + m_3 = n$. In the proof of $A_i \vdash B$, A_i was used in a linear way, since the φ_j's for $j > i$ don't depend on any hypothesis. So by the theorem 3, we can obtain a proof of $\vdash A_i \supset B$ in $O(m_3)$ symbols, say in $d \cdot m_3$ symbols. Also by the induction hypothesis, there are proofs of $A \supset A_{i-1}$ and $A \supset \varphi_i$ of length $c(3m_1 - |A| - |A_{i-1}|)$ and $c(3m_2 - |A| - |\varphi_i|)$ respectively, since A occurs less than l times in those subproofs. We can obtain a proof of $A \supset B$ the following way:

$$
\begin{array}{ll}
\left.\begin{array}{c} \vdots \\ A \supset A_{i-1} \end{array}\right\} & c(3m_1 - |A| - |A_{i-1}|) \\
\left.\begin{array}{c} \vdots \\ A \supset \varphi_i \end{array}\right\} & c(3m_2 - |A| - |\varphi_i|) \\
\left.\begin{array}{c} \vdots \\ A_i \supset B \end{array}\right\} & dm_3 \\
\left.\begin{array}{c} A_{i-1} \wedge \varphi_i \supset A_i \\ A \supset A_i \\ A \supset B \end{array}\right\} & d_1(|A_{i-1}| + |\varphi_i| + |A_i| + |A| + |B|)
\end{array}
$$

Note that if $A_i = A_s = B$, then we don't have to do the work in the third section.

Adding up all the work we get a proof with at most

$$c(3m_1 + 3m_2) + (d_1|A| - 2c|A|) + (d_1|A_{i-1}| - c|A_{i-1}|) + (d_1|\varphi_i| - c|\varphi_i|) + (dm_3 + d_1|B| + d_1|A_i|)$$

number of symbols. Since $dm_3 + d_1|A_i| + d_1|B| \leq m_3(d + d_1)$, taking $c \geq d + d_1$, we get that the number of symbols in the proof is

$$\begin{aligned} &\leq c(3m_1 + 3m_2) - c|A| + cm_3 \\ &\leq c(3m_1 + 3m_2 + 3m_3 - 2m_3 - |A|) \\ &\leq c(3n - |B| - |A|). \end{aligned}$$

So the result follows. □

3 Balanced and Pseudobalanced Trees and Formulas

Definition 4 *We say a tree is* binary *if each node has degree at most two. We say a binary tree is* complete *if each node has degree either zero or two, and all the leaves (i.e. nodes of degree zero) are at the same depth.*

Definition 5 *A tree T is* balanced *if for every subtree T' of T of n leaves, the left immediate subtree of T' has $\lceil (n-1)/2 \rceil$ leaves, and the right immediate subtree of T' has $\lfloor (n-1)/2 \rfloor$ leaves.*

Definition 6 *A tree T is* pseudobalanced *if for every subtree T' of T, the left immediate subtree of T' is a complete binary tree, and the right immediate subtree of T' has at most as many nodes as the left immediate subtree.*

Fact: A complete binary tree is balanced and pseudobalanced.

Parenthesis on formulas can be associated in several different ways. In this paper, we are going to do it in a pseudobalanced way, following the tree model. In fact we can view a formula as a tree, and depending on how the parenthesis are associated, the tree that it represents can be more or less balanced. But before we look at ways to parenthesise a formula, let us see how formulas are represented by trees. Given a formula F,

i) if F is atomic, $F = p$, then the tree has one node and zero edges. The node is labeled as p.

ii) if F is $(A \wedge B)$, $(A \supset B)$ or $(A \vee B)$, then F represents a tree that has a root labeled $\wedge$, $\supset$ or $\vee$, and A represents the left immediate subtree, and B the right immediate subtree. If F is $\neg(A)$, then the root of the tree is labeled $\neg$, and A represents the left immediate subtree of the tree rooted at $\neg$. Such a tree won't have a right immediate subtree.

Definition 7 *A formula A is binary, if it only contains connectives $\wedge$, $\vee$ and $\supset$.*

Definition 8 *The depth of a formula is the height of the tree that represents it.*

Definition 9 *A binary formula is completely balanced if the tree that represents is a complete binary tree. Thus, a completely balanced binary formula has 2^n atomic formulas and $2^n - 1$ connectives, where n can be any natural number. A formula is balanced, if the tree that represents it is balanced. A formula is pseudobalanced, if the tree that represents it is pseudobalanced.*

The definitions of balanced and pseudobalanced formula make sense for binary formulas only, since the presence of $\neg$ makes the formula harder to balance. Although, one way to balance non-binary formulas would be to push $\neg$'s down to the leaves, and then pseudobalance the formula.

Let us now give some examples of balanced and pseudobalanced:

Example 1: A balanced formula like

$$[((A_1 \wedge A_2) \wedge A_3) \wedge ((A_4 \wedge A_5) \wedge A_6)] \wedge [((A_7 \wedge A_8) \wedge A_9) \wedge (A_{10} \wedge A_{11})]$$

produces a highly balanced tree:

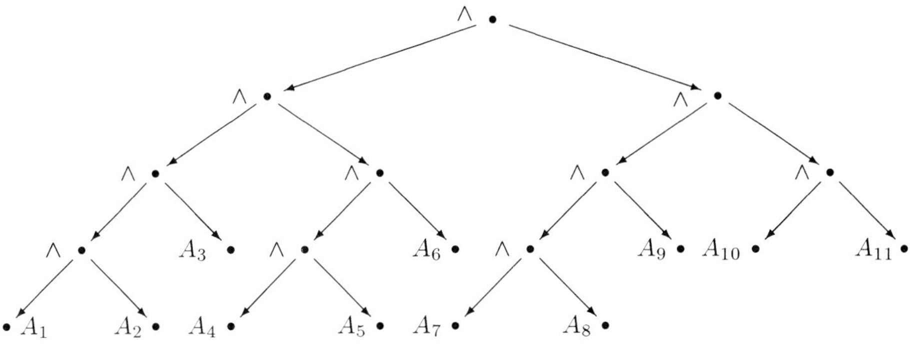

Figure 2

Example 2: A pseudobalanced formula like

$$[((A_1 \wedge A_2) \wedge (A_3 \wedge A_4)) \wedge ((A_5 \wedge A_6) \wedge (A_7 \wedge A_8))] \wedge ((A_9 \wedge A_{10}) \wedge A_{11})$$

produces the following tree:

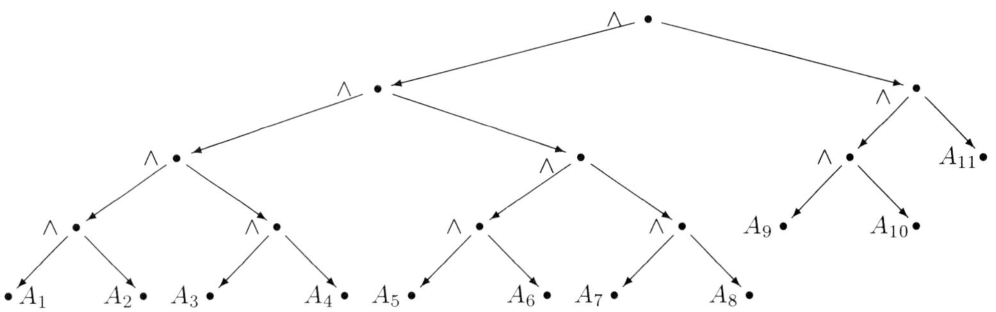

Figure 3

Notice that the depth of the tree is actually the same in the balanced tree.

Now let us prove some facts about what it takes to restore the pseudobalanced property on trees and formulas. Given two pseudobalanced formulas, we will study how many symbols it takes to prove that the conjunction of them implies a new pseudobalanced formula. And if we take two pseudobalanced trees and make a new tree by making each one be an immediate subtree of the root, we will see how can we restore the pseudobalanced property in the new tree. Before we prove these facts, let us define the notion of operations on trees, and what it means to pseudobalance a tree:

Definition 10 *There are two kinds of tree operations:*

1. *Given a node X with left child Y and right child Z, then the operation of switching T_Y and T_Z is a permutation operation.*

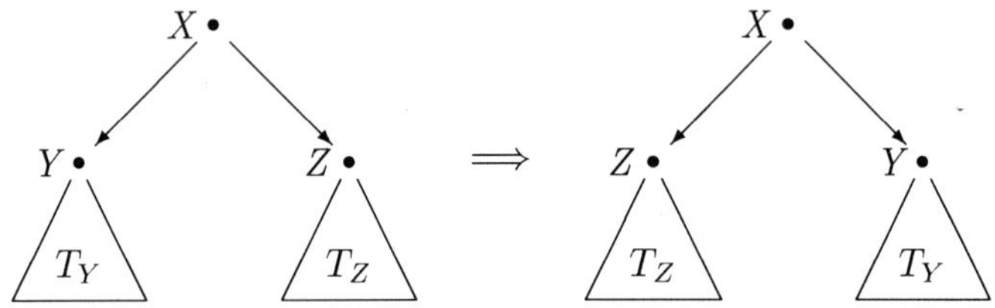

Figure 4

2. *The second kind is called a rotation operation. The notion is best explained by the figure below:*

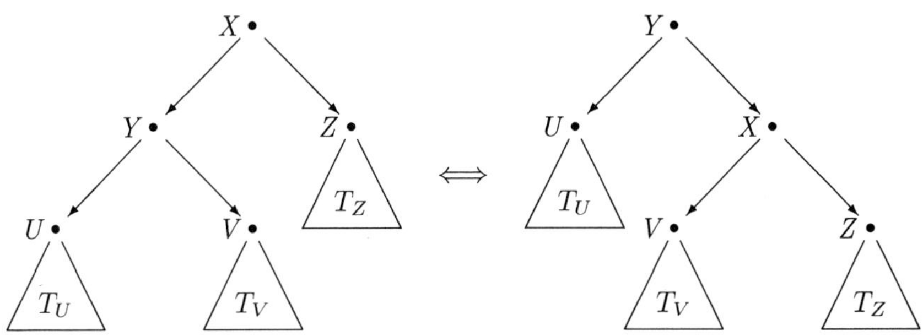

Figure 5

When we go from the picture on the left to the one on the right, we say we perform a right rotation on Y. When we go from the picture on the right to the one on the left, we say we perform a left rotation on X.

The second type of operation is a well-known one. For more information see [9]. Since we can represent formulas as binary tree, let us see the effect on formulas and on the association of parenthesis that these two operations have. The permutation operation corresponds to:

$$(A \wedge B) \Longrightarrow (B \wedge A)$$

The rotation operation corresponds to:

$$((A \wedge B) \wedge C) \Longrightarrow (A \wedge (B \wedge C))$$

and

$$(A \wedge (B \wedge C)) \Longrightarrow ((A \wedge B) \wedge C)$$

Definition 11 *Let T be a binary tree. To pseudobalance T means to obtain a pseudobalance tree T' by doing a succession of tree operations on T. Note that the leaves might be permuted doing the tree operations.*

Theorem 6 *Let T_1 and T_2 be two pseudobalanced binary trees, and let T be the binary tree such that T_1 is the left immediate subtree, and T_2 is the right immediate subtree of the root of T. Let n be the number of leaves in T. Then T can be pseudobalanced with $O(n)$ tree operations.*

Proof by induction on the number of leaves. Suppose the theorem holds for all numbers $< n$. We will prove it for n. Say T_1 has n_1 leaves, and T_2 has n_2 leaves ($n = n_1 + n_2$). Also say $n_1 = 2^r + a$ and $n_2 = 2^s + b$, where $1 \leq a \leq 2^r$ and $1 \leq b \leq 2^s$. Assume w.l.o.g. that $n_1 \geq n_2$. We will label the root of T as X, the root of T_1 as X_1, the root of T_2 as X_2. Also, the roots of the left and right immediate subtrees of T_1 and T_2 are X_{i_L} and X_{i_R} for $i = 1, 2$. The tree T looks as in the figure below.

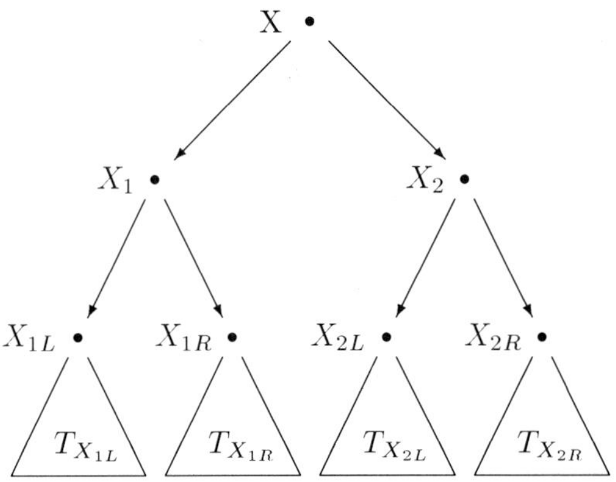

Figure 6

$T_{X_{1_L}}$ and $T_{X_{1_R}}$ have 2^r and a leaves respectively, and $T_{X_{2_L}}$ and $T_{X_{2_R}}$ have 2^s and b leaves respectively.

Case 1: $r = s$. We want to interchange the trees $T_{X_{1_R}}$ and $T_{X_{2_L}}$. For that we have to do the following sequence of 5 operations: right rotation on X_1, left rotation on X_2, permutation of $T_{X_{1_R}}$ and $T_{X_{2_L}}$, right rotation on X, left rotation on X.

Now the left immediate subtree of X is a complete binary tree since $r = s$, and we need to make the right immediate subtree of X a pseudobalanced tree. For that we will use the induction hypothesis on $a + b$. With say $c(a + b)$ operations we can pseudobalance T_{X_2}. At this point T is pseudobalanced, and we have used $c(a + b) + 5$ operations. $c(a + b) + 5 \leq c \cdot n$ for $c \geq 2$.

Case 2: $r > s$. Doing a right rotation on X_1 (see figure 6) we obtain the tree:

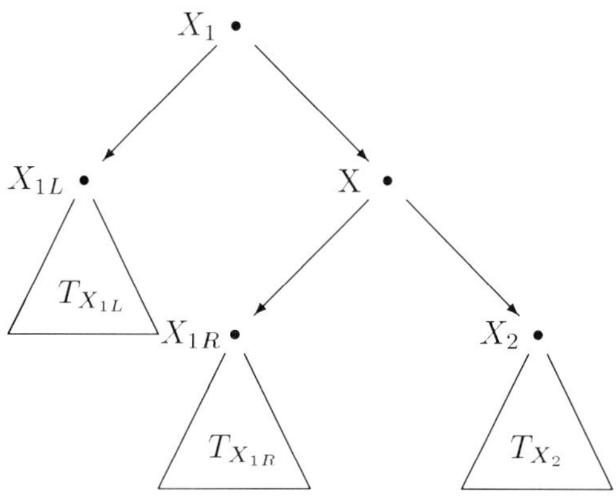

Figure 7

We want to pseudobalance the new subtree rooted at X of $a+b+2^s$ leafs. By the induction hypothesis, we can do that with $c(a+b+2^s)$ operations, obtaining a new tree rooted at X. If $a+b+2^s \leq 2^r$, then we are finished at this point, since the whole tree is then pseudobalanced. But if $a+b+2^s > 2^r$, we are not finished. Since $b \leq 2^s$ and $s < r$, then $b+2^s \leq 2 \cdot 2^s < 2^r$. Also, $a \leq 2^r$, so $a+b+2^s < 2^{r+1}$. So $a+b+2^s = 2^r + c$ for $c < 2^r$. So if $a+b+2^s > 2^r$, the tree looks like:

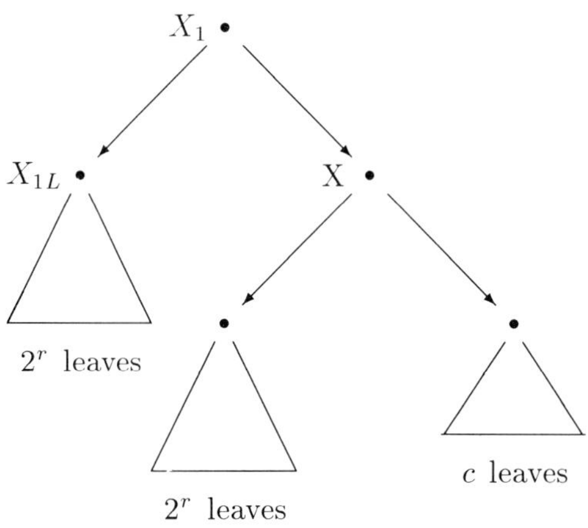

Figure 8

At this point, doing a left rotation on X, we obtain a pseudobalanced tree. We have done at most $c(a+b+2^s)+2$ operations. It is clear that $c(a+b+2^s)+2 \le c \cdot n$ for $c \ge 1$.

The result follows for $c = 2$. □.

The former theorem can be improved to theorem 7. In theorem 7 we don't count number of operations, but number of symbols.

Theorem 7 *Let* $\bigwedge_{i=1}^{t} A_i$ *and* $\bigwedge_{i=t+1}^{n} A_i$ *be pseudobalanced formulas. Then there is a permutation* $i \to j_i$ *(for* $i = 1, \ldots, n$*) and a pseudobalanced formula* $\bigwedge_{i=1}^{n} A_{j_i}$ *such that the formula*

$$\bigwedge_{i=1}^{t} A_i \wedge \bigwedge_{i=t+1}^{n} A_i \supset \bigwedge_{i=1}^{n} A_{j_i}$$

has a proof in $O(n)$ *symbols.*

Proof by induction on n. Suppose the theorem holds for all numbers $< n$. Say $t = 2^r + a$, $n - t = m$, and $m = 2^s + b$, where $0 \le a \le 2^r$, $0 \le b \le 2^s$, and say $r \ge s$.

Case 1: $r = s$. The formula

$$\bigwedge_{i=1}^{t} A_i \wedge \bigwedge_{i=t+1}^{n} A_i \supset [\bigwedge_{i=1}^{2^r} A_i \wedge \bigwedge_{i=t+1}^{t+2^r} A_i] \wedge [\bigwedge_{i=2^r+1}^{t} A_i \wedge \bigwedge_{i=t+2^r+1}^{t+m} A_i] \quad (3)$$

is provable in $O(n)$ symbols. We can apply the induction hypothesis to $a + b \le n$, and get

$$\bigwedge_{i=2^r+1}^{t} A_i \wedge \bigwedge_{i=t+2^r+1}^{t+m} A_i \supset \bigwedge_{i=1}^{a+b} A_{s_i} \quad (4)$$

in $O(a+b)$ symbols, say $c(a+b)$ symbols, where $i \to s_i$ is a permutation for $2^r + 1 \le i \le t$ or $t + 2^r + 1 \le i \le t + m$. Also

$$[\bigwedge_{i=1}^{2^r} A_i \wedge \bigwedge_{i=t+1}^{t+2^r} A_i] \wedge [\bigwedge_{i=2^r+1}^{t} A_i \wedge \bigwedge_{i=t+2^r+1}^{t+m} A_i] \supset [\bigwedge_{i=1}^{2^r} A_i \wedge \bigwedge_{i=t+1}^{t+2^r} A_i] \wedge \bigwedge_{i=1}^{a+b} A_{s_i} \quad (5)$$

is provable with $O(n)$ extra symbols from formula (4). Finally we get

$$[\bigwedge_{i=1}^{2^r} A_i \wedge \bigwedge_{i=2^r+1}^{t} A_i] \wedge [\bigwedge_{i=t+1}^{t+2^r} A_i \wedge \bigwedge_{i=t+2^r+1}^{t+m} A_i] \supset [\bigwedge_{i=1}^{2^r} A_i \wedge \bigwedge_{i=t+1}^{t+2^r} A_i] \wedge \bigwedge_{i=1}^{a+b} A_{s_i}$$

with extra $O(n)$ symbols from formulas (3) and (5). This last formula is what we wanted to show. The total number of symbols we have used is: $c(a+b) + d \cdot n$ for some constant d. We are going to show that if we pick $c \geq 2 \cdot d$, then $c(a+b) + d \cdot n \leq c \cdot n$, and the result follows for this case. If $d \leq c/2$, then

$$\begin{aligned} c(a+b) + d \cdot (a+b+2^r \cdot 2) &\leq c(a+b) + d \cdot (2^r \cdot 2 + 2^r \cdot 2) \\ &\leq c(a+b) + c/2 \cdot (2^r \cdot 2 + 2^r \cdot 2) \\ &\leq c \cdot n \end{aligned}$$

Case 2: $s < r$. By the induction hypothesis applied to $a + b + 2^s$, we get a proof of

$$\bigwedge_{i=2^r+1}^{t} A_i \wedge \bigwedge_{t+1}^{t+m} A_i \supset \bigwedge_{i=1}^{a+m} A_{j_i} \tag{6}$$

in $O(a+m)$ symbols, where $i \to j_i$ is a permutation for $2^r < i \leq t+m$. From formula (6) we get with $O(n)$ extra symbols:

$$\bigwedge_{i=1}^{2^r} A_i \wedge [\bigwedge_{i=2^r+1}^{t} A_i \wedge \bigwedge_{t+1}^{t+m} A_i] \supset \bigwedge_{i=1}^{2^r} A_i \wedge \bigwedge_{i=1}^{a+m} A_{j_i} \tag{7}$$

Since we can prove in $O(n)$ symbols

$$[\bigwedge_{i=1}^{2^r} A_i \wedge \bigwedge_{i=2^r+1}^{t} A_i] \wedge \bigwedge_{t+1}^{t+m} A_i \supset \bigwedge_{i=1}^{2^r} A_i \wedge [\bigwedge_{i=2^r+1}^{t} A_i \wedge \bigwedge_{t+1}^{t+m} A_i], \tag{8}$$

then from (7) and (8) with $O(n)$ extra symbols we prove

$$\bigwedge_{i=1}^{t} A_i \wedge \bigwedge_{i=t+1}^{t+m} A_i \supset \bigwedge_{i=1}^{2^r} A_i \wedge \bigwedge_{i=1}^{a+m} A_{j_i} \tag{9}$$

If $a+m \leq 2^r$, then we are finished at this point, since $\bigwedge^{2^r}(\ ) \wedge \bigwedge^{a+m}(\ )$ is then pseudobalanced. But if $a+m > 2^r$, we are not finished. Since $b < 2^s$ and $s < r$, then $m = b + 2^s < 2 \cdot 2^s < 2^r$. i.e., $m < 2^r$.

Also, $a < 2^r$, so $a + m < 2^{r+1}$. So $a + m = 2^r + e$ for $e < 2^r$. So formula (9) is actually the formula

$$\bigwedge_{i=1}^{t} A_i \wedge \bigwedge_{i=t+1}^{t+m} A_i \supset \bigwedge_{i=1}^{2^r} A_i \wedge [\bigwedge_{i=1}^{2^r} A_{j_i} \wedge \bigwedge_{i=2^r+1}^{e} A_{j_i}] \tag{10}$$

where $\bigwedge_{i=2^r+1}^{c} A_{j_i}$ is pseudobalanced. With $O(n)$ symbols we get from formula (10)

$$\bigwedge_{i=1}^{t} A_i \wedge \bigwedge_{i=t+1}^{t+m} A_i \supset [\bigwedge_{i=1}^{2^r} A_i \wedge \bigwedge_{i=1}^{2^r} A_{j_i}] \wedge \bigwedge_{i=2^r+1}^{e} A_{j_i} \tag{11}$$

which is what we wanted to show.

We have used a total of $c(a+m) + d' \cdot n$ symbols, where d' is a constant. We need to show that

$$c(a+m) + d' \cdot n \leq c \cdot n.$$

Taking $c \geq 3d'$, since $a + m < 2^{r+1}$,

$$\begin{aligned} d' \cdot n + c(a+m) &\leq d'(a+m+2^r) + c(a+m) \\ &\leq d' \cdot 3 \cdot 2^r + c(a+m) \\ &\leq c \cdot 2^r + c(a+m) = c \cdot n \end{aligned}$$

So the result follows taking $c = 3 \cdot \max(d, d')$. □

We can prove a variant of theorem 7 which states how many symbols are required to prove that an arbitrary conjunction implies a pseudobalanced conjunction.

Theorem 8 *Let* $\bigwedge_{i=1}^{n} A_i$ *be a non-pseudobalanced formula of depth* d. *Then, there is a permutation* $i \to j_i$ *(for* $i = 1, \ldots, n$*) and a pseudobalanced formula* $\bigwedge_{i=1}^{n} A_{j_i}$ *such that*

$$\bigwedge_{i=1}^{n} A_i \supset \bigwedge_{i=1}^{n} A_{j_i}$$

has a proof in $O(n \cdot d)$ *symbols.*

Proof by induction on n. Suppose the fact is true for all numbers $< n$.

Say $\bigwedge_{i=1}^{n} A_i = \bigwedge_{i=1}^{t} A_i \wedge \bigwedge_{i=t+1}^{n} A_i$, and the depth of $\bigwedge_{i=1}^{t} A_i$ is n_1, the depth of $\bigwedge_{i=t+1}^{n} A_i$ is n_2 and w.l.o.g., $n_1 \geq n_2$. Then the depth of $\bigwedge_{i=1}^{n} A_i$ is $n_1 + 1$. Now, by the induction hypothesis

$$\bigwedge_{i=1}^{t} A_i \supset \bigwedge_{i=1}^{t} A_{s_i}$$

and

$$\bigwedge_{i=t+1}^{n} A_i \supset \bigwedge_{i=t+1}^{n} A_{s_i}$$

can be proved with say ctn_1 and $c(n-t)n_2$ symbols respectively, for the same constant c. The consequents of those formulas are a permutation of the antecedents pseudobalanced. Also with say $d_1 \cdot n$ symbols, we can get

$$\bigwedge_{i=1}^{t} A_i \wedge \bigwedge_{i=t+1}^{n} A_i \supset \bigwedge_{i=1}^{t} A_{s_i} \wedge \bigwedge_{i=t+1}^{n} A_{s_i}.$$

By theorem 7 in say $d_2 \cdot n$ symbols we can show

$$\bigwedge_{i=1}^{t} A_{s_i} \wedge \bigwedge_{i=t+1}^{n} A_{s_i} \supset \bigwedge_{i=1}^{n} A_{j_i}.$$

i.e. we obtain a pseudobalanced formula from the conjunction of two pseudobalanced ones. Finally, with say $d_3 \cdot n$ symbols more we prove

$$\bigwedge_{i=1}^{n} A_i \supset \bigwedge_{i=1}^{n} A_{j_i}$$

The total number of symbols used is $ctn_1 + c(n-t)n_2 + (d_1 + d_2 + d_3)n$. If we take $c \geq d_1 + d_2 + d_3$, then

$$\begin{aligned} ctn_1 + c(n-t)n_2 + (d_1 + d_2 + d_3)n &\leq ctn_1 + c(n-t)n_2 + cn \\ &\leq ctn_1 + c(n-t)n_1 + cn \\ &\leq cnn_1 + cn = cn(n_1 + 1) \end{aligned}$$

The result follows. □

4 Number of Symbols in Proofs of Permutation Formulas

In this last two sections we are going to talk about two types of propositional formulas, that seem most likely to produce an almost quadratic speed up when using the deduction rule. The first formula to consider is

$$\bigwedge_{i=1}^{n} p_i \supset \bigwedge_{i=1}^{n} p_{j_i} \tag{12}$$

where $i \to j_i$ is any permutation for i such that $1 \leq i \leq n$, and both antecedent and consequent are pseudobalanced. Formula 12 means that a conjunction of atomic formulas implies any permutation of the atoms. We will study the lengths of proofs (counting number of symbols) of such formulas with and without the deduction rule.

Let us first give a sketch of theorem due to Buss which implies that any proof of formula (12) requires $\Omega(n \log n)$ symbols. If A is a tautology and B is a subformula, then we write $A(B/\top)$ and $A(B/\bot)$ to denote the formulas obtained by replacing every occurence of B in A by $\top$ (a fixed tautology) and by $\bot$ (i.e., $\neg\top$), respectively. The subformula B is *needed* in A iff at least one of $A(B/\top)$ or $A(B/\bot)$ is not a tautology. Now suppose A is a tautology with proof P and let X be the set of formulas which are needed in A. Then we claim that the number of symbols in P has the lower bound $\Omega(\sum_{B\in X}|B|)$. To prove this, consider any axiom φ in P; φ is of course an instance of an axiom schema ψ: we say that a subformula B of φ is *affected* by the axiom φ iff some occurence of B in φ has principle connective corresponding to a connective from the schema ψ. It is easy to see that any formula B needed in A must either be affected by some axiom in the proof P or be used as the (second) hypothesis of a modus ponens inference in P; since otherwise every occurence of B in P could be replaced by $\top$ or $\bot$ to yield proofs of $A(B/\top)$ and $A(B/\bot)$, respectively. Since an axiom can affect only a bounded number of formulas, the lower bound on the size of P follows.

For a special case of this lower bound, suppose that A is a tautology, that no non-atomic subformula occurs twice in A and that every non-atomic subformula of A is needed in A. Then any proof P of A requires $\Omega(s)$ symbols where

$$s = \sum \{depth(q) : q \text{ is an } \textit{occurence} \text{ of a variable in } A\}.$$

As a special case, Buss's lower bound shows that, in general, formula (12) will require $\Omega(n \log n)$ symbols. An important observation is that the same lower bound applies (by the same argument as above) to proofs of (12) with the deduction rule.

With the deduction rule we can easily produce an optimal length proof of (12), with $O(n \log n)$ symbols, the following way: first assume $\bigwedge_{i=1}^{n} A_i$. For simplicity, we will assume that n is a power of 2. We obtain

$$\bigwedge_{i=1}^{n/2} A_i \quad \text{and} \quad \bigwedge_{i=(n/2)+1}^{n} A_i.$$

Then we get

$$\bigwedge_{i=1}^{n/4} A_i, \quad \bigwedge_{i=(n/4)+1}^{n/2} A_i, \quad \bigwedge_{(n/2)+1}^{3n/4} A_i, \quad \text{and} \quad \bigwedge_{i=(3n/4)+1}^{n} A_i$$

etc., until finally we obtain

$$A_1, \quad A_2, \quad \ldots, \quad A_n.$$

This process takes $O(n \log n)$ symbols and $O(n)$ lines. We reverse the process to obtain $\bigwedge_{i=1}^{n} A_{j_i}$ in $O(n \log n)$ symbols and $O(n)$ lines more. With the deduction rule we obtain (12) with $O(n \log n)$ symbols and $O(n)$ lines.

If we try to prove the same formula in the Frege system, using the methods of theorem 1 we would obtain a proof of $O(n^2)$ symbols, since the proof with the deduction rule has n lines, and $\bigwedge_{i=1}^{n} A_i$ has size n (see equations 1 and 2). But, can we give a Frege proof of (12) with less than $O(n^2)$ symbols? The answer is yes. We are going to produce a proof of (12) of $O(n \log^2 n)$ symbols without the deduction rule. So we obtain a $O(n/\log^2 n)$ speedup over theorem 1. It is still open whether there is a Frege-proof of (12) with $O(n \log n)$ symbols. Only in some cases we were able to obtain proofs of permutation formulas in $O(n \log n)$ without the deduction rule. Let us see one example:

Lemma 9 *For every m, the formula*

$$\bigwedge_{i=1}^{2^m} q_i \supset \bigwedge \{q_1, q_{(2^m/2)+1}, q_2, q_{(2^m/2)+2}, \ldots, q_{2^m/2}, q_{2^m}\}$$

where antecedent and consequent are completely balanced, has a proof in $O(n \log n)$ symbols, where $n = 2^m$.

Proof by induction on m. Suppose the result is true for all numbers $< m$.

By the induction hypothesis, the formulas

$$\bigwedge_{i=1}^{n/4} q_i \wedge \bigwedge_{i=(n/2)+1}^{3n/4} q_i \supset \bigwedge\{q_1, q_{(n/2)+1}, q_2, q_{(n/2)+2}, \ldots, q_{n/4}, q_{3n/4}\} \quad (13)$$

and

$$\bigwedge_{(n/4)+1}^{n/2} q_i \wedge \bigwedge_{(3n/4)+1}^{n} q_i \supset \bigwedge\{q_{(n/4)+1}, q_{(3n/4)+1}, q_{(n/4)+2}, q_{(3n/4)+2}, \ldots, q_{n/2}, q_n\} \quad (14)$$

have proofs of $O((n/2)\log(n/2))$ symbols.

Say that $E = \bigwedge\{q_1, q_{(n/2)+1}, q_2, q_{(n/2)+2}, \ldots, q_{n/4}, q_{3n/4}\}$ and $F = \bigwedge\{q_{(n/4)+1}, q_{(3n/4)+1}, q_{(n/4)+2}, q_{(3n/4)+2}, \ldots, q_{n/2}, q_n\}$. From (13) and (14) we can obtain

$$[\bigwedge_{i=1}^{n/4} q_i \wedge \bigwedge_{i=(n/2)+1}^{3n/4} q_i] \wedge [\bigwedge_{(n/4)+1}^{n/2} q_i \wedge \bigwedge_{(3n/4)+1}^{n} q_i] \supset E \wedge F \quad (15)$$

with $O(n)$ extra symbols. From (15) we can prove

$$[\bigwedge_{i=1}^{n/4} q_i \wedge \bigwedge_{(n/4)+1}^{n/2} q_i] \wedge [\bigwedge_{i=(n/2)+1}^{3n/4} q_i \wedge \bigwedge_{(3n/4)+1}^{n} q_i] \supset E \wedge F$$

in $O(n)$ extra symbols, which is what we wanted to show.

So we have used a total of say $2c(n/2)\log(n/2) + dn$ symbols. Since $2c(n/2)\log(n/2) + dn = c \cdot n \cdot \log n - c \cdot n + d \cdot n$, the result follows taking $c \geq d$. □

In general, for any possible permutation, there is a proof in $O(n \log^2 n)$ symbols. But before we prove this, let us prove the following lemmas:

Lemma 10 *Let* $\bigwedge_{i=1}^{m} q_i$ *be a pseudobalanced formula such that* $\bigwedge_{i=1}^{m} q_i = \bigwedge_{i=1}^{2^r} q_i \wedge \bigwedge_{i=2^r+1}^{m} q_i$. *Let* n *be the smallest power of* 2 *bigger than* m. *Let*

$\bigwedge\!\!\!\bigwedge_{i=1}^{n} A_i$ be a completely balanced formula where each A_i is either $\top$ (where $\top$ is any fixed tautology) if $i > m$ or q_i if $i \le m$. Then we can prove the formulas

$$\bigwedge\!\!\!\bigwedge_{i=1}^{m} q_i \supset \bigwedge\!\!\!\bigwedge_{i=1}^{n} A_i \quad \text{and} \quad \bigwedge\!\!\!\bigwedge_{i=1}^{n} A_i \supset \bigwedge\!\!\!\bigwedge_{i=1}^{m} q_i$$

in $O(n \log n)$ symbols.

Proof by induction on n. The proof is left to the reader.

Definition 12 *Let A be a binary formula, and let T be the tree that represents A. An extract of A is a formula represented by a tree T' obtained by the following process:*

1. *Removing from T any subset of the leaves of T.*
2. *If a node x labeled $\wedge$, $\vee$ or $\supset$ has only one child, we remove x and connect the child to the ancestor of x. Also we remove the edge from the ancestor of x to x.*
3. *If a non leaf node has no children, we remove it as well as the edge going to it.*

Operations 2 and 3 are to be performed iteratively as long as possible.

So an extract of a formula A, is a formula where we pulled out some atomic subformulas and extra connectives and parenthesis, but keeping the structure of A on what is left over.

Lemma 11 *Let $\bigwedge\!\!\!\bigwedge_{i=1}^{n} q_i$ be a completely balanced formula, and for $m \le n$, $\bigwedge\!\!\!\bigwedge_{i=1}^{m} A_i$ is an extract of $\bigwedge\!\!\!\bigwedge_{i=1}^{n} q_i$, possibly unbalanced formed by removing some of the q_i's from $\bigwedge\!\!\!\bigwedge_{i=1}^{n} q_i$. Then,*

$$\bigwedge\!\!\!\bigwedge_{i=1}^{n} q_i \supset \bigwedge\!\!\!\bigwedge_{i=1}^{m} A_i$$

has a proof in $O(n \log n)$ symbols.

Proof by induction on n. Suppose the fact is true for all powers of two $< n$. By the induction hypothesis,

$$\bigwedge_{i=1}^{n/2} q_i \supset \bigwedge_{i=1}^{t} A_i$$

and

$$\bigwedge_{i=(n/2)+1}^{n} q_i \supset \bigwedge_{i=t+1}^{m} A_i$$

have proofs of say $c(n/2)\log(n/2)$ symbols where t is the number of q_i's that we want to keep from the first half of the conjunction. With extra $O(n)$ symbols we get

$$\bigwedge_{i=1}^{n} q_i \supset \bigwedge_{i=1}^{t} A_i \wedge \bigwedge_{i=t+1}^{m} A_i$$

which is what we wanted to show. □

Theorem 12 *Let* $\bigwedge_{i=1}^{n} q_i$ *and* $\bigwedge_{i=1}^{n} q_{j_i}$ *be completely balanced formulas, where* n *is a power of two, say* $n = 2^m$, *and* $i \to j_i$ *is any permutation for* $i = 1, \ldots, n$. *Then the formula*

$$\bigwedge_{i=1}^{n} q_i \supset \bigwedge_{i=1}^{n} q_{j_i}$$

has a proof in $O(n \log^2 n)$ *symbols.*

Proof by induction on n. Suppose the theorem is true for all powers of 2 less than n. We will prove it for $n = 2^m$.

Let $\bigwedge_{i=1}^{n/2} A_i$ be an extract of $\bigwedge_{i=1}^{n} q_i$ where all the q_{j_i}'s from $i = (n/2)+1$ to n have been removed and the parentheses kept as in $\bigwedge_{i=1}^{n} q_i$, and $\bigwedge_{i=(n/2)+1}^{n} A_i$ be an extract of $\bigwedge_{i=1}^{n} q_i$ where all the q_{j_i}'s from $i = 1$ to $n/2$ have been removed and the parentheses kept the same. Then by lemma 11 we can show

$$\bigwedge_{i=1}^{n} q_i \supset \bigwedge_{i=1}^{n/2} A_i \tag{16}$$

and

$$\bigwedge_{i=1}^{n} q_i \supset \bigwedge_{i=(n/2)+1}^{n} A_i \tag{17}$$

with $O(n \log n)$ symbols. The depth of the formulas $\bigwedge_{i=1}^{n/2} A_i$ and $\bigwedge_{i=(n/2)+1}^{n} A_i$ is at most the depth of $\bigwedge_{i=1}^{n} q_i$. i.e. at most $\log n$. By the theorem 8 the consequents of 16 and 17 can be rebalanced in $O(n \log n)$ symbols. So we can prove

$$\bigwedge_{i=1}^{n/2} A_i \supset \bigwedge_{i=1}^{n/2} A_{k_i} \tag{18}$$

and

$$\bigwedge_{i=(n/2)+1}^{n} A_i \supset \bigwedge_{i=(n/2)+1}^{n} A_{k_i} \tag{19}$$

with $O(n \log n)$ symbols, where $\bigwedge_{i=1}^{n/2} A_{k_i}$ and $\bigwedge_{i=(n/2)+1}^{n} A_{k_i}$ are completely balanced and the A_{k_i}'s are a permutation of the A_i's. Now by the induction hypothesis we can prove

$$\bigwedge_{i=1}^{n/2} A_{k_i} \supset \bigwedge_{i=1}^{n/2} q_{j_i} \tag{20}$$

and

$$\bigwedge_{i=(n/2)+1}^{n} A_{k_i} \supset \bigwedge_{i=(n/2)+1}^{n} q_{j_i} \tag{21}$$

with say $c(n/2)\log^2(n/2)$ symbols each. With $O(n)$ symbols we can prove

$$\bigwedge_{i=1}^{n} q_i \supset \bigwedge_{i=1}^{n/2} q_{j_i}$$

and

$$\bigwedge_{i=1}^{n} q_i \supset \bigwedge_{i=(n/2)+1}^{n} q_{j_i}$$

from 16, 18 and 20, and from 17, 19 and 21. Finally, with $O(n)$ more symbols we prove

$$\bigwedge_{i=1}^{n} q_i \supset \bigwedge_{i=1}^{n} q_{j_i}$$

which is what we wanted to show.

Let us now look at the total number of symbols used in this proof. Say we have used $c2(n/2)\log^2(n/2) + d_1 n \log n + d_2 n$ symbols. we will show that by taking $c \geq d_1 + d_2$, $c2(n/2)\log^2(n/2) + d_1 n \log n + d_2 n \leq cn \log^2 n$. Since $\log^2(n/2) = \log^2 n + 1 - 2\log n$,

$$\begin{aligned} c2(n/2)\log^2(n/2) &+ d_1 n \log n + d_2 n \\ &\leq cn \log^2 n + cn - 2cn \log n + n \log n(d_1 + d_2) \\ &\leq cn \log^2 n - cn \log n + n \log n(d_1 + d_2) \\ &\leq cn \log^2 n \end{aligned}$$

The result follows. □

Corollary 13 *The formula*

$$\bigwedge_{i=1}^{n} q_i \supset \bigwedge_{i=1}^{n} q_{j_i}$$

where $\bigwedge_{i=1}^{n} q_i$ *and* $\bigwedge_{i=1}^{n} q_{j_i}$ *are pseudobalanced, and* $\bigwedge_{i=1}^{n} q_{j_i}$ *is any permutation of* $\bigwedge_{i=1}^{n} q_i$, *has a proof in* $O(n \log^2 n)$ *symbols.*

Proof note that in the corollary we don't require n to be a power of 2. The result follows by two uses of lemma 10, and one use of theorem 12. □

5 Number of Symbols in Proofs of Transitive Closure Formulas

In this last section we are going to be working with closure formulas of the form

$$\bigwedge_{i=1}^{n}(q_{i-1} \supset q_i) \supset \bigwedge_{i=1}^{m}(q_{s_i} \supset q_{t_i}) \tag{22}$$

where both $\bigwedge\!\!\!\bigwedge_{i=1}^{n}(q_{i-1} \supset q_i)$ and $\bigwedge\!\!\!\bigwedge_{i=1}^{m}(q_{s_i} \supset q_{t_i})$ are pseudobalanced, and $\bigwedge\!\!\!\bigwedge_{i=1}^{m}(q_{s_i} \supset q_{t_i})$ is a conjunction of implications $q_{s_i} \supset q_{t_i}$ where $s_i < t_i$.

Again we are going to be studying the length of proofs of such formulas with and without the deduction rule. As in the case of permutation formulas, by the informal argument given there, any proof of (22) has at least $\Omega((n+m)\log n)$ symbols (since $m \leq n^2$, $\log m \leq \log(n^2) = 2\log n$ and $O(n\log n + m\log m) = O((n+m)\log n)$). With the deduction rule, we can get a proof of (22) with $O((n+m)\log n)$ symbols the following way: first assume $\bigwedge\!\!\!\bigwedge_{i=1}^{n}(q_{i-1} \supset q_i)$. Again for simplicity let n be a power of 2. Then we get

$$\bigwedge\!\!\!\bigwedge_{i=1}^{n/2}(q_{i-1} \supset q_i) \quad \text{and} \quad \bigwedge\!\!\!\bigwedge_{i=(n/2)+1}^{n}(q_{i-1} \supset q_i)$$

We undo (22) until we obtain

$$q_0 \supset q_1, q_1 \supset q_2, \ldots, q_{n-2} \supset q_{n-1}, q_{n-1} \supset q_n$$

This process takes $O(n\log n)$ symbols and $O(n)$ lines. By the work of Bonet-Buss [1, 4, 3], we prove

$$q_{s_1} \supset q_{t_1}, q_{s_2} \supset q_{t_2}, \ldots, q_{s_m} \supset q_{t_m}$$

in $O((n+m)\alpha(n))$ symbols and $O((n+m)\alpha(n))$ lines, where α is the inverse of the Ackerman function. Finally we obtain $\bigwedge\!\!\!\bigwedge_{i=1}^{m}(q_{s_i} \supset q_{t_i})$ in $O(m\log m)$ symbols or $O(m)$ lines. So with a total of $O(m\log m + n\log n)$ symbols and $O((n+m)\alpha(n))$ lines we obtain the formula 22.

If we go back to theorem 1, we see that we can prove (22) with at most $O((n+m)(n+m)\alpha(n))$ symbols without the deduction rule (see (1) and (2)). The main question is now whether we can prove (22) without the deduction rule with less than $O((n+m)^2\alpha(n))$ symbols. The answer is yes. We will prove (22) with $O((n+m)\log^3 n)$ symbols without the deduction rule. So we obtain a $O((n+m)\alpha(n)/\log^3 n)$ speedup over theorem 1. It is still open whether (22) has a Frege-proof of $O((n+m)\log n)$ symbols.

Before we prove the main theorem, let us obtain the following lemmas:

Lemma 14 *The formulas*

$$\bigwedge\!\!\!\bigwedge_{i=1}^{n}(q_{i-1} \supset q_i) \supset \bigwedge\!\!\!\bigwedge_{i=1}^{n}(q_0 \supset q_i)$$

and

$$\bigwedge_{i=1}^{n}(q_{i-1} \supset q_i) \supset \bigwedge_{i=0}^{n-1}(q_i \supset q_n)$$

where n is a power of 2, the conjunctions are completely balanced, and the q_i's are atomic formulas, have Frege proofs of $O(n \log^2 n)$ symbols.

Proof by induction on n. Suppose the lemma holds for all numbers $< n$. We will prove it for n.

By the induction hypothesis the formulas

$$\bigwedge_{i=1}^{n/2}(q_{i-1} \supset q_i) \supset \bigwedge_{i=1}^{n/2}(q_0 \supset q_i) \tag{23}$$

and

$$\bigwedge_{i=(n/2)+1}^{n}(q_{i-1} \supset q_i) \supset \bigwedge_{i=(n/2)+1}^{n}(q_{n/2} \supset q_i) \tag{24}$$

have proofs of $O((n/2)\log^2(n/2))$ symbols. Also in the next lines we are going to show how to get a proof of

$$(q_0 \supset q_{n/2}) \wedge \bigwedge_{i=(n/2)+1}^{n}(q_{n/2} \supset q_i) \supset \bigwedge_{i=(n/2)+1}^{n}(q_0 \supset q_i) \tag{25}$$

in $O(n \log n)$ symbols. We are going to do it by induction on the size of conjunctions. Assume it holds for all numbers $< n$, and let us show it for n. By the induction hypothesis the formulas

$$(q_0 \supset q_{n/2}) \wedge \bigwedge_{i=(n/2)+1}^{3n/4}(q_{n/2} \supset q_i) \supset \bigwedge_{i=(n/2)+1}^{3n/4}(q_0 \supset q_i)$$

and

$$(q_0 \supset q_{n/2}) \wedge \bigwedge_{i=(3n/4)+1}^{n}(q_{n/2} \supset q_i) \supset \bigwedge_{i=(3n/4)+1}^{n}(q_0 \supset q_i)$$

have proofs of $O((n/2)\log(n/2))$ symbols. So we can prove both of them in $cn\log(n/2) = cn\log n - cn$ symbols, where c is a constant. With extra $O(n)$ symbols we can prove

$$(q_0 \supset q_{n/2}) \wedge [\bigwedge_{i=(n/2)+1}^{3n/4}(q_{n/2} \supset q_i) \wedge \bigwedge_{i=(3n/4)+1}^{n}(q_{n/2} \supset q_i)] \supset$$

$$[\bigwedge_{i=(n/2)+1}^{3n/4}(q_0 \supset q_i) \wedge \bigwedge_{i=(3n/4)+1}^{n}(q_0 \supset q_i)]$$

which is what we wanted to prove. It is easy to see that we can prove (25) with $O(n \log n)$ symbols, since we used $cn \log n - cn + O(n)$ symbols.

Also we need to prove the formula

$$\bigwedge_{i=1}^{n/2}(q_0 \supset q_i) \supset (q_0 \supset q_{n/2}). \tag{26}$$

Such formula has a proof with $O(n)$ symbols. The proof by induction is left to the reader.

From (25) and (26) we can obtain the formula

$$\bigwedge_{i=1}^{n/2}(q_0 \supset q_i) \wedge \bigwedge_{(n/2)+1}^{n}(q_{n/2} \supset q_i) \supset \bigwedge_{i=(n/2)+1}^{n}(q_0 \supset q_i) \tag{27}$$

with $O(n)$ additional symbols. Also with $O(n)$ extra symbols we prove

$$\bigwedge_{i=1}^{n}(q_{i-1} \supset q_i) \supset \bigwedge_{i=1}^{n}(q_0 \supset q_i)$$

from (23), (24) and (27). The total number of symbols used is $cn\log^2(n/2) + e_1 n \log n + e_2 n$. If $c \geq e_1 + e_2$ then $cn\log^2(n/2) + e_1 n \log n + e_2 n \leq cn \log^2 n$. □

Lemma 15 *The formula*

$$\bigwedge_{i=1}^{l}(p_i \supset X) \wedge \bigwedge_{i=1}^{l}(X \supset q_i) \supset \bigwedge_{i=1}^{l}(p_i \supset q_i)$$

has a proof with $O(l \cdot \log l)$ symbols, where X is an atomic formula.

Proof by induction on l. The proof is left to the reader.

Lemma 16 *Let $\bigwedge_{i=1}^{n} q_i$ be a completely balanced formula for n a power of 2, and $\langle j_1, \ldots, j_s \rangle$ be a sequence from the set $\{1, \ldots, n\}$. Then there is* *some permutation* $\langle k_1, \ldots, k_s\rangle$ *of the sequence* $\langle j_1, \ldots, j_s\rangle$, *and there is a pseudobalanced formula* $\bigwedge_{i=1}^{s} q_{k_i}$ *(q_i's can be omitted or repeated) such that the formula*

$$\bigwedge_{i=1}^{n} q_i \supset \bigwedge_{i=1}^{s} q_{k_i}$$

can be proved with $O((n+s)\log n)$ *symbols.*

Proof by induction on n. Suppose it is true for all numbers $< n$. We'll prove it for n. By the induction hypothesis the formulas

$$\bigwedge_{i=1}^{n/2} q_i \supset \bigwedge_{i=1}^{t} q_{a_i}$$

and

$$\bigwedge_{i=(n/2)+1}^{n} q_i \supset \bigwedge_{i=t+1}^{s} q_{a_i}$$

have proofs of $O(((n/2)+t)\log(n/2))$ and $O(((n/2)+(s-t))\log(n/2)$ symbols, where $\langle a_1, \ldots, a_t\rangle$ is a permutation of the j_i's which are $\leq n/2$, and $\langle a_{t+1}, \ldots, a_s\rangle$ is a permutation of the j_i's which are $\geq n/2$. So the proofs have a total of $cn\log(n/2) + cs\log(n/2)$ symbols. i.e., at most $c(n+s)\log n - c(n+s)$ symbols. We can obtain the formula

$$\bigwedge_{i=1}^{n/2} q_i \wedge \bigwedge_{i=(n/2)+1}^{n} q_i \supset \bigwedge_{i=1}^{t} q_{a_i} \wedge \bigwedge_{i=t+1}^{s} q_{a_i}$$

with extra $O(n+s)$ symbols. Finally we can pseudobalance the consequent with $O(n+s)$ symbols (by theorem 7), to obtain

$$\bigwedge_{i=1}^{n} q_i \supset \bigwedge_{i=1}^{s} q_{k_i}$$

The result follows. $\square$.

Theorem 17 *Let* $\bigwedge_{i=1}^{n}(q_{i-1} \supset q_i)$ *be a completely balanced formula where* n *is a power of* 2. *Let* $\langle s_1, \ldots, s_m\rangle$ *be any sequence from the set* $\{0, 1, \ldots, n-$

1}, and $\langle t_1, \ldots, t_m \rangle$ be any sequence from the set $\{1, \ldots, n\}$ such that $s_i < t_i$. Then the formula

$$\bigwedge_{i=1}^{n}(q_{i-1} \supset q_i) \supset \bigwedge_{i=1}^{m}(q_{s_i} \supset q_{t_i})$$

has a proof with $O((n+m)\log^3 n)$ symbols.

Proof by induction on n. Assume it holds for all numbers $< n$. Say $m = l_1 + l_2 + l_3$ where l_1 is the number of implications for which $q_{t_i} \leq n/2$ (i.e. in the first half), l_2 is the number of implications for which $q_{s_i} \geq n/2$ (i.e. in the second half) and l_3 is the number of implications for which $q_{s_i} < n/2$ and $q_{t_i} > n/2$ (i.e. implications across $q_{n/2}$).

Let $\langle s_i t_i \rangle$ be the tuple representing the implication $s_i \supset t_i$. We can form the sequence of tuples, $\langle\langle s_1 t_1 \rangle, \ldots, \langle s_m t_m \rangle\rangle$. We can partition this sequence in three sequences: $\langle\langle a_1 b_1 \rangle, \ldots, \langle a_{l_1} b_{l_1} \rangle\rangle$ representing the implications in the first half, $\langle\langle c_1 d_1 \rangle, \ldots, \langle c_{l_2} d_{l_2} \rangle\rangle$ representing the implications in the second half, and $\langle\langle g_1 j_1 \rangle, \ldots, \langle g_{l_3} j_{l_3} \rangle\rangle$ representing the implications across $q_{n/2}$.

By the induction hypothesis we can prove

$$\bigwedge_{i=1}^{n/2}(q_{i-1} \supset q_i) \supset \bigwedge_{i=1}^{l_1}(q_{a_i} \supset q_{b_i}) \tag{28}$$

and

$$\bigwedge_{i=(n/2)+1}^{n}(q_{i-1} \supset q_i) \supset \bigwedge_{i=1}^{l_2}(q_{c_i} \supset q_{d_i}) \tag{29}$$

with $c((n/2)\log^3(n/2) + l_1 \log^3(n/2)) + c((n/2)\log^3(n/2) + l_2 \log^3(n/2))$ symbols, for some constant c. So with $c\log^3(n/2)(n + l_1 + l_2)$ symbols. With extra $O(n + l_1 + l_2)$ symbols we prove the following three facts:

$$\bigwedge_{i=1}^{n/2}(q_{i-1} \supset q_i) \wedge \bigwedge_{i=(n/2)+1}^{n}(q_{i-1} \supset q_i) \supset \bigwedge_{i=1}^{l_1}(q_{a_i} \supset q_{b_i}) \wedge \bigwedge_{i=1}^{l_2}(q_{c_i} \supset q_{d_i})$$

from (28) and (29),

$$\bigwedge_{i=1}^{l_1}(q_{a_i} \supset q_{b_i}) \wedge \bigwedge_{i=1}^{l_2}(q_{c_i} \supset q_{d_i}) \supset \bigwedge_{i=1}^{l_1+l_2}(q_{e_i} \supset q_{f_i})$$

where the consequent is now pseudobalanced by theorem 7, and the e_i's are a permutation of the a_i's and c_i's, and the f_i's are the corresponding permutation of the b_i's and d_i's, and finally,

$$\bigwedge_{i=1}^{n}(q_{i-1} \supset q_i) \supset \bigwedge_{i=1}^{l_1+l_2}(q_{e_i} \supset q_{f_i}) \tag{30}$$

from the latter two statements. Say we have used $d_1(n + l_1 + l_2)$ symbols. Now we need to prove such an implication for l_3.

By lemma 14 we can prove

$$\bigwedge_{i=1}^{n/2}((q_{i-1} \supset q_i) \supset \bigwedge_{i=0}^{(n/2)-1}(q_i \supset q_{n/2}) \tag{31}$$

and

$$\bigwedge_{i=(n/2)+1}^{n}(q_{i-1} \supset q_i) \supset \bigwedge_{i=(n/2)+1}^{n}(q_{n/2} \supset q_i) \tag{32}$$

with say $d_2(n/2)\log^2(n/2)$ symbols each. So with a total of at most $d_2 n \log^2 n$. By lemma 16 we can prove the formulas

$$\bigwedge_{i=0}^{(n/2)-1}(q_i \supset q_{n/2}) \supset \bigwedge_{i=1}^{l_3}(q_{g_i} \supset q_{n/2}) \tag{33}$$

and

$$\bigwedge_{i=(n/2)+1}^{n}(q_{n/2} \supset q_i) \supset \bigwedge_{i=1}^{l_3}(q_{n/2} \supset q_{h_i}) \tag{34}$$

with $O(((n/2) + l_3)\log(n/2))$ symbols each where the h_i's are a permutation of the j_i's. Say with at most $d_3 n \log n + 2d_3 l_3 \log n$ symbols. Also we need to obtain the formula $\bigwedge_{i=1}^{l_3}(q_{g_i} \supset q_{j_i})$. By theorem 12 and lemma 15 we can do all this in at most $O(l_3 \log^2 l_3)$ symbols, obtaining

$$\bigwedge_{i=1}^{l_3}(q_{g_i} \supset q_{n/2}) \wedge \bigwedge_{i=1}^{l_3}(q_{n/2} \supset q_{h_i}) \supset \bigwedge_{i=1}^{l_3}(q_{g_i} \supset q_{j_i}) \tag{35}$$

in say $d_4 l_3 \log^2 l_3$ symbols. And from (33), (34) and (35) we obtain

$$\bigwedge_{i=0}^{(n/2)-1}(q_i \supset q_{n/2}) \wedge \bigwedge_{i=(n/2)+1}^{n}(q_{n/2} \supset q_i) \supset \bigwedge_{i=1}^{l_3}(q_{g_i} \supset q_{j_i}) \tag{36}$$

with $O(n + l_3)$ symbols, as well as from (31), (32) and (35) we can obtain

$$\bigwedge_{i=1}^{n}((q_{i-1} \supset q_i) \supset \bigwedge_{i=1}^{l_3}(q_{g_i} \supset q_{j_i}) \tag{37}$$

with $O(n + l_3)$ symbols. The last two formulas are proved with $d_5(n + l_3)$ symbols, for some constant d_5. From (30) and (37) we get

$$\bigwedge_{i=1}^{n}((q_{i-1} \supset q_i) \supset \bigwedge_{i=1}^{l_1+l_2}(q_{e_i} \supset q_{f_i}) \wedge \bigwedge_{i=1}^{l_3}(q_{g_i} \supset q_{j_i}) \tag{38}$$

with $O(n + m)$ symbols. Say with $d_6(n + m)$ symbols, for some constant d_6. At this point we only need to pseudobalance the consequent of (38) (by theorem 7), and give it the required order (by theorem 12). So we prove

$$\bigwedge_{i=1}^{l_1+l_2}(q_{e_i} \supset q_{f_i}) \wedge \bigwedge_{i=1}^{l_3}(q_{g_i} \supset q_{j_i}) \supset \bigwedge_{i=1}^{m}(q_{s_i} \supset q_{t_i}) \tag{39}$$

with at most say $d_7 m \log^2 m$ symbols, where the s_i's are a permutation of the e_i's and g_i's, and the t_i's are a permutation of the f_i's and j_i's. Finally from (38) and (39) we prove

$$\bigwedge_{i=1}^{n}((q_{i-1} \supset q_i) \supset \bigwedge_{i=1}^{m}(q_{s_i} \supset q_{t_i})$$

with say $d_8(n + m)$ symbols, for some constant d_8.

Now we need to compute the total of symbols used. To simplify that, let us group the following terms. Let

$$e_1(n + m) = d_1(n + l_1 + l_2) + d_5(n + l_3) + d_6(n + m) + d_8(n + m)$$

Also, to group $d_2 n \log^2 n$, $d_4 l_3 \log^2 l_3$ and $d_7 m \log^2 m$, consider the fact that $m \leq n^2$. Then $\log^2 m \leq \log^2(n^2) = 4\log^2 n$. Also, $\log^2 l_3 \leq 4\log^2 n$. Then let

$$\begin{aligned} e_2(n + m)\log^2 n &= d_2 n \log^2 n + 4d_7 m \log^2 n + 4d_4 l_3 \log^2 n \\ &= d_2 n \log^2 n + d_7 m \log^2 m + d_4 l_3 \log^2 l_3 \end{aligned}$$

So now we just need to show that

$$\begin{aligned} c(n + m)\log^3 n \quad &\geq \quad c(n + l_1 + l_2)(\log^3 n - 3\log^2 n + 3\log n - 1) + \\ &\quad\; e_1(n + m) + e_2(n + m)\log^2 n + d_3 n \log n + 2d_3 l_3 \log n \end{aligned}$$

Given that if $c \geq \max(e_1, e_2, 2d_3)$ and $n \geq 4$, then

$$e_1 l_3 + 2d_3 l_3 \log n + e_2 l_3 \log^2 n \leq c l_3 \log^3 n,$$

we only need to show that,

$$\begin{aligned} c(n + l_1 + l_2) \quad \geq \quad & c(n + l_1 + l_2)(\log^3 n - 3\log^2 n + 3\log n - 1) + \\ & e_1(n + l_1 + l_2) + e_2(n + l_1 + l_2)\log^2 n + d_3 n \log n \end{aligned}$$

Since we take $c \geq \max(e_1, e_2, 2d_3)$, that is equivalent to proving that

$$c(n + l_1 + l_2)(\log^2 n + \log n + 1) \leq c(n + l_1 + l_2)(3\log^2 n - 3\log n + 1)$$

i.e. equivalent to proving that

$$\log^2 n + \log n + 1 \leq 3\log^2 n - 3\log n + 1.$$

The equation above is true for $n \geq 4$, therefore the result follows. □

In theorem 17 we assumed n was a power of 2. By padding the formula $\bigwedge_{i=1}^{n}(q_{i-1} \supset q_i)$ with $\top$ to make it have the right length we can prove the result for n not a power of 2.

Corollary 18 *With $O((n + m)\log^3 n)$ symbols we can prove the formula*

$$\bigwedge_{i=1}^{n}(q_{i-1} \supset q_i) \supset \bigwedge_{i=1}^{m}(q_{s_i} \supset q_{t_i})$$

where for all i, $1 \leq i \leq m$, $s_i < t_i$.

References

[1] Maria Luisa Bonet. *The Lengths of Propositional Proofs and the Deduction Rule.* PhD thesis, U.C. Berkeley, 1991.

[2] Maria Luisa Bonet and Samuel R. Buss. The deduction rule and linear and near-linear proof simulations. Submitted for publication.

[3] Maria Luisa Bonet and Samuel R. Buss. On the serial transitive closure problem. Submitted for publication.

[4] Maria Luisa Bonet and Samuel R. Buss. On the deduction rule and the number of proof lines. In *Proceedings Sixth Annual IEEE Symposium on Logic in Computer Science*, pages 286–297, 1991.

[5] Samuel R. Buss. The undecidability of k-provability. *Annals of Pure and Applied Logic*, 53:75–102, 1991.

[6] Stephen A. Cook and Robert A. Reckhow. The relative efficiency of propositional proof systems. *Journal of Symbolic Logic*, 44:36–50, 1979.

[7] Stephen C. Kleene. *Introduction to Metamathematics.* Wolters-Noordhoff and North-Holland, 1971.

[8] Robert A. Reckhow. *On the Lengths of Proofs in the Propositional Calculus.* PhD thesis, Department of Computer Science, University of Toronto, 1976. Technical Report #87.

[9] Thomas A. Standish. *Data Structure Techniques.* Addison-Wesley, 1980.

[10] R. Statman. Complexity of derivations from quantifier-free Horn formulae, mechanical introduction of explicit definitions, and refinement of completeness theorems. In *Logic Colloquium '86*, pages 505–517. North-Holland, 1977.

Maria Luisa Bonet
Department of Mathematics
University of California at San Diego
La Jolla, CA 92093
email: mbonet@math.ucsd.edu

Algorithms for Boolean Formula Evaluation and for Tree Contraction

Samuel R. Buss*

Abstract

We present a new, simpler ALOGTIME algorithm for the Boolean sentence value problem (BSVP). Unlike prior work, this algorithm avoids the use of postfix-longer-operand-first formulas. We also show that tree-contraction can be made ALOGTIME uniform.

The Boolean sentence problem restricted to balanced sentences with only the connectives $\land$ and $\lor$ is in online $O(\log(\log n))$ space. Hence every ALOGTIME predicate is deterministic log-time reducible to $\log(\log n)$ space. This balanced and/or BSVP has logarithmic width, linear length, branching programs.

*Supported in part by NSF grants DMS-8902480 and INT-8914569. Part of this work performed while visiting the Czechoslovakian Academy of Sciences.

1 Introduction

A Boolean sentence is an expression formed from propositional connectives $\wedge$, $\vee$, $\neg$, etc., from constants 1 and 0 (for 'True' and 'False', respectively), and from parentheses. The *Boolean sentence value problem* (BSVP) is the decision problem of, given a Boolean sentence, to determine if the value of the sentence is 1 (i.e., True). The Boolean sentence problem has been shown to be in alternating logarithmic time (ALOGTIME) by Buss [4] and Buss-Cook-Gupta-Ramachandran [5]. Hence the Boolean *formula* value problem is also in ALOGTIME (a formula may have free variables—the problem is to find the truth value of a formula given truth values for its variables).

The prior algorithms of [4] and [5] were based on a construction which necessitated converting an infix Boolean formula to Postfix-Longer-Operand-First (PLOF) form and then using complicated methods for choosing breakpoints in a semi-oblivious fashion. One of the main contributions of this paper is to give a new and simpler ALOGTIME algorithm for the BSVP. In our new algorithm, the sentence remains in infix notation and the method for selecting breakpoints is conceptually simpler.

Just as in [5], the method for solving the BSVP can be adapted to give logarithmic depth, polynomial size, arithmetic circuits for arithmetic formula evaluation. More generally, in this paper we analyse a tree-contraction algorithm of Abrahamson-Dadoun-Kirkpatrick-Przytycka [1] and show that it is ALOGTIME uniform.

Thirdly, we consider a special case of the Boolean sentence problem where the sentence is restricted to be balanced (or at least have logarithmic depth) and to have only the connectives $\wedge$ and $\vee$. We show that this 'Balanced, and/or BSVP' has an $O(\log\log n)$ space algorithm. Indeed there is an $O(\log\log n)$ space algorithm for this problem which scans the input from left to right without ever backing up and which has as memory only a single counter which may be incremented, decremented and tested for being equal to zero. To compare this with prior work, recall that Barrington showed that the BSVP has constant-width, quadratic length branching programs. Our result shows that the 'balanced and/or BSVP' has $\log\log n$ width, linear length branching programs. Unlike Barrington's construction, the branching programs we construct correspond to online computations in that the input symbols controlling the branching program are taken in their input order.

From the above we obtain the surprising result that every problem in ALOGTIME is deterministic log time reducible to $\log\log n$ space (by a

Turing machine with random access).

2 The ALOGTIME algorithm for BSVP

In this section we present the ALOGTIME algorithm for the Boolean sentence value problem. We presume basic familiarity with the notion of ALOGTIME; however, we briefly review the definitions and relevant properties. For more information see the introductory sections of [4,5] and the references cited therein.

An ALOGTIME Turing machine is an alternating, multitape Turing machine which runs in logarithmic time; since the runtime is sublinear, the input is accessed via an index tape so that if an integer i is written in binary notation on the index tape then the i-th symbol of the input string is accessible to the finite state control of the Turing machine. Without loss of generality the Turing machine may be constrained so that it only reads a single input symbol (and then immediately halts) on any computation path This is because the Turing machine may be designed to guess input symbols as needed and then branch universally to either verify the guess or to continue the computation.

It is well-known that ALOGTIME corresponds to a uniform version of NC^1; more precisely, the set of predicates recognizable in ALOGTIME is the same as the set of predicates which have U_{E^*}-uniform, log depth, bounded fan-in circuits.

A function f is said to be in ALOGTIME if its bit graph is in ALOGTIME; i.e., if

$$\{\langle x, i, c\rangle : i\text{-th symbol of } f(x) \text{ is } c\}$$

is in ALOGTIME. Some prominent examples of functions known to be in ALOGTIME are counting, multiplication and vector addition. With the aid of counting, it can be seen that any usual parsing operation on parenthesized infix formulas can in performed in ALOGTIME — for example, determining if a formula is properly parenthesis-balanced or identifying matching parentheses can be performed in ALOGTIME by counting parentheses.

It is convenient to picture Boolean sentences as trees with interior nodes labelled with propositional connectives and leaves labelled with 0/1 values. Hence a *node* in a Boolean sentence is (an occurrence of) a non-parenthesis symbol in the sentence and a *leaf* is (an occurrence of) a constant symbol in the sentence. Concepts such as 'depth', 'child', 'father', 'ancestor', 'descendent', etc. that apply to trees thus also apply to Boolean sentences.

Furthermore, by usual methods of parenthesis counting, all these concepts are ALOGTIME definable.

The leaves of a Boolean sentence are numbered from left to right beginning with 1. The *rank* of a leaf is the largest integer d such that 2^d divides the number of the leaf. Both the number and the rank are readily seen to be computable in ALOGTIME.

If U and V are nodes, we write $U \rhd V$ to mean that U is an ancestor of V. Similarly, $U \unrhd V$ means that $U = V$ or $U \rhd V$. The least common ancestor of U and V is denoted $lca(U, V)$ and is the lowest node W such that $W \unrhd U$ and $W \unrhd V$. (Trees, by convention, have root at the top so 'lowest' means of greatest depth.) The least common ancestor function is clearly computable in ALOGTIME.

We shall henceforth assume that all Boolean sentences do not contain negation signs and hence correspond to binary trees. This assumption can be made without loss of generality since there is an ALOGTIME algorithm which converts any Boolean formula containing negation signs to an equivalent one without negation signs. We also assume w.l.o.g. that there are exactly $2^{d+1} - 1$ leaves in a Boolean sentence: this can always be ensured by adding extra $(\cdots \wedge 1)$'s to the sentence — this will at most double the length of the formula. (The only reason for the latter assumption is to simplify the exposition of the algorithm below.)

The rest of this section is devoted to the proof of:

Theorem 1 *The BSVP is in ALOGTIME.*

The ALOGTIME algorithm for the BSVP will be presented as a pebbling game played by two players called the Pebbler and the Challenger. The pebbling game is described next in a top-down fashion.

In the pebbling game, the Pebbler places "pebbles" on nodes of the Boolean sentence: each pebble is labelled with a truth value 0 or 1 and represents an assertion by the Pebbler that the subformula rooted at the pebbled node has that truth value. After the Pebbler has placed some pebbles, the Challenger must respond by challenging one of the pebbled positions: this represents an assertion by the Challenger that the pebble value is incorrect, and furthermore, by challenging a pebble position U, the Challenger asserts that every pebble position $V \lhd U$ is correctly labeled. As an extra condition, we require that the Challenger may not challenge any node W such that either $W \rhd U$ where U has already been challenged or $W \trianglelefteq V \triangleleft U$ for some previously challenged U and some V that was already pebbled when U was challenged.

The pebbling game is devised so that the Pebbler has a winning strategy if the Boolean sentence has value 1 and otherwise the Challenger has a winning strategy. A lot of the rules of the game (such as the 'extra condition' above) are designed so that a play of the game can be run and evaluated in alternating log time. The reader should not be misled by the fact that rules seem somewhat arbitrary and strict — the point is that the strictness of the rules helps with the alternating log time calculation while preserving the property that the correct player has a winning strategy.

The pebbling game is organized into rounds: during each round, the Pebbler pebbles a set of nodes (the set of nodes is specified below) and then the Challenger either challenges a newly pebbled node or rechallenges the previously challenged node. A player loses the game by making a clearly false assertion; e.g., asserting an incorrect value for a leaf, asserting incompatible input and output values for a Boolean gate, etc. A player may also lose the game by violating the rules detailed below.

The pebbling game will yield an ALOGTIME algorithm for the BSVP since the following conditions will hold:

(1) The Pebbler has a winning strategy iff the Boolean sentence has value 1.

(2) The game lasts $\leq \lceil \log n \rceil$ rounds, where n is the number of leaves in the formula.

(3) Complete information about the Pebbler's and Challenger's moves can be given with a constant number of bits per round (12 bits per round in the implementation below).

(4) There is an ALOGTIME algorithm which, given the Boolean sentence and the $O(\log n)$ bits specifying the moves of the players, determines the winner of the game.

The conversion of the pebbling game into an alternating log time Turing machine M is quite straightforward. On input a Boolean sentence φ, M computes $O(\log n)$ bits describing a play of the game using existential guesses for the Pebbler's moves and universal choices for the Challenger's moves; M then accepts iff the Pebbler won this play of the game. By (2), (3) and (4), M runs for $O(\log n)$ time, and by (1) and the definition of alternation, M accepts φ iff φ has value 1.

Condition (3) that only $O(1)$ many bits of information may be given per round is rather stringent. It is also crucial for obtaining an ALOGTIME algorithm. Since a constant number of bits is not sufficient for specifying an arbitrary pebble point (that would require $O(\log n)$ bits), it is necessary that the rounds of the game be at least partly "oblivious" to the previous rounds. We say that a game is *open* at round i if the winner of the game is not determined by the players' moves before round i. The game, as implemented by the Turing machine M, lasts $O(\log n)$ rounds even though the game may not be open for all the rounds. The game will be defined so that the loser of the game is the first player is make an 'obvious' mistake (as defined below) and thus players's moves after the first such mistake are irrelevant to the outcome of the game.

We now define in complete detail the rules of the Pebbling game. In round $i \geq 1$ there will be leaves L_i, C_i, R_i and nodes A_i and B_i (see Figure 1). As a useful mnemonic think of A, B, L, C and R as standing for 'Above', 'Below', 'Left', 'Center' and 'Right'. The following four conditions will be satisfied for as long as the game is open.

(α) $A_i \rhd B_i \unrhd C_i$.

(β) A_i is the lowest (and last) challenged note. B_i is the highest pebbled position satisfying (α). (There will always be such a node B_i if the game is still open at round i.) Informally, the players have agreed at B_i but disagree at A_i.*

(γ) L_i and R_i are leaves of rank $d-i$ and C_i is of rank $> d-i$.

(δ) Every leaf D which is a descendent of A_i but is not a descendent of B_i has number in the range

$$(L_i - \delta_i, L_i + \delta_i) \cup (R_i - \delta_i, R_i + \delta_i)$$

where $\delta_i = 2^{d-i-1}$. Here we are identifying leaves with their numbers in left-to-right order; the rounded parentheses indicate open intervals. Thus $L_i - \delta_i$ and $L_i + \delta_i$ are the leaves closest to L_i of rank exactly $d-i-1$, and similarly for $R_i - \delta_i$ and $R_i + \delta_i$.

---

* For readers familiar with prior expositions of pebbling games: the rules of the game force the "subformula of disagreement" to be 1-scarred; i.e., if $D \lhd A_i$ is a pebbled node, then $D \unlhd B_i$.

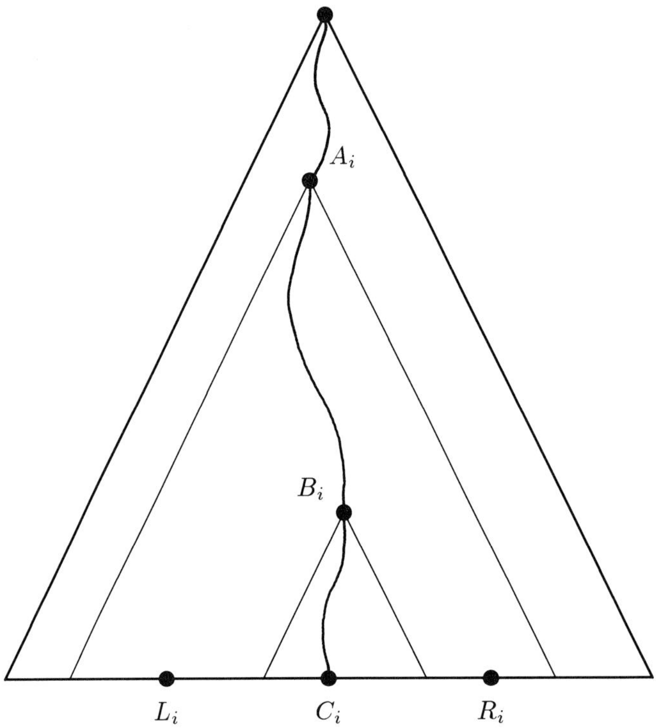

Figure 1

The triangles delineate subtrees. In general, it is not necessary for L_i and R_i to be descendents of A_i. On the other hand, it is also permissable for L_i and R_i to be descendents of B_i.

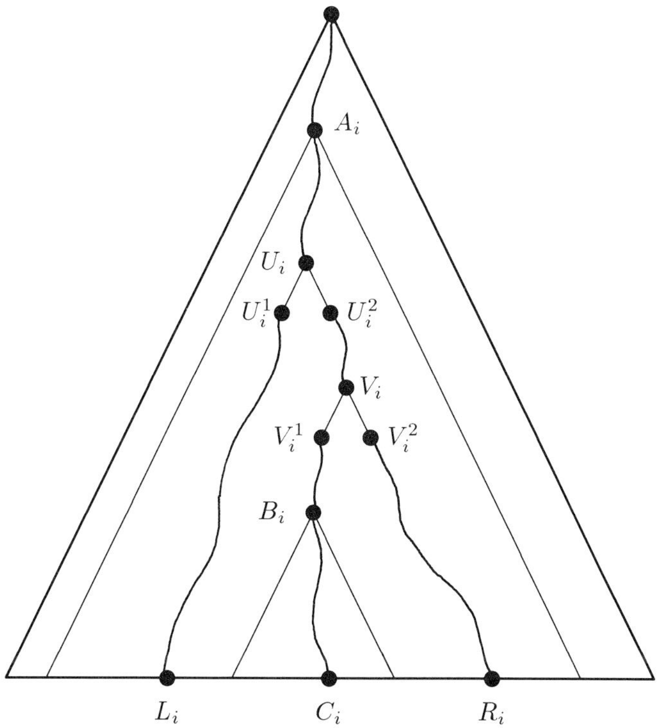

Figure 2

Figure 2 illustrates a possible configuration for the breakpoints chosen in round i. Handdrawn lines indicate paths in the tree.

Challenged Node	Pebbler says $U_i \rhd V_i$	Pebbler says $V_i \rhd U_i$
V_i^1	$C_{i+1} = C_i$ $R_{i+1} = R_i - 2^{\delta_i}$ $L_{i+1} = L_i + 2^{\delta_i}$	$C_{i+1} = C_i$ $R_{i+1} = R_i - 2^{\delta_i}$ $L_{i+1} = L_i - 2^{\delta_i}$
V_i^2	$C_{i+1} = R_i$ $R_{i+1} = R_i + 2^{\delta_i}$ $L_{i+1} = R_i - 2^{\delta_i}$	$C_{i+1} = R_i$ $R_{i+1} = R_i + 2^{\delta_i}$ $L_{i+1} = R_i - 2^{\delta_i}$
U_i^1	$C_{i+1} = L_i$ $R_{i+1} = L_i + 2^{\delta_i}$ $L_{i+1} = L_i - 2^{\delta_i}$	$C_{i+1} = L_i$ $R_{i+1} = L_i + 2^{\delta_i}$ $L_{i+1} = L_i - 2^{\delta_i}$
U_i^2	$C_{i+1} = C_i$ $R_{i+1} = R_i + 2^{\delta_i}$ $L_{i+1} = L_i + 2^{\delta_i}$	$C_{i+1} = C_i$ $R_{i+1} = R_i - 2^{\delta_i}$ $L_{i+1} = L_i + 2^{\delta_i}$
U_i or V_i	Game ends	Game ends

Challenged Node: A_i			
Pebbler says $A_i \rhd U_i, V_i$	Pebbler says $U_i, V_i \unrhd A_i$	Pebbler says $U_i \unrhd A_i \rhd V_i$	Pebbler says $V_i \unrhd A_i \rhd U_i$
$C_{i+1} = C_i$ $R_{i+1} = R_i + 2^{\delta_i}$ $L_{i+1} = L_i - 2^{\delta_i}$	$C_{i+1} = C_i$ $R_{i+1} = R_i - 2^{\delta_i}$ $L_{i+1} = L_i + 2^{\delta_i}$	$C_{i+1} = C_i$ $R_{i+1} = R_i + 2^{\delta_i}$ $L_{i+1} = L_i + 2^{\delta_i}$	$C_{i+1} = C_i$ $R_{i+1} = R_i - 2^{\delta_i}$ $L_{i+1} = L_i - 2^{\delta_i}$

Tables 1 and 2

The pebbling game proceeds as follows:

Round 0: The Pebbler must pebble the root node with value 1 and also also pebble the unique leaf of rank d (recall that there are exactly $2^{d+1}-1$ leaves). The Challenger must challenge the pebble at the root. In preparation for round 1, set A_1 equal to the root, set $B_1 = C_1$ equal to the leaf numbered 2^d, and set L_1 and R_1 equal to the leaves numbered 2^{d-1} and $2^d + 2^{d-1}$.

Round i: Let U_i be $lca(L_i, C_i)$ and V_i be $lca(C_i, R_i)$. Note that U_i and V_i are distinct. Let U_i^1 and U_i^2 be the left and right children of U_i, respectively, and let V_i^1 and V_i^2 be the left and right children of V_i, respectively (see Figure 2). The Pebbler must do the following two things:

(a) State whether $A_i \rhd U_i \rhd V_i$ or $A_i \rhd V_i \rhd U_i$ or $U_i \unrhd A_i \rhd V_i$ or $V_i \unrhd A_i \rhd U_i$ or $U_i, V_i \unrhd A_i$. (Since A_i, U_i and V_i are ancestors of C_i, exactly one will hold; so three bits of information specifies which one.)

(b) Give six truth values which are placed on pebbles on the nodes U_i, V_i, U_i^1, U_i^2, V_i^1, V_i^2 (six bits of information).†

Then the Challenger must challenge one of the pebbled nodes A_i, U_i, V_i, U_i^1, U_i^2, V_i^1, V_i^2 (there are seven possibities so three bits of information suffice). The Challenger must choose to challenge a node W such that $A_i \unrhd W$ and such that $B_i \not\unrhd W$.

For round $i+1$, the values of L_{i+1}, C_{i+1}, R_{i+1} are set according to Tables 1 and 2. And A_{i+1} is the node challenged by the Challenger in round i and B_{i+1} is the highest pebbled node below A_{i+1} (there will always be a unique choice for B_{i+1} if the game is still open).

Remark: There are a number of degenerate cases for round i. For example, it is possible that $U_i^2 = V_i$ and that the Pebbler pebbles this node twice in round i. If the Pebbler has a winning strategy, it must require that the Pebbler consistently give the same value to a repebbled node. As another degenerate case, we may have $B_i \unrhd U_i, V_i$. In this case,

† A slightly more complicated definition of the game could save two bits of information by not having the Pebbler pebble U_i and V_i. This would still preserve the essential properties of the game since the truth values of U_i and V_i are determined by the truth values of their inputs.

no matter what values are pebbled on the six nodes, the Challenger must rechallenge A_i (or lose the game). Likewise, if all six pebble positions are ancestors of A_i then the Challenger must repebble A_i.

To understand the way that the values of L_{i+1}, C_{i+1} and R_{i+1} are set in Table 1, consider as an example the case where $U_i \rhd V_i$ and the Challenger has challenged U_i^2. If $U_i^2 = V_i$ then the winner of the game has been determined by this move because the Pebbler has specified values to the inputs V_i^1 and V_i^2 and to the output $U_i^2 = V_i$ of a Boolean gate and because the Challenger has challenged the value of the output, thereby accepting the Pebbler's values for the inputs. So in this case either the Pebbler or the Challenger has made an obviously false assertion. On the other hand, suppose $U_i^2 \rhd V_i$ so the game may still be open. Referring to Table 1, this implies $C_{i+1} = C_i$, $L_{i+1} = L_i + 2^{d-i-1}$, $R_{i+1} = R_i + 2^{d-i-1}$, $A_{i+1} = U_i^2$, and $B_{i+1} = V_i$. It is easily seen that every leaf W such that $U_i^2 \unrhd W$ and $V_i \not\unrhd W$ has number in the set

$$\left(L_i, L_i + 2^{d-i}\right) \cup \left(R_i, R_i + 2^{d-i}\right).$$

This is because condition (δ) held at the beginning of round i. Of course, this set is equal to

$$\left(L_{i+1} - 2^{d-i-1}, L_{i+1} - 2^{d-i-1}\right) \cup \left(R_{i+1} - 2^{d-i-1}, R_{i+1} - 2^{d-i-1}\right).$$

Thus condition (δ) holds again for round $i+1$. To see that condition (γ) holds at round $i+1$, note that it is trivial that if L_i and R_i have rank $d-i$, then $L_i \pm 2^{d-i-1}$ and $R_i \pm 2^{d-i-1}$ have rank $d-i-1$.

The rest of the cases in Table 1 where $U_i \rhd V_i$ are easily checked in same fashion with the aid of Figure 2. The cases $V_i \rhd U_i$ are reflections of the cases where $U_i \rhd V_i$. We leave it to the reader to verify the rest of Tables 1 and 2.

Now we define precisely who the winner of a play of the game is — this is done by defining the loser of the game by formalizing what an 'obvious mistake' or 'obviously false assertion' is. It also must be shown that (a) someone loses the game by the end of Round $d = \lfloor \log_2 n \rfloor$, and (b) Pebbler has a winning strategy iff the Boolean sentence has value 1, and (c) the loser of a particular play of the game can be determined in ALOGTIME.

Definition The 'obvious mistakes' that cause a player to lose the game are:

- Asserting incompatible values to the inputs and output of a gate. For the Pebbler this occurs when the input nodes are either leaves or are pebbled and the output node has been pebbled incompatibly. For the Challenger, this occurs when he challenges the output of a gate whose input nodes are either leaves or pebbled.
- Asserting an incorrect value for a leaf. For the Pebbler, this can occur if she pebbles a leaf incorrectly; for the Challenger, if he challenges a correctly pebbled leaf.
- (Pebbler only) Being incorrect about whether $U_i \rhd V_i$, $A_i \rhd U_i$, $A_i \rhd V_i$.
- (Pebbler only) Pebbling a node with both 0 and 1.
- (Challenger only) Challenging a node which is above a previously challenged node.
- (Challenger only) Challenging a node which is at or below a previously agreed upon pebble value. A pebble value is defined to be 'previously agreed upon' if the pebble was placed in an earlier round and in that round the Challenger challenged an ancestor of the pebble.

By condition (δ), there are at most 2^{d-i+1} leaves which are below A_i but not at or below B_i. It follows that in round d there are at most two such leaves and thus, since the sentence contains only binary connectives, one of the players will be forced into an 'obvious mistake' before round d. Furthermore, the Pebbler (respectively, the Challenger) has a winning stategy iff the Boolean sentence has value 1 (respectively, 0); since, for the Pebbler the winning strategy is to always tell the truth and, for the Challenger, the winning strategy is to always challenge a lowest incorrectly pebbled node which is not below a previously agreed upon node. It remains to show that the winner of a play of the game can be determined in ALOGTIME. For this, it will suffice to show that the positions L_i, C_i, R_i, A_i and B_i can be obtained in ALOGTIME since then the winner can be determined in ALOGTIME by nondeterministically guessing the first 'obvious mistake', checking that it is indeed an obvious mistake, and then branching universally to verify that there was no earlier obvious mistake. We first explain how to obtain L_i and R_i in ALOGTIME: in the description of the game, we always have $L_{i+1} = L_i \pm \delta_i$ or $L_{i+1} = R_i - \delta_i$ and have $R_{i+1} = R_i \pm \delta_i$ or $R_{i+1} = L_i + \delta_i$, where $\delta_i = 2^{d-i-1}$. Assignments of the form $L_{i+1} = R_i - \delta_i$ and $R_{i+1} = L_i + \delta_i$ are called *mixed*. We say that the next mixed assignment after round i occurs in round j if there is a mixed assignment at round j and no mixed assignment in any round k with $i < k < j$; if the mixed assignment in round j is of the form $L_{j+1} = R_j - \delta_j$ then we define LR_i to be equal to R_i, otherwise the mixed assignment is $R_{j+1} = L_j + \delta_j$ and LR_i is defined to be equal to L_i. If there is no mixed assignment after round i, then LR_i is arbitrarily defined to be L_i. Since it is easy to tell whether an assignment is mixed in a round k of a play of the pebbling game by referring to Table 1—the question of whether LR_i equals R_i or L_i is easily resolved in ALOGTIME. Now LR_{i+1} is always equal to $LR_i \pm \delta_i$ and whether it is plus or minus δ_i is easily determined from the play of the game and the rules of Table 1. Thus LR_i can be determined by summing a vector of values $\pm\delta_i$; this is a particularly easy vector summation and is in ALOGTIME. To find the value of L_i, first find the least value of j such that for all $j \leq k < i$, L_{k+1} is defined by an unmixed assignment $L_{k+1} = L_k + \epsilon_k\delta_k$ with $\epsilon_k = \pm 1$ (so $j = 0$ or L_j is given by a mixed assignment). Then, if $j = 0$, $L_i = 2^{d-1} + \sum_{k=0}^{i-1} \epsilon_k\delta_k$ and if $j > 0$, $L_i = LR_j + \sum_{k=0}^{i-1} \epsilon_k\delta_k$. This vector summation provides an ALOGTIME algorithm to compute L_i. Essentially the same construction gives an ALOGTIME algorithm for determining R_i.

To calculate C_i, one determines the maximum value $j \leq i$ such that $C_j = L_{j-1}$ or R_{j-1} according to Table 1, or, if no such j exists, use $j = 0$; and then $C_i = C_j$. To calculate U_i, U_i^1, U_i^2, V_i, V_i^1, and V_i^2 is now easy since they are the least common ancestors of L_i, R_i and C_i and the children of the least common ancestors. Finally, A_i and B_i may be found from the values of the U's and V's and the play of the game according to Table 1.

That completes the proof that there is an alternating logarithmic time algorithm for the Boolean sentence value problem.

3 ALOGTIME Tree Contraction

The algorithm of the previous section is based on a method of splitting a tree (representing a formula) into subtrees so that the problem of finding the value of the tree is reduced to finding values of the subtrees. The subtrees are generated by picking breakpoints (the nodes U_i and V_i). There is an elegant way to describe the breakpoints which are picked in the i-th round; namely, they are least common ancestors of the leaves of rank $\geq d - i$. The algorithm for the BSVP was presented as a Pebbling game — in essence, this is a top-down procedure. However, in parallel processing applications it is often useful to have a bottom-up algorithm which can be executed efficiently in parallel by multiple processors. Such algorithms have been given by Brent [3] Miller-Reif [8] and Abrahamson, et. al. [1]; these algorithms allow $n/\log n$ many processors to evaluate a Boolean sentence in time $O(\log n)$. All of these algorithms are based on tree contraction algorithms; the tree contraction algorithm of Abrahamson, et. al. is particularly elegant and direct, and in this section we show that their tree contraction algorithm is describable (computable) in alternating log time.

We first briefly review Abrahamson, et. al.'s tree contraction algorithm. They construct a tree by applying "prune" and "bypass" procedures which transform a binary tree as pictured:

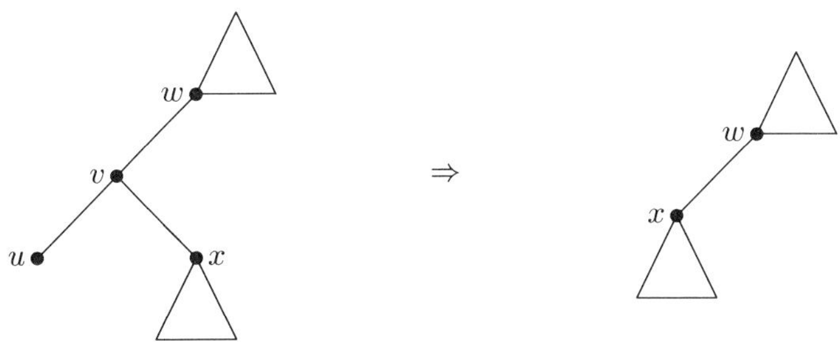

The Prune-Bypass Operation Applied to u
Figure 3

In Figure 3 the leaf node u is removed in the prune operation and then its parent node v is removed (bypassed) and v's other child becomes the child of v's parent w (the triangles represent arbitrary subtrees). Clearly the prune-bypass operation preserves the property of being a binary tree. If x and y are nodes in a binary tree then we say x is a *right (left) descendent* of y iff x is the right (left) child of y or x is a descendent of that child. It is clear that if x and y are not removed in a prune-bypass operation, then the property of x being a left (right) descendent of y is preserved.

The Abrahamson, et. al. tree contraction algorithm is:

Input: A binary, ordered tree T_0 with n leaves
Step (1): Number the leaves from 1 to n in left-to-right order
Step (2): Loop with $i = 1, \ldots, \lfloor \log n \rfloor$:
Apply the prune-bypass operation to all leaves of odd number
Divide the remaining leaf numbers by 2 and call the resulting tree T_i.
Endloop

We have made some inessential modifications to the algorithm. First, nodes are numbered beginning with 1 instead of 0. Second, we allow the root node to be bypassed (in Figure 3, this would make x the new root). Third, the prune-bypass operations may be carried out in any order since we are only concerned with describing the trees T_i; however, Abrahamson, et. al.'s technique of pruning and bypassing first the leaves of number congruent to 1 mod 4 and then the leaves of number congruent to 3 mod 4 still works to give an efficient parallel algorithm.

Clearly each T_i is a binary tree and T_i has exactly $\lfloor n/2^i \rfloor$ leaves. Thus $T_{\lfloor \log n \rfloor}$ has exactly one node and the tree contraction is completed at this stage. The internal nodes in T_i are also internal nodes of T_0; likewise every leaf of T_i was a leaf of T_0. The T_0*-number* and T_0*-rank* of a leaf are defined to be the number and rank that the leaf originally has in T_0.

Theorem 2

(a) *The leaves of T_i are precisely the leaves x of T_0 which have T_0-rank $\geq i$.*

(b) *If x and y are nodes in T_i then their least common ancestor in T_i is the same as their least common ancestor in T_0.*

(c) *An internal node u of T_0 is a node in T_i iff u is the least common ancestor of two leaves of ranks $\geq i$.*

Proof (a) is trivial from the definition of the tree-contraction algorithm. (b) is easily proved by noting that least common ancestors are preserved by a prune-bypass operation. For half of (c), note that (a) and (b) imply that the least common ancestor of two leaves of ranks $\geq i$ is a node in T_i. For the other half of (c), suppose it is not the case that u is the least common ancestor of two leaves of ranks $\geq i$. In this case, it must be the case that either every right descendent leaf of u has T_0-rank $< i$ or (dually) every left descendent leaf of u in T_0 has T_0-rank $< i$. But since leaves of T_0-rank $< i$ do not appear in T_i, it is impossible for u to be a node in T_i since u would have no right (respectively, left) descendent leaf in T_i contradicting the fact that T_i is a binary tree with u an interior node.

It follows immediately from Theorem 2 that the trees T_i are ALOGTIME computable from T_0. If x is a node in T_0, it can be determined if x is a node in T_i by either checking if x is a leaf of rank $\geq i$ or guessing two leaves y and x with T_0-rank $\geq i$ and verifying that x is the least common ancestor of y and z.

4 Balanced BSVP is in log-log space

In this final section we give a very simple algorithm for the BSVP when the logical connectives are restricted to be $\wedge$ and $\vee$. This algorithm runs in logspace but is simpler than N. Lynch's log space algorithm for the BSVP with arbitrary binary connectives [7]. When our algorithm is applied to balanced Boolean sentences, it uses only $O(\log\log n)$ space on inputs of length n. This yields the rather surprising consequences: (1) every ALOGTIME predicate is deterministic log time reducible to a $\log\log n$ space problem, and (2) there are linear length, log width branching programs for the Balanced BSVP restricted to connectives $\wedge$ and $\vee$. Consequence (1) follows since it is known that the Balanced BSVP restricted to $\wedge$ and $\vee$ is complete for ALOGTIME under deterministic log time reductions. It is open whether the $\log\log n$ space bound is optimal.

Since $\log\log n$ space is not space constructible [6], it is useful to think of a $\log\log n$ space Turing machine being initialized with a workspace of size $\log\log n$; however, we shall only consider inputs on which the Turing machine will use only $\log\log n$ space and, in particular, the Turing machine gains no advantage from being able to sense the ends of its worktape. A $\log\log n$-space Turing machine accesses its input via a usual tape head which may be moved left or right one square at a time. However, the $\log\log n$ space algorithm below will only move the tape head obliviously from left-to-right with no backing up.

Theorem 3 *There is a* $\log\log n$ *space algorithm for the Balanced BSVP restricted to connectives* $\wedge$ *and* $\vee$. *Moreover, this algorithm scans the input* *from left-to-right only and its only memory is a single counter which may by incremented by one, decremented by one and tested for zero.*

A Boolean sentence of n symbols is said to be balanced if it has depth $\lceil \log_2 n \rceil$; the algorithm of Theorem 3 actually uses space $\log d$ on inputs which are Boolean sentences of depth d.

The essential idea of the algorithm is based on the following property of the connectives $\wedge$ and $\vee$: either the value of the first operand of an $\wedge$ (or $\vee$) determines the output value or the value of the second operand is equal to the output value (and the value of the first operand determines which case holds). Thus, in the first case, the second operand may be ignored, while in the second case, the value of first operand and the gate type may be forgotten since the second operand's value is equal to the output value.

Algorithm for BSVP with connectives $\wedge$ and $\vee$:

```
Input: a Boolean sentence φ.
Initialize: Ignore_Flag = 0
            Ignore_Depth = 0
            Input tape head at leftmost symbol of φ
Loop:
  Read current symbol α from input tape
  If Ignore_Flag = 1
        If α = "(", increment Ignore_Depth by 1
        If α = ")"
              decrement Ignore_Depth by 1
              if Ignore_Depth = 0, set Ignore_Flag = 0
        Endif
  Endif
  If Ignore_Flag = 0
        If α = "0" set Val = 0
        If α = "1" set Val = 1
        If α = "∧" and Val = 0, set Ignore_Flag = 1
        If α = "∨" and Val = 1, set Ignore_Flag = 1
  Endif
  Move input tape head right one square
Endloop
Output: Val
```

This algorithm is easily implemented on a Turing machine which scans its input from left-to-right. The one bit registers *Val* and *Ignore_Flag* may be remembered by the finite state control. The *Ignore_Depth* register may be incremented, decremented and tested for zero; its maximum value is the depth of the Boolean formula. Thus if the Boolean formula is balanced, *Ignore_Depth* may be stored in $\log\log n$ space.

Corollary 4 *Every ALOGTIME predicate is deterministic log time, many-one reducible to* $\log\log n$ *space.*

This corollary is immediate from Theorem 3 since in [4,5] it is shown that the Balanced BSVP restricted to connectives $\wedge$ and $\vee$ is complete for ALOGTIME under deterministic log time reductions. Since $\log\log n$ is not space constructible, it is useful to define a many-one reduction of a decision problem $A \subset \Sigma^*$ to $\log\log n$ space to be a function f and a Turing machine M such that, for all $x \in \Sigma^*$, $x \in A$ iff $f(x)$ is accepted by M and such that, for all $x \in \Sigma^*$, M uses at most $\log\log|f(x)|$ space on input $f(x)$.

Theorem 5 *The Balanced BSVP, with* $\wedge$ *'s and* $\vee$ *'s only, has logarithmic width, linear length branching programs.*

See Barrington [2] for information on branching programs. Theorem 5 should be compared with Barrington's theorem that the Balanced BSVP has quadratic length, constant width branching programs.

Proof Let $0 < d \leq n$. We shall show how to contruct a branching program of width $4 \cdot d$ and length n which recognizes true Boolean sentences of length n and depth $\leq d$. Taking $d = \lceil \log n \rceil$ this proves the theorem. The branching program can be viewed as a graph on $n+1$ columns of height $4d$. Each of the $4d$ rows corresponds to a particular set of values for *Ignore_Flag*, *Val*, and *Ignore_Depth*. The transition rules from the i-th column to the $i+1$-st column is based on the i-th symbol of the input in the obvious way based on the above algorithm. □

A natural question to ask is whether Theorem 3 can be used to separate the class P of polynomial time predicates from ALOGTIME; since, after all, both $\log \log n$ space and deterministic log time are proper subclasses of P. However, it is not possible to apply Theorem 3 to provide a straightforward polynomial time diagonalization of ALOGTIME: the problem is that the deterministic log time reductions may be functions of arbitrarily high degree polynomial growth rate and these can not be computed by a single polynomial time algorithm.

Acknowledgements

I wish to thank Pierre McKenzie for bringing the tree-contraction algorithm of Abrahamson-Dadoun-Kirkpatrick-Przytycka to my attention; this was the catalyst for the work of this paper. I benefitted greatly from long-ago conversations with Vijaya Ramachandran about formula evaluation. Pavel Pudlák suggested an improvement which has been incorporated into the $\log \log n$ space algorithm.

References

[1] K. Abrahamson, N. Dadoun, D. G. Kirkpatrick, and T. Przytycka, *A simple parallel tree contraction algorithm*, Journal of Algorithms, 10 (1989), pp. 287–302.

[2] D. A. M. Barrington, *Bounded-width polynomial-size branching programs recognize exactly those languages in NC^1*, J. Comput. System Sci., 38 (1989), pp. 150–164.

[3] R. P. Brent, *The parallel evaluation of general arithmetic expressions*, J. Assoc. Comput. Mach., 21 (1974), pp. 201–206.

[4] S. R. Buss, *The Boolean formula value problem is in ALOGTIME*, in Proceedings of the 19-th Annual ACM Symposium on Theory of Computing, May 1987, pp. 123–131.

[5] S. R. Buss, S. A. Cook, A. Gupta, and V. Ramachandran, *An optimal parallel algorithm for formula evaluation.* To appear in SIAM Journal on Computation.

[6] A. R. Freedman and R. E. Ladner, *Space bounds for processing contentless inputs*, Journal of Computer and System Sciences, 11 (1975), pp. 118–128.

[7] N. A. Lynch, *Log space recognition and translation of parenthesis languages*, J. Assoc. Comput. Mach., 24 (1977), pp. 583–590.

[8] G. L. Miller and J. H. Reif, *Parallel tree contraction and its application*, in Proceedings of the 26th IEEE Symposium on Foundations of Computer Science, IEEE Computer Society, 1985, pp. 478–489.

Samuel R. Buss
Department of Mathematics
University of California, San Diego
La Jolla, California 92093
email: sbuss@ucsd.edu

Provably Total Functions in Bounded Arithmetic Theories R_3^i, U_2^i and V_2^i

Samuel R. Buss[*] [†]
Jan Krajíček[*] [‡]
Gaisi Takeuti[*] [§]

Abstract

This paper investigates the provably total functions of fragments of first- and second-order Bounded Arithmetic. The (strongly) Σ_i^b-definable functions of S_3^{i-1} and R_3^i are precisely the (strong) $\mathrm{FP}_3^{\Sigma_{i-1}^p}[wit, log^{O(1)}]$ functions. The $\Sigma_i^{1,b}$-definable functions of V_2^{i-1} and U_2^i are the $\mathrm{EXPTIME}^{\Sigma_{i-1}^{1,p}}[wit, poly]$ functions and the $\Sigma_i^{1,b}$-definable functions of V_2^i are the $\mathrm{EXPTIME}^{\Sigma_i^{1,p}}$-functions. We give witnessing theorems for these theories and prove conservation results for R_3^i over S_3^{i-1} and for U_2^i over V_2^{i-1}.

1 Introduction

This paper discusses the Σ_i^b-definable functions of the first-order theories R_3^i and S_3^{i-1} and the closely related $\Sigma_i^{1,b}$-definable functions of U_2^i and V_2^{i-1}. In addition, we characterize the $\Sigma_i^{1,b}$-definable functions of V_2^i. We give new witnessing theorems for the appropriate fragments of these theories and prove several conservation results.

[*] All three authors supported in part by NSF-ČSAV grant INT-8914569.
[†] Supported in part by NSF Grant DMS-8902480.
[‡] Work performed while visiting San Diego and on leave at Urbana-Champaign.
[§] Supported in part by NSF grant DMS-8800314.

Buss [3] provided a characterization of the Σ_i^b-definable functions of S_2^i as the set of functions which are polynomial time computable with an oracle from the class Σ_{i-1}^p of the polynomial time hierarchy. Later, he characterized the Σ_i^b-definable functions of T_2^{i-1} by showing that S_2^i is conservative over T_2^{i-1} with respect to $\forall\Sigma_i^b$-consequences [4]. In this paper we establish similar characterizations of the Σ_i^b-definable functions of theories S_3^{i-1} and R_3^i. Recall that the theory R_3^i was introduced in various forms by Allen [1], Clote-Takeuti [7] and Takeuti [18]. In analogy with earlier results, we show that R_3^i and S_3^{i-1} have the same Σ_i^b-definable functions and that R_3^i is conservative over S_3^{i-1} with respect to $\forall\Sigma_i^b$-consequences.

The above characterization of provably total functions of R_3^i uses the witness function method but also requires the introduction a new notion of oracle computation: we define a *witness oracle* to be an oracle which when presented with an existential question, either responds 'No' or responds 'Yes' and provides a witness to the truth of the question; i.e., provides a instance or value of the existential quantifier proving that the answer is 'Yes'.[1] It is shown that the Σ_i^b-definable functions of R_3^i and S_3^{i-1} are precisely the functions that can be computed in time $2^{(\log n)^{O(1)}}$ time with witness oracle from the class Σ_{i-1}^p of the polynomial time hierarchy. In addition, we consider a notion of 'strongly Σ_i^b-definable' functions and also characterize the strongly Σ_i^b-definable functions of R_3^i and S_3^{i-1} as being precisely the functions that can be "strongly" computed in time $2^{(\log n)^{O(1)}}$ time with witness oracle from the class Σ_{i-1}^p of the polynomial time hierarchy. Unfortunately, we have not been able to accomplish a similar result for R_2^i with polynomial time computations; it is open whether such a theorem holds. For S_2^{i-1} there is such a theorem known: Krajicek [10] shows that the Σ_i^b-definable functions of S_2^{i-1} are precisely the functions which can be computed in polynomial time with a witness oracle from Σ_{i-1}^p.

It turns out that this investigation of first-order systems is entirely analogous to investigating the $\Sigma_i^{1,b}$-definable functions of V_2^{i-1} and U_2^i. For these systems, we prove a old conjecture of the first author [3] regarding the class of functions with first-order values which can be $\Sigma_i^{1,b}$-defined by U_2^i and V_2^i; however, the method of proof is rather different from what the first author had in mind when making the conjecture. In addition

[1] Our use of witness oracles is closely related to Kreisel's nocounterexample interpretation as well as to the use of Herbrand's theorem in [12, 15, 11]. See also [2, 9] for recent applications of the the use of witness oracles to Peano arithmetic.

we characterize the $\Sigma_i^{1,b}$-definable functions of these theories that have second-order values.

The outline of this paper is as follows: in section 2, we introduce the computational complexity classes using witness oracles and prove several fundamental closure properties for them. In section 3, we briefly review the fragments of Bounded Arithmetic needed and prove that various complexity classes of functions can be defined in these theories. In section 4, we review the witness predicate and prove the witnessing lemma and various corollaries for the first-order systems. Section 5 is a translation of the results of section 4 to the second-order systems. The reader who is interested primarily in first order-systems may safely omit all the sections that pertain to second-order objects and second order-systems (sections 2.2, 3.3 and 5). However, the reader interested in second-order systems must read the entire paper since we frequently omit proofs in the second-order case. A summary of the main results can be found in sections 4.3 and 5.3. A couple of open questions are also mentioned in section 4.3.

2 Computational Complexity

2.1 Witness Oracles and Function Complexity Classes

Complexity classes such as P, NP, the polynomial time hierarchy classes Σ_i^p and Π_i^p, PSPACE, EXPTIME, and LOGSPACE are classes of predicates; i.e., are classes of problems which are computed by resource-bounded Turing machines which provide Yes/No answers. In this section we define related classes of functions. The inputs and outputs of our functions are integers which, by standard coding methods, is equivalent to using strings of characters over a finite alphabet. The length of an integer x is the length of its binary representation and is denoted $|x|$.

The classes of functions we define below are computed with Turing machines with *witness oracles*. A witness oracle is a generalized form of an oracle: when a witness oracle is asked an existential question "$(\exists x)\varphi(x)$?", it responds either with the answer "No" or with a value for x making $\varphi(x)$ true. Since there may be multiple x's making $\varphi(x)$ true this allows the witness oracle of a degree of non-determinism. Because of this non-determinism we shall allow our functions to be multivalued. A multivalued k-ary function is a relation on $\mathbb{N}^k \times \mathbb{N}$; we write $f(\vec{x}) = y$ for $(\vec{x}, y) \in f$; we shall always assume f is total.

To motivate these complexity classes, let's consider a couple of examples of functions that use a witness oracle for an NP predicate. First, let $f(x)$

be the following multi-valued function of values x coding propositional formulas:

$$f(x) = \begin{cases} y & \text{if } y \text{ codes a satisfying assignment for } x \\ 0 & \text{if } x \text{ is not satisfiable} \end{cases}$$

The function $f(x)$ can be easily computed with a single call to a witness oracle for SAT (the set of satisfiable propositional formulas). Second, let $g(x)$ be defined to the multivalued function such that if x codes a graph G then $g(x)$ codes a clique of maximal size in G. To compute $g(x)$, find the maximal clique size using binary search with $O(\log|x|)$ many queries to an NP predicate; then ask a witness oracle for NP for a clique of that maximal size. Both $f(x)$ and $g(x)$ are in the class $\mathrm{FP}^{\Sigma_1^p}[wit, log]$ defined next.

Loosely speaking, the class $\mathrm{FP}^{\Sigma_i^p}[wit, log]$ contains the functions which are polynomial time computable with a witness oracle for Σ_i^p and with the restriction that the oracle may be queried only $O(\log n)$ times. Recall that Σ_i^p and Π_i^p are classes in the polynomial time hierarchy with $\Sigma_0^p = \Pi_0^p = P$ and $\Sigma_1^p = \mathrm{NP}$ and $\Pi_1^p = \mathrm{coNP}$, etc.

Definition $\mathrm{FP}^{\Sigma_i^p}[wit, log]$ is the class of multivalued functions f for which there is a Turing machine M such that the following hold:

(1) M has an input x of length n and M runs in polynomial time. The value x may be a single integer or a vector of integers.

(2) M has Σ_i^p witness oracle for (w.l.o.g.) a predicate of the form

$$\Omega(q) \Leftrightarrow (\exists z, |z| < |q|^k) R(z, q)$$

where $R \in \Pi_{i-1}^p$ and k is a constant. The oracle is accessed with a query tape, an oracle response tape, a query state and oracle accept and reject states. When M enters the query state with q written on the query tape, the next configuration of M is either (i) in the oracle reject state if $\Omega(q)$ is false or (ii) in the oracle accept state with a value z written on the oracle response tape such that $R(z, q)$ holds and such that $|z| < |q|^k$. In case (1) the response tape is blank and in case (2) the tape head is at the leftmost symbol of z and the rest of the tape is blank.

(3) M makes only $O(\log n)$ many queries in any computation.

(4) At the end of the computation $M(x)$ outputs a value y such that $f(x) = y$. However, it is not necessarily the case that for any value $y = f(x)$ there is some sequence of valid oracle answers such that $M(x)$ outputs y.

The restriction that the witness oracle only be called $O(\log n)$ many times is necessary for the use of witness oracles to be meaningful: if polynomially (i.e., arbitrarily) many calls to the witness oracle were allowed, then M could use an ordinary (non-witness) oracle to get witnesses by asking a witness value one bit at a time.

Remark 1: That condition (4) allows $f(x) = y$ even if it is impossible for $M(x)$ to output y may seem surprising at first — especially since this allows the relation $f(x) = y$ to be non-recursive.[2] However, one should think of the problem of computing $f(x)$ as being the problem of searching for a y such that $f(x) = y$. From this point of view, it makes sense to say that M can compute $f(x)$, i.e., solve the search problem, even though M may not have the potential of outputting each y such that $f(x) = y$.

Let $func_M$ be the multivalued problem defined by $func_M(x) = y$ if and only if $M(x)$ *can* output y. An alternative definition of $\mathrm{FP}^{\Sigma_i^p}[wit, log]$ is that it is the class of functions f such that $f \supseteq func_M$ for M satisfying (1)-(3). It is also useful to consider the class of functions of the form $func_M$; accordingly we define:

Definition A function f is in *strong*-$\mathrm{FP}^{\Sigma_i^p}[wit, log]$ if and only if there is a Turing machine satisfying conditions (1)-(3) such that $f = func_M$.

Remark 2: It is possible to modify the definition of $\mathrm{FP}^{\Sigma_i^p}[wit, log]$ so that the witness oracle does not provide a witness until the final oracle call. This would not change the power of the witness oracle since M as defined above can be simulated by a Turing machine M' which runs the following algorithm:

---

[2] To construct a non-recursive f, pick A to be any non-recursive set and let $f(x) = 0$ hold for all x and let $f(x) = 1$ hold iff $x \in A$. The function f is clearly in $\mathrm{FP}^{\Sigma_i^p}[wit, log]$ since M need merely output 0 on all inputs.

Input: x

For $k = 1, \ldots, c \cdot \log n$ /* $c \cdot \log n$ = max. number of queries */

Ask oracle: "Is there a valid computation of M such that M's first $k-1$ queries are answered by $\alpha_1, \ldots, \alpha_{k-1}$ and such that the k-th query is answered 'Yes'?"

If so, set α_k ='Yes',

Else set α_k ='No'.

End for

Ask oracle for a witness to the true statement "There is a computation of M in which the oracle answers are $\alpha_1, \ldots, \alpha_{c \cdot \log n}$".

Output the value y which is on the output tape of the final configuration of the computation of M returned by the witness oracle.

To properly understand the above algorithm we must see why the oracle queries are Σ_i^p queries. Suppose that $\alpha_{i_1}, \ldots, \alpha_{i_r}$ are the ones among $\alpha_1, \ldots, \alpha_{k-1}$ that are equal to 'Yes'. Then the query for a computation of M can be phrased as the following Σ_i^p query about x and $i_1, \ldots, i_r$:

> "Is there (an encoding of) a computation of M such that during the computation M asks some oracle queries $q_1, \ldots, q_{k-1}, q_k$ and receives 'Yes' answers for exactly the queries q_{i_j} for $j = 1, \ldots r$ such that the witness responses β_{i_j} to the queries with 'Yes' answers satisfy $|\beta_{i_j}| < |q_{i_j}|^k$ and $R(\beta_{i_j}, q_{i_j})$?"

This is a Σ_i^p query since the predicate $R(-,-)$ is a Π_{i-1}^p property. Note that the oracle does not have to check if the negative responses by the oracle are correct (this would would make the query too complex anyway) since the α_j's are chosen greedily to be 'Yes' if possible. This is because the sequence $\alpha_1, \ldots, \alpha_{k-1}$ is the lexicographically largest possible sequence of Yes/No oracle answers (taking 'Yes' as greater than 'No' for the lexicographic ordering); hence if the 'Yes' answers are correct with correct witnesses β_{i_j} then the 'No' answers are necessarily correct.

One consequence of this remark is that the machine M may be restricted to use only $O(\log n)$ space until the final witness oracle query (for the purposes of measuring space, the query tape is write-only, is erased after each query, and is not counted in the space computation); the response to the final query is a polynomial size witness which w.l.o.g.

contains the output of M as a substring. Thus M may operate in $O(\log n)$ space until its final query, at which point it merely copies the output from (part of) the response tape to the output tape. Accordingly, another possible name for this function complexity class is $FL^{\Sigma_i^p}[wit, log]$. It can be shown using well-known techniques that the restriction that only $O(\log n)$ queries may be made to the witness oracle may be dropped for FL function classes, and thus this class is the same as $FL^{\Sigma_i^p}[wit]$ (see [8, 6, 19] for these techniques).

It is interesting to note that $func_{M'}$ may not be the same as $func_M$; since not every valid computation of M receives the lexicographically largest sequence of possible Yes/No answers. Part of our reason for using the class $\mathrm{FP}^{\Sigma_i^p}[wit, log]$ instead of strong-$\mathrm{FP}^{\Sigma_i^p}[wit, log]$ is to make Remark (2) hold.

Remark 3: It is also possible to allow M unlimited queries to a Σ_{i-1}^p oracle without changing the class $\mathrm{FP}^{\Sigma_i^p}[wit, log]$. This is because a polynomial time computation with unlimited queries to Σ_{i-1}^p may be simulated by making a single query to a witness oracle for Σ_i^p; namely, ask the witness oracle if there is a correct computation of M with correct answers to the Σ_{i-1}^p queries; of course, there is always a unique correct computation and the witness oracle returns it on its response tape.

Remark 4: It would be possible to define a class of predicates $P^{\Sigma_i^p}[wit, log]$ by considering the class of 0/1-valued functions in $\mathrm{FP}^{\Sigma_i^p}[wit, log]$. However, using the method of Remark 2, it is easy to see that this would be the class $P^{\Sigma_i^p}[log]$ which uses a regular (non-witness) oracle. This is the class of predicates polynomial time truth-table reducible to Σ_i^p (see Krentel [14], Buss-Hay [6], Wagner [19]). Krajíček [10] shows that these are precisely the predicates Δ_{i+1}^b-definable in S_2^i.

Similar considerations show that if a function in $\mathrm{FP}^{\Sigma_i^p}[wit, log]$ is constrained to output only values of length $O(\log n)$ bits, then it is in the class $\mathrm{FP}^{\Sigma_i^p}[log]$ which is defined as above but with a (non-witness) oracle for Σ_i^p.

Remark 5: Krentel [14] gave the original definition of the class $\mathrm{FP}^{\Sigma_i^p}[log]$ of functions. Our function class of strong-$\mathrm{FP}^{\Sigma_i^p}[wit, log]$ with witness oracles provides a seemingly different and possibly more natural function class. For example, the multivalued function defined by $f(x) = y$ if and only if either $y = 0$ or x codes a Boolean formula with y a satisfying assignment, is clearly in strong-$\mathrm{FP}^{\mathrm{NP}}[wit, log]$ since M can ask the witness oracle for a satisfying assignment of x. However, Krentel showed that this function is in $\mathrm{FP}^{\mathrm{NP}}[log]$ if and only if $\mathrm{P} = \mathrm{NP}$. On the other hand, our function class defined in terms of witness oracles seems to have the inherent disadvantage of containing multivalued functions.

For use with the theory R_3^i, we need to a slight modification of the above function class to reflect the presence of the $\#_3$ function in the language:

Definition $\mathrm{FP}_3^{\Sigma_i^p}[wit, log^{O(1)}]$ is the class of multivalued functions defined in exactly the same way as $\mathrm{FP}^{\Sigma_i^p}[wit, log]$ except that the runtime of the Turing machine is bounded by $2^{(\log|x|)^{k_1}}$ and the number of oracle queries is bounded by $(\log|x|)^{k_2}$ for some constants k_1, k_2.

The runtime bound $2^{(\log n)^{O(1)}}$ on inputs of length n is called "$\#_3$ time".

2.2 Functions of second-order objects

We next consider higher-order analogues of $\mathrm{FP}_3^{\Sigma_i^p}[wit, log^{O(1)}]$. There are essentially two modifications: first, the computational complexity will be exponential time or polynomial space and, second, there will two different kinds (called *orders*) of inputs — first-order inputs of length n and second-order inputs of length $2^{n^{O(1)}}$. In our applications to second-order theories U_2^i and V_2^i, these two kinds of inputs correspond to first- and second-order variables.

The class $\Sigma_i^{1,p}$ is the class of predicates which can be defined by a $\Sigma_i^{1,b}$ formula; in terms of Turing machines, this is the class of the predicates that can be recognized by a $2^{n^{O(1)}}$-time Turing machine which has i blocks of existential and universal alternations beginning with an existential block.

Definition $\mathrm{EXPTIME}^{\Sigma_i^{1,p}}$ is the class of single-valued functions f which are computed by a Turing machine M such that

1. M has first-order input x of length n (x may be a vector of values, in which case n is the total length of the first-order inputs). And M has second-order inputs $\vec{\varphi}$. Each second-order input is a string of symbols written on its own input tape.

2. M has run time bounded by 2^{n^k} for some constant k. Thus M can access only 2^{n^k} many squares of the second-order input tapes and M can ask oracle queries of length up to 2^{n^k} symbols.

3. M has a (nonwitness) oracle for a predicate in Σ_i^p. Since exponentially long queries are allowed, this corresponds to asking $\Sigma_i^{1,p}$ queries about the first-order inputs.

4. M outputs either a second-order value or a first-order value. Any first-order output must have length bounded by $n^{k'}$ for some constant k'.

The computational power $\mathrm{EXPTIME}^{\Sigma_i^{1,p}}$ would not be significantly changed if M was allowed to use a witness oracle; this is because the number of oracle queries by M is not restricted and M can ask repeated oracle queries to obtain witnesses one bit at a time.

Definition $\mathrm{EXPTIME}^{\Sigma_i^{1,p}}[\mathit{wit}, \mathit{poly}]$ is a class of multivalued functions; it is defined to be the set of functions such that there is a Turing machine satisfying the conditions 1.-4. above, except that, firstly, the third condition is modified so that M has a witness oracle for Σ_i^p and M may only make n^{k_0} queries to the witness oracle for some constant k_0, and secondly, if $M(x, \vec{\varphi})$ can output y or ψ then $f(x, \vec{\varphi}) = y$ or $f(x, \vec{\varphi}) = \psi$ (but not necessarily conversely).

Strong-$\mathrm{EXPTIME}^{\Sigma_i^{1,p}}[\mathit{wit}, \mathit{poly}]$ is the class of functions func_M for M satisfying the modified conditions 1.-4.

The remarks above about $\mathrm{FP}^{\Sigma_i^p}[\mathit{wit}, \mathit{log}]$ above apply also to $\mathrm{EXPTIME}^{\Sigma_i^{1,p}}[\mathit{wit}, \mathit{poly}]$. For example, the Turing machine M for $\mathrm{EXPTIME}^{\Sigma_i^{1,p}}[\mathit{wit}, \mathit{poly}]$ may be restricted so that only its final oracle query returns a witness; likewise M may be restricted to use only polynomial space until after the final witness query it copies part of the response tape to the output tape. As before, the write-only query tape is erased after each query and is not considered in the space computation. Also, an $\mathrm{EXPTIME}^{\Sigma_i^{1,p}}[\mathit{wit}, \mathit{poly}]$ Turing machine M may make unrestricted queries to a Σ_{i-1}^p oracle. If M outputs only first-order values then a usual oracle for $\Sigma_i^{1,b}$ suffices and the use of the witness oracle is unnecessary.

Finally, one should note that $\mathrm{EXPTIME}^{\Sigma_i^{1,p}}$ and $\mathrm{EXPTIME}^{\Sigma_i^{1,p}}[wit, poly]$ may be regarded as being the same classes as $\mathrm{FP}_3^{\Sigma_i^p}$ and $\mathrm{FP}_3^{\Sigma_i^p}[wit, log^{O(1)}]$ if the second-order inputs are reinterpreted as being first order inputs. This is because $\#_3$-time of inputs of length $2^{n^{O(1)}}$ is the same thing as time $2^{n^{O(1)}}$. This is related to the 'RSUV isomorphism' discussed in the next section.

3 Fragments of Bounded Arithmetic

3.1 Preliminaries

In this section we review the fragments of Bounded Arithmetic that are used in this paper. We shall assume familiarity with Buss [3] and shall need some theorems from Buss [4]. The systems we shall deal with are R_3^i, S_3^i, S_2^i, T_2^i, U_2^i and V_2^i. The subscript indicates the growth rate of function symbols in the language; the subscript 2 indicates that 0, S, $+$, $\cdot$, $\#$, $\lfloor \frac{1}{2}x \rfloor$ and $|x|$ are function symbols and the subscript 3 indicates that the $\#_3$ function is also in the language where $x \#_3 y = 2^{|x|\#|y|}$. The function $\#_2$ has polynomial growth rate and the growth rate of $\#_3$ is superpolynomial and subexponential: the $\#_2$ function allows formation of terms with growth rate $2^{|x|^{O(1)}}$ and the $\#_3$ function allows formation of terms with growth rate $2^{2^{|(|x|)|^{O(1)}}}$. In addition R_3^i has function symbols $\dot{-}$ and MSP for subtraction and "most significant part". (This choice of functions symbols avoids problem with bootstrapping and makes R_3^0 a useful theory. We shall not discuss the details of bootstrapping in this paper; since we are dealing primarily with R_3^i with $i > 1$, it is only necessary to show that R_3^i contains S_3^{i-1} and then the well-known bootstrapping for S_3^1 applies.)

The definition of the classes Σ_i^b and Π_i^b of bounded formulas is as usual, counting alternations of bounded quantifiers $(Qx \leq t)$ but ignoring sharply bounded quantifiers of the form $(Qx \leq |t|)$. The terms in bounded quantifiers may contain the $\#_3$ function if it is in the language. Recall that Σ_i^b and Π_i^b formulas represent precisely the predicates in the corresponding level of the polynomial time hierarchy when the language does not contain $\#_3$; if the language does contain $\#_3$ then these formulas can represent precisely the predicates in the corresponding levels of the $\#_3$-time hierachy.

S_2^i and S_3^i are axiomatized with the Σ_i^b-PIND rule and T_2^i is axiomatized with the the Σ_i^b-IND rule. The definitions of R_2^i and R_3^i are based on the work of Allen [1] who dealt with a theory D^i which is equivalent to R_2^i and on the independent work of Clote-Takeuti [7]. This paper uses the definition for R_2^i and R_3^i from Takeuti [18]. The axioms of R_2^i and R_3^i are the BASIC axioms which define the function symbols and the Σ_i^b-LBIND rules:

$$\frac{A(\lfloor \frac{1}{2}a \rfloor), \Gamma \longrightarrow \Delta, A(a)}{A(0), \Gamma \longrightarrow \Delta, A(|t|)}$$

where Γ and Δ are arbitrary cedents of formulas and $A \in \Sigma_i^b$ and the *eigenvariable* a must not occur in the lower sequent. This is equivalent to the Σ_i^b-LLIND rule

$$\frac{A(a), \Gamma \longrightarrow \Delta, A(Sa)}{A(0), \Gamma \longrightarrow \Delta, A(||t||)}$$

Allen [1] showed that R_2^i and R_3^i prove the Δ_i^b-comprehension axioms. Clote-Takeuti [7] and Takeuti [18] showed that R_2^i and R_3^i prove the Π_i^b-separation axioms. Takeuti and Allen also showed that R_3^i contains the theory S_2^{i-1}, for $i \geq 1$ (the method of proof is similar to the proof that S_2^i contains T_2^{i-1}). The next theorem, due to Allen, generalizes these three results since Σ_i^b-replacement implies S_2^{i-1} is shown by Buss [3] and since it is easy to see directly that Σ_i^b-replacement implies Π_i^b-separation.

Theorem 1 (Allen [1]) *Σ_i^b-replacement is a consequence of R_2^i and R_3^i.*

Proof Recall that the Σ_i^b-replacement axioms can be stated as

$$(\forall x \leq |t|)(\exists y \leq s)A(x,y) \rightarrow \\ (\exists w \leq SqBd(s,t))(\forall x \leq |t|)(A(x, \beta(Sx,w)) \wedge \beta(Sx,w) \leq s)$$

where $A(x,y)$ is a Σ_i^b-formula, possibly with other free variables besides x and y, and where w.l.o.g. the term s does not contain x. Here $SqBd(s,t)$ is a term that bounds the size of a minimal Gödel number of a sequence of $|t|+1$ numbers of values $\leq s$ [3]. Let X and Y be the hypothesis and the conclusion (respectively) of the above replacement axiom and let $Z(j)$ be the formula

$$(\forall u \leq |t|)(\exists w \leq SqBd(s,t))(\forall x \leq |t|) \\ [(x \leq j \wedge u + x \leq |t|) \rightarrow A(u+x, \beta(Sx,w)) \wedge \beta(Sx,w) \leq s].$$

Now it is trivial that R_2^i and R_3^i can prove $X \to Z(0)$ and it is not hard to see that they also prove $Z(\lfloor \frac{1}{2} j \rfloor) \to Z(j)$ and also that they prove $Z(|t|) \to Y$. By Σ_i^b-LBIND on Z, they prove $Z(0) \to Z(|t|)$ and hence R_2^i and R_3^i prove the Σ_i^b-replacement axiom. □

U_2^i and V_2^i are second-order systems axiomatized with the comprehension rule for bounded first-order ($\Sigma_0^{1,b}$) properties and with induction rules $\Sigma_i^{1,b}$-PIND and $\Sigma_i^{1,b}$-IND, respectively [3]. It is easy to see that $U_2^{i+1} \supseteq V_2^i$ by the well-known methods (analogously to the proof that S_2^{i+1} contains T_2^i, or more precisely, to the proof that R_3^{i+1} contains S_3^i). There is a sharp analogy between the second-order theories U_2^i and V_2^i and the first-order theories R_3^i and S_3^i, respectively. This analogy is called the "RSUV isomorphism and is developed by [17, 18]. The basic idea of the RSUV isomorphism is that bounded second-order objects in one of the second-order theories correspond to first-order objects in the first-order theory and that first-order objects in the second-order theory correspond to lengths of objects in the appropriate first-order theory. By a bounded second-order object, we mean a predicate φ on the integers $< x$ for some first-order object x; to make the correspondence between a bounded second-order object φ and a first-order object y, we interpret the truth values of $\varphi(i)$ for $i < x$ as the bits in the binary representation of y. This makes a second-order quantifier correspond to a (bounded) first-order quantifier and makes a bounded first-order quantifier correspond to a sharply bounded quantifer. Since the second-order theory has $\#_2$, the corresponding first-order theory has the *lengths* of integers closed under $\#_2$, i.e., the first-order theory must have integers closed under $\#_3$. In addition, the $\Sigma_i^{1,b}$-PIND axioms of the theory U_2^i correspond to Σ_i^b-LBIND axioms of R_3^i. Similarly, the $\Sigma_i^{1,b}$-IND axioms of the theory U_2^i correspond to Σ_i^b-LIND axioms of S_3^i. The analogies are summarized by the table below:

Second-order Theory	First-order Theory
Second-order object (predicate)	First-order object (integer)
First-order object (integer)	Length of an integer
$\#_2$ on first-order objects	$\#_3$ on first-order objects
$\Sigma_i^{1,b}$	Σ_i^b
PIND/LIND	LBIND/LLIND
IND	LIND/PIND
$\text{EXPTIME}^{\Sigma_i^{1,p}}$	$FP_3^{\Sigma_i^p}$
$\text{EXPTIME}^{\Sigma_i^{1,p}}[wit, poly]$	$\text{FP}_3^{\Sigma_i^p}[wit, log^{O(1)}]$
U_2^i	R_3^i
V_2^i	S_3^i

The Σ_i^b-definable functions of some of these first-order theories can be characterized: For S_2^i and T_2^{i-1}, the Σ_i^b-definable functions are precisely the $\square_i^p$ functions, i.e., the $\text{FP}^{\Sigma_{i-1}^b}$ functions [3, 4]. For R_2^1 and $i = 1$, they are precisely the NC functions, see Allen [1] and Clote-Takeuti [7]. The $\Sigma_1^{1,b}$-definable functions of U_2^1 and V_2^1 were shown by Buss [3] to be precisely the polynomial space and exponential time computable functions, respectively. Buss [3] made a conjecture about the $\Sigma_i^{1,b}$-definable first-order-valued functions of U_2^i and V_2^i. We prove this conjecture, and we also characterize the functions with second-order values that can be $\Sigma_i^{1,b}$-defined by U_2^i and V_2^i; namely, they are the $\text{EXPTIME}^{\Sigma_i^{1,p}}[wit, poly]$ and $\text{EXPTIME}^{\Sigma_i^{1,p}}$ functions, respectively.

Also in this paper we characterize the Σ_i^b-definable functions of S_3^{i-1} and R_3^i as being precisely the $\text{FP}_3^{\Sigma_i^p}[wit, log^{O(1)}]$ functions, for $i > 1$.

The prior characterizations for R_2^1, S_2^i, U_2^1 and V_2^1 concerned the definability of *single-valued* functions. However, to work with the classes $\text{FP}_3^{\Sigma_i^p}[wit, log^{O(1)}]$ we also require a notion of Σ_i^b-definition of a multivalued function.

Definition Let f be a multivalued function (i.e. a relation). Then we say f is Σ_i^b-defined by a theory T if and only if for some Σ_i^b-formula $A(x, y)$,

(1) $T \vdash (\forall x)(\exists y)A(x, y)$, and

(2) Whenever $A(n, m)$ is true (in the standard model), then $f(n) = m$ is true.

The fact that condition (2) of the definition of Σ_i^b-definability is an "if ... then ..." is perhaps somewhat surprising; however, this matches the condition (4) of the definition of $\mathrm{FP}^{\Sigma_{i-1}^p}[wit, log]$. There is also a "strong" version of Σ_i^b-definability:

Definition A multivalued function f is *strongly* Σ_i^b*-definable* iff f is Σ_i^b-definable and for all $m, \vec{n}$, $A(\vec{n}, m)$ holds iff $f(\vec{n}) = m$.

We shall prove that the class of functions strongly Σ_i^b-definable by R_3^i (and by S_3^{i-1}) is precisely strong-$\mathrm{FP}_3^{\Sigma_{i-1}^p}[wit, log^{O(1)}]$.

3.2 Some Σ_i^b-Definitions of Functions in R_3^i and S_3^{i-1}

As part of characterizing the Σ_i^b-definable functions of R_3^i and S_3^{i-1} we need to give some intensional Σ_i^b-definitions of functions in these two theories. Later, we shall prove that all functions Σ_i^b-definable in R_3^i and S_3^{i-1} are captured by the next theorem. We say that M is an explicit $\mathrm{FP}_3^{\Sigma_i^p}[wit, log^{O(1)}]$ Turing machine if M has built-in 'clocks' that limit the run time to $2^{(\log n)^k}$ steps and the number of oracle queries to $\leq (\log n)^k$ on inputs of length n, for some constant k. This means that if M exceeds the runtime or oracle query limits then M aborts and outputs some constant (say 0). Given a description of a Turing machine with a witness oracle for a Σ_i^b-predicate Ω the property "w codes a valid execution of M on input x" can be expressed as a $\Sigma_{i+1}^b \cap \Pi_{i+1}^b$ formula $Run_M(x, w)$. Loosely speaking $Run_M(x, w)$ states that w completely codes a computation of $M(x)$ and that for each query q made to the witness oracle ω either (1) $\Omega(q)$ is false and M entered the oracle reject state after the query, or (2) M entered the oracle accept state with the response tape containing a value z witnessing the truth of $\Omega(q)$.

Theorem 2 *Fix* $i \geq 1$. *Let* M *be an explicit* $\mathrm{FP}_3^{\Sigma_i^p}[wit, log^{O(1)}]$ *Turing machine.*

(a) $R_3^{i+1} \vdash (\forall x)(\exists w) Run_M(x, w)$.

(b) $S_3^i \vdash (\forall x)(\exists w) Run_M(x, w)$.

Corollary 3 *Let* $i > 1$. R_3^i *and* S_3^{i-1} *can each* Σ_i^b*-define every* $\mathrm{FP}_3^{\Sigma_{i-1}^p}[wit, log^{O(1)}]$ *function.*

It is an open question whether R_3^1 can Σ_1^b-define all $\#_3$-time functions (this would be $i = 1$ of part (a) of the theorem). What is known is that R_3^1 can Σ_1^b-define exactly the polylog-space computable functions; this follows from the 'RSUV isomorphism' and from the characterization of the provably total functions of U_2^1 as the polynomial space functions.

Proof Since $R_3^{i+1} \vdash S_3^i$ it will suffice to prove (b). We fix M an $\mathrm{FP}_3^{\Sigma_i^p}[wit, log^{O(1)}]$ machine, which has a fixed oracle $\Omega \in \Sigma_i^p$ and runs in time $2^{(\log n)^k}$ and asks $(\log n)^k$ queries on inputs of length n. The oracle $\Omega(q)$ is of the form $(\exists z \leq t(q))B(z, q)$ for some $B \in \Pi_{i-1}^p$ and some term t in the language of S_3^i. We reason inside S_3^i. We say that w codes a *precomputation* of M if w is a sequence of configurations of M's execution respect to an unspecified oracle; that is, w being a precomputation implies nothing about whether the oracle answers in w are correct. A Q-computation is a precomputation in which all the 'Yes' answers are correct for the oracle Ω (but the 'No' answers may be incorrect). Let $QComp_M(w, x, v)$ be the following formula which states that w codes a Q-computation of M on input x such that all the 'Yes' answers are correct with a correct response tape contents and such that, for all j, the j-th most significant bit of v is a "1" if and only if the j-th query of $M(x)$ to the witness oracle yields a 'Yes' answer.

$$
\begin{aligned}
&QComp_M(w, x, v) \;\Leftrightarrow \\
&\qquad w \text{ codes an precomputation of } M(x) \text{ and} \\
&\qquad (\forall j \leq (\log|x|)^k)[YesAns(w, j) \to Bit((\log|x|)^k - j, v) = 1 \text{ and} \\
&\qquad (\forall j \leq (\log|x|)^k)[Bit((\log|x|)^k - j, v) = 1 \to CorrectYes(w, j)]
\end{aligned}
$$

where the formula $YesAns(w, j)$ asserts the j-th oracle query in the precomputation w receives a 'Yes' answer and the formula $CorrectYes(w, j)$ asserts that the j-th oracle query in the precomputation w yields a 'Yes' answer and that $B(z, q) \wedge z \leq t(q)$ holds, where q is the j-th oracle query in w and z is the response tape contents after the j-th oracle query. It is easy to see that $CorrectYes$ and $QComp_M$ are Π_{i-1}^b formulas, since B is (unless $i = 1$ in which case, they are Δ_1^b formulas). $YesAns$ is, of course, always a Δ_1^b-formula.

By coding precomputations efficiently, it can be presumed that any precomputation w for $M(x)$ can be bounded by some term $r(x)$; this is because M runs in $\#_3$ time. Likewise the v in $QComp_M(w,x,v)$ has length $\leq (\log|x|)^k$ and hence $v \leq |s(x)|$ for some term s in the language of S_3^i. Thus the formula $(\exists w)QComp(w,x,v)$ is (equivalent to) a Σ_i^b-formula. Since S_3^1 can prove that deterministic, non-oracle, $\#_3$-time Turing machines always halt, it follows that S_3^i proves that $(\exists w)QComp(w,x,0)$; namely, S_3^i proves that there is precomputation of $M(x)$ with all the oracle queries answered 'No'. And since S_3^i admits the "length maximization axioms" Σ_i^b-LMAX, S_3^i can prove that there exists a maximum $v < |s(x)|$ such that $(\exists w)QComp(w,x,v)$. Let v now denote this maximum value; it follows that S_3^i can prove that if $QComp(w,x,v)$ then w codes a computation with all witness oracle answers correct. To prove this in S_3^i, one argues that all 'Yes' answers must be correct since $QComp_M$ holds and that, for any j, if the j-th 'No' answer were incorrect v would not be maximum since one could obtain a larger value v' by changing the j-th most significant bit of v to "1" and setting all lower bits to zero. Then from $(\exists w)QComp(w,x,v)$, S_3^1 proves that $(\exists w')QComp(w',x,v')$ by letting w' code the precomputation corresponding to w up to the j-th query, then coding a 'Yes' answer with a valid witness on the response tape for the j-th query and subsequently coding M's computation with all oracle queries returning 'No' answers.

Thus S_3^i can prove that $M(x)$ always has at least one valid computation and S_3^i can Σ_{i+1}^b-define the function which M computes. The formula which Σ_{i+1}^b-defines the function $M(x) = y$ is the formula

$$(\exists y)(\exists w)[Run_M(x,w) \text{ and computation } w \text{ outputs } y]$$

which asserts that there is a w encoding a valid computation of $M(x)$ which outputs the value y. Note that w and y can be bounded by terms involving x.
Q.E.D. Theorem 2

Corollary 4 *Let $i > 1$. R_3^i and S_3^{i-1} can each strongly Σ_i^b-define every strong $\mathrm{FP}_3^{\Sigma_{i-1}^p}[wit, log^{O(1)}]$ function.*

Proof This follows immediately from Theorem 2 since the Σ_i^b-definition of $func_M$,

$$(\exists y)(\exists w)[Run_M(x,w) \text{ and computation } w \text{ outputs } y],$$

is also a strong Σ_i^b-definition of $func_M$. □

It should be noted that the proof of Theorem 2 involved formalizing, in S_3^i, the algorithm of Remark 2 above.

It is easy to see that R_3^i and S_3^{i-1} prove that the class $\mathrm{FP}_3^{\Sigma_i^p}[wit, log^{O(1)}]$ is closed under composition. Next we show that R_3^i and S_3^{i-1} are also able to prove that the class $\mathrm{FP}_3^{\Sigma_i^p}[wit, log^{O(1)}]$ is closed under *limited logarithmic recursion on notation.* We say that f is *defined from g and h by limited logarithmic recursion on notation with bound k* if and only if f is defined by the following:

$$f(x,\vec{y}) \;=\; f^*(|x|,\vec{y})$$

where

$$\begin{aligned} f^*(0,\vec{y}) &= g(\vec{y}) \\ f^*(z,\vec{y}) &= \min\{h(z,\vec{y},f^*(\lfloor \tfrac{1}{2}z \rfloor,\vec{y})),k(\vec{y})\} \qquad \text{for } z \neq 0. \end{aligned}$$

The definition of limited logarithmic recursion on notation is phrased so that the only purpose of k is to bound the size of the values of $f^*(z,\vec{y})$. Thus we shall henceforth consider only bounds $k(\vec{y})$ of the form $k_c(\vec{y}) = 2^{2^{||(|\vec{y}|)|^c}}$ where c is a constant.

Theorem 5 *Let $i \geq 1$. Suppose M_g and M_h are explicitly $\mathrm{FP}_3^{\Sigma_i^p}[wit, log^{O(1)}]$ Turing machines which compute multivalued functions g and h and suppose $c > 0$. Then there is a canonical, explicitly $\mathrm{FP}_3^{\Sigma_i^p}[wit, log^{O(1)}]$ Turing machine M computing the function f obtained by limited logarithmic recursion from g and h with bound k_c such that S_3^i can prove that the function computed by M_f satisfies the defining equations for f in terms of g and h.*

Proof It is easy enough for S^i_3 to construct M_f from M_g, M_h and c so that M_f computes f in the obvious straightforward manner. It is also easy for S^i_3 to prove that M_f has $\#_3$-runtime and only asks $(\log(|x|+|\vec{y}|))^{O(1)}$ oracle queries, since M_g and M_h also satisfy such bounds and the computation of f consists of computing g once and iterating h only $(\log|x|)$ many times. □

Next we show that $\mathrm{FP}_3^{\Sigma_i^p}[wit, log^{O(1)}]$ is closed under a form of parallel computation.

Definition Suppose $n \geq 1$ and f is an n-ary multivalued function. Then $\overline{f}$ is the multivalued function defined by:

$$\overline{f}(m,\vec{x}) \quad = \quad \langle f(0,\vec{x}), f(1,\vec{x}), \ldots, f(|m|-1,\vec{x})\rangle. \tag{1}$$

Theorem 6 *Fix $i \geq 1$.*

(a) *Suppose f is a $\mathrm{FP}_3^{\Sigma_i^p}[wit, log^{O(1)}]$ function. Then $\overline{f}$ is also a $\mathrm{FP}_3^{\Sigma_i^p}[wit, log^{O(1)}]$ function.*

(b) *Let M_f be an explicitly $\mathrm{FP}_3^{\Sigma_i^p}[wit, log^{O(1)}]$ Turing machine computing a function f. Then there is a canonical $\mathrm{FP}_3^{\Sigma_i^p}[wit, log^{O(1)}]$ Turing machine $M_{\overline{f}}$ computing the function $\overline{f}$ such that S^i_3 can prove that for every value of $func_{M_{\overline{f}}}(m,\vec{x})$ there are values of $f(0,\vec{x}),\ldots,f(|m|-1,\vec{x})$ that satisfy equation (1).*

Remark: Theorem 6 can be strengthened by additionally requiring in (b) that S^i_3 can prove that for all values of $f(0,\vec{x}),\ldots,f(|m|-1,\vec{x})$, there is a value of $func_{M_{\overline{f}}}(m,\vec{x})$ which makes equation (1) true. An indirect way to prove this is to use Theorem 20 below and the fact that R^i_3 proves Σ^b_i-replacement.

Proof (The main ideas of this proof may be found already in [8, 5, 19].) We shall prove (a) and leave it to the reader to show that for fixed M_f, the proof can be formalized in S^i_3. Let M_f be an $\mathrm{FP}_3^{\Sigma_i^p}[wit, log^{O(1)}]$ Turing machine that runs in time $2^{(\log n)^k}$ and asks $(\log n)^k$ queries to a Σ_i^p witness oracle $\Omega(q) = (\exists z \le t(q))B(q,z)$ on inputs of length n. Let $\overline{M}$ be the Turing machine which computes $\overline{f}$ in the straightforward way by running M on the inputs $0, \vec{x}$, the inputs $1, \vec{x}$, etc., up to $|m|-1, \vec{x}$. The runtime of $\overline{M}$ is $O(|m| \cdot 2^{(\log n)^k})$ which is $O(2^{(\log n)^{k+1}})$; however, the number of witness oracle queries made by $\overline{M}$ is only bounded by $|m| \cdot (\log n)^k$ which is $O(n(\log n)^k)$. So $\overline{M}$ has the desired runtime but makes far too many oracle queries and we need to devise a more sophisticated Turing machine $M_{\overline{f}}$ which is in $\mathrm{FP}_3^{\Sigma_i^p}[wit, log^{O(1)}]$ and computes the same function as $\overline{M}$ (or, more correctly, $func_{M_{\overline{f}}} \subseteq func_{\overline{M}}$).

By Remark 2 we may assume that M_f ignores the contents of its oracle response tape until after the final query to the witness oracle. It follows that, for any fixed input values, there are at most $2^{(\log n)^k} - 1$ many possible oracle queries which can be made in any precomputation of M on those inputs (recall that a precomputation need not have correct oracle answers). This is because the i-th query of M will depend only on the inputs and on the prior Yes/No answers of the oracle and because M asks at most $(\log n)^k$ many oracle queries. Thus, for an fixed input values $m, \vec{x}$, there is a set containing at most $|m| \cdot (2^{(\log n)^k} - 1)$ queries which contains all the queries that $\overline{M}$ may ask in any precomputation. This set of queries can be indexed by pairs (i,j) where $i < |m|$ and $0 < j < 2^{(\log n)^k}$; namely, let $\ell = |j| - 1$ and define $q_{i,j}$ to be the $(\ell+1)$-st query in a precomputation of $M(i, \vec{x})$ if, for all $k < \ell$, the $(k+1)$-st query in the pre-computation was answered 'Yes' iff $Bit(k,j) = 1$. In other words, $q_{i,j}$ is the next query $M(i, \vec{x})$ will ask if the prior queries were answered as specified by the bits of the binary representation of j. Form the array

$$\left(q_{i,j} : 0 \le i < |m|, 0 < j < 2^{(\log n)^k}\right)$$

of all possible queries in any precomputation of $\overline{M}(m, \vec{x})$.

We are now ready to describe the Turing machine $M_{\overline{f}}$. First, $M_{\overline{f}}$ computes all the entries (queries) $q_{i,j}$ in the array. Second, $M_{\overline{f}}$ uses a binary search procedure to find the number of entries q in the array such that $\Omega(q)$ holds. This is accomplished by asking queries of the following form, for p an integer:

"Do there exist at least p many entries q such that $\Omega(q)$?"

Since $\Omega \in \Sigma_i^p$, these queries are also Σ_i^p properties (of p and the array of queries). The response to such a query either is a 'No' answer or is a 'Yes' answer and an array of values z which witness at least p of the queries satisfying Ω. After $|m| + (\log n)^k$ many queries, $M_{\overline{f}}$ has ascertained the precise number of entries which for which $\Omega(q)$ is true and on the response tape there is any array of values $z_{i,j}$ which either indicate that $\Omega(q_{i,j})$ is false or which are a witness to the truth of $\Omega(q_{i,j})$. Third, $M_{\overline{f}}$ simulates the execution of $\overline{M}$ except that whenever $\overline{M}$ would query the oracle, $M_{\overline{f}}$ instead looks up the answer (which is already on $M_{\overline{f}}$'s response tape).

That completes the proof of part (a). The reader should convince him- or herself that this argument can be formalized in S_3^i. It might be useful to remember that S_3^i admits PIND for $\Sigma_{i+1}^b \cap \Pi_{i+1}^b$ predicates [4]. For instance, the formula expressing the property that there are $\geq p$ and $< p'$ many entries $q_{i,j}$ which satisfy $\Omega(q_{i,j})$ is a Boolean combination of Σ_i^b-formulas and S_3^i admits PIND for such a formula.
Q.E.D. Theorem 6

Theorem 6 is quite useful in a variety of situations. As one application, let $P(x, \vec{y}, z)$ be a predicate; we say that $h(\vec{y}, z)$ is defined by *length-bounded minimization* from P if

$$h(\vec{y}, z) = \begin{cases} \text{the least } x \leq |z| \text{ such that } P(x, \vec{y}, z) \text{ if such an } x \text{ exists} \\ |z| + 1 \quad \text{otherwise} \end{cases}$$

We use $h(\vec{y}, z) = (\mu x \leq |z|)P(x, \vec{y}, z)$ as a compact notation for definition by length-bounded minimization.

Theorem 7 *($i \geq 1$) Let $P(x, \vec{y}, z)$ be a Π_i^p-predicate. Then the function $h(\vec{y}, z)$ defined from P by length-bounded minimization is in $\mathrm{FP}_3^{\Sigma_i^p}[wit, log^{O(1)}]$. Furthermore, there is a canonical, explicitly $\mathrm{FP}_3^{\Sigma_i^p}[wit, log^{O(1)}]$ Turing machine M such that S_3^i can prove the the function h computed by M satisfies the above defining equation for length-bounded minimization.*

Proof Let $f(x, \vec{y}, z)$ be the function which is equal to 1 if $P(x, \vec{y}, z)$ holds and is equal to 0 otherwise. Clearly f is in $\mathrm{FP}_3^{\Sigma_i^p}[wit, log^{O(1)}]$

since it can be computed with a single (non-witness) oracle query to the Π_i^p-predicate P. By Theorem 6, it follows that the function $\overline{f}$ is also in $\mathrm{FP}_3^{\Sigma_i^p}[wit, log^{O(1)}]$. And h can be easily computed in polynomial time without any further oracle queries from $\overline{f}(2z+1, \vec{x}, z)$ since $|2z+1| = |z|+1$. □

3.3 Some Function Definitions in U_2^i and V_2^i

In this section we shall begin the investigation of the functions which are $\Sigma_{i+1}^{1,b}$-definable in the three theories V_2^i, U_2^{i+1} and V_2^{i+1} where $i \geq 1$. It is shown in this section that every $\mathrm{EXPTIME}^{\Sigma_i^{1,p}}$ function is $\Sigma_{i+1}^{1,b}$-definable in V_2^{i+1} and that every $\mathrm{EXPTIME}^{\Sigma_i^{1,p}}[wit, poly]$ function is $\Sigma_{i+1}^{1,b}$-definable in the theories V_2^i and U_2^{i+1}. Recall that Buss [3] has shown that the $\Sigma_1^{1,b}$-definable functions of U_2^1 and V_2^1 are precisely the polynomial space and exponential time computable functions, respectively (in [3], this was only shown for functions with first-order values, but the methods immediately extend to functions with second-order values). Because of the "RSUV isomorphism", the theories U_2^i and V_2^i are in some sense equivalent to R_3^i and S_3^i; thus we shall often just outline or omit proofs because they will precisely parallel the proofs already given for R_3^i and S_3^i.

A Turing machine M is said to be an explicit $\mathrm{EXPTIME}^{\Sigma_i^{1,p}}$ Turing machine or an explicit $\mathrm{EXPTIME}^{\Sigma_i^{1,p}}[wit, poly]$ Turing machine if it has builtin 'clocks' that limit the run time to 2^{n^k} step and, in the latter case, the number of oracle queries to n^k for some constant k. Here n is the total length of the first-order inputs to M; these runtime bounds will limit M to accessing at most the first 2^{n^k} symbols (i.e., bits) of its oracles; thus if a second-order input has a length $> 2^{n^k}$ the excess symbols are ignored.

If M has a Σ_i^p (witness) oracle let the formula $Run_M(x, \alpha, \zeta)$ state that the second-order object ζ codes a correct computation of M on input x, α: it is easy to see that Run_M is a $\Sigma_{i+1}^{1,b} \cap \Pi_{i+1}^{1,b}$ formula. (In the sequel, x and α may be vectors of first- and second-order objects, respectively.)

Theorem 8 *For $i \geq 0$ and M an explicit $\mathrm{EXPTIME}^{\Sigma_i^{1,p}}$ Turing machine,*

$$V_2^{i+1} \vdash (\forall x)(\forall \alpha)(\exists \zeta) Run_M(x, \alpha, \zeta)$$

Hence V_2^{i+1} *can* $\Sigma_{i+1}^{1,b}$*-define every* $\text{EXPTIME}^{\Sigma_i^{1,p}}$*-function.*

The proof of Theorem 8 is entirely analogous to the proof of the the fact that S_3^{i+1} can Σ_{i+1}^b-define every function $\#_3$-time computable with an oracle for Σ_i^p. To prove this directly for V_2^{i+1}, one can modify the proof of Theorem 10.1 of [3].

Theorem 9 *Let* $i \geq 1$ *and* M *be an explicit* $\text{EXPTIME}^{\Sigma_i^{1,p}}[wit, poly]$ *Turing machine. Then*

(a) $V_2^i \vdash (\forall x)(\forall \alpha)(\exists \zeta) Run_M(x, \alpha, \zeta)$

(b) $U_2^{i+1} \vdash (\forall x)(\forall \alpha)(\exists \zeta) Run_M(x, \alpha, \zeta)$

Hence V_2^i *and* U_2^{i+1} *can* $\Sigma_{i+1}^{1,b}$*-define every* $\text{EXPTIME}^{\Sigma_i^{1,p}}[wit, poly]$*-function.*

The proof of Theorem 9 is analogous to the proof of Theorem 2. It suffices to prove the theorem for V_2^i since it is a subtheory of U_2^{i+1}. The notion of a precompuation of M is defined analogously as before and likewise a formula $QComp_M(\zeta, x, \alpha, v)$ is defined which says that ζ codes a precomputation of M in which the i-th oracle query returns a (correct) yes answer iff the i-th most significant bit of the binary representation of v is a "1". Note that since only polynomially many queries may be made by M, v is a first-order object. Now V_2^i can prove that there exists a maximum value for v such that $(\exists \zeta) M(\zeta, x, \alpha, v)$; this maximum value must give a true, correct computation of M. We leave it to the reader to supply the rest of the details of the proof.

It is obvious that the classes $\text{EXPTIME}^{\Sigma_i^{1,p}}$ and $\text{EXPTIME}^{\Sigma_i^{1,p}}[wit, poly]$ are closed under composition, and provably so in the theories V_2^{i+1} and the theories V_2^i and U_2^{i+1}, respectively. We next discuss the closure of these classes under limited forms of primitive recursion.

Definition Let g and h be (possibly multivalued) functions which have second-order values. We say that f is defined from g and h by *first-order* *recursion* iff f is defined by

$$\begin{aligned} f(0, \vec{y}, \vec{\alpha}) &= g(\vec{y}, \vec{\alpha}) \\ f(x+1, \vec{y}, \vec{\alpha}) &= h(x, \vec{y}, \vec{\alpha}, f(x, \vec{y}, \vec{\alpha})) \end{aligned}$$

We say f is defined by *first-order recursion on notation* iff f is defined by

$$\begin{aligned} f(0, \vec{y}, \vec{\alpha}) &= g(\vec{y}, \vec{\alpha}) \\ f(x, \vec{y}, \vec{\alpha}) &= h(x, \vec{y}, \vec{\alpha}, f(\lfloor \tfrac{1}{2} x \rfloor, \vec{y}, \vec{\alpha})) \end{aligned}$$

Because the second-order objects have length exponential in the length of the first-order objects, one can think of "first-order" as a synonym for "logarithmic"; hence the notion of *first-order recursion on notation* is analogous to the notion of *logarithmic recursion on notation.* One important thing to note about the definition of f by first-order recursion (on notation) is that g and h and hence f must take on second-order values. Because of this there is no need to limit to growth rate of the values of f by a function k as we did in the definition of limited logarithmic recursion on notation. For this, it is important that the runtime bounds and number of queries bounds for computation in $\mathrm{EXPTIME}^{\Sigma_i^{1,p}}$ and $\mathrm{EXPTIME}^{\Sigma_i^{1,p}}[wit, poly]$ are in terms of the total length n of the first-order inputs.

Theorem 10 *Let $i \geq 0$. Suppose M_g and M_h are explicit $\mathrm{EXPTIME}^{\Sigma_i^{1,p}}$ Turing machines computing functions of the appropriate number of first- and second-order arguments with second-order outputs. Then there is a canonical explicit $\mathrm{EXPTIME}^{\Sigma_i^{1,p}}$ Turing machine M_f computing the function f defined from g and h by first-order recursion such that V_2^{i+1} proves that the function computed by M_f satisfies the defining equation for f in terms of g and h.*

Theorem 11 *Let $i \geq 0$. Suppose M_g and M_h are explicit $\mathrm{EXPTIME}^{\Sigma_i^{1,p}}[wit, poly]$ Turing machines computing multivalued functions g and h of the appropriate number of first- and second-order arguments with second-order outputs. Then there is a canonical explicit $\mathrm{EXPTIME}^{\Sigma_i^{1,p}}[wit, poly]$ Turing machine M_f computing the multivalued* *function f defined from g and h by first-order recursion on notation such that V_2^i proves that the function computed by M_f satisfies the defining equation for f in terms of g and h.*

The proofs of Theorems 10 and 11 are based on the fact that the straightforward computation of f in terms of g and h satisfies the proper runtime bounds and, for the second theorem, makes only polynomially many oracle queries.

Next we shall define a new version of $\overline{f}$ and prove an analogue of Theorem 6. Since f will generally have second-order values, we need to define a notion of sequences of second-order object; a sequence of second-order objects can be coded by a single second-order object using the $\beta\!\!\!/$ and $\big\langle \cdots \big\rangle$ conventions from [3]. Recall that $\beta\!\!\!/(a, \alpha)$ is the abstract $\{x\}\alpha(\langle a, x\rangle)$ and that if α_i are second-order objects then

$$\big\langle \alpha_1, \ldots, \alpha_n \big\rangle$$

is the second-order object such that $\beta\!\!\!/\,\big(a, \big\langle \vec{\alpha} \big\rangle\big)$ is α_a for all $a \le n$.

Definition Let $f(x, \vec{y}, \vec{\alpha})$ be a function which takes second-order values. The function $\overline{f}$ is defined by

$$\overline{f}(x, \vec{y}, \vec{\alpha}) \quad = \quad \big\langle f(0, \vec{y}, \vec{\alpha}), f(1, \vec{y}, \vec{\alpha}), \ldots, f(x, \vec{y}, \vec{\alpha}) \big\rangle. \qquad (2)$$

Note that $\overline{f}$ may be multivalued if f is.

Theorem 12 *Fix $i \ge 0$.*

(a) *Suppose f is a* $\mathrm{EXPTIME}^{\Sigma_i^{1,p}}$ *-function. Then $\overline{f}$ is also a* $\mathrm{EXPTIME}^{\Sigma_i^{1,p}}$ *-function.*

(b) *Suppose M_f is an explicit* $\mathrm{EXPTIME}^{\Sigma_i^{1,p}}$ *Turing machine computing a function f. There there is a canonical* $\mathrm{EXPTIME}^{\Sigma_i^{1,p}}$ *Turing machine that computes the function $\overline{f}$ such that V_2^{i+1} can prove that the function computed by the Turing machine satisfies equation (2).*

Theorem 12 is proved by noting that $\overline{f}$ can be computed from f by straightforwardly computing each requisite value of f — this can be easily formalized in V_2^{i+1}.

Theorem 13 *Fix* $i \geq 1$.

(a) *Suppose* f *is a* $\mathrm{EXPTIME}^{\Sigma_i^{1,p}}[wit, poly]$*-function (multivalued). Then* $\overline{f}$ *is also a* $\mathrm{EXPTIME}^{\Sigma_i^{1,p}}[wit, poly]$*-function (multivalued).*

(b) *Suppose* M_f *is an explicit* $\mathrm{EXPTIME}^{\Sigma_i^{1,p}}[wit, poly]$ *Turing machine computing a function* f. *There there is a canonical* $\mathrm{EXPTIME}^{\Sigma_i^{1,p}}[wit, poly]$ *Turing machine that computes the function* $\overline{f}$ *such that* V_2^i *can prove that the function computed by the Turing machine satisfies equation (2).*

Theorem 13 is proved by a rather complicated construction analogous to the proof of Theorem 6. We omit this proof.

Finally we consider *first-order minimization.* Let $P(x, \vec{y}, z, \vec{\alpha})$ be a predicate where $x, \vec{y}, z$ are first-order arguments; $h(\vec{y}, z, \vec{\alpha})$ is defined by *first-order minimization* from P if

$$h(\vec{y}, z, \vec{\alpha}) = \begin{cases} \text{the least } x \leq z \text{ such that } P(x, \vec{y}, z, \vec{\alpha}) \text{ if such an } x \text{ exists} \\ z + 1 \quad \text{otherwise} \end{cases}$$

Note the function h defined by first-order minimization has a first-order value. We use $h(\vec{y}, z, \vec{\alpha}) = (\mu x \leq z) P(x, \vec{y}, z, \vec{\alpha})$ as a compact notation for definition by first-order minimization.

Theorem 14 *($i \geq 1$) Let* $P(x, \vec{y}, z, \vec{\alpha})$ *be a* $\Pi_i^{1,p}$*-predicate. Then the function* $h(\vec{y}, z, \vec{\alpha})$ *defined from* P *by first-order minimization is in* $\mathrm{EXPTIME}^{\Sigma_i^{1,p}}[wit, poly]$. *Furthermore, there is a canonical, explicitly* $\mathrm{FP}_3^{\Sigma_i^p}[wit, log^{O(1)}]$ *Turing machine* M *such that* V_2^i *can prove the the function* h *computed by* M *satisfies the above defining equation for first-order minimization.*

Theorem 14 is proved similarly to Theorem 7; namely, apply Theorem 13 to the characteristic function of P.

4 The Σ_i^b-Definable Functions of R_3^i and S_3^{i-1}

In this section, the Σ_i^b-definable functions of R_3^i and S_3^{i-1} are characterized. We have already shown that every $\mathrm{FP}_3^{\Sigma_{i-1}^p}[wit, log^{O(1)}]$ function is Σ_i^b-definable in these theories. It will suffice to establish the converse for the stronger theory R_3^i: we shall do this by proving a 'witnessing theorem' which states that every sequent of Σ_i^b-formulas provable in R_3^i is 'witnessed' by a $\mathrm{FP}_3^{\Sigma_{i-1}^p}[wit, log^{O(1)}]$ function (provably in S_3^i). This not only characterizes the Σ_i^b-definable functions of R_3^i and S_3^{i-1}, but also gives a conservation result between the two theories and proves that S_3^{i-1} admits Δ_i^b-PIND.

4.1 The Witness Formula

We next review briefly a definition from [3] which is necessary for the the characterization of the Σ_i^b-definable functions of S_3^{i-1} and R_3^i. For the rest of this section, $i \geq 1$ will be a fixed integer; the applications in this paper only need $i > 1$. Let $A(\vec{a})$ be a Σ_i^b-formula. A formula $Witness_A^{i,\vec{a}}(w, \vec{a})$ is defined which has limited quantifier complexity and which states that w is a number 'witnessing' the truth of $A(\vec{a})$.

Definition Suppose $A(\vec{a}) \in \Sigma_i^b$ and $\vec{a}$ is a vector of variables including all those free in A. The formula $Witness_A^{i,\vec{a}}$ is defined below, inductively on the complexity of A:

(1) If $A \in \Pi_{i-1}^b$ then $Witness_A^{i,\vec{a}}$ is just A itself.

(2) If A is $B \wedge C$ then define

$$Witness_A^{i,\vec{a}}(w, \vec{a}) \iff Witness_B^{i,\vec{a}}(\beta(1, w), \vec{a}) \wedge Witness_C^{i,\vec{a}}(\beta(2, w), \vec{a}).$$

(3) If A is $B \vee C$ then define

$$Witness_A^{i,\vec{a}}(w, \vec{a}) \iff Witness_B^{i,\vec{a}}(\beta(1, w), \vec{a}) \vee Witness_C^{i,\vec{a}}(\beta(2, w), \vec{a}).$$

(4) If A is $B \rightarrow C$ and is not in $\Pi_{i-1}^{1,b}$ then we define

$$Witness_A^{i,\vec{a}}(w, \vec{a}) \iff Witness_{\neg B}^{i,\vec{a}}(\beta(1, w), \vec{a}) \vee Witness_C^{i,\vec{a}}(\beta(2, w), \vec{a}).$$

(5) If $A \notin \Pi^b_{i-1}$ and $A(\vec{a})$ is $(\forall x \leq |s(\vec{a})|)B(\vec{a},x)$ then define

$$\begin{aligned} Witness^{i,\vec{a}}_A(w,\vec{a}) \iff\ & Seq(w) \wedge Len(w) = |s(\vec{a})| + 1 \wedge \\ & \wedge (\forall x \leq |s(\vec{a})|)\, Witness^{i,\vec{a},b}_{B(\vec{a},b)}(\beta(x+1,w),\vec{a},x). \end{aligned}$$

In words, w witnesses $A(\vec{a})$ if $w = \langle w_0,\ldots,w_{|s|}\rangle$ and each w_i witnesses $B(\vec{a},i)$. The formula $Seq(w)$ says w is a valid Gödel number of a sequence and $Len(w)$ is a function giving the number of entries in the sequence w.

(6) If $A \notin \Pi^b_{i-1}$ and A is $(\exists x \leq t(\vec{a}))B(\vec{a},x)$ then define

$$\begin{aligned} Witness^{i,\vec{a}}_A(w,\vec{a}) \iff\ & Seq(w) \wedge Len(w) = 2 \wedge \beta(1,w) \leq t(\vec{a}) \wedge \\ & \wedge\, Witness^{i,\vec{a},b}_{B(\vec{a},b)}(\beta(2,w),\vec{a},\beta(1,w)). \end{aligned}$$

So w witnesses $A(\vec{a})$ if $w = \langle n,v\rangle$ where $n \leq t(\vec{a})$ and v witnesses $B(\vec{a},n)$.

(7) If $A \notin \Pi^b_{i-1}$ and A is $\neg B$ then use prenex operations to push the negation sign into the formula so that it can be handled by cases (1)–(6).

The purpose of defining *Witness* is to give a canonical way of verifying that $A(\vec{a})$ is true. It is easy to see that $(\exists w)\, Witness^{i,\vec{a}}_A(w,\vec{a})$ is equivalent to $A(\vec{a})$. The next propositions express some properties of *Witness*; these are proved mostly by induction on the complexity of A.

Proposition 15 *For $i \geq 2$, and $A \in \Sigma^b_i$, $Witness^{i,\vec{a}}_A$ is a Π^b_{i-1}-formula.*

Proposition 16 $(i \geq 1)$. *Let $A(\vec{a})$ be a Σ^b_i-formula. Then*

$$S^{i-1}_3 \vdash Witness^{i,\vec{a}}_A(w,\vec{a}) \rightarrow A(\vec{a})$$

and there is a term $t_A(\vec{a})$ such that

$$S^{i-1}_3 + \Sigma^b_i\text{-replacement} \vdash A(\vec{a}) \leftrightarrow (\exists w \leq t_A)\, Witness^{i,\vec{a}}_A(w,\vec{a}).$$

Also there is a Σ^b_1-defined function $g_A(w)$ such that

$$S^1_2 \vdash Witness^{i,\vec{a}}_A(w,\vec{a}) \rightarrow Witness^{i,\vec{a}}_A(g_A(w),\vec{a}) \wedge g_A(w) \leq t_A.$$

Proposition 16 also holds for S_2^{i-1} and is proved by induction on the complexity of A exactly as in the proofs of the corresponding theorems in Buss [3, 4].

Proposition 17 ($i \geq 1$). *Let A be a Σ_i^b-formula. The predicate $Witness_A^{i,\vec{a}}$ is a Π_{i-1}^p-predicate.*

4.2 The Witnessing Theorem for S_3^{i-1} and R_3^i

In this section we give the proof that every Σ_i^b-definable function of S_3^{i-1} and R_3^i is in $\mathrm{FP}_3^{\Sigma_{i-1}^p}[wit, log^{O(1)}]$. The proof is a proof-theoretic 'witnessing theorem' and hence is constructive; however, it uses cut elimination and is not feasibly constructive (since cut-elimination involves superexponential growth rate). We now consider the theories of Bounded Arithmetic formalized in a Gentzen-style sequent calculus: each line in a proof is a *sequent* of the form

$$A_1, \ldots, A_k \longrightarrow B_1, \ldots, B_\ell$$

where each A_j and B_j is a formula. The intended meaning of this sequent is that the conjunction of the *antecedent* $A_1, \ldots, A_k$ implies the disjunction of the *succedent* $B_1, \ldots, B_\ell$. Note that the sequent connective symbol $\longrightarrow$ is distinct from the logical connective $\rightarrow$. Capital Greek letters $\Gamma, \Delta, \Pi, \Lambda, \ldots$ will be used to denote a series of formulae separated by commas, these are called *cedents*.

There are about 23 rules of inference for the sequent calculus; in addition, there are induction rules which replace the induction axioms. The *initial sequents* (i.e., axioms) of a sequent calculus proof must be equality axioms, logical axioms or non-logical axioms. The theories S_3^{i-1} and R_3^i each have a finite set of open (i.e., quantifier free) sequents as initial sequents. There are no induction axioms as initial sequents since induction rules are used instead. An important theorem (due to Gentzen, but see also Takeuti [16]) concerning the sequent calculus is that many instances of the cut rule may be eliminated from proofs — more precisely, all *free cuts* may be eliminated from a proof. Rather than define precisely what a free cut is, let us merely say that for a proof of a Σ_i^b-formula in a theory S_3^{i-1} or R_2^i, we may assume that every formula appearing in the proof is a Σ_i^b- or a Π_i^b-formula. For more information on the sequent calculus see Takeuti [16] and for information on the sequent calculus for theories of Bounded Arithmetic, consult chapter 4 of [3].

If Γ is a cedent we write $\bigwedge\Gamma$ and $\bigvee\Gamma$ to denote the conjunction and disjunction, respectively, of the formulae in Γ. Conjunction and disjunction associate from right to left; for example, if Γ is A, B, C then $\bigwedge\Gamma$ denotes $A \wedge (B \wedge C)$.

The theory R_3^i has the function and predicate symbols β, Seq, and Len which manipulate Gödel numbers of sequences ($i > 1$). These function symbols can be Σ_1^b-defined by S_3^{i-1}, for $i > 1$, and w.l.o.g. we assume that the language of S_3^{i-1} is enlarged to contain these function symbols. We use $\langle a_1, \ldots, a_n \rangle$ to denote the Gödel number of the sequence $a_1, \ldots, a_n$. Also, $*$ is a binary function defined so that

$$\langle a_1, \ldots, a_n \rangle * a_{n+1} = \langle a_1, \ldots, a_n, a_{n+1} \rangle.$$

Finally $\langle\!\langle a_1, \ldots, a_n \rangle\!\rangle$ is equal to $\langle a_1, \langle a_2, \ldots, \langle a_{n-1}, a_n \rangle \ldots \rangle \rangle$.

These conventions allow us to conveniently discuss witnessing a cedent. For example, suppose Γ is $A_1, \ldots, A_n$ and that $w = \langle\!\langle w_1, \ldots, w_n \rangle\!\rangle$. Then $Witness^{i,\vec{a}}_{\bigwedge\Gamma}(w, \vec{a})$ holds if and only if $Witness^{i,\vec{a}}_{A_j}(w_j, \vec{a})$ holds for all $1 \leq j \leq n$.

Theorem 18 (The Witnessing Lemma for S_3^{i-1} and R_3^i)
Fix $i > 1$. Suppose the sequent $\Gamma, \Pi \longrightarrow \Delta, \Lambda$ is a theorem of R_3^i and each formula in $\Gamma \cup \Delta$ is Σ_i^b and each formula in $\Pi \cup \Lambda$ is Π_i^b. Let $c_1, \ldots, c_p$ be the free variables in the sequent and let G and H be the formulae

$$G = \left(\bigwedge\Gamma\right) \wedge \bigwedge\{\neg C : C \in \Lambda\}$$

and

$$H = \left(\bigvee\Delta\right) \vee \bigvee\{\neg C : C \in \Pi\}.$$

Then there is a $\mathrm{FP}_3^{\Sigma_{i-1}^p}[wit, log^{O(1)}]$ function f which is Σ_i^b-defined by S_3^{i-1} such that

$$S_3^{i-1} \vdash Witness^{i,\vec{c}}_G(w, \vec{c}) \rightarrow Witness^{i,\vec{c}}_H(f(w, \vec{c}), \vec{c}).$$

Furthermore, S_3^{i-1} defines f as being equal to $func_M$ for some explicit $\mathrm{FP}_3^{\Sigma_{i-1}^p}[wit, log^{O(1)}]$ Turing machine.

Proof

A formula is said to be in *negation normal form* if every negation sign ($\neg$) has an atomic formula in its scope. Since any formula is logically equivalent to a formula in negation normal form, we may, without loss of generality, restrict our attention to proofs in which every formula is in negation normal form. In particular, the induction formulas and every formula in the endsequent are restricted to being in negation normal form. This simplifies the notation considerably since now Π and Λ may, without loss of generality, be presumed to be the empty cedent.

By the free-cut elimination theorem there is a R_3^i-proof P of $\Gamma \longrightarrow \Delta$ such that every formula in the proof is in negation normal form and every cut in P has a Σ_i^b principal formula and such that P is in free variable normal form (see [3] for definitions). The proof of Theorem 18 is by induction on the number of sequents in the proof P.

To begin, consider the case where P has no inferences and consists of a single sequent. This sequent must be a nonlogical axiom of R_3^i or a logical axiom or an equality axiom. In any event, it contains only atomic formulae and is also a consequence of S_3^{i-1}. For atomic formulae A, $Witness_A^{i,\vec{c}}$ is just A itself; hence this case is completely trivial.

The argument for the induction step splits into thirteen cases depending on the final inference of P. Since the general form of this induction is by now quite familiar, we shall omit the easier cases and discuss only the more difficult cases of the induction step.

Case (1): ($\vee$:left) Suppose the last inference of P is

$$\frac{B, \Gamma^* \longrightarrow \Delta \qquad C, \Gamma^* \longrightarrow \Delta}{B \vee C, \Gamma^* \longrightarrow \Delta}$$

Let D be the formula $B \wedge (\bigwedge \Gamma^*)$ and let E be $C \wedge (\bigwedge \Gamma^*)$ and let F be $(B \vee C) \wedge (\bigwedge \Gamma^*)$. By the induction hypothesis, there are $\mathrm{FP}_3^{\Sigma_{i-1}^p}[wit, log^{O(1)}]$ functions g and h such that

$$S_3^{i-1} \vdash Witness_D^{i,\vec{c}}(w, \vec{c}) \rightarrow Witness_{\bigvee\Delta}^{i,\vec{c}}(g(w, \vec{c}), \vec{c})$$

and

$$S_3^{i-1} \vdash Witness_E^{i,\vec{c}}(w, \vec{c}) \rightarrow Witness_{\bigvee\Delta}^{i,\vec{c}}(h(w, \vec{c}), \vec{c}).$$

Let the function k be defined by

$$k(v, w, \vec{c}) = \begin{cases} v & \text{if } Witness_{\bigvee\Delta}^{i,\vec{c}}(v, \vec{c}) \\ w & \text{otherwise} \end{cases}$$

By Proposition 17, $k(w, a, b, \vec{c})$ can be computed with a single call to a Σ^p_{i-1}-oracle; thus k is a $\mathrm{FP}_3^{\Sigma^p_{i-1}}[wit, log^{O(1)}]$ function. Let f be the function

$$f(w, \vec{c}) = k\big(g^*(w, \vec{c}), h^*(w, \vec{c}), \vec{c}\big)$$

where

$$g^*(w, \vec{c}) = g(\langle \beta(1, \beta(1, w)), \beta(2, w)\rangle, \vec{c})$$

and

$$h^*(w, \vec{c}) = h(\langle \beta(2, \beta(1, w)), \beta(2, w)\rangle, \vec{c}).$$

Since f is defined as the composition of $\mathrm{FP}_3^{\Sigma^p_{i-1}}[wit, log^{O(1)}]$ functions, f is itself in $\mathrm{FP}_3^{\Sigma^p_{i-1}}[wit, log^{O(1)}]$. Clearly f can be intensionally defined by S_3^{i-1} and the conditions of Theorem 18 are satisfied.

Case (2): ($\wedge$:right). Suppose the last inference of P is

$$\frac{\Gamma \longrightarrow B, \Delta^* \qquad \Gamma \longrightarrow C, \Delta^*}{\Gamma \longrightarrow B \wedge C, \Delta^*}$$

The argument for this case is similar to that of case (1). Let D be the formula $B \vee (\bigvee \Delta^*)$, let E be $C \vee (\bigvee \Delta^*)$ and let F be $(B \wedge C) \vee (\bigvee \Delta^*)$. The induction hypothesis is that there are $\mathrm{FP}_3^{\Sigma^p_{i-1}}[wit, log^{O(1)}]$ functions g and h so that

$$S_3^{i-1} \vdash \mathit{Witness}^{i,\vec{c}}_{\bigwedge \Gamma}(w, \vec{c}) \rightarrow \mathit{Witness}^{i,\vec{c}}_{D}(g(w, \vec{c}), \vec{c})$$

and

$$S_3^{i-1} \vdash \mathit{Witness}^{i,\vec{c}}_{\bigwedge \Gamma}(w, \vec{c}) \rightarrow \mathit{Witness}^{i,\vec{c}}_{E}(h(w, \vec{c}), \vec{c}).$$

Let k be the function such that

$$k(v, w, \vec{c}) = \begin{cases} v & \text{if } \mathit{Witness}^{i,\vec{c}}_{\bigvee \Delta^*}(v, \vec{c}) \\ w & \text{otherwise.} \end{cases}$$

By Proposition 17, $\mathit{Witness}^{i,\vec{c}}_{\bigvee \Delta^*}$ is a Π^b_{i-1}-predicate; hence k is a $\mathrm{FP}_3^{\Sigma^p_{i-1}}[wit, log^{O(1)}]$ function. Let f be the function

$$f(w, \vec{c}) = \big\langle \langle \beta(1, g(w, \vec{c})), \beta(1, h(w, \vec{c}))\rangle, k(\beta(2, g(w, \vec{c})), \beta(2, h(w, \vec{c})), \vec{c})\big\rangle.$$

As in case (1), it is clear that f satisfies the desired conditions.

Case (3): ($\exists\leq$:left). Suppose the last inference of P is

$$\frac{a\leq s, B(a), \Gamma^* \longrightarrow \Delta}{(\exists x\leq s)B(x), \Gamma^* \longrightarrow \Delta}$$

The free variable a is the *eigenvariable* and appears only as indicated. Let D be the formula $a\leq s\wedge\big(B(a)\wedge(\bigwedge\Gamma^*)\big)$ and let E be $(\exists x\leq s)B(x)\wedge(\bigwedge\Gamma^*)$. By the induction hypothesis, there is a $\mathrm{FP}_3^{\Sigma_{i-1}^p}[wit, log^{O(1)}]$ function g such that

$$S_3^{i-1}\vdash Witness_D^{i,\vec{c},a}(w,\vec{c},a)\rightarrow Witness_{\bigvee\Delta}^{i,\vec{c}}(g(w,\vec{c},a),\vec{c}).$$

(Note that the variable a can be omitted from the right hand side of the implication since it does not appear free in Δ.)

We wish to define a function $h(w,\vec{c})$ which produces a value for a such that $B(a)$ holds: we can apply the function g to this to get a witness for $\bigvee\Gamma$. To define h, we must consider three subcases: first, if $(\exists x\leq s)B$ is in $\Sigma_i^b\setminus\Pi_{i-1}^b$, let $h(w,\vec{c})=\beta(1,\beta(1,w))$; if w is a witness for E then $h(w,\vec{c})$ is a value for a such that $B(a)$ holds and such that $\beta(2,\beta(1,w))$ is a witness for $B(a)$. Second, if $(\exists x\leq s)B\in\Sigma_{i-1}^b\cap\Pi_{i-1}^b$, $h(w,\vec{c})$ is the (possibily multivalued) function which is computed by asking a witness oracle for $(\exists x\leq s)B(x)$ for a value for x; i.e., $h(w,\vec{c})=a$ iff $a\leq s$ and $B(a)$ holds. Third, if $(\exists x\leq s)B\in\Pi_{i-1}^b\setminus\Sigma_{i-1}^b$, then the quantifier $(\exists x\leq s)$ must be sharply bounded with $s=|r|$ for some term r. Define h by $h(w,\vec{c})=(\mu x\leq|r|)B(x,\vec{c})$; by Theorem 7, h is a $\mathrm{FP}_3^{\Sigma_{i-1}^p}[wit, log^{O(1)}]$ function definable by S_3^{i-1}.

In each case, we have that

$$S_3^{i-1}\vdash Witness_E^{i,\vec{c}}(w,\vec{c})\rightarrow B(h(w,\vec{c}),\vec{c})\wedge h(w,\vec{c})\leq s(\vec{c})$$

and, indeed, that

$$S_3^{i-1}\vdash Witness_E^{i,\vec{c}}(w,\vec{c})\rightarrow Witness_B^{i,\vec{c},a}(\beta(2,\beta(1,w)),\vec{c},h(w,\vec{c})).$$

The desired $\mathrm{FP}_3^{\Sigma_{i-1}^p}[wit, log^{O(1)}]$ function $f(w,\vec{c})$ is given by

$$f(w,\vec{c})=g(\langle\!\langle 0,\beta(2,\beta(1,w)),\beta(2,w)\rangle\!\rangle,\vec{c},h(w,\vec{c}))$$

and it is can easily be checked that all the conditions of Theorem 18 hold.

Case (4): ($\forall \leq$:right). Suppose the last inference of P is:

$$\frac{a \leq s, \Gamma \longrightarrow B(a), \Delta^*}{\Gamma \longrightarrow (\forall x \leq s)B(x), \Delta^*}$$

(As usual this is one of the hardest cases.) The free variable a is the *eigenvariable* and must appear only as indicated. Let D be the formula $a \leq s \wedge (\bigwedge \Gamma)$, let E be $B(a) \vee (\bigvee \Delta^*)$ and let F be $(\forall x \leq s)B(x) \vee (\bigvee \Delta^*)$. The induction hypothesis is that there is a $\mathrm{FP}_3^{\Sigma_{i-1}^p}[wit, log^{O(1)}]$ function g such that

$$S_3^{i-1} \vdash Witness_D^{i,\vec{c},a}(w, \vec{c}, a) \rightarrow Witness_E^{i,\vec{c},a}(g(w, \vec{c}, a), \vec{c}, a).$$

First, consider the case where $(\forall x \leq s)B(x)$ is in Π_{i-1}^b. We shall define a function f such that

$$S_3^{i-1} \vdash Witness_D^{i,\vec{c}}(w, \vec{c}) \rightarrow Witness_F^{i,\vec{c}}(f(w, \vec{c}), \vec{c}).$$

by informally describing how to compute f. It will be clear that f is a $\mathrm{FP}_3^{\Sigma_{i-1}^p}[wit, log^{O(1)}]$ function since g is. To compute $f(w, \vec{c})$, first ask a Σ_{i-1}^p witness oracle if there exists a value $a \leq s(\vec{c})$ such that $\neg B(a, \vec{c})$ holds. If such an value exists, the oracle returns a value a, and $f(w, \vec{c}) = g(\langle 0, w \rangle, \vec{c}, a)$. If no such value exists, then $f(w, \vec{c}) = \langle 0, 0 \rangle$. In the latter case, $\langle 0, 0 \rangle$ is a witness for $F(\vec{c}, s)$ since 0 is a witness for the true Π_{i-1}^b formula $(\forall x \leq s)B(x)$.

Second, consider the case where $(\forall x \leq s)B(x)$ is in $\Sigma_i^b \setminus \Pi_{i-1}^b$. Then it must be that $(\forall x \leq s)$ is sharply bounded and $s = |r|$ for some term r. Let k be the function defined by

$$k(w, \vec{c}, a) = \begin{cases} \langle 0, \beta(1, g(w, \vec{c}, a)) \rangle & \text{if } Witness_B^{i,\vec{c},a}(\beta(1, g(w, \vec{c}, a)), \vec{c}, a) \\ \langle 1, \beta(2, g(w, \vec{c}, a)) \rangle & \text{otherwise.} \end{cases}$$

To understand the defintion of k; note that the function $g(w, \vec{c}, a)$ provides a witness for E; such a witness is an ordered pair $\langle v_1, v_2 \rangle$ such that either v_1 is a witness for $B(a)$ or v_2 is a witness for $\bigvee \Delta^*$. By definition, $k(w, \vec{c}, a)$ is equal to $\langle 0, v_1 \rangle$ in the former case and to $\langle 1, v_2 \rangle$ otherwise. It is clear that k is a $\mathrm{FP}_3^{\Sigma_{i-1}^p}[wit, log^{O(1)}]$ function since g is and by Proposition 17. Now let $\overline{k}(w, \vec{c}, a)$ be the $\mathrm{FP}_3^{\Sigma_{i-1}^p}[wit, log^{O(1)}]$ function defined from k as in Theorem 6. And define the function $f(w, \vec{c})$ in terms of $\overline{k}$ by letting

$$f(w, \vec{c}) = \begin{cases} \langle \langle w_0, \ldots, w_{|s|} \rangle, 0 \rangle & \text{if } a_i = 0 \text{ for all } i \\ \langle 0, w_i \rangle & \text{for the least } i \text{ s.t. } a_i = 1, \text{ otherwise} \end{cases}$$

where $\overline{k}(2s(\vec{c})+1, \vec{x}, a) = \langle\langle a_0, w_0\rangle, \ldots, \langle a_{|s|}, w_{|s|}\rangle\rangle$. Clearly f is defined by a polynomial function of the value of $\overline{k}$ and, by construction, f satisfies the desired conditions for Theorem 18.

Case (5): (Σ_i^b-LLIND). Suppose the last inference of P is

$$\frac{B(\lfloor \frac{1}{2}a \rfloor), \Gamma^* \longrightarrow B(a), \Delta^*}{B(0), \Gamma^* \longrightarrow B(|t|), \Delta^*}$$

where a is the *eigenvariable* and must not appear in the lower sequent.

Let D be the formula $B(\lfloor \frac{1}{2}a \rfloor) \wedge (\bigwedge \Gamma^*)$, let $E(\vec{c}, a)$ be $B(a) \vee (\bigvee \Delta^*)$, let F be $B(0) \wedge (\bigwedge \Gamma^*)$ and let A be $B(|t|) \vee (\bigvee \Delta^*)$. The induction hypothesis is that there is a $\mathrm{FP}_3^{\Sigma_{i-1}^p}[wit, log^{O(1)}]$ function g such that

$$S_2^{i-1} \vdash Witness_D^{i+1,\vec{c},a}(w, \vec{c}, a) \rightarrow Witness_E^{i+1,\vec{c},a}(g(w, \vec{c}, a), \vec{c}, a).$$

By Proposition 16 there is a polynomial time computable function g_E which is Σ_1^b-definable in S_3^{i-1} and a term t_E such that S_3^{i-1} can prove that if w is a witness for $E(\vec{c}, a)$ then $g_E(w)$ is $\leq t_E(\vec{c}, a)$ and is also a witness for $E(\vec{c}, a)$. Define the function h by

$$h(0, w, \vec{c}) = g_E(\langle \beta(1, w), 0\rangle)$$

$$h(a, w, \vec{c}) = \begin{cases} h(\lfloor \frac{1}{2}a \rfloor, w, \vec{c}) & \text{if } Witness_E^{i,\vec{c},a}(h(\lfloor \frac{1}{2}a \rfloor, w, \vec{c}), \vec{c}, a) \\ g_E(g(\langle \beta(1, h(\lfloor \frac{1}{2}a \rfloor, w, \vec{c})), \beta(2, w)\rangle, \vec{c}, a)) & \text{otherwise} \end{cases}$$

for $a > 0$. Also define $f(w, \vec{c}) = h(|t|, w, \vec{c})$. Because of the use of the function g_E, the values of $h(a, w, \vec{c})$ are bounded by $t_E(\vec{c}, a)$ and thus f is defined by limited logarithmic recursion on notation from functions in $\mathrm{FP}_3^{\Sigma_{i-1}^p}[wit, log^{O(1)}]$ and by Theorem 5, f is also in $\mathrm{FP}_3^{\Sigma_{i-1}^p}[wit, log^{O(1)}]$ and S_2^{i-1} can prove that f can be computed by a canonical Turing machine.

Thus, S_3^{i-1} can prove that, for any computation of the value of $f(w, \vec{c})$, there is a sequence of values

$$h(0, w, \vec{c}), \cdots, h(\lfloor |t|/2^j \rfloor, w, \vec{c}), \cdots, h(|t|, w, \vec{c})$$

obtained during the computation. Now it is easy for S_3^{i-1} to prove that $h(\lfloor |t|/2^j \rfloor, w, \vec{c})$ is a witness for $E(\vec{c}, \lfloor |t|/2^j \rfloor)$ by Σ_{i-1}^b-LBIND since the formula $Witness_E^{i,\vec{c},a}$ is Π_{i-1}^b (in fact, Σ_{i-1}^b-LIND is available). Thus

$$S_3^{i-1} \vdash Witness_F^{i,\vec{c}}(w, \vec{c}) \rightarrow Witness_A^{i,\vec{c}}(f(w, \vec{c}), \vec{c}).$$

Q.E.D. Theorem 18

4.3 Some Corollaries

Theorem 19 $(i > 1)$ *The* Σ_i^b*-definable functions of* R_3^i *and of* S_3^{i-1} *are precisely the* $\mathrm{FP}_3^{\Sigma_{i-1}^p}[wit, log^{O(1)}]$ *functions.*

Proof It is immediate from Theorem 18 that every Σ_i^b-definable function of R_3^i is an $\mathrm{FP}_3^{\Sigma_{i-1}^p}[wit, log^{O(1)}]$ function. Since $R_3^i \vdash S_3^{i-1}$, the same is true of every function Σ_i^b-definable function of S_3^{i-1}. Finally, by Theorem 3, every $\mathrm{FP}_3^{\Sigma_{i-1}^p}[wit, log^{O(1)}]$ function is Σ_i^b-definable by S_3^{i-1}. □

Theorem 20 *The functions which are strongly* Σ_i^b*-definable by* R_3^i *(or, by* S_3^{i-1}*) are precisely the strong* $\mathrm{FP}_3^{\Sigma_{i-1}^p}[wit, log^{O(1)}]$ *functions.*

Proof Corollary 4 gives one direction. Conversely, suppose $R_3^i \vdash (\forall\vec{x})(\exists y)A(\vec{x}, y)$ where $A \in \Sigma_i^b$ and let f be the function such that $f(\vec{x}) = y$ iff $A(\vec{x}, y)$. We need to show that f is in strong $\mathrm{FP}_3^{\Sigma_{i-1}^p}[wit, log^{O(1)}]$. By Theorem 18, there is an explicit $\mathrm{FP}_3^{\Sigma_{i-1}^p}[wit, log^{O(1)}]$ Turing machine M such that $S_3^{i-1} \vdash (\forall\vec{x})A(\vec{x}, func_M(\vec{x}))$. Without loss of generality, assume $A(\vec{x}, y)$ is of the form $(\exists z \le t')B(\vec{x}, y, z)$ with $B \in \Pi_{i-1}^b$ and with t' a term in the variables $\vec{x}$ only. Construct a Turing machine M' which runs the following algorithm:

> **Input:** $\vec{x}$
>
> **(1)** Simulate $M(\vec{x})$ until an output y_0 is obtained.
>
> **(2)** Ask the witness oracle query "$(\exists y \le t)(y = y)$?";
> Then ask the witness oracle query "$(\exists z \le t')(z = z)$?"
> Let y_1 and z_1 be the oracle responses.
>
> **(3)** Ask the Σ_{i-1}^b-query "$\neg B(\vec{x}, y_1, z_1)$?"
> If answer is Yes (so B is false), then output y_0 and halt.
> Otherwise, answer is No: output y_1 and halt.

Clearly M' is explicitly $\mathrm{FP}_3^{\Sigma_{i-1}^p}[wit, log^{O(1)}]$ since M is. Step (2) of the algorithm consists of asking known-to-be-true queries for the sole purpose of generating nondeterministically values for y_1 and z_1. If these values for y_1 and z_1 happen to witness the truth of $(\exists y \leq t)A$, then y_1 is output, otherwise y_0 is output. Having y_0 ensures that at least one value of $f(\vec{x})$ can be found; and nondeterministically guessing y_1 and z_1 makes it possible for any $y = f(\vec{x})$ to be output. □

Theorem 21 $(i > 1)$ *R_3^i is $\forall\Sigma_i^b$-conservative over S_3^{i-1}.*

Proof This is also immediate from Theorem 18: if R_3^i proves a $(\forall\vec{x})A(\vec{x})$ with $A \in \Sigma_i^b$, then R_3^i proves the sequent $\longrightarrow A(\vec{c})$ and, by Theorem 18, S_3^{i-1} proves $(\exists w)\,\mathit{Witness}_A^{i,\vec{c}}(w,\vec{c})$. Thus, Proposition 16, S_3^{i-1} proves $(\forall\vec{x})A(\vec{x})$. □

The class of formulas which are Boolean combinations of Σ_i^b-formulas is denoted $\mathcal{B}(\Sigma_i^b)$; and $\forall\mathcal{B}(\Sigma_i^b)$ is the set of universal generalizations of Boolean combinations of Σ_i^b-formulas.

Theorem 22 *R_3^i is $\forall\mathcal{B}(\Sigma_i^b)$-conservative over $S_3^{i-1} + \Sigma_i^b$-replacement.*

Recall that $R_3^i \supseteq S_3^{i-1} + \Sigma_i^b$-replacement; however, we have no indication as to whether these theories are distinct.

Proof It suffices, of course, to show that any Boolean combination of Σ_i^b-formulas (with free variables) provable by R_3^i is provable in $S_3^{i-1} + \Sigma_i^b$-replacement. Since any Boolean combination can be written in conjunctive normal form, it suffices to show that any disjunction of Σ_i^b- and Π_i^b-formulas provable in R_3^i is provable in $S_3^{i-1} + \Sigma_i^b$-replacement. By rewriting a disjunction as an implication in the form $\bigwedge A_j \to \bigvee B_j$ with each A_j and B_j in Σ_i^b, it suffices to show that if R_3^i proves a sequent $A \longrightarrow B$ with $A, B \in \Sigma_i^b$ then $S_3^{i-1} + \Sigma_i^b$-replacement proves the sequent too. If the sequent is provable by R_3^i then by Theorem 18, S_3^{i-1} proves

$$(\exists w)\,\mathit{Witness}_A^{i,\vec{c}}(w,\vec{c}) \longrightarrow (\exists w)\,\mathit{Witness}_B^{i,\vec{c}}(c,\vec{c}).$$

By Proposition 16, $S_3^{i-1} + \Sigma_i^b$-replacement proves $(\exists x)\, Witness_A^{i,\vec{c}}(w, \vec{c})$ is equivalent to $A(\vec{c})$ and likewise for B. Thus $S_3^{i-1} + \Sigma_i^b$-replacement proves $A \rightarrow B$. □

Recall that a formula is Δ_{i+1}^b with respect to a theory if and only if the theory proves that the formula is equivalent to a Σ_{i+1}^b-formula and to a Π_{i+1}^b-formula.

Theorem 23 $(i \geq 1)$ S_3^i *admits* Δ_{i+1}^b*-PIND.*

Proof For any formula A which is Δ_{i+1}^b w.r.t. S_3^i, the induction axiom for A is a $\forall\Sigma_{i+1}^b$-sentence. Since R_3^{i+1} proves Δ_{i+1}^b-PIND, it follows that S_3^i does too. □

Actually there is another proof of the previous theorem which also applies to S_2^i; this gives the following stronger result which answers a question from Buss [4].

Theorem 24 $(i \geq 1)$ S_2^i *admits* Δ_{i+1}^b*-PIND.*

Proof Based on a theorem of Ressayre it was shown in [4] that $S_2^i + \Sigma_{i+1}^b$-replacement is Σ_{i+1}^b-conservative over S_2^i. Since the Δ_{i+1}^b-PIND axioms are equivalent to $\forall\Sigma_{i+1}^b$-formulas, it suffices to show that $S_2^i + \Sigma_{i+1}^b$-replacement proves the Δ_{i+1}^b-PIND axioms. It is easy to see that Σ_{i+1}^b-replacement implies 'comprehension' for Δ_{i+1}^b formulas; i.e., if $A(u)$ is Δ_{i+1}^b then $S_2^i + \Sigma_{i+1}^b$-replacement proves

$$(\exists w)(\forall u \leq |t|)(Bit(u, w) = 1 \leftrightarrow A(u))$$

where t may be any term not involving u, w and $A(u)$ may have other variables besides u as parameters. Now LIND for A follows easily from the existence of w (just do LIND on $Bit(u, w) = 1$). □

We conclude this section with a couple open questions regarding the theories above results.

First, it would be interesting to know if any Σ_i^b-defined function of R_3^i (equivalently, S_3^{i-1}) can be strenthened to be a provably single-valued function. That is, if R_3^i proves $(\forall x)(\exists y)A(x,y)$ with $A \in \Sigma_i^b$, must there be a formula $A^*(x,y) \in \Sigma_i^b$ such that R_3^i proves $A^*(x,y) \to A(x,y)$ and such that R_3^i proves $(\forall x)(\exists! y)A^*(x,y)$? It would also be nice to know this for S_3^{i-1} as well (this does not seem to automatically follow from the statement for R_3^i). Note that this capability of strenthening a function definition does exist for prior-studied theories such as S_2^i.

Second, it is open whether the conservation results of Theorems 21 and 22 hold for the theories R_2^i and S_2^{i-1} in place of R_3^i and S_3^{i-1}. Likewise, it would be quite interesting to know what the Σ_i^b-definable functions of R_2^i are. It seems that R_3^i is somehow a much more natural theory than R_2^i; as if the growth rate of the $\#_3$ function is somehow naturally linked to the LLIND and LBIND axioms (at least at our present state of knowledge). Note that Krajíček [10] has shown that the Σ_i^b-definable functions of S_2^{i-1} are precisely the functions in $\mathrm{FP}^{\Sigma_i^p}[wit, log]$.

5 The $\Sigma_i^{1,b}$-Definable Functions of Three Second-Order Systems

In this section, the $\Sigma_i^{1,b}$-definable functions of the theories V_2^{i-1}, U_2^i and V_2^i are characterized. We have already shown that every $\mathrm{EXPTIME}^{\Sigma_{i-1}^{1,p}}[wit, poly]$ function is $\Sigma_i^{1,b}$-definable in these theories in the first two theories and that every $\mathrm{EXPTIME}^{\Sigma_i^{1,p}}$ function is $\Sigma_i^{1,b}$-definable in the third theory. To prove the converses, we shall prove 'witnessing theorems' regarding sequents of $\Sigma_i^{1,b}$-formulas provable in these theories.

5.1 The Witness Formula

In [3] a predicate *Witness2* was defined for the purpose of witnessing $\Sigma_1^{1,b}$-formulas; we must generalize this definition to handle witnessing $\Sigma_i^{1,b}$-formulas. For the rest of this section, $i \geq 1$ will be a fixed integer; the applications in this paper only need $i > 1$. Let $A(\vec{a}, \vec{\alpha})$ be a $\Sigma_i^{1,b}$-formula. A formula $Witness2_A^{i,\vec{a},\vec{\alpha}}(\gamma, \vec{a}, \vec{\alpha})$ is defined which has limited quantifier complexity and which states that γ is a second-order object 'witnessing' the truth of $A(\vec{a}, \vec{\alpha})$.

Definition Suppose $A(\vec{a},\vec{\alpha}) \in \Sigma_i^{1,b}$ and $\vec{a},\vec{\alpha}$ is a vector of variables including all those free in A. The formula $Witness2_A^{i,\vec{a},\vec{\alpha}}$ is defined below, inductively on the complexity of A:

(1) If $A \in \Pi_{i-1}^{1,b}$ then $Witness2_A^{i,\vec{a},\vec{\alpha}}$ is just A itself.

(2) If A is $B \wedge C$ then define

$$Witness2_A^{i,\vec{a},\vec{\alpha}}(\gamma,\vec{a},\vec{\alpha}) \iff Witness2_B^{i,\vec{a},\vec{\alpha}}(\beta(1,\gamma),\vec{a},\vec{\alpha}) \wedge Witness2_C^{i,\vec{a},\vec{\alpha}}(\beta(2,\gamma),\vec{a},\vec{\alpha}).$$

(3) If A is $B \vee C$ then define

$$Witness2_A^{i,\vec{a},\vec{\alpha}}(\gamma,\vec{a},\vec{\alpha}) \iff Witness2_B^{i,\vec{a},\vec{\alpha}}(\beta(1,\gamma),\vec{a},\vec{\alpha}) \vee Witness2_C^{i,\vec{a},\vec{\alpha}}(\beta(2,\gamma),\vec{a},\vec{\alpha}).$$

(4) If A is $B \to C$ and is not in $\Pi_{i-1}^{1,b}$, then we define

$$Witness2_A^{i,\vec{a},\vec{\alpha}}(\gamma,\vec{a},\vec{\alpha}) \iff Witness2_{\neg B}^{i,\vec{a},\vec{\alpha}}(\beta(1,\gamma),\vec{a},\vec{\alpha}) \vee Witness2_C^{i,\vec{a},\vec{\alpha}}(\beta(2,\gamma),\vec{a},\vec{\alpha}).$$

(5) If $A \notin \Pi_{i-1}^{1,b}$ and $A(\vec{a},\vec{\alpha})$ is $(\forall x \le s)B(x,\vec{a},\vec{\alpha})$ then define

$$Witness2_A^{i,\vec{a},\vec{\alpha}}(\gamma,\vec{a},\vec{\alpha}) \iff (\forall x \le s)\, Witness2_{B(b,\vec{a},\vec{\alpha})}^{i,b,\vec{a},\vec{\alpha}}(\beta(x+1,\gamma),x,\vec{a},\vec{\alpha}).$$

(6) If $A \notin \Pi_{i-1}^{1,b}$ and $A(\vec{a},\vec{\alpha})$ is $(\exists x \le s)B(x,\vec{a},\vec{\alpha})$ then define

$$Witness2_A^{i,\vec{a},\vec{\alpha}}(\gamma,\vec{a},\vec{\alpha}) \iff (\exists x \le s)\, Witness2_{B(b,\vec{a},\vec{\alpha})}^{i,b,\vec{a},\vec{\alpha}}(\gamma,x,\vec{a},\vec{\alpha}).$$

(7) If $A \notin \Pi_{i-1}^{1,b}$ and A is $(\exists\varphi)B(\vec{a},\vec{\alpha},x)$ where φ is a unary second-order predicate, then define

$$Witness2_A^{i,\vec{a},\vec{\alpha}}(\gamma,\vec{a},\vec{\alpha}) \iff Witness2_{B(b,\vec{a},\vec{\alpha})}^{i,\vec{a},\vec{\alpha},\varphi}(\beta(2,\gamma),\vec{a},\vec{\alpha},\beta(1,\gamma)).$$

The assumption that φ is unary is sufficient for proving the witnessing lemmas below. For k-ary predicates we would replace $\beta(1,\gamma)$ with $ARY_k(\beta(1,\gamma))$ as in [3].

(8) If $A \notin \Pi_{i-1}^{1,b}$ and A is $\neg B$ then use prenex operations to push the negation sign "into" the formula so that it can be handled by cases (1)–(6).

The purpose of defining *Witness2* is to give a canonical way of verifying that $A(\vec{a},\vec{\alpha})$ is true. It is easy to see that $(\exists\gamma)\, \mathit{Witness2}_A^{i,\vec{a},\vec{\alpha}}(\gamma,\vec{a},\vec{\alpha})$ is equivalent to $A(\vec{a},\vec{\alpha})$. The next propositions express some properties of *Witness*; these are are analogous to Propositions 15-17 and are proved similarly.

Proposition 25 *For* $i \geq 1$, *and* $A \in \Sigma_i^{1,b}$, $\mathit{Witness2}_A^{i,\vec{a},\vec{\alpha}}$ *is a* $\Pi_{i-1}^{1,b}$*-formula.*

Proposition 26 $(i \geq 1)$. *Let* $A(\vec{a},\vec{\alpha})$ *be a* $\Sigma_i^{1,b}$*-formula. Then*

$$V_2^{i-1} \vdash \mathit{Witness2}_A^{i,\vec{a},\vec{\alpha}}(\gamma,\vec{a}) \rightarrow A(\vec{a})$$

and

$$V_2^{i-1} + \Sigma_i^{1,b}\text{-replacement} \vdash A(\vec{a},\vec{\alpha}) \leftrightarrow (\exists\gamma)\, \mathit{Witness2}_A^{i,\vec{a},\vec{\alpha}}(\gamma,\vec{a},\vec{\alpha}).$$

Recall that $V_2^{i-1} + \Sigma_i^{1,b}$-replacement is a subtheory of U_2^i (by Theorem 9.16 of [3]).

Proposition 27 $(i \geq 1)$. *Let* A *be a* $\Sigma_i^{1,b}$*-formula. The predicate represented by* $\mathit{Witness2}_A^{i,\vec{a},\vec{\alpha}}$ *is a* $\Pi_{i-1}^{1,p}$*-predicate.*

5.2 Two Witnessing Theorems for Second-Order Theories

In this section we state two witnessing theorems regarding the definability and computability of witness functions for sequents of $\Sigma_i^{1,b}$-formulas. The first, for V_2^i, is the easiest to prove since it does not depend on the use of witness oracles.

Theorem 28 (The Witnessing Lemma for V_2^i)
Fix $i \geq 1$. Suppose the sequent $\Gamma, \Pi \longrightarrow \Delta, \Lambda$ is a theorem of V_2^i and each formula in $\Gamma \cup \Delta$ is $\Sigma_i^{1,b}$ and each formula in $\Pi \cup \Lambda$ is $\Pi_i^{1,b}$. Let $\vec{c}, \vec{\alpha}$ be the free variables in the sequent and let G and H be the formulae

$$G = \left(\bigwedge \Gamma\right) \wedge \bigwedge\{\neg C : C \in \Lambda\}$$

and

$$H = \left(\bigvee \Delta\right) \vee \bigvee\{\neg C : C \in \Pi\}.$$

Then there is a $\mathrm{EXPTIME}^{\Sigma_{i-1}^{1,p}}$ function f which is $\Sigma_i^{1,b}$-defined by V_2^i such that

$$V_2^i \vdash \mathit{Witness2}_G^{i,\vec{c},\vec{\alpha}}(w, \vec{c}, \vec{\alpha}) \rightarrow \mathit{Witness2}_H^{i,\vec{c},\vec{\alpha}}(f(w, \vec{c}, \vec{\alpha}), \vec{c}, \vec{\alpha}).$$

The case $i = 1$ of Theorem 28 is already proved in Buss [3]; the proof of the general case is exactly the same as the proof of the case $i = 1$ except that all exponential time computations are relative to an oracle which is a complete predicate of Σ_{i-1}^p.[3] Another way to think about Theorem 28 is to use the 'RSUV isomorphism' to see that the theorem corresponds to a witnessing theorem for S_3^i — the witnessing theorem for S_3^i is completely idential to the witnessing theorem for S_2^i except that $\#_3$-time is used in place of polynomial time. Because the proof of Theorem 28 is so similar to earlier proofs, we omit it here.

Theorem 29 (The Witnessing Lemma for V_2^{i-1} and U_2^i)
Fix $i > 1$. Suppose the sequent $\Gamma, \Pi \longrightarrow \Delta, \Lambda$ is a theorem of U_2^i and each formula in $\Gamma \cup \Delta$ is $\Sigma_i^{1,b}$ and each formula in $\Pi \cup \Lambda$ is $\Pi_i^{1,b}$. Let $\vec{c}, \vec{\alpha}$ be the free variables in the sequent and let G and H be the formulae

$$G = \left(\bigwedge \Gamma\right) \wedge \bigwedge\{\neg C : C \in \Lambda\}$$

and

$$H = \left(\bigvee \Delta\right) \vee \bigvee\{\neg C : C \in \Pi\}.$$

[3] Recall that exponentially long queries may be made to the Σ_{i-1}^p-oracle so, effectively, the computation may ask any $\Sigma_i^{1,p}$-query.

Then there is a $\mathrm{EXPTIME}^{\Sigma_{i-1}^{1,p}}[wit, poly]$ *function* f *which is* $\Sigma_i^{1,b}$*-defined by* V_2^{i-1} *such that*

$$V_2^{i-1} \vdash Witness2_G^{i,\vec{c},\vec{\alpha}}(w,\vec{c},\vec{\alpha}) \rightarrow Witness2_H^{i,\vec{c},\vec{\alpha}}(f(w,\vec{c},\vec{\alpha}),\vec{c},\vec{\alpha}).$$

The proof of Theorem 29 is entirely analogous to the proof of Theorem 18. For the case of $\forall$*:left* inferences, we use the definability of first-order minimization in place of length-bounded minimization; and for the case of $\Sigma_i^{1,b}$-PIND, we use the closure of $\mathrm{EXPTIME}^{\Sigma_{i-1}^{1,p}}[wit, poly]$ under first-order recursion on notation in place of the closure of $\mathrm{FP}_3^{\Sigma_{i-1}^{p}}[wit, log^{O(1)}]$ under limited, logarithmic recursion on notation. We omit the details. It is also possible to prove Theorem 29 as a corollary of Theorem 18 using the RSUV isomorphism.

5.3 Main Results for Second-Order Theories

The following theorems are all corollaries of the witnessing lemmas; these are proved similarly to the corresponding theorems for first-order theories in section 4.3 above.

Theorem 30 (See [3]) $(i \geq 1)$ *The* $\Sigma_i^{1,b}$*-definable functions of* V_2^i *are precisely the* $\mathrm{EXPTIME}^{\Sigma_{i-1}^{1,p}}$ *functions.*

Theorem 31 $(i > 1)$ *The* $\Sigma_i^{1,b}$*-definable functions of* U_2^i *and of* V_2^{i-1} *are precisely the* $\mathrm{EXPTIME}^{\Sigma_{i-1}^{1,p}}[wit, poly]$ *functions.*

The strongly $\Sigma_i^{1,b}$*-definable functions of* U_2^i *and of* V_2^{i-1} *are precisely the strong* $\mathrm{EXPTIME}^{\Sigma_{i-1}^{1,p}}[wit, poly]$ *functions.*

The $\Sigma_i^{1,b}$*-definable functions of* U_2^i *and of* V_2^{i-1} *which have first-order values are precisely the functions in* $\mathrm{PSPACE}^{\Sigma_{i-1}^{1,p}}$ *(equivalently, in* $\mathrm{EXPTIME}^{\Sigma_{i-1}^{1,p}}[poly]$*).*

Theorem 32 $(i > 1)$ U_2^i *is* $\forall\Sigma_i^{1,b}$*-conservative over* V_2^{i-1}.

Theorem 33 *U_2^i is $\forall\mathcal{B}(\Sigma_i^{1,b})$-conservative over $V_2^{i-1} + \Sigma_i^{1,b}$-replacement.*

A formula is $\Delta_{i+1}^{1,b}$ with respect to a theory if and only if the theory proves that the formula is equivalent to a $\Sigma_{i+1}^{1,b}$-formula and to a $\Pi_{i+1}^{1,b}$-formula.

Theorem 34 $(i \geq 1)$ *V_2^i admits $\Delta_{i+1}^{1,b}$-PIND.*

Theorem 34 follows from Theorem 32 since a $\Delta_{i+1}^{1,b}$-formula is equivalent to a $\forall\Sigma_{i+1}^{1,b}$-formula.

The *bounded $\Delta_{i+1}^{1,b}$ comprehension* axioms are the formulas

$$(\forall a)(\exists\varphi)(\forall x \leq a)\,(\varphi(x) \leftrightarrow A(x))$$

where A is $\Delta_{i+1}^{1,b}$ and may contain other free variables in addition to x.

Theorem 35 $(i \geq 1)$ *$V_2^i \vdash$ bounded $\Delta_{i+1}^{1,b}$ comprehension.*

Theorem 35 improves a result of Takeuti that V_2^i proves $\Sigma_i^{1,b}$-bounded comprehension [17].

Proof First we show that, for $i \geq 0$, U_2^{i+1} proves bounded $\Delta_{i+1}^{1,b}$ comprehension (this extends Theorem 2.1 of [13]). Let A be $\Delta_{i+1}^{1,b}$ with respect to U_2^{i+1} and let $B(z,a)$ be the formula

$$(\forall x \leq a)(\exists\varphi)(\forall y \leq z)(\varphi(x+y) \leftrightarrow A(x+y));$$

note that B is a $\Sigma_{i+1}^{1,b}$-formula. Now we can use the usual "doubling trick" (see Theorems 2.22 or 9.16 of [3]): it is easy to see that $U_2^{i+1} \vdash B(\lfloor \frac{1}{2}z \rfloor, a) \rightarrow B(z,a)$, from whence, by $\Sigma_{i+1}^{1,b}$-PIND, $U_2^{i+1} \vdash B(a,a)$, which readily implies that U_2^{i+1} proves the bounded $\Delta_{i+1}^{1,b}$-comprehension axiom for $A(x)$.

Now suppose $i \geq 1$ and $A(x)$ is $\Delta_{i+1}^{1,b}$ with respect to V_2^i. The bounded $\Delta_{i+1}^{1,b}$ comprehension axiom for A is a $\forall\Sigma_{i+1}^{1,b}$-formula and is provable in U_2^{i+1}. Hence, by Theorem 32, it is provable also by V_2^i.

References

[1] B. ALLEN, *Arithmetizing uniform NC*, Annals of Pure and Applied Logic, 53 (1991), pp. 1–50.

[2] S. R. BUSS, *The witness function method and fragments of Peano arithmetic*. To appear in the Proceedings of the Ninth International Congress on Logic, Methodology and Philosophy of Science.

[3] ——, *Bounded Arithmetic*, Bibliopolis, 1986. Revision of 1985 Princeton University Ph.D. thesis.

[4] ——, *Axiomatizations and conservation results for fragments of bounded arithmetic*, in Logic and Computation, proceedings of a Workshop held Carnegie-Mellon University, 1987, vol. 106 of Contemporary Mathematics, American Mathematical Society, 1990, pp. 57–84.

[5] S. R. BUSS AND L. HAY, *On truth-table reducibility to SAT and the difference hierarchy over NP*, in Proceedings of the Structure in Complexity Conference, June 1988, pp. 224–233.

[6] ——, *On truth-table reducibility to SAT*, Information and Computation, 91 (1991), pp. 86–102.

[7] P. CLOTE AND G. TAKEUTI, *Bounded arithmetics for NC, ALOGTIME, L and NL*. To appear in *Annals of Pure and Applied Logic*.

[8] L. A. HEMACHANDRA, *The strong exponential hierarchy collapses*, in Proceedings 19-th Annual ACM Symposium on Theory of Computing, 1987, pp. 110–122. To appear in JCSS.

[9] R. KAYE, *Using Herbrand-type theorems to separate strong fragments of arithmetic*. Typeset manuscript, August 1991.

[10] J. KRAJÍČEK, *Fragments of bounded arithmetic and bounded query classes*. To appear in Transactions of the A.M.S.

[11] ——, *No counter-example interpretation and interactive computation*, in Logic From Computer Science: Proceedings of a Workshop held November 13-17, 1989, Mathematical Sciences Research Institute Publication #21, Springer-Verlag, 1992, pp. 287–293.

[12] J. Krajíček, P. Pudlák, and G. Takeuti, *Bounded arithmetic and the polynomial hierarchy*, Annals of Pure and Applied Logic, 52 (1991), pp. 143–153.

[13] J. Krajíček and G. Takeuti, *On bounded Σ^1_1 polynomial induction*, in Feasible Mathematics: A Mathematical Sciences Institute Workshop, Ithaca, June 1989, Birkhäuser, 1990, pp. 259–280.

[14] M. W. Krentel, *The complexity of optimization problems*, Journal of Computer and System Sciences, 36 (1988), pp. 490–509.

[15] P. Pudlák, *Some relations between subsystems of arithmetic and the complexity of computations*, in Logic From Computer Science: Proceedings of a Workshop held November 13-17, 1989, Mathematical Sciences Research Institute Publication #21, Springer-Verlag, 1992, pp. 499–519.

[16] G. Takeuti, *Proof Theory*, North-Holland, 2nd ed., 1987.

[17] ——, S^i_3 *and* $\overset{\circ}{V}{}^i_2(BD)$, Archive for Math. Logic, 29 (1990), pp. 149–169.

[18] ——, *RSUV isomorphisms*. Typeset manuscript, 1991.

[19] K. W. Wagner, *Bounded query classes*, SIAM Journal on Computing, 19 (1990), pp. 833–846.

Samuel R. Buss
Department of Mathematics
University of California, San Diego
La Jolla, California 92093
email: sbuss@ucsd.edu

Jan Krajíček
CSAV, Matematicky Ustav
Mathematics Institute
Czechoslovakian Academy of Sciences
Phaha 1, Zitna Ulice 25
Czechoslovakia 11587
email: krajicek@csearn.bitnet

Gaisi Takeuti
Department of Mathematics
University of Illinois
1409 West Green Street
Urbana, IL 61801
email: takeuti@symcom.math.uiuc.ed

On polynomial size Frege proofs of certain combinatorial principles

Peter Clote*

Abstract

Using a theorem of Barrington, we give a new recursion theoretic characterization of the class of $ALOGTIME$ computable functions and then introduce a corresponding free-variable equational calculus ALV'. The system ALV', more elegant than an earlier system ALV, allows straightforward definitions of parity and other counting predicates, as well as the proof of correctness of these predicates. One thus easily obtains polynomial size Frege proofs for certain combinatorial principles such as the pigeonhole principle (S.R. Buss) and the equipartition principle (A. Goerdt).

1 Introduction

In [6] certain parallel complexity classes of functions were characterized in a recursion-theoretic manner as the smallest class of functions containing certain initial functions and closed under certain variants of primitive recursion. Based on the function algebra N_0 which in [6] characterized the class of $ALOGTIME$ computable functions, in [7] a free-variable equational calculus ALV was defined and the a result similar to S. Cook's ER Simulation theorem[1] [9] was proved. This result, which we here call the F simulation theorem,[2] states that if $ALV \vdash f = g$, then the family $\langle |f = g|_n : n \in \mathbf{N} \rangle$

*Part of this research supported by NSF Grant # INT-9014569 and CCR-9102896.
Key words and phrases: Frege proofs, propositional calculus, parallel complexity, NC, free variable equational logic.

[1] for extended resolution simulation.

[2] for Frege simulation.

has polynomial size Frege proofs, where $|f = g|_n$ is a propositional formula expressing the equality of f and g on inputs of length at most n.

In [11] G. Takeuti asked for an explicit definition in N_0 of

$$PARITY(x) = x_0 \oplus x_1 \oplus \cdots \oplus x_{n-1}.$$

Because of its technical nature, this question is answered in the appendix. As it seems quite complicated to prove within AVL that our explicit definition of $PARITY$ satisfies the required properties, using a theorem of Barrington [3, 4] we give an alternative characterization N_0' of $ALOGTIME$ computable functions. We define the associated free-variable equational calculus ALV', in which $PARITY$ and other counting predicates are easily defined, and which provably in ALV' satisfy the required properties. As ALV' also enjoys the F simulation theorem, it is straightforward to prove the existence of polynomial size Frege proofs for combinatorial principles such as the parity principle, k-equipartition principle, pigeonhole principle, etc.

The plan of the paper is as follows. In the next section, we introduce certain initial functions and recursion schemes which will be used in the sequel. In section 3, a new characterization of $ALOGTIME$ computable functions is given, the free variable equational calculus ALV' is defined, and the F simulation theorem is stated. As well, proofs in ALV' of the pigeonhole principle and the equipartition principle are given. Finally, in the appendix, our rather technical construction answering G. Takeuti's question is given. The technique therein may prove useful in an attempt to prove the equivalence of the systems AVL and AVL'.

2 Preliminaries

We assume familiarity with Turing machines (with random access to admit sublinear runtimes), the polynomial time hierarchy PH, the logtime hierarchy LH, etc. The bitgraph of a function f is the binary predicate $A_f(x, i)$ which holds iff the i-th bit of $f(x)$ is 1. The length of the binary representation of an integer x is denoted by $|x|$. A function f has polynomial growth if there is a polynomial p for which $|f(x)| \leq p(|x|)$. If $\mathcal{C}$ is a complexity class, then $\mathcal{FC}$ denotes the class of all functions of polynomial growth, whose bitgraph belongs to $\mathcal{C}$.

Definition 1 $o(x) = 0$,$s(x) = x+1$, $s_0(x) = 2x$,$s_1(x) = 2x+1$, $i_k^n(x_1, \ldots, x_n) = x_k$, $|x| = \lceil log_2(x+1) \rceil$, $x * y = 2^{|y|} \cdot x + y$, $x \# y = 2^{|x| \cdot |y|}$, $tr(x) = \lfloor x/2 \rfloor$, $parity(x) = x - 2 \cdot tr(x)$, $bit(x, i) = parity(\lfloor x/2^i \rfloor)$, $trunc(x, y) = \lfloor x/2^{|y|} \rfloor$.

For $x > 1$, $tree(x)$ is defined to be the 0, 1 value obtained by evaluating a perfect *and/or* tree on the $4^{\lfloor log_4(|x|-1)\rfloor}$ many least significant bits of x. The formal definition is given in the appendix. With this definition, if x has binary representation 1101 of length 4, then $tree(x) = 1$, as this is the least significant bit (or parity) of x, whereas if x has binary representation 11101 of length 5, then

$$tree(x) = (1 \wedge 1) \vee (0 \wedge 1) = 1.$$

Definition 2 *The function f is defined by concatenation recursion on notation (CRN) from g, h_0, h_1 if*

$$\begin{aligned} f(0, \vec{u}) &= g(\vec{u}), \\ f(2x, \vec{u}) &= s_{h_0(x,\vec{u})}(f(x, \vec{u})) \text{ if } x \neq 0 , \\ f(2x+1, \vec{u}) &= s_{h_1(x,\vec{u})}(f(x, \vec{u})). \end{aligned}$$

Definition 3 *The algebra A_0 (resp. N_0) is the smallest class of functions containing o, s_0, s_1, i^n_k, bit, #, $|x|$ (resp. and additionally tree) and closed under composition and CRN.*

In [6], it is shown that A_0 characterizes the class of functions in LH, the log-time hierarchy and that N_0 characterizes the class of functions computable in $ALOGTIME$.

Sharply bounded quantifiers are of the form $(\exists x < |t|)$ and $(\forall x < |t|)$.

Lemma 4 *The algebra A_0 is closed under sharply bounded quantification.*

Proof Suppose that $g, h \in A_0$ and that f is defined by

$$f(x) = \begin{cases} 1 & \text{if } \exists i < |g(x)|[h(i,x) = 0] \\ 0 & \text{else.} \end{cases}$$

We must show that $f \in A_0$. Define f_0 by CRN as follows:

$$\begin{aligned} f_0(x, 0) &= 0 \\ f_0(x, s_i(y)) &= \begin{cases} s_0(f(x,y)) & \text{if } h(|y|, x) \neq 0 \\ s_1(f(x,y)) & \text{else} \end{cases} \end{aligned}$$

Then $f(x) = sg(f_0(x, g(x)))$, where the signum function is defined by $sg(0) = 0$ and $sg(x) = 1$ for $x > 0$. In [6], sg was shown to be in A_0. In a similar manner, one can show closure under sharply bounded universal quantification. □

Definition 5 *The function f is defined from g, h by sharply bounded μ-operator if*

$$f(x) = \begin{cases} i_0 & \text{if } i_0 < |g(x)| \wedge h(i_0, x) = 0 \wedge \forall i < i_0(h(i,x) \neq 0) \\ |g(x)| & \text{else.} \end{cases}$$

This is denoted by $f(x) = \mu i < |g(x)|[h(i,x) = 0]$.

Lemma 6 *A_0 is closed under sharply bounded μ-operator.*

Proof Given $g, h \in A_0$, define f_0 by CRN as follows:

$$\begin{aligned} f_0(x, 0) &= 0 \\ f_0(x, s_i(y)) &= \begin{cases} s_0(f_0(x,y)) & \text{if } \forall i \leq |y|[h(i,x) \neq 0] \\ s_1(f_0(x,y)) & \text{else.} \end{cases} \end{aligned}$$

Let $f(x) = |g(x)| - |f_0(x, g(x))|$. Then it is easy to see that

$$f(x) = \mu i < |g(x)|[h(i,x) = 0].$$

In [6] it was shown that modified subtraction

$$x \dot{-} y = \begin{cases} x - y & \text{if } x \geq y \\ 0 & \text{else} \end{cases}$$

belongs to the algebra A_0. It follows that $f \in A_0$. □

Definition 7 *The function f is defined from g, h_0, h_1 by $Wk - BRN$ (weak, k-bounded recursion on notation) if $f(x, \vec{u}) = F(|x|, \vec{u})$, where*

$$\begin{aligned} F(0, \vec{u}) &= g(\vec{u}) \\ F(s_0(x), \vec{u}) &= h_0(x, \vec{u}, F(x, \vec{u})) \textit{if } x \neq 0 \\ F(s_1(x), \vec{u}) &= h_1(x, \vec{u}, F(x, \vec{u})) \end{aligned}$$

provided that $F(x, \vec{u}) \leq k$, for all $x, \vec{u}$.

Lemma 8 *A_0 is closed under $Wk - BRN$.*

Proof Let $g, h_0, h_1 \in A_0$, let k be a fixed integer, and let f be defined from g, h_0, h_1 by weak k-bounded recursion on notation. Then $f(x, \vec{u}) = y$ if and only if there exists a sequence number t encoding the sequence $(t_0, \ldots, t_{||x||})$ such that

(i) $\forall i \leq ||x||(t_i < k)$
(ii) $t_0 = g(\vec{u})$
(iii) $\forall i \leq ||x||((bit(|x|, i) = 0 \rightarrow t_{i+1} = h_0(i, \vec{u}, t_i)) \wedge (bit(|x|, i) = 1 \rightarrow t_{i+1} = h_1(i, \vec{u}, t_i)))$.
(iv) $t_{||x||} = y$.

We define a pairing function as follows. Given integers x, y, let $m = max\{|x|, |y|\}$. Then $pair(x, y)$ is the integer whose binary representation is obtained by concatenating $(2^m + x)$ with $(2^m + y)$. For instance, if $x = 3$ and $y = 5$, then $m = max\{|3|, |5|\} = 3$ and so the binary representation of $pair(3, 5)$ is 10111101.

To encode the sequence $(t_0, \ldots, t_{n-1})$, first find m, the least power of 2 greater than or equal to $|t_i| + 1$, for $i < n$. Then define the sequence number

$$\langle t_0, \ldots, t_{n-1} \rangle$$

encoding the sequence to be $pair(t, 2^m)$, where

$$t = \sum_{i<n} (2^{m-1} + t_i) \cdot 2^{mi}.$$

This manner of encoding sequences first appeared in [8] and can be shown to be in A_0. If $(t_0, \ldots, t_{n-1})$ is a sequence of integers, each less than k, then the length of s, the sequence number encoding this sequence is at most $n \cdot (|k| + 1) + 2^{|k|+1} + 2$, hence $O(n)$. Letting s now encode the sequence $(t_0, \ldots, t_{||x||})$ described at the beginning of this proof, it follows that s has length $O(||x||)$ and so $s < |x|^{\bar{k}}$ for some $\bar{k}$. We have that

$$f(x) = y \iff \exists s < |x|^{\bar{k}} [(i) \wedge (ii) \wedge (iii) \wedge (iv)]$$

and so by lemma 4, f belongs to A_0. □

3 Main Results

In this section, we give a new recursion theoretic characterization of $ALOGTIME$ computable functions, introduce the new free-variable equational calculus

ALV' and give proofs in ALV' of the parity and pigeonhole principles. A variant of the argument given yields proofs in ALV' of the k-equipartition principle, a generalization of the parity principle. Using ALV', other combinatorial principles proved using counting are similarly easy to establish. The proof of the main theorem of [7] can be modified to prove what we here call the F-simulation theorem, which states that if ALV' proves the equality $f = g$ of two functions, then there are polynomial size Frege proofs of a related family $\langle |f = g|_n : n \in \mathbf{N} \rangle$ of propositional formulas expressing the equality of f and g when restricted to inputs of length at most n. This yields a simple approach towards establishing the existence of polynomial size Frege proofs for a variety of combinatorial principles.

The parity and pigeonhole principles were studied by M. Ajtai, who in [1] (resp. [2]) showed that there are no polynomial size constant depth Frege proofs of the pigeonhole principle (resp. parity principle, even assuming the pigeonhole principle). The question of existence of polynomial size Frege proofs for the pigeonhole principle, raised by S.A. Cook and R. Reckhow, was first answered by S.R. Buss [5] by a direct construction. To our knowledge, G. Takeuti first raised the question of polynomial size Frege proofs of the parity principle. This question was answered by A. Goerdt [10], who displayed polynomial size cutting plane proofs, and proved that Frege systems p-simulate cutting plane proofs. Our contribution is to set up machinery to easily yield polynomial size Frege proofs for such types of combinatorial principles: first one proves the principle within ALV', and then one applies the F-simulation theorem.

3.1 The system ALV'

Definition 9 *The function f is defined by k-bounded recursion on notation ($k - BRN$) from g, h_0, h_1 if*

$$\begin{aligned} f(0, \vec{x}) &= g(\vec{x}), \\ f(2n, \vec{x}) &= h_0(n, \vec{x}, f(n, \vec{x})) \text{ if } n \neq 0 \\ f(2n+1, \vec{x}) &= h_1(n, \vec{x}, f(n, \vec{x})), \end{aligned}$$

provided that $f(n, \vec{x}) \leq k$ holds for all $n, \vec{x}$.

Definition 10 *The algebra N_0' is the smallest class of functions containing o, s_0, s_1, i_k^n, bit, $\#$, $|x|$ and closed under composition, CRN and $k - BRN$, for $k \in \mathbf{N}$.*

Theorem 11 $\mathcal{F}ALOGTIME = N_0'$.

Proof sketch Consider first the inclusion from right to left. $\mathcal{F}-ALOGTIME$ is easily seen to be closed under the initial functions of N_0', and the operations of composition and CRN. Suppose that g, h_0, h_1 are $ALOGTIME$ computable and that f is defined by $k - BRN$ from g, h_0, h_1. Given $x, \vec{u}$, define $\alpha(x, \vec{u}) : \{0, \ldots, k\} \to \{0, \ldots, k\}$ by

$$\alpha(x, \vec{u})(i) = h_{parity(x)}(\lfloor x/2 \rfloor, \vec{u}, i).$$

Given $x, \vec{u}$, let m be the least power of 2 greater than or equal to $|x|$. For $1 \leq i \leq |x|$, let the i-th leaf of a perfect binary tree of depth $\log_2(m)$ be $\alpha(\lfloor x/2^i \rfloor, \vec{u})$ and for $|x| < i \leq m$, let the i-th leaf be the identity function $id_k : \{0, \ldots, k\} \to \{0, \ldots, k\}$. Let each interior node of the binary tree have as value the function obtained by composing the functions associated with that node's children. Specifically, if H_ℓ, H_r are $n+2$-place functions associated with the left, right children of a node, then associate the function $H(x, \vec{u}, i) =_{\mathrm{df}} H_r(x, \vec{u}, H_\ell(x, \vec{u}, i))$ to the node. Composition of functions whose domain and range are contained in $\{0, \ldots, k\}$ can be implemented by finite table look-up. Evaluating the binary tree yields a finite function, which when composed with the initial value $g(\vec{u})$, yields the value $f(x, \vec{u})$. From this description, standard techniques allow the construction of an alternating Turing machine which computes the bitgraph of f in logarithmic time.

Consider now the direction from left to right. By work of [3, 4], the word problem for the group S_5 is complete for $ALOGTIME$ under LH reductions. By [6], LH functions are exactly those in A_0. Define the five functions $f_0, \ldots, f_4$ mapping $\mathbf{N}$ into $\{0, 1, 2, 3, 4\}$ by

$$f_i(x) = \sigma_m \circ \cdots \circ \sigma_1(i)$$

for $0 \leq i \leq 4$. Here σ_i is a permutation on the set $\{0, 1, 2, 3, 4\}$, and the input x codes in a natural manner a sequence $(\sigma_1, \ldots, \sigma_m)$ of m permutations in S_5. Now if $A \subseteq \mathbf{N}$ is in $ALOGTIME$, then let $g \in \mathcal{F}LH$ be such that $x \in A$ holds iff $g(x)$ is a sequence of permutations in S_5 evaluating to the identity permutation in S_5. Thus

$$x \in A \iff \bigwedge_{0 \leq i \leq 4} f_i(g(x)) = i.$$

By hypothesis, $g \in \mathcal{F}LH$, so $g \in A_0$ and the f_i are defined by $4 - BRN$. It follows that the characteristic function of A belongs to N_0'. An arbitrary function f of $\mathcal{F}ALOGTIME$ is one whose bitgraph $A_f(x, i) \equiv$

$bit(f(x), i) = 1$ belongs to $ALOGTIME$. Since N_0' is closed under concatenation recursion on notation, it now follows that $f \in N_0'$. ∎

Definition 12 *By induction, we define the function symbols and terms of* ALV'.

(1) The constant 0 *is a term.*
(2) There are infinitely many numerical variables $x_0, x_1, \ldots$ *and each such variable is a term.*
(3) s_0, s_1, tr, $parity$, s, $|x|$ *are* 1*-place function symbols.*
(4) $trunc$, pad, $\#$ *are* 2*-place function symbols.*
(5) $cond$ *is a* 3*-place function symbol.*
(6) If f *is an* n*-place function symbol and* $t_1, \ldots, t_n$ *are terms, then* $f(t_1, \ldots, t_n)$ *is a term.*
(7) If t *is a term and* $x_1, \ldots, x_n$ *is a list of variables including* <u>*all*</u> *variables in* t*, then* $[\lambda x_1 \ldots x_n.t]$ *is an* n*-place function symbol.*
(8) If f, g, h *are respectively* n*-place,* $n+1$*-place,* $n+1$*-place function symbols, then* $R[f, g, h]$ *is an* $n+1$*-place function symbol.*
(9) If f, g, h *are respectively* n*-place,* $n+2$*-place,* $n+2$*-place function symbols, then* $R_\ell[f, g, h]$ *is an* $n+1$*-place function symbol.*

The intended interpretation of the function symbols is as follows: s_0, s_1, tr, $parity$, $trunc$, $\#$ are as in definition 1. The *conditional* function satisfies

$$cond(x, y, z) = \begin{cases} y & \text{if } x = 0 \\ z & \text{else} \end{cases}$$

and the *padding* function satisfies $pad(x, y) = 2^{|y|} \cdot x$. The recursor $R[f, g, h]$ satisfies

$$R[f, g, h](x, \vec{u}) = \begin{cases} f(\vec{u}) & \text{if } x = 0 \\ 2 \cdot R[f, g, h](\lfloor x/2 \rfloor, \vec{u}) + sg(g(x, \vec{u})) & \text{if } x > 0 \text{ even} \\ 2 \cdot R[f, g, h](\lfloor x/2 \rfloor, \vec{u}) + sg(h(x, \vec{u})) & \text{else} \end{cases}$$

and the recursor $R_\ell[f, g, h]$ defines a function from f, g, h by $k - BRN$, where $k = 2^\ell - 1$. We remark that at the expense of an increase in notational difficulty, we could have defined $R_\ell[f, g, h]$ to correspond to $\ell - BRN$ and obtained similar results.

Definition 13 *The axioms of ALV' are as follows:*

(A1) $s_0(0) = 0$.

(A2a) $parity(s_0(x)) = 0$.
(A2b) $parity(s_1(x)) = 1$.

(A3a) $tr(s_0(x)) = x$.
(A3b) $tr(s_1(x)) = x$.

(A4a) $cond(0, y, z) = y$.
(A4b) $cond(s_0(x), y, z) = cond(x, y, z)$.
(A4c) $cond(s_1(x), y, z) = z$.

(A5a) $pad(x, s_0(y)) = cond(y, x, s_0(pad(x, y)))$.
(A5b) $pad(x, s_1(y)) = s_0(pad(x, y))$.

(A6a) $trunc(x, s_0(y)) = cond(y, x, tr(trunc(x, y)))$.
(A6b) $trunc(x, s_1(y)) = tr(trunc(x, y))$.

(A7a) $x\#s_0(y) = cond(y, 1, pad(x\#y, x))$.
(A7b) $x\#s_1(y) = pad(x\#y, x)$.

(A8a) $s(0) = s_1(0)$.
(A8b) $s(s_0(x)) = s_1(x)$.
(A8c) $s(s_1(x)) = s_0(s(x))$.

(A9a) $|0| = 0$.
(A9b) $|s_0(x)| = cond(x, 0, s(|x|))$.
(A9c) $|s_1(x)| = s(|x|)$.

(A10) $[\lambda x_1, \ldots, x_n.t](x_1, \ldots, x_n) = t$.

(A11) $R[f, g, h](x, \vec{u}) = cond(x, f(\vec{u}), cond(parity(x), U, V))$,

where

$$U = cond(g(tr(x), \vec{u}), s_0(t), s_1(t))$$

and

$$V = cond(h(tr(x), \vec{u}), s_0(t), s_1(t))$$

and $t = R[f, g, h](tr(x), \vec{u})$, *which corresponds to term formation using* CRN,

(A12) $R_\ell[f, g, h](x, \vec{u}) = cond(x, f(\vec{u}), cond(parity(x), U_\ell(x, \vec{u}), V_\ell(x, \vec{u})))$, *where*

$$U_\ell(x, \vec{u}) = g(tr(x), \vec{u}, R_\ell[f, g, h](tr(x), \vec{u})) \bmod 2^\ell$$

and

$$V_\ell(x, \vec{u}) = h(tr(x), \vec{u}, R_\ell[f, g, h](tr(x), \vec{u})) \bmod 2^\ell.$$

In the above definition, in order to express the unary function $f_\ell(x) = x \bmod 2^\ell$ for fixed ℓ, we proceed as follows. Noting that $bit(x, |y|) = parity(trunc(x, y))$, define $F(x, y, z) = cond(parity(trunc(x, y)), s_0(z), s_1(z))$. Then

$$f_\ell(x) = F(x, 0, F(x, 1, F(x, 2, F(x, 2^2, F(\cdots F(x, 2^{\ell-2}, 0)\cdots,))).$$

Definition 14 *The rules of inference of* ALV' *are as follows.*

(R1) $t = u \vdash u = t$.
(R2) $t = u, u = v \vdash t = v$.
(R3) $t_1 = u_1, \ldots, t_n = u_n \vdash f(\vec{t}) = f(\vec{u})$.
(R4) $t = u \vdash t[v/x] = u[v/x]$, *$x$ a variable, v any term.*
(R5) Induction on notation.

$$\begin{array}{cc} & f[0/x] = g[0/x], \\ f[s_0(x)/x] = h_0[f/a] & f[s_1(x)/x] = h_1[f/a] \\ g[s_0(x)/x] = h_0[g/a] & g[s_1(x)/x] = h_1[g/a] \end{array}$$

$$f = g$$

for any terms f, g, h_0, h_1 *and distinct variables* x, a.

Suppose that p is a polynomial, and that f, g are functions in ALV'. Then $|f = g|^n_{p(n)}$ is a propositional formula which expresses that for all x of length at most n, the function values $f(x)$,$g(x)$ have length at most $p(n)$ and $f(x) = g(x)$. The explicit formulation of this propositional formula, which is of size polynomial in n, is given in [7]. With this notation, we have the following F-simulation theorem.

Theorem 15 *If $ALV' \vdash f = g$, then for some polynomial p there are polynomial size Frege proofs of $\langle |f = g|^n_{p(n)} : n \in \mathbf{N}\rangle$.*

Since the proof is but a small variant of the long and tedious proof given in [7], it will not be given.

3.2 Applications

Definition 16 *The k-equipartition principle, denoted $k - EQ$, states that if an integer n is not evenly divisible by k, then there is no partition of $\{0, \ldots, n-1\}$ into disjoint sets of size k.*

This principle was investigated by M. Ajtai in [2], where it was called the parity principle. The propositional formulation $(k+1)-EQ_n$ of the $(k+1)$-equipartition principle could be stated as follows. For $1 \leq i_1, \ldots, i_{k+1} \leq n$, let the propositional variable $x_{i_1}, \ldots, x_{i_{k+1}}$ assert that $i_1, \ldots, i_{k+1}$ all belong to the same partition class. Let $J = \{(j_1, \ldots, j_k) : 1 \leq j_1 < \cdots < j_k \leq n\}$. For $\vec{\jmath} \in J$, we write $i \in \vec{\jmath}$ (resp. $i \notin \vec{\jmath}$) to mean that there exists (resp. does not exist) $1 \leq \ell \leq k$ such that $i = j_\ell$. Suppose that $n \not\equiv 0 \bmod k+1$. Then the negation of $(k+1) - EQ_n$ can be expressed by

$$\bigwedge_{1 \leq i \leq n} \bigvee_{\vec{\jmath} \in J, i \notin \vec{\jmath}} x_{i,\vec{\jmath}} \wedge \bigwedge_{1 \leq i \leq n} \bigwedge_{\vec{\jmath}, \vec{\jmath}' \in J, i \notin \vec{\jmath}, \vec{\jmath}'} (\overline{x}_{i,\vec{\jmath}} \vee \overline{x}_{i,\vec{\jmath}'}).$$

Alternatively, we will later define alternating logtime functions F, G, such that $F = G$ expresses $(k+1) - EQ$, and could take $(k+1) - EQ_n$ to be $|F = G|_n$. Though tedious to show, there are polynomial size Frege proofs of the equivalence between both propositional formulations. This remark applies to all propositional formulations of combinatorial principles in this paper.

In [10], A. Goerdt gave polynomial size "cutting planes" proofs of the parity principle, and showed that Frege systems p-simulate cutting planes.

It follows that the parity principle has polynomial size Frege proofs. Below, we prove within ALV' a free-variable equational formulation of the parity principle. The existence of polynomial size Frege proofs of $2-EQ_n$ follows from the F-simulation theorem for ALV'.

Let $reversebit(x,i) = bit(x^R, i)$, where x^R designates the integer whose binary representation is the string obtained by reversing the binary representation of x. In [6], *reversebit* is explicitly defined in the algebra A_0. Define the predicate $REVBIT(x,i)$ to hold iff $reversebit(x,i) = 1$. We encode a partition as follows. Given integer y and integers $x, i < |y|$, if we temporarily denote the partition encoded by y as f, then we think of $f(x) = i$ as holding iff x is in class i iff $REVBIT(y, \langle x, i\rangle)$ holds. With this convention, note that if y encodes a partition for which $f(x) = i$, and y is an initial segment of another integer z (in particular, if $z = 2 \cdot y$ or $z = 2 \cdot y + 1$), then $f(x) = i$ holds for the partition encoded by z. This property will be exploited in our proof in ALV' of the parity property.

Define the predicate $PART_2(y)$ to hold iff

$$(\forall x_1, x_2, x_3, i < |y|)(\bigwedge_{1\leq j<k\leq 3} x_j \neq x_k \rightarrow \neg \bigwedge_{1\leq j\leq 3} REVBIT(y, \langle x_j, i\rangle))$$

Then $PART_2(y)$ holds exactly when y encodes a partition, all of whose equivalence classes have at most 2 elements. Let

$$F(y) = \begin{cases} 2 & \text{if } \neg PART_2(y) \\ |A_y| \bmod 2 & \text{else} \end{cases}$$

where A_y is the set

$$\{i < |y| : (\exists x < |y|)(\forall x' < |y|)(REVBIT(y, \langle x, i\rangle) \wedge \neg REVBIT(y, \langle x', i\rangle))\}.$$

Thus, if y encodes a partition, then $F(y)$ counts modulo 2 the number of equivalence classes of the partition encoded by y having exactly 1 element.

Let

$$G(y) = \begin{cases} 2 & \text{if } \neg PART_2(y) \\ |B_y| \bmod 2 & \text{else} \end{cases}$$

where B_y is the set

$$\{i < |y| : (\exists x < |y|)(REVBIT(y, \langle x, i\rangle)\}.$$

Thus, if y encodes a partition, then $G(y)$ counts modulo 2 the elements in the partition.

Using the recursor R_1, it is straightforward to define the predicate $PART_2$ and the functions F, G in ALV' and to prove their basic properties by induction on notation. We now have the result.

Theorem 17

$$ALV' \vdash F = G.$$

Proof The proof is by induction on notation.

Base case $y = 0$.
In this case, the partition is empty, so there is nothing to prove.

First inductive case $y = 2 \cdot v$.
In this case, the partition encoded by y is as the partition encoded by v. Apply the inductive hypothesis.

Second inductive case $y = 2 \cdot v + 1$.
By the inductive hypothesis, $F(v) = G(v)$ holds.

Subcase 1 $\neg PART_2(y)$.
In this case, $F(y) = 2 = G(y)$.

Subcase 2 $PART_2(y)$ and $|y| - 1 = \langle x, i \rangle$ for some x, i and

$$(\exists x' < |y|)(x' \neq x \wedge REVBIT(y, \langle x', i \rangle)).$$

In this case, y encodes a partition f for which $f(x) = i$ and $f(x') = i$. Since $PART_2(y)$ and $F(v) = G(v)$ hold, it must be that the number of elements in the partition encoded by v is odd and so the number of elements in the partition encoded by y must be even. Thus $F(y) = G(y)$.

Subcase 3 $PART_2(y)$ and $|y| - 1 = \langle x, i \rangle$ for some x, i and

$$(\neg \exists x' < |y|)(x' \neq x \wedge REVBIT(y, \langle x', i \rangle)).$$

In this case, y encodes a partition f for which $f(x) = i$ and $f(x') \neq i$ for all distinct x'. Since $PART_2(y)$ and $F(v) = G(v)$ hold, it must be that the number of elements in the partition encoded by v is even and so the number of elements in the partition encoded by y is odd. Thus $F(y) = G(y)$.

By induction on notation, we have established $F = G$ in ALV'. Thus any partition of elements in disjoint classes, all of size 2, must have an even number of elements. ∎

Application of the F simulation theorem for ALV' now implies the

Corollary 18 *There are polynomial size Frege proofs of the parity principle.*

It is straightforward, modifying the above proof, to exhibit polynomial size Frege proofs of the k-equipartition principle, for any k.

In [5], S.R. Buss gave polynomial size Frege proofs of the pigeonhole principle. Reasoning in a manner similar to the above proof, we can similarly demonstrate the existence of polynomial size Frege proofs of the pigeonhole principle.

Let MAP_y be the following predicate

$$(\forall i, i', j, j' < |y|)[(j \neq j' \rightarrow \neg REVBIT(y, \langle i, j\rangle) \vee \neg REVBIT(y, \langle i, j'\rangle))$$
$$\wedge (i \neq i' \rightarrow \neg REVBIT(y, \langle i, j\rangle) \vee \neg REVBIT(y, \langle i', j\rangle))].$$

Thus MAP_y holds exactly if y encodes a one-to-one mapping. Define

$$f(y) = \begin{cases} 2 & \text{if } \neg MAP_y \\ |C_y| \bmod 2 & \text{else} \end{cases}$$

where C_y is the set

$$\{i < |y| : (\exists j < |y|) REVBIT(y, \langle i, j\rangle)\}.$$

If y codes a partial injection ϕ, then $f(y)$ is the number mod 2 of elements in the domain of ϕ.

Define

$$g(y) = \begin{cases} 2 & \text{if } \neg MAP_y \\ |D_y| \bmod 2 & \text{else} \end{cases}$$

where D_y is the set

$$\{j < |y| : (\exists i < |y|) REVBIT(y, \langle i, j\rangle)\}.$$

If y codes a partial injection ϕ, then $g(y)$ is the number mod 2 of elements in the range of ϕ. A similar proof by induction on notation shows that $ALV' \vdash f = g$, thus establishing the pigeonhole principle.

Theorem 19 $ALV' \vdash f = g$.

By the F simulation theorem we have the following corollary, first proved by S.R. Buss [5].

Corollary 20 *The pigeonhole principle has polynomial size Frege proofs.*

References

[1] M. Ajtai, "The complexity of the pigeonhole principle", *Proc. 29th IEEE Symp. on Foundations of Computer Science* (1988), 346-355.

[2] M. Ajtai, "Parity and the pigeonhole principle", in *Feasible Mathematics – A Mathematical Sciences Institute Workshop*, Ithaca, New York, June 1989 eds. S.R. Buss, P. Scott, Birkhäuser, 1-24 (1990)

[3] D.A. Barrington", "Bounded-width polynomial-size branching programs recognize exactly those languages in NC^1", *Journal of Computer and System Science* **38** (1989) 150-164, Preliminary version in *Proceedings of the 18th Annual ACM Symposium on Theory of Computing*(1986) 1-5.

[4] D. Mix Barrington, N. Immerman and H. Straubing, "On uniformity in NC^1, *J. Comp. Syst. Sci.* **41** (3), 274-306 (1990), Preliminary version appeared in *Structure in Complexity Theory: 3rd Annual Conference*, IEEE Computer Society Press (1988), 47-59.

[5] S.R. Buss, "Polynomial size proofs of the propositional pigeonhole principle", *Journal of Symbolic Logic* **52** (1987) 916-927.

[6] P. Clote, "Sequential, machine-independent characterizations of the parallel complexity classes $ALOGTIME$, AC^k, NC^k and NC", in *Feasible Mathematics – A Mathematical Sciences Institute Workshop*, Ithaca, New York, June 1989 eds. S.R. Buss, P. Scott, Birkhäuser, 49-70 (1990)

[7] P. Clote, "ALOGTIME and a conjecture of S.A. Cook", in *Proceedings of IEEE Symposium on Logic in Computer Science*, Philadelphia, (June 1990). Journal version to appear in *Annals of Mathematics and Artificial Intelligence*.

[8] P. Clote, G. Takeuti, "Bounded arithmetics for NC, $ALOGTIME$, L and NL", to appear in *Annals of Pure and Applied Logic* (1992).

[9] S.A. Cook, "Feasibly constructive proofs and the propositional calculus", *Proc. 7th ACM Symp. on the Theory of Computation* (1975) 83-97.

[10] A. Goerdt, "Cutting plane versus Frege proof systems", Proceedings of *Computer Science Logic 1990*, Springer Lecture Notes in Computer Science **553**, 174-194, ed. E. Börger (1992).

[11] G. Takeuti, private correspondence dated 12 August 1990.

4 Appendix: An explicit definition of $PARITY$

In this section, we state a formal definition of the function *tree*, used in [6] to characterize alternating logtime computable functions[3] as those of the algebra N_0. We explicitly define $PARITY$ in N_0, answering a question of G. Takeuti, and hope that this construction may be of use in a possible proof of the equivalence of ALV and ALV'.

Definition 21 *The function* <u>*tree*</u> *taking values* 0, 1 *is defined from the auxilary functions and(x), or(x), as follows.*

$$\begin{aligned}
and(0) &= 0\\
and(1) &= 1\\
and(2) &= 1\\
and(3) &= 1\\
and(s_0(s_0(x))) &= s_0(and(x)) \text{ if } x > 0\\
and(s_0(s_1(x))) &= s_0(and(x)) \text{ if } x > 0\\
and(s_1(s_0(x))) &= s_0(and(x)) \text{ if } x > 0\\
and(s_1(s_1(x))) &= s_1(and(x)) \text{ if } x > 0
\end{aligned}$$

$$\begin{aligned}
or(0) &= 0\\
or(1) &= 1\\
or(2) &= 1\\
or(3) &= 1\\
or(s_0(s_0(x))) &= s_0(or(x)) \text{ if } x > 0\\
or(s_0(s_1(x))) &= s_1(or(x)) \text{ if } x > 0\\
or(s_1(s_0(x))) &= s_1(or(x)) \text{ if } x > 0\\
or(s_1(s_1(x))) &= s_1(or(x)) \text{ if } x > 0
\end{aligned}$$

$$tree(x) = \begin{cases} parity(x) & \text{if } x < 16\\ tree(or(and(x))) & \text{else.}\end{cases}$$

To define $PARITY$ in N_0, we would like first to show the existence of a function $t \in A_0$ such that $|t(x)| = 2^{2\cdot||x||}$. To this end, we define auxiliary

[3] There is a minor typographic mistake in the definition in [6] which is corrected here.

functions $t_0, t_1, \ldots, t_7$ as follows. Let

$$t_0(x) = (|x|\#1)\#(|x|\#1).$$

In [6] it was shown that $trunc(x, y) =_{\text{def}} \lfloor x/2^{|y|} \rfloor$ belongs to A_0. Let

$$t_1(x) = trunc(t_0(x), |x|\#|x|).$$

Then

$$t_0(x) = 2^{||x||^2+2\cdot||x||+1}$$

and

$$\begin{aligned} t_1(x) &= \lfloor \frac{2^{||x||^2+2\cdot||x||+1}}{2^{||x||^2+1}} \rfloor \\ &= 2^{2\cdot||x||}. \end{aligned}$$

Now let

$$\begin{aligned} t_2(x, 0) &= 1 \\ t_2(x, s_i(y)) &= \begin{cases} s_0(t_2(x, y)) & \text{if } |y| \leq t_1(x) \\ s_1(t_2(x, y)) & \text{else.} \end{cases} \end{aligned}$$

For y such that $|y| \geq t_1(x)$, it is clear that the binary reprentation of $t_2(x, y)$ is

$$1\underbrace{0\cdots0}\underbrace{1\cdots1}$$

where there are $t_1(x)$ many 0's and $|y| - t_1(x)$ many 1's. Let

$$t_3(x) = (x\#x) * (x\#x) * (x\#x) * (x\#x).$$

Claim. $|t_3(x)| \geq 2^{2\cdot||x||}$.
Proof of claim. For $x > 0$, $||x|| = log_2(|x|) + 1$, so $2^{||x||} \leq 2 \cdot |x|$. Thus

$$\begin{aligned} (2^{||x||})^2 &\leq 4 \cdot |x|^2 \\ &\leq |2^{4\cdot|x|^2+3}| \\ &\leq |t_3(x)|.\Box \end{aligned}$$

Let

$$t_4(x) = t_2(x, t_3(x))$$

and

$$t_5(x) = \mu i < |t_4(x)|[bit(t_4(x), i) = 0].$$

By lemma 6, $t_5 \in A_0$. The binary representation of $t_4(x)$ is

$$1\underbrace{0\cdots0}\underbrace{1\cdots1}$$

where there are $2^{2\cdot||x||}$ many 0's and $|t_3(x)| - 2^{2\cdot||x||}$ many 1's. As well, $t_5(x)$ is the position of the rightmost 0 occurring in the binary representation of $t_4(x)$. In [6], it was shown that the function $shift\_right(x, i) = \lfloor x/2^i \rfloor$ belongs to A_0. Now define the desired function t by

$$t(x) = shift\_right(t_4(x), s_1(t_5(x))).$$

Then $t \in A_0$ and the binary representation of $t(x)$ is

$$1\underbrace{0\cdots0}$$

where there are $2^{2\cdot||x||-1}$ many 0's. Let

$$t_6(x) = \mu i < |x|[bit(x, i) = 1]$$

and

$$t_7(x) = \begin{cases} 1 & \text{if } t_6(x) = |x| \dot{-} 1 \\ 0 & \text{else.} \end{cases}$$

Then $t_6, t_7 \in A_0$ and

$$t_7(x) = 1 \iff x \text{ is a power of 2.}$$

Let

$$t_8(x) = \lfloor t_1(x)/4 \rfloor.$$

Using t_8 in the place of t_1 in the development leading to the definition of the function t, we obtain the function $t' \in A_0$, where

$$t'(x) = 2^{2^{2\cdot||x||-2}-1}.$$

Note that $|t'(x)| = 2^{2\cdot||x||-2} = |t(x)|/4$. Now define

$$\bar{t}(x) = \begin{cases} x & \text{if } x \in \{0, 1, 2\} \\ t'(x) & \text{if } x \text{ is power of 2 and } x \notin \{0, 1, 2\} \\ t(x) & \text{else.} \end{cases}$$

It follows that $|\bar{t}(x)|$ is the square of the least power of 2 greater than or equal to the length of x.

Here are some sample values:

$\|x\|$	$\|\|x\|\|$	$\|t(x)\|$	$\|\bar{t}(x)\|$
1	1	4	1
2	2	16	4
3	2	16	16
4	3	64	16
5	3	64	64

Let x have binary representation $x_{n-1}, \ldots, x_0$. Our goal is to define the function

$$\begin{aligned} PARITY(x) &= x_0 \oplus x_1 \oplus \cdots \oplus x_{n-1} \\ &= (\sum_{i<n} x_i) \bmod 2 \end{aligned}$$

in the algebra N_0.

For the sake of clarity, we give two examples before proceeding with the construction.

Example 22 *Suppose that the binary representation of x is $x_1 x_0$. Recall that*

$$tree(x) = \begin{cases} x & \text{if } x < 16 \\ tree(or(and(x))) & \text{else.} \end{cases}$$

Letting $\bar{x}_i$ designate $1 \dot{-} x_i$, we have

$$\begin{aligned} PARITY(x) &= x_0 \oplus x_1 \\ &= tree(1 x_0 \bar{x}_1 \bar{x}_0 x_1) \end{aligned}$$

Example 23 *Suppose that the binary representation of x is $x_3 x_2 x_1 x_0$. Then*

$$\begin{aligned} PARITY(x) &= x_0 \oplus x_1 \oplus x_2 \oplus x_3 \\ &= tree(1 x_0 \bar{x}_1 \bar{x}_0 x_1 x_2 x_3 \bar{x}_2 \bar{x}_3 x_0 x_1 \bar{x}_0 \bar{x}_1 x_2 \bar{x}_3 \bar{x}_2 x_3) \end{aligned}$$

and the reader should apply a perfect binary tree of depth 2 with alternating AND *and* OR *gates to compute the output of this example.*

For the example with x of length 4, our explicit definition of $PARITY(x)$ is as follows:

(i) compute $\bar{t}(x)$, such that $|\bar{t}(x)|$ is the square of the least power of 2 greater than or equal to $|x|$,

(ii) using $\bar{t}(x)$, which in this example satisfies $|\bar{t}(x)| = 16$, construct the integer with binary representation

$$1x_0x_1x_0x_1x_2x_3x_2x_3x_0x_1x_0x_1x_2x_3x_2x_3$$

(iii) using $W2 - BRN$, define whether the above bits are to be complemented or not, thus producing the integer with binary representation

$$1x_0\bar{x}_1\bar{x}_0x_1x_2x_3\bar{x}_2\bar{x}_3x_0x_1\bar{x}_0\bar{x}_1x_2\bar{x}_3\bar{x}_2x_3$$

(iv) apply *tree* to the result of (iii).
Steps (i),(ii),(iii) can be carried out within A_0. Only step (iv) requires the function *tree*.

Construction for step (ii)
Let

$$\begin{aligned}\bar{f}(0, m, z) &= 0 \\ \bar{f}(s_i(y), m, z) &= \bar{f}(y, m, z) * bit(|z|, 2 \cdot (|m| \dot{-} 1 \dot{-} |y|))\end{aligned}$$

Let $f(m, z) = \bar{f}(m, m, z)$. Then $f \in A_0$ and

$$f(m, z) = \sum_{i<|m|} 2^i \cdot bit(|z|, 2i).$$

Let

$$\begin{aligned}F(m, x, 0) &= 1 \\ F(m, x, s_i(z)) &= F(m, x, z) * bit(x, f(m, z)).\end{aligned}$$

Define $G(x) = F(|x|, x, \bar{t}(x))$. We claim that $G(x)$, ostensibly in A_0, produces the result of step (ii). Before proving this, we illustrate the computation with a previous example.

Example 24 *Let x have binary representation x_1x_0. Then $G(x) = F(2, x, 8)$.*

$$\begin{aligned}F(2, x, 0) &= 1 \\ F(2, x, 1) &= 1 * bit(x, f(2, 0)) \\ &= 1 * bit(x, 0) \\ &= 1x_0 \\ F(2, x, 2) &= 1x_0 * bit(x, f(2, 1))\end{aligned}$$

$$\begin{aligned}
&= 1x_0x_1 \\
F(2,x,4) &= 1x_0x_1 * bit(x, f(2,2)) \\
&= 1x_0x_1x_0 \\
F(2,x,8) &= 1x_0x_1x_0 * bit(x, f(2,4)) \\
&= 1x_0x_1x_0x_1
\end{aligned}$$

Example 25 *Let x have binary representation $x_3x_2x_1x_0$. Then $G(x) = F(4, x, 2^{15})$.*

$$\begin{aligned}
F(4,x,0) &= 1 \\
F(4,x,1) &= 1 * bit(x, f(4,0)) \\
&= 1 * bit(x,0) \\
&= 1x_0 \\
F(4,x,2) &= 1x_0 * bit(x, f(4,1)) \\
&= 1x_0x_1 \\
F(4,x,4) &= 1x_0x_1 * bit(x, f(4,2)) \\
&= 1x_0x_1x_0 \\
F(4,x,8) &= 1x_0x_1x_0 * bit(x, f(4,4)) \\
&= 1x_0x_1x_0x_1 \\
F(4,x,16) &= 1x_0x_1x_0x_1 * bit(x, f(4,8)) \\
&= 1x_0x_1x_0x_1x_2 \\
F(4,x,32) &= 1x_0x_1x_0x_2 * bit(x, f(4,16)) \\
&= 1x_0x_1x_0x_1x_2x_3 \\
F(4,x,64) &= 1x_0x_1x_0x_2x_3 * bit(x, f(4,32)) \\
&= 1x_0x_1x_0x_1x_2x_3x_2 \\
&\textit{etc.}
\end{aligned}$$

Define

$$l(x) = \sum_{i < \frac{|\bar{t}(x)|^{\frac{1}{2}}}{2}} 2^i \cdot bit(x, \frac{|\bar{t}(x)|^{\frac{1}{2}}}{2} + i)$$

and

$$r(x) = \sum_{i < \frac{|\bar{t}(x)|^{\frac{1}{2}}}{2}} 2^i \cdot bit(x, i).$$

Clearly, $l(x)$ and $r(x)$ respectively are the "left" and "right" parts of x.[4]

Letting P temporarily denote $PARITY$, notice that P can be defined as follows:

$$P(x) \quad = \quad (P(l(x)) \wedge \neg P(r(x))) \vee (\neg P(l(x)) \wedge P(r(x)))$$

and

$$\neg P(x) \quad = \quad (P(l(x)) \wedge P(r(x))) \vee (\neg P(l(x)) \wedge \neg P(r(x)))$$

These considerations lead to the definition

$$
\begin{array}{rcl}
g(x,0) & = & 0 \\
g(x,s_0(y)) & = & \left\{ \begin{array}{lll}
0 & \text{if } g(x,y) = 0 & \\
-\ + & \text{if } g(x,y) = + & \\
-\ - & \text{if } g(x,y) = - & \\
- & \text{if } g(x,y) = + & - \\
+ & \text{if } g(x,y) = - & + \\
+ & \text{if } g(x,y) = + & + \\
- & \text{if } g(x,y) = - & -
\end{array} \right. \\
g(x,s_1(y)) & = & \left\{ \begin{array}{lll}
+ & \text{if } g(x,y) = 0 & \\
+\ - & \text{if } g(x,y) = + & \\
+\ + & \text{if } g(x,y) = - & \\
+ & \text{if } g(x,y) = + & - \\
- & \text{if } g(x,y) = - & + \\
+ & \text{if } g(x,y) = + & + \\
- & \text{if } g(x,y) = - & -
\end{array} \right.
\end{array}
$$

Then for

$$|\bar{t}(x)| \leq y < 2 \cdot |\bar{t}(x)|$$

we have $g(x,y) \in \{+,-\}$ and $g(x,y)$ is the sign associated with

$$bit(G(x), y - |\bar{t}(x)|).$$

We urge the reader to work through the above definition for x of length 2 (resp. 4) and $4 \leq y < 8$ (resp. $16 \leq y < 32$).

Note that y is bounded by a function in the length of x, hence $g(x,y)$ can be defined by $Wk - BRN$, and so belongs to A_0.

Construction for step (iii)

---

[4] The intent of $l(x)$ and $r(x)$ is to take the left and right parts of the "word" obtained from x by possibly padding x with preceding 0's to produce a word of length the smallest power of 2 greater than or equal to $|x|$.

As in the construction for step (ii), let

$$f(m, z) = \sum_{i<|m|} 2^i \cdot bit(|z|, 2i).$$

Let

$$\begin{aligned} \bar{F}(m, x, 0) &= 1 \\ \bar{F}(m, x, s_i(z)) &= \bar{F}(m, x, z) * h(m, x, z) \end{aligned}$$

where

$$h(m, x, z) = \begin{cases} bit(x, f(m, z)) & \text{if } g(x, |\bar{t}(x)| + z) = + \\ 1 \dot{-} bit(x, f(m, z)) & \text{if } g(x, |\bar{t}(x)| + z) = - \end{cases}$$

Define $\bar{G}(x) = \bar{F}(|x|, x, \bar{t}(x))$. Working through the definitions, we see that $\bar{G}(x)$, ostensibly in A_0, produces the result of step (iii). It then follows that

$$PARITY(x) = tree(\bar{G}(x)).$$

Peter Clote
Department of Computer Science
Boston College
Chestnut Hill, MA 02167
email: clote@bcuxs1.bc.edu

Interpretability and fragments of arithmetic

Petr Hájek

Abstract

We investigate prominent fragments of first and second order arithmetic and show how they factorize with respect to interpretability.

1 Introduction.

Q is Robinson's arithmetic and PA is Peano arithmetic (Q plus full induction schema). Prominent fragments of PA are $I\Sigma_n$ (Q plus induction for Σ_n formulas) and $B\Sigma_{n+1}$ ($I\Sigma_0$ plus collection for Σ_{n+1} formulas). It is well-known that the theories

$$I\Sigma_0, B\Sigma_1, I\Sigma_1, B\Sigma_2, \ldots I\Sigma_n, B\Sigma_{n+1}, \ldots, PA$$

from a strictly increasing hierarchy. It is customary to call the fragments containing $I\Sigma_1$ *strong* and proper subsystems of $I\Sigma_1$ *weak*. The most prominent weak fragment is $I\Sigma_0$; further prominent weak fragments result by adding to $I\Sigma_0$ an axiom postulating that a given primitive recursive function is total, notably:

$I\Sigma_0 + \Omega_n$
(Ω_n says that $\omega_n(x)$ is total, where $\omega_1(x) = x^{|x|}$ and $\omega_{n+1}(x) = x^{\omega_n(|x|)}$; $|x|$ is the length of the binary expression of x.)

$I\Sigma_0 + exp$ (exp says that 2^x is total).

I refer to the book [HP] for detailed treatment of these fragments, including the questions of interpretability (Chapter III Sect. 3-4, Chapter V Sect.5; for references to original authors see Bibliographical remarks in the book). One important topic not covered by the book is the investigation of the lattice of interpretability types of finite (or: axiomatizable) extensions of PA, due to Lindström [Li] and Švejdar [Šv]. (A generalization of their results to extensions of a given fragment remains as a task for the future.)

ACA_0 is the second order arithmetic extending PA conservatively: it has variables for numbers, variables for classes, usual operations on numbers, Σ^0_∞-comprehension, extensionality and a single induction axiom. (Note that Σ^0_0-formulas are formulas of RCA_0 whose only quantifiers are bounded number quantifiers; Σ^0_n, Π^0_n formulas result in the obvious way.) RCA_0 is a fragment of ACA_0 having only Δ^0_1-comprehension (for $\sigma \in \Sigma^0_1, \pi \in \Pi^0_1$):

$$(\forall x)(\sigma(x) \equiv \pi(x)) \rightarrow (\exists Z)(\forall x)(x \in Z \equiv \sigma(x)))$$

but has the $I\Sigma^0_1$-induction schema.

WKL_0 extends RCA_0 by the single axiom saying that each infinite dyadic tree has an infinite branch.

Note that RCA_0, WKL_0, ACA_0 are the first three of five prominent fragments of the second-order arithmetic; these five theories play a crucial role in Simpson's reverse mathematics. All necessary information on them may be found in the book [Simp]. [Sieg 85] is also very informative. We shall also consider strengthenings of RCA_0 or WKL_0 by the schemas $I\Sigma^0_{n+1}, B\Sigma^0_{n+1}$.

Our *aim* is to study interpretability among those prominent fragments. We shall attempt to prove all interpretabilities by exhibiting a simple, well comprehensible interpretation. In the case of interpretations of a second-order theory in a first-order theory we shall produce ω-interpretations, i.e. numbers and operations on numbers will be interpreted identically (absolutely).

As a result we obtain a complete classification of our prominent fragments with respect to interpretability; this will improve our understanding of the hierachy of fragments. Some corollaries will be obtained and several open problems will be formulated. We shall complete this introduction by a survey of known facts.

1.1. **Theorem.** (Wilkie, Solovay; see [HP], [Wil]).

(a) The theories $Q, I\Sigma_0, I\Sigma_0 + \Omega_1, \ldots, I\Sigma_0 + \Omega_n, \ldots$ are all mutually interpretable.

(b) $I\Sigma_0 + exp$ is not interpretable in $I\Sigma_0$ (and thus not in any of the above theories).

(c) $I\Sigma_1$ is not interpretable in $I\Sigma_0 + exp$ (this is trivial since $I\Sigma_1 \vdash Con(I\Sigma_0 + exp)$).

1.2. **Theorem.** $I\Sigma_{n+1}$ is not interpretable in $B\Sigma_{n+1}$.

Proof. $I\Sigma_{n+1}$ proves $Con(B\Sigma_{n+1})$ (see [HP] or [CHP]); if $I\Sigma_{n+1}$ were interpretable in $B\Sigma_{n+1}$ we would have $I\Sigma_1 \vdash Con(B\Sigma_{n+1}) \to Con(I\Sigma_{n+1})$ (since $I\Sigma_{n+1}$ is finitely axiomatizable) and therefore $I\Sigma_{n+1} \vdash Con(I\Sigma_{n+1})$. This would contradict Gödel's Second incompleteness theorem.

1.3. **Theorem.** RCA_0 has an ω-interpretation in $I\Sigma_1$.

(Obvious, cf. [Ign]: in $I\Sigma_1$, just take Δ_1-sets for classes in the sense of the interpretation, interpreting numbers, addition and multiplication by themselves.)

In this paper we shall produce two kinds of interpretations: an interpretation of $T + B\Sigma_n$ in T for a first-order T and an appropriate n, and an ω-interpretation of WKL_0 in $I\Sigma_1$. This will be done in Sections 2-3 respectively, Section 4 contains a summary of results and a list of problems.

A substantial part of the results was obtained during my one-month stay in San Diego (June-July 1990); I acknowledge the extremely pleasing and creative atmosphere of the workshop. I referred on the results of this paper also on the 9.International Congres of Logic, Methodology and Philosophy of Science (Uppsala, Sweden, August 1991). I greatly acknowledge a grant from the organizers that made my participation possible.

2 Interpreting collection.

We shall present an easy method of interpreting collection. It uses two things: the trivial observation that if $S \lhd (T + \varphi)$ and $S \lhd (T + \neg\varphi)$ then $S \lhd T$ (read $\lhd$ as "is interpretable in") and a modification of a model-theoretic lemma ([HP]IV.1.22) saying that if $M \models I\Sigma_0$ and M has a proper end-extension $M \prec_{e,k} K \models I\Sigma_k$ then $M \models B\Sigma_{k+1}$. Let us recall the following:

2.1 **Definition.** $J(x)$ is a T-definable cut (where $T \supseteq I\Sigma_0$) if $T \vdash J(0) \,\&\, (\forall x)(J(x) \to J(x+1)) \,\&\, (\forall x, y)(J(x) \,\&\, y < x \to J(y)))$. If J_1, J_2 are T-definable cuts then J_1 is shortes than J_2 if $T \vdash J_1(x) \to J_2(x)$. Recall Solovay's method of shortening cuts: each cut J has a shortening J_1 which is provably closed under $+, *$, (i.e. $T \vdash J_1(x) \,\&\, J_1(y) \to J_1(x + y) \,\&\, J_1(x * y)$); moreover, for each given n you may request J_1 to be closed under ω_n.

The promised modification of the above lemma follows.

2.2 **Lemma.** Let $J(x)$ be a definable cut in $I\Sigma_k$ closed under $+, *$ and if $k > 0$ assume J to be k-elementary with respect to the universe, i.e. $I\Sigma_k$ proves that for each Σ_k-formula $z \in J$ and each evaluation e of free variables of z by elements of J we have

$$Sat_{\Sigma_k}(z,e) \equiv Sat^J_{\Sigma_k}(z,e)$$

(definable satisfaction at the right hand side restricted to J). Then the interpretation given by restricting all quantifiers to J is an interpretation of $B\Sigma_{k+1}$ in ($I\Sigma_k + J$ is proper).

Proof. Imitate the proof of [HP]IV.1.22.

We shall present various applications of this lemma.

2.3 **Theorem.** $B\Sigma_1$ is interpretable in $I\Sigma_0$.

Proof. In $I\Sigma_0$ distinguish two cases: if $I\Sigma_1$ holds (Case 1, this is a single axiom, due to finite axiomatizability of $I\Sigma_1$) then interpret everything absolutely (recalling $I\Sigma_1 \vdash B\Sigma_1$); otherwise (Case 2) we define a proper cut (closed under operations, which suffices).

Subcase 1. If *exp* holds, i.e. exponentiation is total, use the Σ_1-definition of satisfaction for Σ_1 formulas and define J_1 to be the intersection of all Σ_1-cuts; since we are in Case 2, this intersection is non-empty and is a proper cut. If necessary, shorten to get a cut closed on $+, *$.

Subcase 2. If exp does not hold, let $J(x)$ say "2^x exists"; this is a proper cut closed under $+$ and may be shortened to achieve closedness under $*$.

2.4. **Theorem.** $B\Sigma_1 + exp$ is interpretable in $I\Sigma_0 + exp$.

Proof. Now we have to guarantee that $J(x)$ is closed under exponentiation. Let 2_x be the superexponential function ($2_0 = 1$, $2_{x+1} = 2^{2_x}$) and let J_0 be the domain of 2_x: $x \in J_0$ iff 2_x exists. Clearly, J_0 is a definable cut (in $I\Sigma_0 + exp$).

Case 1: $I\Sigma_1$ holds in the sense of J_0. Then interpret by J_0 (recalling that $I\Sigma_1 \vdash B\Sigma_1 + exp$).

Case 2: Otherwise J_0 has a proper shortening J_1 (reason as above); define J as the cut given by the image of J_1 in 2_x, i.e.

$$J(x) \equiv (\exists y)(J_1(y) \,\&\, x < 2^y).$$

Clearly, J is closed under exponentiation and is a proper cut; therefore it satisfies $B\Sigma_1$.

2.5. **Theorem.** For each n, $B\Sigma_1 + \Omega_n$ is interpretable in $I\Sigma_0 + \Omega_n$.

Proof: completely analogous to the proof of 2.4.

2.6 **Theorem.** For each $n \geq 1$, $B\Sigma_{n+1}$ is interpretable in $I\Sigma_n$.

Proof. We proceed in $I\Sigma_n$. For each x, let $E_n(x)$ be the least y such that, for each Σ_n-formula $\Theta = (\exists u)\nu, \Theta < x$ and each evaluation $e < x$ of its free variables, if e satisfies Θ then there is a witness a for this beneath y (i.e. e extended by a as the interpretation of u satisfies ν). Clearly, E_n is a total function (cf. [HP]IV.1.63). Define $M(0) = 0, M(x+1) = 1 + E_n(M(x))$. We cannot prove (in $I\Sigma_n$) that M is total; but we can prove that the domain J_0 of M is a cut. Thus we may continue as above: *Case 1.* If J_0 satisfies $I\Sigma_{n+1}$ then define the interpretation using J_0. *Case 2.* otherwise shorten J_0 to a proper subcut J_1 and define J by

$$J(x) \equiv (\exists y)(J_1(y) \ \& \, x < M(y)).$$

Clearly, J is a k-elementary proper cut (cf.2.2). Thus $B\Sigma_{n+1}$ holds in the sense of J.

2.7 **Remark.** Pudlák and Buss called my attention to the fact that some interpretations claimed to exist in 2.4-2.6 may be obtained by proof-theoretic methods by showing that $I\Sigma_0 + exp$ proves the corresponding relative Herbrand consistency (cf. [Pu] and Friedman's results in [Smo]). Here we have presented rather simple and transparent particular interpretations.

3 An ω-interpretation.

An interpretation of a second-order arithmetic in a first-order arithmetic is an *ω-interpretation* if first-order notions are interpreted absolutely (by themselves); thus an ω-interpretation is just given by a definition of classes (and membership). Our aim is to prove the following

3.1 **Theorem.** WKL_0 has an ω-interpretation in $I\Sigma_1$.

The proof is given below; first we shall elaborate some properties of *very low* classes in $I\Sigma_1$ (cf [HK] and [HP]). The basic means is a possibly strong version of the *low basis Theorem* in $I\Sigma_1$. (The Low Basis Theorem was proved in [JS]; Clote showed [Cl 1,2] that the proof formalizes in $B\Sigma_2$.

A version of the Low Basis Theorem was proved in $I\Sigma_1$ in the paper [HK]; further results are in [HP]. Here we need even more.).

Remark. Recall that in $I\Sigma_1$ a class X is called *piecewise coded (p.c.)* if for each x there is a finite sequence s of zeros and ones of length x such that $(\forall i < x)((s)_i = 1 \equiv i \in X)$. ($s$ is called a *piece* of X.) There is a relativized definition of satisfaction $Sat_{0,X}$ for $\Sigma_0(X)$-formulas such that

$$I\Sigma_1 \vdash (X \text{ is p.c.} \rightarrow Sat_{0,X} \text{ satisfies Tarski's truth conditions}).$$

This extends to a definition of $\Sigma_0^H(X)$-formulas where H is a function; quantifiers $(\forall x \leq H(y))$, $(\exists x \leq H(y))$ are allowed. The corresponding satisfaction is denoted by $Sat_{0,X}^H$; $I\Sigma_1$ proves the following: [if X is p.c. and H is a total Δ_1 increasing function then $Sat_{0,X}^H$ satisfies Tarski's truth conditions; moreover, each $\Sigma_0^H(X)$ class is p.c.]

In ($I\Sigma_1$), a class is $\Sigma_0^*(\Sigma_1)$ if it is a $\Sigma_0^H(X)$-class for some Σ_1-class X and some Δ_1 total increasing function H. We have a uniform definition of satisfaction for all $\Sigma_0^*(\Sigma_1)$ classes; each $\Sigma_0^*(\Sigma_1)$-class is p.c. and therefore we have induction for all $\Sigma_0^*(\Sigma_1)$ formulas. (Using satisfaction, induction for these formulas is just a single statement of $I\Sigma_1$; needless to say, it implies each particular instance of the corresponding axiom schema.)

3.2 **Definition** ($I\Sigma_1$). A $\Sigma_0^*(\Sigma_1)$ class X is *very low* (or LL_1) if each $\Sigma_1(X)$-class is a $\Sigma_0^*(\Sigma_1)$-class; thus $\Sigma_1(X) \subseteq \Sigma_0^*(\Sigma_1)$. (Note that the definition is meaningful since X is p.c.)

3.3 **Lemma ($I\Sigma_1$).** For each $X \in LL_1$, $\Delta_1(X) \subseteq LL_1$, i.e. if a class is recursive in a LL_1-class it is itself LL_1. (See [HP]I.2.79.)

Caution. Note that we do not claim that LL_1 classes are closed under Turing join; probably they are not, for the analogous family of low Δ_2 classes is not closed under Turing join [Pos-Rob]. I have not tested if the proof of [Pos-Rob] formalizes in $I\Sigma_1$ as a theorem on LL_1-classes.

3.4 **Theorem ($I\Sigma_1$).** Each LL_1 infinite dyadic tree has an infinite LL_1 branch. (See [HP]I.3.10.)

3.5 **Remark.** (1) As usual, "dyadic" may be replaced by "LL_1-estimated", meaning that there is an $F \in LL_1$ such that, for each x and each $s \in T$ of length x we have $s \leq F(x)$.

(2) We shall strengthen the theorem even more. We state the unrelativized form as a lemma.

3.6 **Lemma** ($I\Sigma_1$). There is a $B^* \in LL_1$ such that each Δ_1 dyadic infinite tree has an infinite branch $B \in \Delta_1(B^*)$.

Proof. We shall replace all Δ_1 trees by a single Π_1 tree as follows: Assume that each number codes a (possibly empty) Turing machine. For each m let T^{Σ}_m be the class of all s such that there is an accepting computation z of m with the input s; similarly, let T^{Π}_m be the set of all s such that all computations z on m with the input s are accepting. Clearly, $T^{\Sigma}_m \in \Sigma_1$, $T^{\Pi}_m \in \Pi_1$, $T^{\Sigma}_m \subseteq T^{\Pi}_m$ and the equality holds iff m defines a total function on strings. Let (i,j) be the usual pairing function; if s is a string of length $x > (i,j)$ then s_i is the sequence $((s)_{(i,0)}, (s)_{(i,1)}, \ldots, (s)_{(i,k)})$ where k is maximal such that $(i,k) < x$; $(s)_i \dagger j$ is $(s)_{(i,0)}, \ldots, (s)_{(i,j)}$ (restriction of s_i to j). Let $t \in\in U$ (t is a *strong element* of U) mean that each initial segment of t is in U. Let, for each string s,

$$s \in T \equiv (\forall (i,j) < lh(s))((\exists t \in\in T^{\Sigma}_i, lh(t) = j) \rightarrow (s_i \dagger j) \in T^{\Pi}_i).$$

Observe that T is Π_1, T is a dyadic tree and is infinite: think of s as $(s_0, \ldots, s_p)$ and observe that to show that there is an $s \in T$ of arbitrary given length it suffices to show for each i that there is an arbitrarily long s_i satisfying the condition

$$(\forall j < lh(s_i))((\exists t \in\in T^{\Sigma}_i, lh(t) = j) \rightarrow (s_i) \dagger j) \in T^{\Pi}_i.$$

But this is evident: either T^{Σ}_i has arbitrarily long strong elements (and then all of them are in T^{Π}_i) or there is a maximal x such that T^{Σ}_i has a strong element s_0 of length x; then any s prolonging s_0 satisfies the above condition. Evidently if a total function Z with $range(Z) \subseteq \{0,1\}$ is an infinite branch through T and m is a total Turig machine defining a Δ_1 infinite dyadic tree then Z_i defined by $Z_i(j) = Z((i,j))$ is an infinite branch through T_m. Thus it remains to prove that T has an infinite branch $B^* \in LL_1$. This follows from the following

3.7 **Lemma** ($I\Sigma_1$). Each infinite dyadic Π_1 tree has in infinite LL_1 branch.

Proof. One can check that the proof of the low basis theorem for Δ_1 trees works also for Π_1 trees as it stands. Alternatively, one may easily show that for each Π_1 tree T there is a Δ_1 tree T' having the same infinite branches (this was observed by Pudlák).

3.8 **Theorem** ($I\Sigma_1$). For each $X \in LL_1$ there is a set $B^*(X) \in LL_1$ such that each infinite dyadic tree $T \in \Delta_1(X)$ has an infinite branch $B \in \Delta_1(B^*(X))$.

Proof by relativization.

3.9 **Construction.** Observe that the proof of 3.8 gives a definition (in $I\Sigma_1$) of an operation B^* that assigns to each $X \in LL_1$ the corresponding class $B^*(X)$. We can start with $\emptyset$ and investigate sequences q such that $(q)_0 = \emptyset$, for each $i < lh(q)$, $(q)_i \in LL_1$ and if $i > 0$ then $(q)_i = B^*((q)_{i-1})$. For a moment, call each such sequence a *chain.* Let $i \in J$ iff there is a unique chain of length i. Clearly, J is a definable cut in $I\Sigma_1$; we define

$$Class(X) \equiv (\exists i \in J)(\text{for the unique chain } q = (B_0, \ldots, B_i),\ X \in \Delta_1(B_i)).$$

Furthermore, put $Number(x) \equiv (x = x)$; $\in$ is the membership in the sense of LL_1-classes.

3.10 **Theorem.** The predicates $Number$, $Class$, $\in$ together with $S, +, *, \leq, 0$ define an ω-interpretation of WKL_0 in $I\Sigma_1$.

Proof. We verify the axioms of WKL_0 in $I\Sigma_1$. Note that each class X is a particular LL_1-class; thus we have induction for X. Moreover, if $Class(X_1), \ldots, Class(X_n)$ and $Z \in \Delta_1(X_1, \ldots, X_n)$ then for some $x \in J$ we have $Z \in \Delta_1(B_x)$; thus $Class(Z)$. This proves Δ_1-comprehension. Since we have $I\Sigma_1(LL_1)$, we have Σ^0_1-induction. And clearly, if $Class(T)$ and T is and unbounded dyadic tree then, for some $x \in J$, $T \in \Delta_1(B_x)$ and hence T has an unbounded branch $B \in \Delta_1(B_{x+1})$; $Class(B)$ is immediate. This completes the proof.

3.11. **Corollary.** (1) (Harrington's theorem.) WKL_0 is a conservative extension of $I\Sigma_1$.

(2) (cf. [Ign]). WKL_0 has only polynomial speed-up over $I\Sigma_1$.

3.12 **Remark.** Harrington's result was obtained later but independently by Ratajczyk, see [Rat]. A proof can be found in [Simp] and in [Sieg 87]; the last paper also contains generalization to other fragments. Below we generalize 3.10 and obtain conservation results of [CHP] as corollaries. (Note that in [CHP], the reference to [Sieg 87] was incomplete and the reference to [Sieg 85] was erroneously omitted.)

3.13 **Theorem.** For each $n \geq 1$, the interpretation of 3.10 is an interpretation of

$$\text{(a)} \qquad WKL_0 + B\Sigma^0_{n+1} \text{ in } B\Sigma_{n+1},$$
$$\text{(b)} \qquad WKL_0 + I\Sigma^0_{n+1} \text{ in } I\Sigma_{n+1}.$$

Proof. One easily shows that, for each $n \geq 1$, $I\Sigma_1$ proves $\Sigma_{n+1}(LL_1) = \Sigma_{n+1}$; Σ^0_{n+1} classes of the interpretation are some $\Sigma_{n+1}(LL_1)$ classes, thus some Σ_{n+1} classes.

3.14 **Corollary.** 3.12 generalizes to the pairs of theories in 3.13, thus: $WKL_0 + B\Sigma^0_{n+1}$ is conservative and has only a polynomial speed up over $B\Sigma_{n+1}$; similarly for B replaced by I.

4 Summary of results; some problems.

Fig.1 visualizes the factorization of investigated theories with respect to degrees of interpretability.

In the following problems, Ω_∞ is the set of all axioms $\Omega_n, (n \leq 1)$; PRA is Primitive Recursive Arithmetic.

(1) Is $B\Sigma_1 + \Omega_\infty$ interpretable in $I\Sigma + \Omega_\infty$?

(2) Is $I\Sigma_0 + \Omega_\infty$ interpretable in $I\Sigma_0$? (Note that there is no interpretation of $I\Sigma_0 + \Omega_\infty$ in $I\Sigma_0$ defined by a cut, see [Pu] Corollary 3.7.)

(3) Is $PRA + B\Sigma_1$ interpretable in PRA?

(4) Does WKL_0 have an ω-interpretation in RCA, i.e. can we define in RCA_0 an interpretation of WKL_0 by interpreting first order notions by themselves and defining new classes as all old classes plus possibly some further classes definable from the old ones?

(5) Is WKL^*_0 (as WKL_0 but with $I\Sigma^0_1$ replaced by $I\Sigma^0_0 + exp$) interpretable in $I\Sigma_0 + exp$? (Cf. [Simp].)

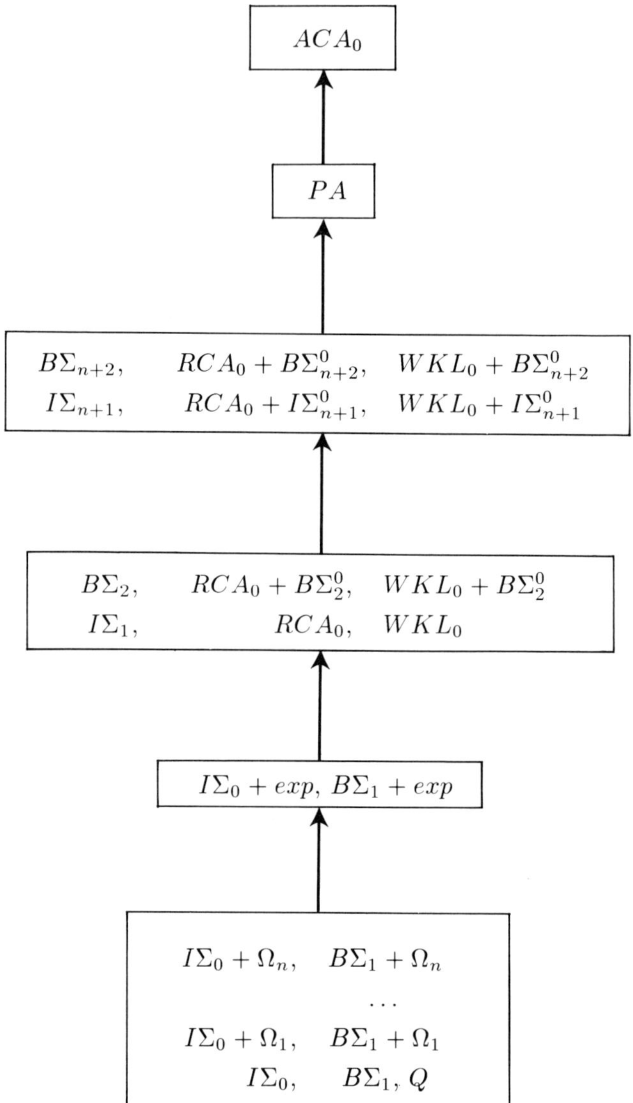
ACA_0
PA
$B\Sigma_{n+2}$, $RCA_0 + B\Sigma^0_{n+2}$, $WKL_0 + B\Sigma^0_{n+2}$
$I\Sigma_{n+1}$, $RCA_0 + I\Sigma^0_{n+1}$, $WKL_0 + I\Sigma^0_{n+1}$
$B\Sigma_2$, $RCA_0 + B\Sigma^0_2$, $WKL_0 + B\Sigma^0_2$
$I\Sigma_1$, RCA_0, WKL_0
$I\Sigma_0 + exp$, $B\Sigma_1 + exp$
$I\Sigma_0 + \Omega_n$, $B\Sigma_1 + \Omega_n$
$\dots$
$I\Sigma_0 + \Omega_1$, $B\Sigma_1 + \Omega_1$
$I\Sigma_0$, $B\Sigma_1$, Q

Fig. 1

References

[Cl1] Clote P.: Applications of the Low Basis Theorem in arithmetic. In: Recursion theory week Oberwolfach 1984, Lect. Notes Math 1141, Springer-Verlag 1985, ed. H.D. Ebbinghaus, G.H. Müller, G. E. Sacks, pp. 65-88.

[Cl2] Clote P.: Partition relations in arithmetic. In: Methods in Mathematical Logic (Proc. 6th Latin American Symposium on Mathematical Logic, Caracas 1983), Lecture Notes in Mathematics 1130, Springer-Verlag 1985, ed. C. A. DiPrisco, pp. 32-68.

[CHP] Clote P., Hájek P., Paris J.: On some formalized conservation results in arithmetic, Arch. Math. Log. 30 (1990), 201-218.

[HK] Hájek P., Kučera A.: On recursion theory in $I\Sigma_1$. Journ. Symb. Log. 54 (1989) 576-584.

[HP] Hájek P., Pudlák P.: Metamathematics of first-order arithmetic, Springer-Verlag (in print).

[Ign] Ignjatović, A.: Fragments of first and second order arithmetic and lengths of proofs. Ph.D. Thesis, Berkeley 1990.

[JS] Jockusch C. G., Soare R. I.: Π_1-classes and degrees of theories. TAMS 173 (1972) 33-56.

[Li] Lindström P.: Some results on interpretability, Proc. 5th Scand. Symp. (Jensen et al., ed.) Aalborg 1979, 329-361.

[Pos-Rob] Posner D. B., Robinson R. W.: Degrees joining to $0'$. Journ. Symb. Log. 46 (1981) 714–722.

[Pu] Pudlák P.: Cuts, consistency statements and interpretations, Journal Symb. Log. 50 (1985), 423-441.

[Rat] Ratajczyk Z.: abstract, Journ. Symb. Log. 50 (1985) 273-274.

[Sieg85] Sieg W.: Fragments of arithmetic, Annals of pure and appl. logic 28 (1985) 33-71.

[Sieg87] Sieg W.: Provably recursive functionals of theories with König's lemma, Rend. Sem. Mat. Univ. Politechn. Torino, Fasc. Speciale 1987, 75-92

[Simp] Simpson S.: Subsystems of second order arithmetic, book in preparation. (Chapter IX: Non-ω-models, Pennsylvania State University, Dept. of Math. Research Report No 88013, 1988).

[Smo] Smoryński C.: Non-standard models and related developments. In: Harvey Friedman's research in foundations of mathematics, North Holland 1985, 179-229.

[Šv] Švejdar V.: Degrees of interpretability, Comm. Math. Univ. Carol. 19 (1978), 789-813.

[Wi] Wilkie A.: On sentences interpretable in systems of arithmetic, Logic Colloquium '84, North-Holland 1986, pp. 331-337.

Petr Hájek
Institute of Computing and Information Science
ČSAV
Pod vodarenskou vezi 2
182 07 Prague, Czechoslovakia
e-mail: HAJEK@CSPGCS11.BITNET

Abbreviating proofs using metamathematical rules

Petr Hájek Franco Montagna Pavel Pudlák

0 Introduction.

There are examples of sentences which are provable in theory T but which have only very long proofs. All the known examples have the property that at the same time there is a simple metamathematical argument which shows that they are provable; thus, they have short proofs in some stronger theory whose soundness is as acceptable as the soundness of the original one. Such a phenomenon, which one calls speed up by provability, was first studied by Parikh. He proved in [Par71] that, if T is a recursively enumerable extension of Peano Arithmetic PA, then there is a sentence φ such that the formalization of "φ is provable in T" has a much shorter proof than φ. This result is generalized in [DM89] and in [Mon90].

In order to learn more about the speed-up-phenomenon, one should consider the following questions:

1) How much can we abbreviate proofs by proving only the provability of a sentence?

2) Are there mathematically interesting examples of such sentences?

3) What are other metamathematical rules that can abbreviate proofs and how much can they abbreviate proofs?

Before describing our results, we have to recall some definitions and introduce notation.

We shall work in fragments of PA $I\Sigma_n$, $n \leq 1$ (these are PA with the induction schema restricted to Σ_n formulas) and theories T such that $I\Delta_0 + \Omega_1 \subseteq T \subseteq I\Delta_0 + \mathrm{Exp}$ (In the following, $I\Delta_0$ denotes Robinson's Ω plus induction for Δ_0 formulas, thus $I\Delta_0$ is $I\Sigma_0 + \mathrm{Exp}, I\Delta_0 + \Omega_1, I\Delta_0 +$ Supexp are $I\Delta_0$ plus the totality of $\mathrm{Exp}(x) = 2^x$, of the function $\omega_1(x) = 2^{|x|^2}$, where $|x|$ denotes the binary length of x, and of the superexponential function 2_x defined inductively by $2_0 = 1, 2_{x+1} = 2^{2_x}$ respectively; cf e.g. [HP92] or [WP87] for the definitions).

$\mathrm{Con}_T(x)$ denotes the formalization of "there is no proof of contradiction in T of size (i.e., number of symbols)$\leq x$"; note that this formula very much depends on the particular representation of the axioms. We shall always assume that the axiom set is in the complexity class NP, since this ensures some good properties, cf [Pud86]. We shall use $nm(x)$ for the dyadic numeral of x, $\ulcorner\varphi\urcorner$ for the numeral of the Gödel number of φ, $\mathrm{Proof}_T(x, y)$ for the formalization of "x codes a proof of the formula y in T", $\mathrm{Prov}_T(y)$ for $\exists x \mathrm{Proof}_T(x, y)$. We shall use the sign $\vdash^n$ for provability by a proof of size $\leq n$.

We also consider the modal logic of provability L, with modality $\Box$, whose axioms and rules are stated in Section 1. An *interpretation* of L in an NP axiomatized arithmetical system $T \supseteq I\Delta_0 + \Omega_1$ (shortly: a T *interpretation*) is a mapping * sending modal formulas into sentences which commutes with Boolean connectives and such that for each modal formula A, $(\Box A^* \equiv \mathrm{Prov}_T(\ulcorner A^*\urcorner)$. If A, B are modal formulas, adding the *modal rule* A/B to T means that, for each T interpretation *, we are allowed to derive B^* whenever we have already a proof of A^*. Of course, we restrict ourselves to rules which can be used (i.e., such that, for some interpretation *, T proves A^*) but do not produce new theorems (these rules will be called *reasonable*). We are now ready to state our results:

1) We shall show a superexponential speed up by provability for the sentences $\mathrm{Con}_T(2_n)$ and "$2_{2_{nm(n)}}$ exists" (the former for $T \supseteq I\Delta_0 + \Omega_1$, the latter for $T = I\Delta_0 + \Omega_1$). These results are easy corollaries of the result [Pud86]; they are interesting since they match the upper bound in the case of $T = I\Delta_0 + \Omega_1$.

2) We shall show also a superexponential *upper* bound for speed up by provability for $T = I\Delta_0 + \Omega_1$.

3) Given a reasonable rule A/B such that $L \not\vdash \Box A \rightarrow \Box B$ and a theory $T \supseteq I\Delta_0 + \Omega_1$ proving the schema $\varphi \rightarrow \mathrm{Prov}_T(\ulcorner\varphi\urcorner)$ for all Σ_1 sentences φ, we shall show that for infinitely many T interpretations *, A^* has a superexponentially shorter proof than B^*, but that proofs of B^* are always bounded (w.r. to proofs of A^*) by a finite iteration of the superexponential function if $T \subseteq I\Delta_0 + \mathrm{Exp}$.

4) For reasonable rules in general, and for T as in 3), $T \subseteq I\Delta_0 + \mathrm{Exp}$, we can only prove a supersuperexponential upper bound; however, we shall prove that, at least in some cases, this upper bound can not be substantially improved.

5) We shall prove that any reasonable modal rule can be polynomially simulated by the rule $\Box p/p$ provided that $T \supseteq I\Delta_0 + \Omega_1$ and T proves the schema $\varphi \rightarrow \mathrm{Prov}_T(\ulcorner\varphi\urcorner)$ for all Σ_1 sentences φ.

6) We shall prove some analogues of the previous results for stronger theories as $I\Sigma_n$ (n=1,2,3,... etc) or PA. In particular, we prove that a

lower bound for speed up by provability in $I\Sigma_n$ (PA respectively) is given by a natural envelope for $I\Sigma_n$ (PA) provably total functions; this result is extended to consistency statements such as in 1. An analogue of 3) also holds, but with the superexponential function replaced by a natural envelope for $I\Sigma_n$ (PA respectively) provably total functions.

Results 3),4),5) and 6) are a nice example of application of modal analysis of theories. Modal logic enables us to investigate an interesting class of metamathematical rules which can be used to abbreviate proofs. Of course, there are other rules which can not be analyzed in this way. For example, we get a superexponential speed up by taking predicative extensions of theories as PA and ZF, cf [Pud86].

Acknowledgement. The last named author would like to acknowledge the excellent conditions provided by the University of Siena under which a part of results was proved.

1 Preliminaries.

In this Section, we recall some definitions and results which will be used throughout the rest of the paper. In the following, *even without explicit mention, we assume that each theory taken into consideration is NP axiomatized and Σ_1 sound.*

Definition 1. L is the modal logic characterized by the following axioms and rules:
Axioms of L:

1) Tautologies
2) $\Box(A \to B) \to (\Box A \to \Box B)$
3) $\Box(\Box A \to A) \to \Box A$

Rules of L: Modus Ponens: $A, A \to B/B$ and Necessitation: $A/\Box A$.

S is the modal logic whose axioms are theorems of L plus the schema $\Box A \to A$. The only rule of S is modus ponens; in particular, S is not closed under $A/\Box A$.

Definition 2 An *interpretation* L in an arithmetical system $T \supseteq I\Delta_0 + \Omega_1$ (shortly: a T-interpretation) is a map * sending modal formulas into T

sentences which commutes with Boolean connectives and such that for each modal formula $A, \Box A^* \equiv \mathrm{Prov}_T(\ulcorner A^* \urcorner)$.

In the following, we use sometimes T-interpretations * such that for every modal formula A, A^* is not a sentence, but a formula in the free variable x (these interpretations will be called *open T interpretations*). If * is an open T-interpretation and $A^* \equiv \varphi(x)$, then $\Box A^* \equiv \mathrm{Prov}_T(\ulcorner \varphi(nm(x)) \urcorner)$, where $\varphi(nm(x))$ is the result of substituting $nm(x)$ for x in $\varphi(x)$. Moreover, we use $A^*(n)$ to denote $\varphi(nm(n))$, where $\varphi(x) \equiv A^*$.

In the sequel, when we use the word "T-interpretation", if we do not explicitly mention that it is open, we mean "T-interpretation according to Definition 2" (hence its values are always sentences).

Proposition 1. (Cf. [Smo85].) Let $T \supseteq I\Delta_0 + \Omega_1$, let C, D be modal formulas, p be a propositional variable occuring in C; let, here and in the following, $\boxdot\, q$ be an abbreviation for $q \wedge \Box q$; then, for each (possibly open) T interpretation *, one has:

$$T \vdash (\boxdot\ (p \leftrightarrow D)^* \rightarrow (C(p) \leftrightarrow C(D))^*.$$

Proposition 2. (Cf. [Sol76] and [DJM91]) If $T \supseteq I\Delta_0 + \Omega_1$ and $T \vdash \varphi \rightarrow \mathrm{Prov}_T(\ulcorner \varphi \urcorner)$ for all sentences $\varphi \in \Sigma_1$ then, for every modal formula A, one has:

1) $L \vdash A$ iff for all *, $T \vdash A^*$,
2) $S \vdash A$ iff for all *, A^* is true the standard model.

The assumption that $T \vdash \varphi \mathrm{Prov}_T(\ulcorner \varphi \urcorner)$ for all sentences $\varphi \in \Sigma_1$ is not needed for the left-to-right implications.

In [Mon90], the notion of a reasonable modal rule is introduced and the problem of shortening of proofs by modal rules is investigated. If A, B are modal formulas, adding the rule A/B to T means that for each T interpretation * one can derive B^* from A^*, provided that A^* has already been deduced. The rule can be used repeatedly with different interpretations.

A rule A/B is called T *sound* iff for every T-interpretation *, $T \vdash A^*$ implies $T \vdash B^*$ i.e. iff, for any *, $\mathrm{Prov}_T(\ulcorner A^* \urcorner) \rightarrow \mathrm{Prov}_T(\ulcorner B^* \urcorner)$ is true in the standard model. If T is as in Proposition 2, then this is the case iff $\vdash \Box A \rightarrow \Box B$. The rule A/B is called T *reasonable* if it is T sound and it can be used, i.e., for some $^*, T \vdash A^*$. If T is as in Proposition 2, this last condition is equivalent to the consistency of $S + \Box A$ (Cf.[Mon90]).

Throughout the whole paper, we make often use of the following results.

Proposition 3. (Cf. [Mon90]) Let $T \supseteq I\Delta_0 + \Omega_1$; then:
(i) if $L \vdash \Box A \to \Box B$, then the rule $R : A/B$ shortens T proofs polynomially; (this means: there is a polynomial $p(x)$ such that for each proof Π of some formula φ from the axioms of T by means of the rule R, there is a T proof Π' of φ in the usual sense such that size $(\Pi') \le p($ size $(\Pi)))$. Moreover, there is a polynomial $p(y)$ such that, for each T-interpretation * and each T proof Π of A^*, there is a T proof Π' of B^* whose size is $\le p(\text{size}(\Pi))$.
(ii) if $L \vdash A(p_1 \dots, p_n)$, then there is a polynomial $q(x_1, \dots, x_n)$ such that, for each T-interpretation *, there is a T proof of $A(p_1, \dots, p_n)^*$ whose size is $\le q(\text{size}(p_1^*), \dots, \text{size}(p_n^*))$.
(iii) If $T \vdash \varphi \to \text{Prov}_T(\lceil\varphi\rceil)$ for every sentence $\varphi \in \Sigma_1, R : A/B$ is a T reasonable rule and $L \nvdash \Box A \to \Box B$, then for each multiexponential function $f(y)$, there are a T-interpretation * and a T proof Π of A^*, such that each T proof of B^* has size $> f(\text{size}(\Pi))$. So, R shortens proofs more than multiexponentially.

Proposition 4. (Cf. [Par71].) Let T be one of $I\Delta_0$, $I\Delta_0 + Exp$, $I\Delta_0 + \Omega_1$, $I\Delta_0 +$Supexp (the last three formulated in a language with symbols for exponentation, for $\omega_1(x)$ and for superexponentation respectively). Then, for each $\Delta_0(T)$-formula $\varphi(x, y)$ (i.e., for each formula whose quantifiers are bounded by a term of T), if $T \vdash \forall x \exists y \varphi(x, y)$, there is a term $t(x)$ of the language of T such that $T \vdash \forall x \exists y \le t(x) \varphi(x, y)$.

Proposition 5 (cf. [Pud86]). There is a polynomial $p(x)$ such that, for each n, the formula $\exists x(2_{nm(n)} = x)$ has a proof in $I\Delta_0 + \Omega_1$ of size $\le p(n)$.

Proposition 6 (cf. [Pud86]). For each $T \supseteq I\Delta_0 + \Omega_1$ there is a positive rational number δ such that for every n, each T proof of $\text{Con}_T(n)$ has size $\le n^\delta$.

Proposition 7 (Cf.[Mon90]) Let T be as in Proposition 2. Let A, B be modal formulas such that A/B is a reasonable rule, and $L \nvdash \Box A \to \Box B$. Then. there are two T-interpretations * and $^\circ$ such that $T \vdash A^*$ (hence, by our assumption $T \vdash B^*$) and $T \vdash \text{Prov}_T(\lceil B^\circ \rceil) \to A^\circ, T \nvdash B^\circ$.

In the rest of the paper, in order to simplify notation, reference to the theory T will be sometimes omitted; thus, we shall often write: "proof" or "interpretation" instead of "T-proof", "T-interpretation", respectively.

2 The relation between proofs of φ and proofs of $\mathbf{Prov}_T(\ulcorner\varphi\urcorner)$.

We start with a simple corollary to Propositions 5 and 6.

Theorem 1. Let $T \supseteq I\Delta_0 + \Omega_1$. Then

(a) There is a positive rational number δ such that, for all n, each proof of $\mathrm{Con}_T(2_n)$ has size $\geq (2_n)^\delta$.
(b) There is a polynomial $p(x)$ such that, for all n, there is a proof of $\mathrm{Prov}_T(\ulcorner\mathrm{Con}_T(2_{nm(n)})\urcorner)$ of size $\leq p(n)$.

Proof. We start from the following Lemma.

Lemma 1. $I\Delta_0 + \Omega_1 \vdash \forall x \forall y[y = 2_x \rightarrow Prov_T(\ulcorner nm(y) = 2_{nm(x)}\urcorner)]$.
The proof is left to the reader as an exercise.

Proof of claim (a). It follows from the proof of Lemma 1 (or even by Lemma 1 and Proposition 4) that for every m, n such that $2_n = m$, there is a proof of $2_{nm(n)} = nm(m)$ of size polynomial in m. Now a proof of $\mathrm{Con}_T(m)$ is easily obtained from a proof of $\mathrm{Con}_T(2_n)$ and one of $2_{nm(n)} = nm(m)$, so the size of the shortest proof of $\mathrm{Con}_T(m)$ differs from the size of the shortest proof of $\mathrm{Con}_T(2_n)$) by a polynomial on $|m|$ at most. The claim follows from Proposition 6.

Lemma 2. $I\Delta_0 + \Omega_1 \vdash \forall z \mathrm{Prov}_T(\ulcorner\mathrm{Con}_T(nm||z||)\urcorner)$.

Proof of Lemma 2. Clearly, it suffices to prove:

$$I\Delta_0 + \Omega_1 \vdash \mathrm{Con}_T \rightarrow \forall z \mathrm{Prov}_T(\ulcorner\mathrm{Con}_T(nm||z||)\urcorner).$$

Each proof of size $\leq ||z||$ has Gödel number bounded by $p(|z|)$ for some polynomial $p(x)$; therefore, it suffices to prove:

$$I\Delta_0 + \Omega_1 \vdash \mathrm{Con}_T \rightarrow \mathrm{Prov}_T(\ulcorner\forall u \leq p(|z|)(\neg\ \mathrm{Proof}\ _T(u, \bot))\urcorner);$$

but this is easy, since

$$I\Delta_0 + \Omega_1 \quad \vdash \quad \mathrm{Con}_T \to \forall u \le p(|z|)$$
$$(\neg \text{ Proof }_T(u, \bot)) \to \mathrm{Prov}_T(\ulcorner \forall u \le p(|z|)(\neg \mathrm{Proof}_T(u, \bot))\urcorner)$$

(the last by the provability of Σ_1^b completeness in $I\Delta_0 + \Omega_1$ and by the fact that the formula $\forall u \le p(|z|)\neg \mathrm{Proof}_T(u, \bot))$ is Σ_1^b). This completes the proof of Lemma 2.

We go on with the proof of Claim (b). Let $T \vdash^n \varphi$ mean that φ has a proof in T of size $\le n$, and let $T \vdash^{n^{0(1)}} \varphi(nm(n))$ mean that there is a polynomial $q(x)$ such that, for each n, $T \vdash^{q(n)} \varphi(nm(n))$. By Lemma 1 and Proposition 5, we get:

$$I\Delta_0 + \Omega_1 \vdash^{n^{0(1)}} \exists z \mathrm{Prov}_T(\ulcorner nm||z|| = 2_{nm(n)}\urcorner)].$$

Thus, by Lemma 2,

$$I\Delta_0 + \Omega_1 \vdash^{n^{0(1)}} \exists z \mathrm{Prov}_T(\ulcorner nm||z|| = 2_{nm(n)} \wedge \mathrm{Con}_T(nm||z||)\urcorner).$$

Therefore,

$$I\Delta_0 + \Omega_1 \vdash^{n^{0(1)}} \exists u \mathrm{Prov}_T(\ulcorner nm(u) = 2_{nm(n)} \wedge \mathrm{Con}_T(nm(u))\urcorner)$$

and

$$I\Delta_0 + \Omega_1 \vdash^{n^{0(1)}} \forall u \forall v[\mathrm{Prov}_T((\ulcorner v = 2_{nm(n)} \wedge u = 2_{nm(n)}) \to u = v)\urcorner)].$$

It follows $I\Delta_0 + \Omega_1 \vdash^{n^{0(1)}} \mathrm{Prov}_T\ulcorner \forall u(u = 2_{nm(n)} \to \mathrm{Con}_T(nm(u)))\urcorner$. This completes the proof of Theorem 1.

We now present simpler formulas whose provability is superexponentially easier to prove (in $I\Delta_0 + \Omega_1$) than the formulas themselves.

Definition 3. A $\Delta_0(\omega)$ *proof in* $I\Delta_0 + \Omega_1$ is a proof in the sequent calculus which uses the induction rule

$$\frac{\Gamma, \varphi(a) \Longrightarrow \varphi(S(a)), \Delta}{\Gamma, \varphi(0) \Longrightarrow \varphi(t), \Delta} \quad \varphi \in \Delta_0(\omega)$$

(where of course, S is the symbol for successor) and has only $\Delta_0(\omega)$ formulas in the cut rule. We shall consider proofs in tree form only.

It follows from the definition that a $\Delta_0(\omega)$ proof of a $\exists\Delta_0(\omega)$ formula Φ contains only $\Delta_0(\omega)$ formulas and (several occurences of) Φ.

Lemma 3. Let $t(x_1, \ldots, x_n)$ be a term of size n in the language of $I\Delta_0+\Omega_1$; then $\mathbf{N} \models t(x_1, \ldots, x_n) \leq 2^{(2+|\max(x_1,\ldots,x_n)|)^{2^n}}$.

Proof. . By induction on n. For $n = 1$, the claim is clear; as for the induction step, assume that the claim holds for n and let t be a term of size n; there are then terms s, u of size $< n$ such that either $t = S(s), t = s+u$ or $t = s \cdot u$ or $t = \omega_1(s)$; we shall consider only the last case and leave the first three to the reader. Using the induction hypothesis, we get: $t(x_1, \ldots, x_n) = \omega_1(s(x_1, \ldots, x_n)) \leq \omega_1(2^{(2+|\max(x_1,\ldots,x_n)|)^{2^n}})^2 = 2^{(2+|\max(x_1,\ldots,x_n)|)^{2^{n+1}}}$.

Definition 4. Let $\Sigma \equiv \Gamma(a, \mathbf{b}) \Longrightarrow \Delta(a, \mathbf{b}), \exists y \varphi(a, y)$ (where $\mathbf{b} \equiv (b_1, \ldots, b_n)$) be a sequent with $\Gamma, \Delta \subseteq \Delta_0(\omega), \varphi \in \Delta_0(\omega)$. A *witnessing function* for Σ is a function f such that $\mathbf{N} \models \Gamma(a, \mathbf{b}) \Longrightarrow \Delta(a, \mathbf{b}), \varphi((a, f(a, \mathbf{b}))$.

Lemma 4. Suppose that a sequent Σ as in Definition 4 is provable in $I\Delta_0 + \Omega_1$ by a $\Delta_0(\omega)$ proof of size $\leq n$. Then, there is a witnessing fuction f for Σ such that $\mathbf{N} \models f(a, \mathbf{b}) \leq 2^{(2+|\max(a,b_1,\ldots,b_n)|)^{2^n}}$.

Proof. By induction on the length l of the proof of Σ; if $l = 1$, the claim is trivial. Let us do the induction step:

Case 1. The last rule is $\exists R$:

$$\frac{\Gamma(a, \mathbf{b}) \Longrightarrow \Delta(a, \mathbf{b})), \varphi(a, s)}{\Gamma(a, \mathbf{b}) \Longrightarrow \Delta(a, \mathbf{b}), \exists y \varphi(a, y))}$$

Simply apply Lemma 3 to s; if $\exists y \varphi((a, y)$ is already in $\Delta(a, \mathbf{b})$, take the maximum between the bound given by Lemma 3 and the bound given by the induction hypothesis.

Case 2. The last rule is the cut rule:

$$\frac{\Gamma(a, \mathbf{b}') \Longrightarrow \Delta(a, \mathbf{b}'), \exists y \varphi(a, y), \psi(\mathbf{b}) \qquad \psi(\mathbf{b}), \Gamma, (a, \mathbf{b}') \Longrightarrow \Delta(a, \mathbf{b}'), \exists y \varphi(a, y)}{\Gamma(a, \mathbf{b}) \Longrightarrow \Delta(a, \mathbf{b}), \exists y \varphi(a, y)}$$

where $\mathbf{b}'$ is a (possibly proper) subset of $\mathbf{b}$ and $\psi \in \Delta_0(\omega)$. By the induction hypothesis, we have a bound for the upper sequents in terms of $a, \mathbf{b}$). So, we can choose the supermum of the two bounds and substitute 0 for all variables in $\mathbf{b} - \mathbf{b}$ in it, thus getting a bound for the lower sequent.

Case 3. The last rule is the induction rule:

$$\frac{\Gamma(a,\mathbf{b}), \psi(b_i) \Longrightarrow \psi(s(b_i)), \Delta(a,\mathbf{b}), \exists y(a,y)}{\Gamma(a,\mathbf{b}), \psi(0) \Longrightarrow \psi(t), \Delta(a,\mathbf{b})), \exists y\varphi(a,y)}$$

where $\mathbf{b} = \mathbf{b}' \cup \{b_i\}$. Let f be a witnessing function for the upper sequent and assume that f is increasing; a witnessing function f' for the lower one is obtained by substituting for b_i in $f(a,\mathbf{b})$ a number m with $0 \leq m < t$ such that $\mathbf{N} \models \psi(m) \wedge \neg\psi(S(m))$ if such a m exists and 0 otherwise; So, we can substitute t in $f(a,\mathbf{b})$ and we get a bound $2^{2+|\max(a,b_1,\ldots,b_{i-1},t,b_{i+1},\ldots,b_n))|^{2^n}}$. Let k denote the size of t; the same argument as in Lemma 3 shows that the value of the last expression is bounded by $2^{2+|\max(a,b_1,\ldots b_{i-1},b_{i+1},\ldots,b_n)|)^{2^n+k}}$. This satisfies our requirements, because the size of the proof increases at least by k.

Case 4. The last rule introduces a bounded quantification, e.g.

$$\frac{b_i \leq t, \Gamma(a,\mathbf{b}) \Longrightarrow \Delta(a,\mathbf{b}), \psi(b_i), \exists y\varphi(a,y)}{\Gamma(a,\mathbf{b}), \forall z \leq t\psi(z), \exists y\varphi(a,y)}$$

Substitute t for b_i in the bound for the upper sequent and argue as in the previous case.

The remaining cases are easy and therefore left to the reader.

Corollary 1. Let $\varphi \in \Delta_0(\omega)$; suppose $I\Delta_0 + \Omega_1 \vdash^n \exists y\varphi(a,y)$ by a $\Delta_0(\omega)$ proof; then $\mathbf{N} \models \forall x \exists y \leq 2^{(2+|x|)^{2^n}} \varphi(x,y)$.

Corollary 2. Let $\varphi \in \Delta_0(\omega)$; suppose $I\Delta_0 + \Omega_1 \vdash^n \exists y\varphi(a,y)$ by an arbitrary proof; then $\mathbf{N} \models \forall x \exists y \leq 2^{(2|x|)^{2^{3n}}} \varphi(x,y)$.

Proof. First use free cut elimination (Cf. [HP92]) and then apply Corollary 1.

Theorem 2. For all n, φ, if $I\Delta_0 + \Omega_1 \vdash^n \mathrm{Prov}_{I\Delta_0+\Omega_1}(\ulcorner\varphi\urcorner)$, then $I\Delta_0 + \Omega_1 \vdash^{2^{(|\varphi|+1)^{2^{3n}}}} \varphi$.

Proof. The claim follows from Corollary 2, letting $\varphi(x,y) \equiv \exists y[x = \ulcorner\varphi\urcorner \rightarrow \mathrm{Proof}_{I\Delta_0+\Omega_1}(y,x)]$.

Theorem 2 gives us an upper bound for the size of proofs (in $I\Delta_0+\Omega_1$) of *any sentence* φ in terms of the size of a proof of $\mathrm{Prov}_{I\Delta_0+\Omega_1}(\ulcorner\varphi\urcorner)$. In order to find a (worse) upper bound for theories T such that $I\Delta_0+\Omega_1 \subseteq T \subseteq I\Delta_0+\mathrm{Exp}$, we can use a result of Visser (cf. [Vis90]), stating that for every Π_2 formula $\varphi(x)$ one has:

(*) $I\Delta_0+\mathrm{Supexp} \vdash \forall x[\mathrm{Prov}_{I\Delta_0+\mathrm{Exp}}(\ulcorner\varphi(nm(x))\urcorner) \rightarrow \varphi(x)]$.
Replacing $\varphi(x)$ by $\mathrm{Prov}_T(x)$ in (*) and using $T \subseteq I\Delta_0+\mathrm{Exp}$, we obtain:
(**) $I\Delta_0+\mathrm{Supexp} \vdash \forall x[\mathrm{Prov}_T(\ulcorner\mathrm{Prov}_T(nm(x))\urcorner) \rightarrow \mathrm{Prov}_T(x)]$

By (**) and by Proposition 4 (Parikh's theorem), we obtain an upper bound which is a finite iteration of the superexponential function.

Returning to $I\Delta_0+\Omega_1$, we use Corollary 2 to derive also a lower bound for the shortening of proofs of $\mathrm{Prov}_{I\Delta_0+\Omega_1}(\ulcorner\varphi\urcorner)$ with respect to proofs of φ in the case where $\varphi \equiv \exists y(y = 2_{2_{nm(n)}})$.

Theorem 3 (a) There is a polynomial $p(x)$ such that, for all n,

$$I\Delta_0+\Omega_1 \vdash^{p(n)} \mathrm{Prov}_{I\Delta_0+\Omega_1}(\ulcorner\exists y(y = 2_{2_{nm(n)}})\urcorner).$$

(b) For n sufficiently large, each proof in $I\Delta_0+\Omega_1$ of $\exists y(y = 2_{2_{nm(n)}})$ has size $\geq (2_n)/4$.

Proof. (a) By Proposition 5, one has:

(i) $I\Delta_0+\Omega_1 \vdash^{n^{0(1)}} \exists y(y = 2_{nm(n+1)})$. Thus, there is a polynomial $q(x)$ such that, for all x, there is a proof of $\exists y(y = 2_{nm|x|)})$ of size $\leq q(|x|)$; it follows that for some term $r(x)$ of $I\Delta_0+\Omega_1$ and for all x, there is a proof of Gödel number $\leq r(x)$ of $\exists y(y = 2_{nm(|x|)})$. Since these proofs can be coded in $I\Delta_0+\Omega_1$, the above argument can be formalized in that theory, yielding:

(ii) $I\Delta_0+\Omega_1 \vdash \forall x \mathrm{Prov}_{I\Delta_0+\Omega_1}(\ulcorner\exists y(y = 2_{nm(|x|)})\urcorner)$. By (i) and (ii) we obtain:

(iii) $I\Delta_0+\Omega_1 \vdash^{n^{0(1)}} \exists x[x = 2_{nm(n+1)}) \cdot \wedge \mathrm{Prov}_{I\Delta_0+\Omega_1}(\ulcorner\exists y(y = 2_{nm(|x|)})\urcorner)]$.

Since we have proofs of size polynomial in $|n|$ of the formulas $x = 2_{nm(n+1} \rightarrow |x| = 2_{nm(n)}$, form Lemma 1 we get:

(iv) $I\Delta_0+\Omega_1 \vdash^{n^{0(1)}} x = 2_{nm(n+1)} \rightarrow \mathrm{Prov}_{I\Delta_0+\Omega_1}(\ulcorner nm(|x|) = 2_{nm(n)}\urcorner)$.
Putting (iii) and (iv) together, we obtain (a).

To prove (b), suppose $I\Delta_0 + \Omega_1 \vdash^m \exists y(y = 2_{2_{nm(n)}})$. The size of $\exists y(y = 2_{2_{2nm(n)}})$ is bounded by a polynomial in n, thus it is $\leq 2_{n-1} - 2$ for n sufficiently large. Thus, by Corollary 2, we get: $\mathbf{N} \models \forall x \exists y \leq 2_{3m+n}(y = 2_{2_{nm(n)}})$.Hence, $2_{2_n} \leq 2_{3m+n}$, i.e. $2_n \leq 3m + n$. If n is sufficiently large, then $n \leq m$, thus $2_n \leq 4 \cdot m$, whence $m \leq (2_n)/4$. This completes the proof of the Theorem.

3 Shortenings by modal rules.

We begin this Section with an improvement of Proposition 3, (iii).

Theorem 4. Suppose $T \supseteq I\Delta_0 + \Omega_1$ and $T \vdash \varphi \rightarrow \text{Prov}_T(\ulcorner\varphi\urcorner)$ for all sentences $\varphi \in \Sigma_1$; let A/B be a T reasonable rule such that $L \not\vdash \Box A \rightarrow \Box B$; then, there are an open interpretation φ and a polynomial $p(x)$ such that for each natural number n, $A^\varphi(n)$ has a proof of size $\leq p(n)$ and $B^\varphi(n)$ has no proof of size $\leq 2_n$. Hence, the rule A/B shortens proofs in T at least by 2_{n^δ} for some positive rational number δ. If $T \supseteq I\Delta_0 + \text{Exp}$, then we can also bound proofs of $A^\varphi(n)$ by $p(|n|)$ instead of $p(n)$, therefore the shortening is by $2_{2_{n^\delta}}$ for some positive rational number δ.

Proof. By Proposition 7, there are two interpretations $^*, ^\circ$, such that $T \vdash A^*, T \vdash B^*, T \vdash \text{Prov}_T(\ulcorner B^\circ \urcorner) \rightarrow A^\circ, T \not\vdash B^\circ$. For each T formula $\xi(x)$ having exactly x free, we denote by ξ the open interpretation given by $p_i^\xi(x) \equiv (\xi(x) \rightarrow p_i^* \wedge (\neg\xi(x) \rightarrow p_i^\circ)$. Moreover, for each modal formula A, we write $A^\xi(x)$ instead of A^ξ to stress that A^ξ has (only) x free.

Let, by self reference, $\varphi(x)$ be such that:
(i) $T \vdash \varphi(x) \leftrightarrow \forall v[2_{x+2} = v \rightarrow \forall w \leq v \neg \text{Proof}_T(w, \ulcorner B^\varphi(nm(x))\urcorner)]$

We claim that the interpretation $^\varphi$ satisfies the requirements of the Theorem.

First note that for all n, $B^\varphi(n)$ has no proof of size $\leq 2_n$, otherwise, for some $n, B^\varphi(n)$ would have a proof of Gödel number $\leq 2_{n+2}$ and from (i) we would easily obtain $T \vdash \neg\varphi(nm(n))$; from this and from the definition of $p_i^\varphi(x)$, we would get, for each propositional variable p_i occuring in A or in $B, T \vdash p_i^\varphi(nm(n)) \leftrightarrow p_i^\circ$, therefore we would conclude $T \vdash B^\varphi(n) \leftrightarrow B^\circ$, and $T \not\vdash B^\varphi(n)$, a contradiction.

Let, for every formula $\tau(x)$, $\hat{\text{Prov}}_T(\ulcorner \tau(nm(x))\urcorner)$ be an abbreviation for $\tau(x) \wedge \text{Prov}_T(\ulcorner \tau(nm(x))\urcorner)$. For each propositional variable p_i occuring in A or in B, we obtain:

$$T \vdash \hat{\text{Prov}}_T(\ulcorner \varphi(nm(x))\urcorner) \rightarrow \hat{\text{Prov}}_T(\ulcorner p_i^\varphi(nm(x)) \leftrightarrow p_i^* \urcorner).$$

From this and Proposition 1), it follows:
(i) $T \vdash \hat{\text{Prov}}_T(\ulcorner\varphi(nm(x))\urcorner) \rightarrow [A^\varphi(x) \leftrightarrow A^*)] \rightarrow A^\varphi(x)$

Moreover, again by Proposition 1, we get, for each propositional variable p_i occuring in A or in B:

$$T \vdash \hat{\text{Prov}}_T(\ulcorner\neg\varphi(nm(x))\urcorner) \rightarrow \hat{\text{Prov}}_T(\ulcorner p_i^\varphi(nm(x)) \leftrightarrow p_i^\circ\urcorner)$$

From this and Proposition 1), it follows:
(iii)

$$\begin{aligned} T \vdash \hat{\text{Prov}}_T(\ulcorner\neg\varphi(nm(x))\urcorner) \rightarrow & \\ \rightarrow \quad & [(A^\varphi(x) \leftrightarrow A^\circ) \wedge \\ & (\text{Prov}_T(\ulcorner B^\varphi(nm(x))\urcorner) \leftrightarrow \text{Prov}_T(\ulcorner b^\circ\urcorner))] \\ \rightarrow \quad & [\text{Prov}_T(\ulcorner B^\varphi(nm(x))\urcorner) \rightarrow A^\varphi(x)]. \end{aligned}$$

We prove now:
(iv) $T \vdash \forall x[\text{“}2_{x+4} \text{ exists”} \rightarrow A^\varphi(x)]$.

If $\psi(v, u_1, \ldots u_n)$ is a formula with only $v, u_1, \ldots u_n$ free whose quantifiers are bounded by a polynomial $p(v)$, then, whenever $u_1 \leq p(v), \ldots, u_n \leq p(v)$ and $\psi(v, u_1, \ldots u_n)$ is true, $\psi(v, u_1, \ldots u_n)$ has a proof whose size is bounded by a polynomial in v. This can be shown by induction on the complexity of ψ. The most difficult case is the case corresponding to bounded universal quantifier; in this case, if $r(v)$ is a bound corresponding to the induction hypothesis, we get a bound for the induction step linear in $p(v) \cdot r(v)$. It follows that there is a polynomial $q(v)$ such that whenever $\psi(v, u_1, \ldots u_n)$ is true and $u_1 \leq p(v), \ldots, u_n \leq p(v)$, there is a proof of $\psi(v, u_1, \ldots u_n)$ whose Gödel number is bounded by $2^{q(v)}$; thus, for v sufficiently large, this Gödel number is bounded by 2^{2^v}. Formalizing (a particular case of) this in $I\Delta_0 + \Omega_1$ we get: if $\psi(x, v)$ has only x, v free, all quantifiers in $\psi(x, v)$ are bounded by a polynomial in v, and $I\Delta_0 + \Omega_1 \vdash \psi(x, v) \rightarrow x \leq v$, then:

$$I\Delta_0 + \Omega_1 \vdash \forall v[(\psi(x, v) \wedge \text{“}2^{2v} \text{ exists”}) \rightarrow \text{Prov}_T(\ulcorner(nm(x), nm(v))\urcorner)].$$

Next, let $\varphi_0(x, v)$ be the formula $2_{x+2} = v \wedge \forall w \leq v \neg\text{Proof}_T(w, \ulcorner B^\varphi(nm(x))\urcorner)$. Note that all quantifiers in $\varphi_0(x, v)$ are bounded by a polynomial in v and $I\Delta_0 + \Omega_1 \vdash \psi(x, v) \rightarrow x \leq v$. Moreover, $T \vdash \forall x[\exists v \varphi_0(x, v) \rightarrow \varphi(x)]$. So,

$$\begin{aligned} T \vdash \forall x\{(\text{“}2_{x+4} \text{ exists”} \wedge \varphi(x)) \rightarrow \quad & \exists v[\varphi_0(x, v) \wedge ?\text{‘}2^{2v} \text{ exists”}]\} \\ \rightarrow \quad & \exists v[\text{Prov}_T(\ulcorner\varphi_0(nm(x), nm(v))\urcorner)]\} \\ \rightarrow \quad & \text{Prov}_T(\ulcorner\exists v \varphi_0(nm(x), v)\urcorner)\} \\ \rightarrow \quad & \text{Prov}_T(\ulcorner\varphi(nm(x))\urcorner)\}. \text{ It follows:} \end{aligned}$$

$$\begin{aligned} \text{(v)} \quad T \vdash \forall x\{(\text{“}2_{x+4} \text{ exists ”} \wedge \varphi(x)) \rightarrow \quad & \hat{\text{Prov}}_T(\ulcorner\varphi(nm(x))\urcorner)\} \\ \rightarrow \quad & A^\varphi(x)\} \text{ (by(ii))}. \end{aligned}$$

Thus, it suffices to prove:
(vi) $T \vdash \forall x\{(\text{“}2_{x+4}\text{ exists”} \wedge \neg\varphi(x)) \rightarrow a^{\varphi}(x)\}$.

We prove first:
(vii) $T \vdash \forall x\{(\text{“}2_{x+4}\text{ exists”} \wedge \neg\varphi(x)) \rightarrow [\mathrm{Prov}_T(\ulcorner B^{\varphi}(nm(x))\urcorner) \rightarrow A^{\varphi}(x)]\}$.

Let $\psi_o(v,x)$ be the formula $2_{x+2} = v \wedge \exists w \leq v \mathrm{Proof}_T(w, \ulcorner B^{\varphi}(nm(x))\urcorner)$. Note that each quantifier.in $\psi_o(v,x)$ is bounded by a polynomial in v, $I\Delta_0 + \Omega_1 \vdash \psi_o(x,v) \rightarrow x \leq v$, and $T \vdash \forall x[\exists v \psi_o(x,v) \rightarrow \neg\varphi(x)]$. So,

$$
\begin{aligned}
T \vdash \forall x\{(\text{“}2_{x+4}\text{ exists”} \wedge \neg\varphi(x)) \rightarrow \quad & \exists v[\psi_o(x,v) \wedge \text{“}2^{2^v}\text{ exists”}]\} \\
\rightarrow \quad & \exists v \mathrm{Prov}_T(\ulcorner \psi_o(nm(x), nm(v))\urcorner)\} \\
\rightarrow \quad & \mathrm{Prov}_T(\ulcorner \exists v \psi_o(nm(x), v)\urcorner)\} \\
\rightarrow \quad & \mathrm{Prov}_T(\ulcorner \neg\varphi(nm(x))\urcorner)\}. \text{ So,} \\
T \vdash \forall x\{(\text{“}2_{x+4}\text{ exists”} \wedge \neg\varphi(x)) \rightarrow \quad & \hat{\mathrm{Prov}}_T(\ulcorner \neg\varphi(nm(x))\urcorner)\} \\
\rightarrow \quad & [\mathrm{Prov}_T(\ulcorner B^{\varphi}(nm(x)))\urcorner) \rightarrow A^{\varphi}(x)]\} \text{ (by (iii)).}
\end{aligned}
$$

This proves (vii). Now, from (i) it easily follows:
(viii) $T \vdash \forall x[(\text{“}2_{x+4}\text{ fexists”} \wedge \neg\varphi(x)) \rightarrow \mathrm{Prov}_T(\ulcorner B^{\varphi}(nm(x))\urcorner)]$.

By claims (vii) and (viii), we get (vi) and, by (v), we conclude
(iv) $T \vdash \forall x\{\text{“}2_{x+4}\text{ exists”} \rightarrow A^{\varphi}(x))\}$.

Now, by Proposition 5, $T \vdash^{n^{0(1)}} \text{“}2_{nm(n)+4}\text{ exists”}$. So, by (iv), there is a polynomial $p(x)$ such that for all n, $A^{\varphi}(n)$ has a proof of size $\leq p(n)$.

If $T \supseteq I\Delta_0 + \mathrm{Exp}$, there is a cut $J(x)$ such that:
(ix) $T \vdash \forall x[J(x) \rightarrow \text{“}2_x\text{ exists”}]$.

As proved e.g. in [Pud86], there is a polynomial $p(x)$ such that for each n, $J(n)$ has a proof of size$\leq p(|n|)$; it folows that there is a polynomial $q(x)$ such that for each n, “2_n exists” has a proof of size $\leq q(|n|)$ (use a proof of size $\leq p(|n|)$ of $J(n)$ and a fixed proof of (ix)). Last, by (iv), we get a polynomial $r(x)$ such that $A^{\varphi}(n)$ has a proof of size $\leq p(|n|)$. This completes the proof of the Theorem.

When dealing with upper bounds for the shortenings by modal rules, two problems naturally arise:

1) Given two modal formulas A,B such that A/B is a reasonable rule, find a function $f(x)$ (as slow as possible) such that for every T interpretation *, if x is the size of a proof of A^*, then there is a proof of B^* of size $\leq f(x)$ (we call such a f an *upper bound* (in T) *for proofs of B from proofs of A).*

2) Given a reasonable rule R: A/B, find a function $f(x)$ (as low as possible) such that, for every proof Π of any formula φ by means of R, there is a proof in the usual sense of φ of size $\leq f(\mathrm{size}(\Pi))$ (we call such a function an *upper bound for shortenings* (in T)*by means of R).*

We start with some observations: first of all, the half of Proposition 2 concerning validity of L and S holds for all Σ_1 sound extensions T of $I\Delta_0 + \Omega_1$ (Cf. Proposition 2). It follows that, if $S \vdash \Box A \to \Box B$, the rule A/B is T sound for all such extensions. The other direction (i.e., if A/B is T sound, then $S \vdash \Box A \to \Box B$) is proved using the other half of Proposition 2, under the additional assumption that $T \vdash \varphi \to \mathrm{Prov}_T(\ulcorner\varphi\urcorner)$ for $\varphi \in \Sigma_1$). The remaining results of this Section (except from Theorem 6) can be proved for all theories T such that $I\Delta_0 + \Omega_1 \subseteq T \subseteq I\Delta_0 + \mathrm{Exp}$ under the assumption that $S \vdash \Box A \to \Box B$. Of course, if $T \vdash \varphi \to \mathrm{Prov}_T(\ulcorner\varphi\urcorner)$ for all sentences $\varphi \in \Sigma_1$, then we can replace the assumption $S \vdash \Box A \to \Box B$ by the weaker assumptions: A/B is a T sound rule. So, these results can be stated and proved indiferently under weaker assumptions about the theory T (i.e. $I\Delta_0 + \Omega_1 \subseteq T \subseteq I\Delta_0 + \mathrm{Exp}$) and stronger assumptions about the rule (i.e. $S \vdash \Box A \to \Box B$) or viceversa (i.e., with the additional assumption that $T \vdash \varphi \to \mathrm{Prov}_T(\ulcorner\varphi\urcorner)$ for all sentences $\varphi \in \Sigma_1$ for the theory, and with the only assumption of T soundness for the rule). We choose the first possibility, but *we point out here and for all that we can choose the second one as well.*

Needless to say, the problem of upper bounds becomes trivial if either the rule A/B can never be applied (in which case the identity is an upper bound) or $L \vdash \Box A \to \Box B$, in which case, by proposition 1, upper bounds for proofs of A and upper bounds for the shortening by the rule A/B can be chosen to be polynomial.

We wish to point out that the problem of upper bounds for shortening by a modal rule A/B is in general different from the problem of upper bounds for proofs of B from proofs of A. Indeed, a modal rule A/B can be used several times in a proof, thus possibly yielding a stronger speed up.

Lemma 5. If $S \vdash \Box A \to \Box B$, and k is the number of $\Box$ subformulas of $\Box A \to \Box B$, then $L \vdash \Box A \to \Box^{k+1} B$.

Proof. . If $L \nvdash \Box A \to \Box^{k+1} B$, there is a finite transitive irreflexive treelike Kripke model $\mathbf{M} \equiv < X, R, \Vdash >$ with root x_0 such that $x_0 \Vdash \Box A$ and $x_0 \nVdash \Box^{k+1} B$ (Cf [Smo85]). It follows that there are nodes $x_0\, R\, x_1\, R\, x_2, \ldots, R\, x_k$ s.t. $x_i \nVdash \Box A, i = 0, \ldots, k, x_1 \Vdash \neg\Box^k B, x_2 \Vdash \neg\Box^{k-1} b, \ldots\ldots, x_k \Vdash \neg\Box B$.

Now, $\Box$ formulas are preserved upwards, the number of nodesnodes $x_0, \ldots, x_k$ is $k + 1$ and the number of $\Box$ subformulas of $\Box a \to \Box B$ is k; thus, there is an $i < k$ s.t. x_i, x_{i+1} force the same $\Box$ subformulas of $\Box A \to \Box B$. Let $\Box C$ be a $\Box$ subformula of $\Box A \to \Box B$. We get:

$$x_{x+1} \Vdash \Box C \Rightarrow x_i \Vdash \Box C \Rightarrow x_{i+1} \Vdash C$$

So, $x_{i+1} \vdash \Box C \to C$. It follows:
$x_{i+1} \vdash \wedge\{(\Box C \to C) : \Box C$ a subformula of $\Box A \to \Box B) \wedge \Box A \wedge \neg\Box B$.
Thus, $L \nvdash \wedge\{(\Box C \to C) : \Box C$ a subformula of $\Box A \to \Box B) \to (\Box A \to \Box B)$.

By the characterization of provability in S in terms of provability in L (Cf. [Smo85]), this is equivalent to $S \nvdash \Box A \to \Box B$.

Theorem 5 (i) Let T be such that $I\Delta_0 + \Omega_1 \subseteq T \subseteq I\Delta_0 + \text{Exp}$ and let A/B be a rule such that $S \vdash \Box A \to \Box B$. Then there is an upper bound in T for proofs of B from proofs of A which is a finite iteration of the function 2_x.

(ii) If A/B is as above and T is $I\Delta_0 + \Omega_1$, then there is an upper bound in T for proofs of B from proofs of A of the form $f^{(k)}(x)$, where $f(x) = 2^{(2+|x|)^2_{3x}}$, k is the number of $\Box$ subformulas of $\Box A \to \Box B$ and $f^{(k)}$ denotes f iterated k times.

(iii) If T and A/B are as in i), there is an upper bound for the shortening by the rule A/B in T which is of the form supsupexp$(n{\cdot}x)$ (for some fixed n), where supsupexp(x) is defined by supsupexp$(0) = 1$; supsupexp$(x+1) = 2_{\text{supsupexp(x)}}$.

Proof. (i) If $S \vdash \Box A \to \Box B$ then $L \vdash \Box A \to \Box^{k+1} B$, where k is the number of $\Box$ subformulas of $\Box A \to \Box B$; therefore, by Proposition 3, for each interpretation *, if $T \vdash A^*$ by proof Π of size x, there is a proof of $(\Box^k B)$ of size polynomial in x. By what we observed at the end of the proof of Theorem 2, there is a finite iteration $g(v)$ of the superexponential function (not depending on the interpretation) such that, for $0 \leq i < k$, whenever $(\Box^{i+1} B)^*$ has a proof of size $v (\Box^i B)^*$ has a proof of size $\leq g(v)$. Using this k times, we obtain a bound which is a k-iteration of g, hence a finite iteration of the superexponential function.

(ii) As in (i), but with g replaced by $f(x) = 2^{(2+|x|)^2_{3x}}$ (Cf Section 2, Theorem 2).

(iii) If a proof Π of some formula φ by means of the rule A/B has size x, then the rule A/B is applied less than x times. Let $A_1/B_1, \ldots, A_y/B_y (y < x)$ be all the instances of application of the rule A/B in the proof. By (i), there is n-iteration $g(v)$ of the superexponential function (some fixed n) such that we can replace each proof of A^* (* any interpretation) of size v by a proof of B^* of size $\leq g(v)$; so we can get a proof of B_1 and one of A_2 of size $\leq g(x)$, a proof of B_2 and one of A_3 of size $\leq g(g(x))$, etc. We get a proof (in the usual sense) of φ which is bounded by $g^{(x)}(x)$, hence by supsupexp$((n \cdot x)$.

The upper bound for rules found in Theorem 5 (iii) seems very bad, because there is a big gap between it and the lower bound given by Theorem 4. We shall prove that, at least in the case where T is $I\Delta_0 + Exp$ and the rule is $\Box p/p$ (or even $\Box^n p/\Box^m p$ with $m < n$), this upper bound cannot be improved substantially, as the following theorem shows.

Theorem 6. For every n, there is a $I\Delta_0 + \mathrm{Exp}$ interpretation * such that for every positive $i \leq n$, $(\Box^{2i+1}p)^*$ has a proof which is superexponentially shorter than any proof of $(\Box^{2i-1}p)^*$. Thus, for all $m < n$, no finite iteration of the superexponential function can be an upper bound for the rule $\Box^n p/\Box^m p$ in $I\Delta_0 + \mathrm{Exp}$.

Proof. We start from the following Lemma.

Lemma 6. $I\Delta_0 + \mathrm{Exp} \vdash \forall x \mathrm{Prov}_{I\Delta_0+\mathrm{Exp}}(\ulcorner 2_{nm(x)} \text{ exists}\urcorner)$.

Proof of Lemma 6 (Sketch) For every x, the formula "2_x exists" has a proof of size polynomial in x, and, therefore, of Gödel number bounded by a term $t(x)$ of $I\Delta_0 + \mathrm{Exp}$ (use proofs in $I\Delta_0 + \mathrm{Exp}$ of "2_0 exists" and of "$\forall u(2_u \text{ exists} \rightarrow 2_{u-1} \text{ exists})$"). Once we have such a bound $t(x)$, it is not difficult to prove the $\Delta_0(\mathrm{Exp})$ formula $\exists y \leq t(x)\mathrm{Proof}_{I\Delta_0+\mathrm{Exp}}(y, \ulcorner 2_{nm(x)} \text{ exists }\urcorner)$ by induction on x. This completes the proof of Lemma 6.

We go on with the proof of Theorem 6. To simplify notation, throughout the whole proof, we shall write $\vdash, \varphi \vdash, \mathrm{Proof}, \mathrm{Prov},$ and $\hat{\mathrm{Prov}}$ for $I\Delta_0 +$ Exp $\vdash, I\Delta_0 + \mathrm{Exp} + \varphi \vdash, \mathrm{Proof}_{i\Delta_0+\mathrm{Exp}}$ and $\hat{\mathrm{Prov}}_{I\Delta_0+\mathrm{Exp}}$ respectively (Cf. the proof of Theorem 4 for the notation $\hat{\mathrm{Prov}}_T$); moreover, for each pair $\exists y\varphi(y), \exists y\psi(y)$ of Σ_1 formulas such that $\varphi(y), \psi(y) \in \Delta_0$, we write $\exists y\varphi(y) <_{\mathrm{Supexp}} \exists y\psi(y)$ for $\exists y\exists z[\varphi(y) \wedge 2_y = z \wedge \forall u < z\neg\psi(u)]$. Let n be given. By self reference, there is an interpretation $^\circ$ such that

$$\text{(i)} \qquad \vdash p^\circ \leftrightarrow \wedge\{(\Box^{2i+1}p)^\circ <_{\mathrm{Supexp}} (\Box^{2i-1}p)^\circ; i = 1, \ldots, n\}.$$

We prove that $\vdash p^\circ$. This is sufficient to prove the claim of Theorem 6, because, if $\vdash p^\circ$, then p° is true in the standard model, therefore, by (i), for each positive $i \leq n$, there is a proof of $(\Box^{2i+1}p)^\circ$ which is superexponentially shorter than any proof of $(\Box^{2i-1}p)^\circ$; in particular, for each h, m, there is a proof of $(\Box^{2(r+h)+1}p)^\circ$ which is shorter than any proof of $(\Box^{2r+1}p)^\circ$ by the h times iterated superexponential function, therefore for no h, the h times iterated superexponential function can be an upper bound for the shortening by the rule $\Box^n p$, $n < m$. In order to prove $\vdash p^\circ$, it sufices to prove that for *every* interpretation *:

$$\text{(ii)} \qquad \vdash \text{Prov}(\ulcorner p^* \leftrightarrow \wedge\{(\Box^{2i+1}p)^* <_{\text{Supexp}} (\Box^{2i-1}p)^*; 1 \le i \le n\}\urcorner) \to (\Box^{2i+1}p)^*$$

(indeed, since the premise of the implication in (ii) is provable, with $^\circ$ in place of *, we get that $(\Box^{2i+1}p)^\circ$ is provable too, whence, by the soundness of $I\Delta_0 + \text{Exp}$, we can conclude that p° is in turn provable. We prove (ii) by induction on n (in the metalanguage). Assume $n = 1$; then,

$$\begin{aligned}
\hat{\text{Prov}}(\ulcorner p^* \leftrightarrow \quad & ((\Box^3 p)^*) <_{\text{Supexp}} (\Box p)^*)\urcorner) \vdash (\Box^3_p)^* \to \exists x \text{Proof}(x, \ulcorner(\Box^2 p)^*\urcorner) \\
\to \quad & \exists x \text{Prov}(\ulcorner \text{Proof}(nm(x), \ulcorner(\Box^2 p)^*\urcorner)\urcorner) \\
\to \quad & \exists x \text{Prov}(\ulcorner \text{Proof}(nm(x), \ulcorner(\Box^2 p)^*\urcorner) \wedge 2_{nm(x)} \text{ exists } \urcorner) \quad \text{(Lemma 6)} \\
\to \quad & \exists x \text{Prov}(\ulcorner p^* \vee (\Box p)^*\urcorner) \quad \text{(by (i) with } n = 1) \\
\to \quad & (\Box^2 p)^*.
\end{aligned}$$

We conclude

$$\hat{\text{Prov}}(\ulcorner p^* \leftrightarrow ((\Box^3 p)^* <_{\text{Supexp}} (\Box p)^*)\urcorner) \vdash (\Box^3 p)^* \to (\Box^2 p)^*.$$

Putting Prov in front of both sides of $\vdash$ and recalling that, for any formula φ, $\text{Prov}(\ulcorner \hat{\text{Prov}}(\ulcorner\varphi\urcorner)\urcorner)$ is equivalent to Prov $(\ulcorner\varphi\urcorner)$, we obtain:

$$\text{Prov}(\ulcorner p^* \leftrightarrow ((\Box^3 p)^* <_{\text{Supexp}} (\Box p)^*)\urcorner) \vdash \text{Prov}(\Box^3 p)^* \to (\Box^2 p)^*.$$

By Löb's Theorem, $\text{Prov}(\ulcorner p^* \leftrightarrow ((\Box^3 p)^* <_{\text{Supexp}} (\Box p)^*)\urcorner) \vdash (\Box^3 p)^*$. We proceed with the induction step; assume that (ii) holds for $n = k$, and let us prove it for $n = k + 1$; in order to simplify notation, let us write, for every $m \ge 1, F_m(p^*)$ for $\wedge\{(\Box^{2i+1}p)^* <_{\text{Supexp}} (\Box^{2i-1}p)^* > 1 \le i \le m\}$. Note that

(iii) $p^* \leftrightarrow F_{k+1}(p^*) \vdash [(\Box^{2k+3}p)^* \to (p^* \leftrightarrow F_k(p^*))$ and that by Σ_1 completeness,

(iv) $(\Box^{2k+3p})^* <_{\text{Supexp}} (\Box^{2k+1}p)^* \vdash \text{Prov}(\ulcorner(\Box^{2k+3}p)^* <_{\text{Supexp}} (\Box^{2k+1}p)^*\urcorner)$.

Now,

$$
\begin{array}{ll}
\hat{\text{Prov}} & (\ulcorner p^* \leftrightarrow F_{k+1}(p^*)\urcorner) \vdash (\Box^{2k+3}p)^* \to \exists x \text{Proof}(x, \ulcorner(\Box^{2k+2}p)^*\urcorner) \\
\to & \exists x \text{Prov}(\ulcorner \text{Proof}(nm(x), \ulcorner(\Box^{2k+2}p)^*\urcorner)\urcorner) \\
\to & \exists x \text{Prov}(\ulcorner \text{Proof}(nm(x), \ulcorner(\Box^{2k+2}p)^*\urcorner) \wedge 2_{nm(x)} \text{ exists } \urcorner) \text{ (Lemma 6)} \\
\to & \exists x \text{Prov}(\ulcorner(\Box^{2k+1}p)^* \vee ((\Box^{2k+3}p)^* <_{\text{Supexp}} (\Box^{2k+1}p)^*))\urcorner) \\
\to & \text{Prov}(\ulcorner(\Box^{2k+1}p)^* \vee \text{Prov}(\ulcorner(\Box^{2k+3}p)^* <_{\text{Supexp}} (\Box^{2k+1}p)^*)\urcorner)\urcorner) \quad \text{by(iv)} \\
\to & \text{Prov}(\ulcorner(\Box^{2k+1}p)^* \vee (\text{Prov}(\ulcorner(p^* \leftrightarrow F_k(p^*))\urcorner)\urcorner) \quad \text{(by(iii))} \\
\to & \text{Prov}(\ulcorner(\Box^{2k+1}p)^* v (\Box^{2k+1}p)^*\urcorner) \quad \text{(by the induction hypothesis)} \\
\to & \text{Prov}(\ulcorner(\Box^{2k+1}p)^*\urcorner) \\
\to & (\Box^{2k+2}p)^*.
\end{array}
$$

We conclude

$$\hat{\text{Prov}}(\ulcorner p^* \leftrightarrow F_{k+1}(p^*)\urcorner) \vdash (\Box^{2k+3}p)^* \to (\Box^{2k+2}p)^*.$$

Putting Prov in front of both sides of $\vdash$ and recalling again that, for any formula φ, $\text{Prov}(\ulcorner \hat{\text{Prov}}(\ulcorner\varphi\urcorner)\urcorner)$ is equivalent to $\text{Prov}(\ulcorner\varphi\urcorner)$, we obtain:

$$\text{Prov}(\ulcorner p^* \leftrightarrow F_{k+1}(p^*)\urcorner) \vdash (\text{Prov}(\ulcorner((\Box^{2k+3}p)^* \to (\Box^{2k+2}p)^*)\urcorner).$$

By this Löb's Theorem, $\text{Prov}(\ulcorner p^* \leftrightarrow F_{k+1}(p^*)\urcorner) \vdash (\Box^{2k+3}p)^*$. This completes the proof.

Among the reasonable modal rules which shorten proofs more than polynomially, the rule $\Box p/p$ seems to be the most natural one. Also for this reason, the first results presented in this paper concern the speed up by provability. The next Theorem says that there is "a posteriori" another reason for concertrating ourselves on this rule; indeed, all sound modal rules are in a sense dominated by the rule $\Box p/p$; using other rules instead we do not shorten proofs more than polynomially.

Theorem 7. Let T be such that $I\Delta_0 + \Omega_1 \subseteq T \subseteq I\Delta_0 + \text{Exp}$, and let $R : A/B$ be a modal rule such that $S \vdash \Box A \to \Box B$. Then, there is a polynomial $p(x)$ such that, to each proof Π using the rule A/B we can associate a proof Π' using the rule $\Box p/p$ instead whose size is $\leq p(\text{size}(\Pi))$.

Proof. By Lemma 5, Ł $\vdash \Box A \to \Box^{k+1}B$. Thus, by Proposition 3 i) and ii) there are polynomials $p'(y), q(x_1, \dots x_n)$ such that:

a) for each interpretation * there is a proof of $(\Box A \to \Box^{K+1}B)^*$ of size $\leq q(\text{size}(p_1^*), \dots, \text{size}(p_n^*))$ $(p_1, \dots, p_n)$ are the variables in A, B);

b) for each interpretation * and each proof Π of A^*, there is a proof Π' of $(\Box^k B)^*$ of size $\leq p(\text{size}(\Pi))$.

Moreover:

c) There is a polynomial $r(y)$ s.t. if Π is a proof of φ, there is a proof of $\text{Prov}_T(\ulcorner\varphi\urcorner)$ of size $\leq r(\text{size}(\Pi))$. (Cf.e.g.[WP87]).

d) There are polynomials $s(x), t(x)$ such that, if Π is a proof of φ from sentences $\varphi_1, \ldots, \varphi_n$, then there is a proof in T of size $\leq s(\text{size}(\Pi))$ of $\varphi_1 \wedge \ldots \wedge \varphi_n \rightarrow \varphi$ and there is a proof of size $\leq t(\text{size}(\Pi))$ of the sentence $\text{Prov}(\ulcorner(\varphi_1 \wedge \ldots \wedge \varphi n)\urcorner) \rightarrow \text{Prov}_T(\ulcorner\varphi\urcorner)$. (We use a proof polynomial in size(Π) of $\text{Prov}_T(\ulcorner((\varphi_1 \wedge \ldots \wedge \varphi) \rightarrow \varphi)\urcorner)$, and a fixed general proof of $\forall x \forall y[\text{Prov}_T(x \rightarrow y) \rightarrow (\text{Prov}_T x \rightarrow \text{Prov}_T y)]$).

If Π is a proof which uses some instance A^*/B^* of the rule A/B, then $\text{size}(\Pi) \leq \text{size}(p_i^*)$; thus, letting $q'(y) = q(y, \ldots, y)$, we get, for all such *, a proof of $(\Box A \rightarrow \Box^{K+1} B)^*$ of size$\leq q'(\text{size}(\Pi))$.

To simplify notation, we consider the sum $P(y) - p(y) + q'(y) + r(y) + s(y) + t(y)$ as a bound for cases a),b),c) and d).

Now, consider a proof Π with the rule A/B; we shall change Π into a proof Π' of size $h \cdot \text{size}(\Pi) \cdot P(\text{size}(\Pi))$, where h is a constant which does not depend on Π.

The proof is obtained as follows. Let $A_1/B_1, \ldots A_n/B_n$ be the occurences of the rule A/B in Π (here A_i/B_i) can correspond to different interpretations of A, B). Note that Π contains derivations (in T) of A_i from $B_i, \ldots, B_{i-1}$. So, by d), there are derivations Π_i of $\text{Prov}_T(\ulcorner\wedge_{j<i} B_j \rightarrow A_i\urcorner)$ of size $\leq P(\text{size}(\Pi))$. By Lemma 5, there are also derivations Σ_i of size $\leq P(\text{size}(\Pi))$ of $\text{Prov}_T(\ulcorner A_i\urcorner) \rightarrow \text{Prov}_T^{k+1}(\ulcorner B_i\urcorner)$.

Last, we have a derivation Δ of $\text{Prov}_T(\ulcorner A_1\urcorner)$ and derivations Δ_i of

$$\text{Prov}_T(\ulcorner\wedge_{j<i} B_j\urcorner) \rightarrow [(\text{Prov}_T(\ulcorner B_i\urcorner) \rightarrow \text{Prov}_T(\ulcorner\wedge_{j\leq i} B_j\urcorner))],$$

which are all of size $\leq P(\text{size}(\Pi))$.

At this point, we can produce the desired proof Π', which is as follows:

$$A_1, \Delta, \text{Prov}_T(\ulcorner A_1\urcorner), \Sigma_1, \text{Prov}_T(\ulcorner A_1\urcorner) \rightarrow \text{Prov}_T^{k+1}(\ulcorner B_1\urcorner),$$

$$\text{Prov}_T^{k+1}(\ulcorner B_1\urcorner), \text{Prov}_T^{k}(\ulcorner B_1\urcorner), \ldots, \text{Prov}_T(\ulcorner B_1\urcorner), B_1, \Pi_1,$$

$$\text{Prov}_T(\ulcorner B_1\urcorner) \rightarrow \text{Prov}_T(\ulcorner A_2\urcorner), \text{Prov}_T(\ulcorner A_2\urcorner), \Sigma_2,$$

$$\text{Prov}_T(\ulcorner A_2\urcorner) \rightarrow \text{Prov}_T^{k+1}(\ulcorner B_2\urcorner), \text{Prov}_T^{k+1}(\ulcorner B_2\urcorner), \text{Prov}_T^{k}(\ulcorner B_2\urcorner), \ldots, \text{Prov}(\ulcorner B_2\urcorner), B_2, \Delta_2,$$

$$\text{Prov}_T(\urcorner B_1\urcorner) \rightarrow (\text{Prov}_T(\ulcorner B_2\urcorner) \rightarrow \text{Prov}_T(\ulcorner B_1 \wedge B_2\urcorner),$$

$$\text{Prov}_T(\ulcorner B_2\urcorner) \rightarrow \text{Prov}_T(\ulcorner B_1 \wedge B_2\urcorner), \text{Prov}_T(\ulcorner B_1 \wedge B_2), \Pi_2,$$

$$\text{Prov}_T(\ulcorner B_1 \wedge B_2\urcorner) \rightarrow \text{Prov}_T(\ulcorner A_3\urcorner), \text{Prov}_T(\ulcorner A_3\urcorner), \Sigma_3,$$

$$\mathrm{Prov}_T(\ulcorner A_3 \urcorner) \to \mathrm{Prov}_T^{k+1}(\ulcorner B_3 \urcorner) \text{ etc...}$$

$$\Sigma_n, \mathrm{Prov}_T(\ulcorner A_n \urcorner) \to \mathrm{Prov}_T^{k+1}(\ulcorner B_n \urcorner), \mathrm{Prov}_T^{k+1}(\ulcorner B_n \urcorner), \mathrm{Prov}_T^{k}(\ulcorner B_n \urcorner), \ldots,$$
$$\mathrm{Prov}_T(\ulcorner B_n \urcorner), B_n, \Pi$$

(note that Π is a derivation in T of its last formula from $B_1, \ldots, B_n$ and $B_1, \ldots, B_n$ are derived in the part preceding Π). The resulting proof has size bounded by $h \cdot n \cdot P(\mathrm{size}(\Pi))$ (for some h not depending on Π) hence by $h \cdot \mathrm{size}(\Pi))$. This completes the proof of Theorem 7.

4 Strong fragments.

In the preceding section we dealt with fragments of arithmetic called *weak*, i.e. subsystems of $I\Delta_0 + \exp$. These fragments are presently in the focus of interest, whereas the interest in stronger fragments like $I\Sigma_n$ seems to decline, at least temporarily. In the present section we shall show that the results of preceding sections have natural analogous for strong fragments and for PA itself. This is not too surprising, but not evident. We concentrate to results that can be obtained more or less easily using or imitating the results obtained so far; some of our results might be strengthened or more elaborted. The important feature of previous results and the main observation for our generalizations is the fact that the speed up we can exhibit is so big that there are lower bounds that are not provably recursive (dominate all provably recursive functions). We shall show that the same holds for strong fragments. To this end we use known characterizations of provably recursive functions, as presented in [HP92].

Definition 5. (1) A function $f : \mathbf{N} \to \mathbf{N}$ is Σ_1-*definable in* T if there is a Σ_1-formula φ defining f in $\mathbf{N}$ and such that T prove that φ defines a (partial) function, i.e.

$$T \vdash \forall x, y, z\big(\varphi(x, y) \& \varphi(x, z) \to y = z\big).$$

The formula φ is a Σ_1-*definition of* f *in* T. The function f is T-*provably recursive* if f has a Σ_1-definition in T such that T proves that φ defines a total function, i.e.

$$T \vdash \forall x \exists y \varphi(x, y).$$

(2) A Σ_1 system $\{f_n(x) | n\}$ of functions is an *envelope* for T-provably recursive functions of each f_n is T-provably recursive and each T-provably recursive function is majorized by some f_n.

Definition 6. (Schwichtenberg-Wainer hierarchy). For $\alpha < \varepsilon_0$,

$$\begin{aligned} f_0(x) &= x+1 \\ f_{\alpha+1}(x) &= f_\alpha^x(x+1) \\ f_x(x) &= f_{<\lambda>(x)}(x+1) \text{ for } \lambda \text{ limit} \end{aligned}$$

($< \lambda > (x)$, $x = 0, 1, \ldots$ is an increasing sequence of ordinals with limit λ, see [HP92] Chap. II Sect. 3).

$$\begin{aligned} G_{k,n}(x) &= f_{\omega_{k-1}^n}(x+1) \\ G_n(x) &= f_{\omega_{n-1}}(x+1) \\ G_\infty(x) &= G_x(x+1) \end{aligned}$$

Lemma 7. (1) For each $k > 1$, the system $\{G_{k,n} | n = 1, 2, \ldots\}$ is an envelope for $I\Sigma_k$-provably recursive functions; $I\Sigma_k$ proves that each $G_{k,n}$ is increasing and if $m > n$ then $G_{k,m}$ dominates $G_{k,n}$. Similarly, $\{G_n | n = 1, 2 \ldots\}$ is an envelope for PA-provably recursive functions; PA proves that each G_n is increasing and if $m > n$ then G_m dominates G_n.

(2) $G_{k+1} = G_{k+1,1}$ satisfies $G_{k+1}(x) = G_{k,x}(x+1)$ for all x and hence is not $I\Sigma_k$-provably recursive. G_∞ is not PA provably recursive.

(3) The relation $\{< y, x, k, n > | y = G_{k,n}(x)\}$ is defined by a Δ_0-formula, provably in $I\Sigma_1$.

(4) Let $k \geq 1$. There is a polynomial $p(x)$ such that, for each n, the formula $(\exists y)\big(y = G_{k+1}(n)\big)$ has an $I\Sigma_k$-proof of size $\leq p(|n|)$. Similarly, there is a polynomial $p(x)$ such that, for each n, the formula $(\exists y)(y = G_\infty(n))$ has a PA-proof of size $\leq p(|n|)$.

Proof. (1)-(3) are known; (4) is the main new device. For (1)-(2) see e.g. [HP92] (Chap. II Sec. 3 and Chap. IV Sec. 3). For (3) see [Som90] 4.1. (His definitions are inessentially different.) We prove (4).

Let us inspect the proof of [HP92] II-3.30 (1), saying that for each n $I\Sigma_{k+1}$ proves $(\forall \alpha \prec \omega_k^n)(f_\alpha$ is total). Recall that $L(\omega_k^n, \Sigma_m)$ is a single statement saying that the least number principle holds for all non-empty Σ_m subsets of $(\preceq \omega_k^n)$. Clearly, $I\Sigma_1$ proves $(\forall \beta \prec \alpha)(f_\beta \text{ total} \rightarrow f_\alpha \text{ total})$; thus $L(\omega_k^n, \Sigma_k)$ implies that for all $\alpha \prec \omega_k^n$ (and hence for all $\alpha \leq \omega_k^n$)), the function f_α is total. Thus II.3.30(1) follows from the fact ([HP92] II

3.18) that $I\Sigma_{k+1} \vdash L(\omega_k^n, \Sigma_k)$ for all n. A complexity analysis shows the following:

$$I\Sigma_{k+1} \vdash (\forall x)(L(\omega_1^x, \Sigma_{k+1}) \to L(\omega_k^x, \Sigma_k), \text{ thus} \qquad (1)$$
$$I\Sigma_{k+1} \vdash^{|n|^{O(1)}} L(\omega_1^n, \Sigma_{k+1}) \to L(\omega_k^n, \Sigma_k).$$

$$I\Sigma_{k+1} \vdash L(\omega_1^1, \Sigma_{k+1}) \text{ and} \qquad (2)$$
$$I\Sigma_{k+1} \vdash \forall y(L(\omega_1^y, \Sigma_{k+1} \to L(\omega_1^{y+1}, \Sigma_{k+1}), \text{ thus}$$
$$I\Sigma_{k+1} \vdash^{|n|^{O(1)}} L(\omega_1^n, \Sigma_{k+1})$$

(this needs a usual trick with a definable cut closed under addition). Furthermore, we have

$$I\Sigma_{k+1} \vdash (\forall x)(L(\omega_k^x, \Sigma_2) \to (\forall \alpha \preceq \omega_k^x)(f_\alpha \text{ total})), \text{ thus} \qquad (3)$$
$$I\Sigma_{k+1} \vdash^{|n|^{O(1)}} L(\omega_k^n, \Sigma_2) \to (\forall \alpha \preceq \omega_k^n)(f_\alpha \text{ total})).$$

This completes the proof.

Theorem 8. Let $T \supseteq I\Sigma_{k+1}, k \geq 1$. Write $d(n)$ for $G_{k+1}(n)$.

(a) There is a positive rational δ and that, for all n, each T-proof of $\mathrm{Con}_T(d(n))$ has size $\geq d(n)^\delta$.

(b) There is a polynomial $p(n)$ such that, for all n, there is a T-proof of $\mathrm{Prov}_T(\ulcorner \mathrm{Con}_T(d(nm(n))) \urcorner)$ of size $\leq p(n)$.

Proof. Similar to the proof of Theorem 1. Observe the following:

(1) $T \vdash^{|n|^{O(1)}} nm(d(n)) = d(nm(n))$ (since $y = d(x)$ is Δ_0),
(2)$T \vdash y = d(x) \to \mathrm{Prov}_T(nm(y)) = d(nm(x))$ (provable Σ_1-completeness),
(3) $T \vdash (\forall z)\mathrm{Prov}_T(\ulcorner \mathrm{Con}_T(nm(z) \urcorner)$ – formalize the usual argument for the fact that for each n, $T \vdash \mathrm{Con}_T(n)$.

Now (a) follows similarly as (a) in Theorem 1. On the other hand:

$$T \vdash y = d(x) \to \mathrm{Prov}_T(nm(y)) = d(nm(x)) \,\&\, \mathrm{Con}_T(nm(y)),$$

thus, by Lemma 7(4),

$$T \vdash^{|n|^{O(1)}} \exists y \mathrm{Prov}_T(nm(y) = d(nm(n)) \,\&\, \mathrm{Con}(nm(y))),$$

which completes the proof.

Theorem 9. Let $T \supseteq I\Sigma_k, k \geq 1$, let A/B be a reasonable rule, let $L \vdash \Box A \rightarrow \Box B$. Then there is an open interpretation φ and a polynomial $p(x)$ such that, for each n, $A^\varphi(n)$ has a T-proof of size $\leq p(n)$ and $B^\varphi(n)$ has no proof of size $\leq \alpha(n)$. Thus the rule A/B shortens the proof at least by $(d(n))^\delta$ for some positive rational δ.

Proof. The proof is a simplification of the proof of Theorem 4. Let φ be such that

$$T \vdash \varphi(x) \equiv (\forall v)(v = 2^{2^{d(x)}} \rightarrow (\forall \omega \leq v) \neg \text{ Proof }_T(\omega, \ulcorner B^\varphi(nm(x))\urcorner)\,.$$

First prove, using provable Σ_k-completeness,

$$T \vdash \forall x(d(x) \text{ exists } \rightarrow A^\varphi(x))$$

and then conclude

$$T \vdash^{n^{O(1)}} A^\varphi(n).$$

On the other hand, clearly $B^\varphi(n)$ has no proof of size $\leq |x|$.

Theorem 10. Let T be $I\Sigma_k, k \geq 1$. There is an n such that the function $G_{k+1,n}$ is an upper bound for proofs of A from proofs of $\Box A$.

Proof. Recall that $I\Sigma_{k+1} \vdash \text{Con}(I\Sigma_k + Tr(\Pi_1))$ (see [HP92]). Thus $I\Sigma_{k+1} \vdash (\forall w \in \Sigma_1 -$ sentences $)(\text{Prov}_T(w) \rightarrow Tr_{\Sigma_1}(w))$. In particular,

$$I\Sigma_{k+1} \vdash (\forall z)(\text{Prov}_T(\ulcorner \text{Prov}_T(z)\urcorner)) \rightarrow \text{Prov}_T(z)).$$

This means that the function associating to each T-proof of $\text{Prov}_T(\varphi)$ a T-proof of φ is provably recursive in $I\Sigma_{k+1}$ and hence majorized by some $G_{k+1,n}$.

Remark. Theorem 9 may be used to imitate Theorem 5; prove the following: Let $T = I\Sigma_k, k \geq 1$, let A/B be a reasonable such such that $S \vdash \Box A \rightarrow \Box B$. If $S \vdash \Box A \rightarrow \Box B$ then for some n, $G_{k+1,n}$ is an upper bound of proofs of B from the proofs of A. (See the proof of Theorem 5.)

References

[DM89] De Jongh D.H.J., Montagna F.: *Much shorter proofs;* Zeitschr. Math. Log. 35 (1989), pp. 247-260

[DJM91] De Jongh D.H.J., Jumelet M., Montagna F.: *On the proof of Solovay's Theorem;* Studia Logica 50 (1991) 51-70

[EM71] Ehrenfeucht A., Mycielski J.: *Abbreviating proofs by adding new axioms,* Bull. AMS 77 (1971) 366-367

[FR79] Ferrante J., Rackoff C.: *The computational comlexity of logical theories,* Lect. Notes Math. 718, Springer Verlag 1979.

[HP92] Hájek P. Pudlák P.: *The metamathematics of first order arithmetic;* Springer Verlag Berlin-Heidelberg-New York (in press)

[Mon90] Montagna F.; *Polynomially and superexponentially shorter proofs in fragments of Arithmetic;* Preprint University of Siena 1990; to appear in the Journal of Symbolic Logic.

[Par71] Parikh R.; *Existence and feasibility in arithmetic;* J. Symb. Log. 36 (1971), pp. 494-508.

[Pud86] Pudlák P.: *On the length of proofs of finitistic consistency statements in first order theories;* Logic Colloquium 1984, J. Paris, A.J. Wilkie, G.M. Wilmers Editors North Holland 1986, pp. 165-196

[Pud] Pudlák P.; *Short formulae defined by iteration;* Unpublished manuscript.

[Pud87] Pudlák P.: *Improved bounds to the length of proofs of finitistic consistency statements;* Contemporary Mathematics 65 (1987) pp. 309-331

[Smo85] Smoryński C.: *Self refernce and modal logic;* Springer Verlag (1985)

[Sol76] Solovay R.M.; *Provability interpretations of modal logic;* Israel Journ. of Math. 25 (1976), pp. 287-304

[Som90] Sommer R.: *Transfinite induction and hierarchies generated by transfinite recursion within Peano arithmetic;* PhD thesis, University of California at Berkeley, 1990

[Vis90] Visser A.; *Interpretability Logic;* In: Mathematical logic (Proceedings of the Heyting Conference, Bulgaria 1988), Plenum Press 1990, pp.175-210

[WP87]; Wilkie A.J., Paris J.B.; *On the scheme of induction for bounded arithmetical formulae;* Annals of Pure Applied Logic 35 (1987), pp. 261-302

Petr Hájek
Institute of Computing and Information Science
ČSAV
Pod vodarenskou vezi 2
182 07 Prague, Czechoslovakia
email: hajek@cspgcs11.bitnet

Franco Montagna
Scuola di Spezializzazione in Logica Matematica
Universita di Siena
Via del Capitano 15
I-53100 SIENA
email: montagna@sivax.cineca.it

Pavel Pudlák
Matematicky Ustav
ČSAV
Zitna Ulice 25
115 87 Prague, Czechoslovakia
email: pudlak@csearn.bitnet

Open induction, Tennenbaum phenomena, and complexity theory

Richard Kaye

1 Introduction

This paper collects together several ideas concerning weak systems of arithmetic related to the theory *IOpen* of Shepherdson (1964, 1965) and the theory IE_1 of Wilmers (1985), and describes some of the author's investigations in this area over the last few years. Its two main themes are encapsulated by the following questions: in what way do these theories relate to the computational complexity of natural problems? And: how might one go about proving independence results in this realm? To set the scene and to discuss these questions in more detail we must first give some definitions.

Definitions

In this paper, $\mathcal{L}_\mathsf{A}$ will denote the language of arithmetic with nonlogical symbols $+, \cdot, \dot{-}, <, 0, 1$. This language differs from the usual one by having an extra symbol for the modified subtraction operation $x \dot{-} y =_{\text{def}} \max(x - y, 0)$, but this is merely a matter of convenience. The theory PA^- is the theory of the nonnegative parts of discretely ordered rings, which can be axiomatized in essentially the same way as described in chapter 2 of Kaye (1991), modified only slightly to allow for the new symbol $\dot{-}$. Note that, in the particular language $\mathcal{L}_\mathsf{A}$ chosen here, PA^- is axiomatized by *universal* sentences. All models considered in this paper will be models for languages extending $\mathcal{L}_\mathsf{A}$ and satisfying at least PA^-.

Now consider some language $\mathcal{L}$ extending $\mathcal{L}_\mathsf{A}$. (We shall assume that the standard model $\mathbb{N}$ has some natural interpretation as an $\mathcal{L}$-structure.) An *open* formula of $\mathcal{L}$ is one that is quantifier-free. We shall also use $E_0(\mathcal{L})$ and $U_0(\mathcal{L})$ to denote the collection of open formulas of $\mathcal{L}$. Inductively, we define

$$E_{n+1}(\mathcal{L}) = \{\ \exists \bar{x} < t(\bar{y})\, \theta(\bar{x}, \bar{y}) : \theta \in U_n(\mathcal{L}), t \text{ an } \mathcal{L}\text{-term}\ \}$$

and

$$U_{n+1}(\mathcal{L}) = \{\ \forall \bar{x} < t(\bar{y})\, \theta(\bar{x}, \bar{y}) : \theta \in E_n(\mathcal{L}), t \text{ an } \mathcal{L}\text{-term}\ \}.$$

The $\mathcal{L}$-theories $IE_n(\mathcal{L})$ and $IU_n(\mathcal{L})$ are defined to be some base theory (to be specified in the relevant context, but always including PA^-, and usually also some further axioms involving the other nonlogical symbols from $\mathcal{L} \setminus \mathcal{L}_{\mathsf{A}}$) together with the axiom scheme of induction

$$\forall \bar{a}, b\, \big((\theta(0, \bar{a}) \wedge \forall x < b\, (\theta(x, \bar{a}) \to \theta(x+1, \bar{a}))) \to \forall x \leqslant b\, \theta(x, \bar{a})\big)$$

for all θ from $E_n(\mathcal{L})$, respectively $U_n(\mathcal{L})$. Similarly, $LE_n(\mathcal{L})$ and $LU_n(\mathcal{L})$ are the theories consisting of some base theory together with the least-number principle

$$\forall \bar{a}, b\, \big(\theta(b, \bar{a}) \to \exists x \leqslant b\, (\theta(x, \bar{a}) \wedge \forall z < x\, \neg\theta(z, \bar{a}))\big)$$

for all θ from $E_n(\mathcal{L})$, or $U_n(\mathcal{L})$. When $n = 0$, we denote $IE_0(\mathcal{L})$ (equivalently $IU_0(\mathcal{L})$) by $I\mathit{Open}(\mathcal{L})$ and $LE_0(\mathcal{L})$ (or $LU_0(\mathcal{L})$) by $L\mathit{Open}(\mathcal{L})$. If $\mathcal{L} = \mathcal{L}_{\mathsf{A}}$ we omit mention of it from the notation. Thus $I\mathit{Open}$ is essentially the theory studied by Shepherdson (1964, 1965), and the hierarchy IE_n, IU_n is that introduced by Wilmers (1985).

If Γ is a class of formulas (for example, E_n or U_n) we denote by $\Gamma^{\mathbb{N}}$ the collection of subsets of $\mathbb{N}^{<\omega}$ defined by some formula in Γ. The collections $E_n^{\mathbb{N}}$ and $U_n^{\mathbb{N}}$ form a hierarchy whose union is the Δ_0-definable subsets of $\mathbb{N}^{<\omega}$; note however that it is not known whether this hierarchy is proper or even whether or not it collapses to its first level, i.e., whether or not $E_1^{\mathbb{N}} = U_1^{\mathbb{N}} = \Delta_0^{\mathbb{N}}$. By Parikh's theorem (Parikh 1971; see also Kaye 1991 p. 73), this last statement is equivalent to the statement that $\Pi_1\text{-Th}(\mathbb{N})$ proves the Matijasevič–Robinson–Davis–Putnam theorem on the diophantine representability of r.e. predicates, $\Pi_1\text{-Th}(\mathbb{N}) \vdash \text{MRDP}$. More precisely, we say a theory T *proves* the MRDP theorem, $T \vdash \text{MRDP}$, if for all Δ_0 formulas $\theta(\bar{x})$ there is a quantifier-free formula $\phi(\bar{x}, \bar{y})$ such that

$$T \vdash \forall \bar{x}\, (\theta(\bar{x}) \leftrightarrow \exists \bar{y}\, \phi(\bar{x}, \bar{y})),$$

and a model M *satisfies* the MRDP theorem, $M \vDash \text{MRDP}$, if the same is true but with '$T \vdash$' replaced by '$M \vDash$'. Although $I\mathit{Open} \nvdash \text{MRDP}$, it is not known if $I\Delta_0 \vdash \text{MRDP}$, or even if $IE_1 \vdash \text{MRDP}$. Note also that, by Parikh's theorem, if IE_1 or $I\Delta_0$ does prove the MRDP theorem, then every U_1 formula is equivalent to some E_1 formula in IE_1 (respectively $I\Delta_0$), and in particular it would follow that $E_1^{\mathbb{N}} = U_1^{\mathbb{N}} = \Delta_0^{\mathbb{N}}$.

Motivation

Historically, the family of theories IE_n, IU_n was introduced before Buss's hierarchy $\Sigma_n^b - LIND$ (Buss, 1986), and the language $\mathcal{L}_A$ (or $\mathcal{L}_A$ without modified subtraction) was chosen simply because it is often the most convenient one for stronger theories such as PA and because it is the most natural one for the study of the nonnegative parts of discretely ordered rings. This said however, this choice of language is important for Shepherdson's work, and it is $\mathcal{L}_A$ that makes IE_1 seem at first glance so weak—and in consequence makes Wilmers' main result (that nonstandard models M of IE_1 have recursively saturated reducts $M{\restriction}+)$ concerning IE_1 so surprising. However, there is little other reason for us to restrict our attention to $\mathcal{L}_A$, and the main objective of the current paper is to investigate in general terms the effects of expanding this language in some definable way.

In broad terms, the E_n/U_n hierarchy is closely associated with the linear time hierarchy of Wrathall (1978). She proved that there is a fixed $k \in \mathbb{N}$ such that $E_n^{\mathbb{N}} \subseteq \Sigma_{k+n}^{\mathsf{lin}}$ and $\Sigma_n^{\mathsf{lin}} \subseteq E_{k+n}^{\mathbb{N}}$ for all n, and indeed this result is unaffected by expansion of the language $\mathcal{L}_A$, as long as we restrict ourselves to adding functions or predicates from the linear time hierarchy. It would apparently follow that the study of the hierarchies of theories $IE_n(\mathcal{L})$ is intimately linked with that of the linear time hierarchy, just as Buss's theories are associated with the polynomial time hierarchy, but connections between the theories IE_n and the linear time hierarchy (or even the other natural hierarchy $E_k^{\mathbb{N}}$ ($k \in \mathbb{N}$) of Δ_0 sets) are not worked out in so much detail, and indeed there are reasons for suspecting that such connections may not be so tight. Thus the only result known in the IE_n setting analogous to Buss's results connecting P and $\Sigma_1^b - LIND$ only works in one direction—one can demonstrate that if certain fragments of IE_1 (say) can prove a sentence of a specific restricted form, then that sentence can be 'witnessed' by a function of bounded computational complexity—but these fragments of IE_1 do not seem to capture any particularly natural complexity class. (The result just alluded to was worked out by the author in Manchester several years ago, but appears in this paper below for the first time in printed form.) Nor is there to date any result analogous to that of Krajíček, Pudlák and Takeuti (1991) relating the possible collapse of the hierarchy of theories $\Sigma_1^b - LIND$ to the collapse of the polynomial time hierarchy.

The problems insinuated in the last paragraph may not turn out to be so bad after all, however. The IE_n hierarchy (especially at its lowest levels) has important algebraic and number-theoretic characteristics which could in principle be utilized to prove separation results. This may in turn have a direct bearing on complexity-theoretic issues via results on the diophantine representation of sets of given complexity class. This indeed was Wilmers' motivation for studying IE_1, but the algebraic nature of $IOpen$ is already clear from Shepherdson's work on the remarkable theory $IOpen$, and also later work by Wilkie (1978), van den Dries (1980), Macintyre and Marker (1989). Two related features of $IOpen$ are particularly conspicuous: the theory's tight relationship with real closed fields, enabling Shepherdson to prove for example that induction for open $\mathcal{L}_{\mathsf{A}}$ formulas is equivalent to the least number principle for such formulas; and the presence of recursive nonstandard models of the theory. But the analogous question whether IU_1 proves LU_1 is a well known open problem from Wilmers' 1985 paper (and turns out to be equivalent to asking if IE_1 proves IE_2 as we will see below), and the construction of recursive nonstandard models of interesting theories stronger than $IOpen$ is apparently much harder than was first thought. Two fairly well known open problems in the latter vein are: do recursive nonstandard models of $IOpen$ plus the *normality* axioms exist? (for a discussion of normality see van den Dries 1980) and: do recursive nonstandard models of $IOpen$ having a infinite prime (irreducible, etc.) exist? The second question is a rather embarrassing gap in our knowledge first noticed by Macintyre and Marker (1989), and the first question seems to be still open despite van den Dries' faith that $IOpen$ + normality is a 'weak fragment of arithmetic' (according to his definition) worthy of study by Logicians, Algebraists and Number theorists alike (see van den Dries 1980 paragraph 5.4).

2 Between $IOpen$ and IE_1

Shepherdson (1964, 1965) characterized models M of open induction (for the usual language $\mathcal{L}_{\mathsf{A}}$) in terms of their *real closure* $\mathcal{RC}(M)$: M satisfies $IOpen$ iff every $r > 0$ from $\mathcal{RC}(M)$ has an integral part in M. More precisely,

Result 1 (Shepherdson, 1964) $M \models IOpen$ iff $M \models \mathsf{PA}^-$ and for all $x > 0$ in $\mathcal{RC}(M)$, there exists y in M with $(|y - x| < 1)$.

I would like to sketch the proof of this result since it highlights very nicely the difficulties one might expect in trying to extend it to other languages $\mathcal{L}$ in place of $\mathcal{L}_{\mathsf{A}}$.

For one direction, suppose at first that Q is the field of fractions corresponding to M, $M \models IOpen$, $\bar{a} \in Q$, and $r \in \mathcal{RC}(M)$ satisfies $p(x, \bar{a}) = 0$.

By multiplying out by a common denominator we may suppose $\bar{a} \in Z$, the discretely ordered ring associated with M. Note also that r is bounded above and below by elements of Z. Since $p(x, \bar{a})$ has only finitely many roots in $\mathcal{RC}(M)$ there is $b \in Z$ with $b < r$ such that for any root s of $p(x, \bar{a})$ in the interval (b, r), $(s, r) \cap Z$ is empty; there is also $q \in Q$ with $q > r$ such that $p(b, \bar{a})$ and $p(q, \bar{a})$ have different signs. Then, by the sign change property, there is a quantifier-free formula $\phi(y, \bar{a})$ expressing that 'there is no root s of $p(x, \bar{a})$ in the interval $(b, \min(y, q))$'. By induction on ϕ it follows that there is $t_1 \in Z$ such that $p(x, \bar{a})$ changes sign in the interval $(t_1, \min(t_1 + 1, q))$. Thus t_1 is the integral part of at least one of the roots of $p(x, \bar{a})$; but there are only finitely many such roots, so their integral parts (and, in particular, the integral part of r) may be found in turn by the same method. Finally, if $\bar{a}$ is not rational, this tuple may be approximated sufficiently closely by some $\bar{a}' \in Q$ to give the same result.

For the other direction, consider a quantifier-free formula $\phi(y, \bar{a})$. It is straightforward to see that ϕ is equivalent to a boolean combination of formulas of the form $p(y, \bar{a}) > 0$, where $p(y, \bar{x}) \in \mathbb{Z}[y, \bar{x}]$, and (since each such polynomial has at most finitely many roots in $\mathcal{RC}(M)$) each formula $p(y, \bar{a}) > 0$ is equivalent to a boolean combination of formulas $y < r$ for certain $r \in \mathcal{RC}(M)$. It follows that the set of $y \in M$ satisfying $\phi(y, \bar{a})$ is a union of finitely many intervals in M whose end points are defined by elements of $\mathcal{RC}(M)$. But every $r \in \mathcal{RC}(M)$ has an integral part in M, so $\{y \in M : M \vDash \phi(y, \bar{a})\}$ is a union of finitely many intervals with end points in M, so induction (and also the least-number principle) holds for the formula ϕ.

COROLLARY 2 (Shepherdson, 1964) $I\mathit{Open} \vdash L\mathit{Open}$.

As might be expected from its proof (which uses properties of polynomials in an essential way), the corollary does not work for other languages, as we shall see.

Shepherdson's main application of the characterization of models of $I\mathit{Open}$ was the construction of nonstandard models:

RESULT 3 (Shepherdson, 1964) $I\mathit{Open}$ has nonstandard recursive models.

Shepherdson's model consists of elements of the form

$$a_p X^{p/q} + a_{p-1} X^{(p-1)/q} + \cdots + a_1 X^{1/q} + a_0$$

where $p, q \in \mathbb{N}$, $q > 0$, $a_p, a_{p-1}, \ldots, a_1$ are real algebraic, and a_0 is integral. Addition and multiplication are defined in this model in the obvious way, and the model is made into a discretely ordered ring by making X 'infinitely large'. The model (and also its nonnegative part) is recursive (by coding real algebraic numbers in some suitable way, and using the decidability of real closed fields) and satisfies *IOpen* by the preceding criterion, since its real closure is readily seen to be the field of Puiseux series

$$a_p X^{p/q} + a_{p-1} X^{(p-1)/q} + \cdots + a_1 X^{1/q} + a_0 + a_{-1} X^{-1/q} + \cdots$$

the a_is being real algebraic.

The existence of recursive nonstandard models is an indication that the theory *IOpen* is rather weak, and examination of Shepherdson's model shows that *IOpen* cannot prove even some very basic number-theoretic facts such as the irrationality of $\sqrt{2}$ as well as stronger assertions not known to be provable in more powerful theories.

Despite a great deal of effort by many people, very few interesting recursive models of fragments of arithmetic have been constructed. If we attempt to develop Shepherdson's ideas further to languages $\mathcal{L}$ extending $\mathcal{L}_{\mathsf{A}}$, we must find some criterion by which we can demonstrate that a model satisfies induction for quantifier-free formulas of $\mathcal{L}$. Put another way, we must be able to understand the structure of sets defined by quantifier-free formulas. Unfortunately, this structure becomes extremely complicated even for rather modest extensions of $\mathcal{L}_{\mathsf{A}}$, and recursive nonstandard models for the *least number principle* $L\,Open(\mathcal{L})$ cannot exist for such languages $\mathcal{L}$. (There is the 'back-door' possibility that recursive nonstandard models for the *induction* scheme $I\,Open(\mathcal{L})$ might exist, but I have no bright ideas concerning how one might construct them.) These difficulties are illustrated nicely by the next result.

THEOREM 4 (Wilmers and Kaye) If $\mathcal{L}$ contains the function $\lfloor x/y \rfloor$ and the base theory contains

$$\forall x, y\, (y \neq 0 \rightarrow y \cdot \lfloor x/y \rfloor \leqslant x < y \cdot (\lfloor x/y \rfloor + 1))$$

then $L\,Open(\mathcal{L}) \vdash IE_1(\mathcal{L})$. For $n \geqslant 1$ (and no restrictions on $\mathcal{L}$ or the base theory) we have $LU_n(\mathcal{L}) \vdash IE_{n+1}(\mathcal{L})$.

Proof: We show that $LOpen(\mathcal{L}) \vdash LE_1(\mathcal{L})$. The equivalence of $IE_1(\mathcal{L})$ and $LE_1(\mathcal{L})$ is proved exactly as in Wilmers 1985 for $\mathcal{L} = \mathcal{L}_A$.

Consider for simplicity the $E_1(\mathcal{L})$-formula $\exists y < t(x, \bar{a})\, \theta(x, y, \bar{a})$, where θ is quantifier-free, and suppose r, s exist satisfying $s < t(r, \bar{a}) \wedge \theta(r, s, \bar{a})$ in some model of $LOpen(\mathcal{L})$. Let b be greater than $t(x, \bar{a})$ for all $x \leqslant r$. Then the model satisfies

$$\exists w \left(w - b\lfloor w/b\rfloor < t(\lfloor w/b\rfloor, \bar{a}) \wedge \theta(\lfloor w/b\rfloor, w - b\lfloor w/b\rfloor, \bar{a})\right).$$

If w_0 is the least element satisfying this, then it is clear that $r_0 = \lfloor w_0/b\rfloor$ is least satisfying $\exists y < t(r_0, \bar{a})\, \theta(r_0, y, \bar{a})$, as required. All the other cases are similar. ■

Recall from Wilmers 1985 that all nonstandard models of IE_1 are nonrecursive. In fact, Wilmers proved a much stronger result: all nonstandard models M of IE_1 have recursively saturated reduct $M\restriction+$. Thus the structure of quantifier-free definable sets in nonstandard models of $LOpen(\lfloor x/y\rfloor)$ (i.e., $LOpen(\mathcal{L}_A \cup \{\lfloor x/y\rfloor\})$) is very complicated—at least as far as the least-number principle is concerned.

Notice also that $IOpen$ defines $\lfloor x/y\rfloor$ in such a way that $\lfloor x/y\rfloor$ satisfies the sentence given in the statement of the last theorem, but that there is no obvious translation of open formulas of $\mathcal{L}_A \cup \lfloor x/y\rfloor$ into open formulas of $\mathcal{L}_A$. We shall see later that $IOpen$ and $IOpen(\lfloor x/y\rfloor)$ are not equivalent.

3 Open induction and complexity theory

The main theorem in this section is the 'witnessing theorem' alluded to in the introduction. Its most general form will be given first.

THEOREM 5 (Witnessing theorem for $IOpen(\mathcal{L})$)

Let $\mathcal{L}$ extend $\mathcal{L}_A$ by various relations and functions, and suppose that all new functions are of linear growth rate, i.e., for any such function $f(\bar{x})$ there is a polynomial $t(\bar{x})$ such that (in the natural interpretation of $\mathbb{N}$ as an $\mathcal{L}$-structure)

$$\forall \bar{x}\, (f(\bar{x}) \leqslant t(\bar{x})).$$

Take as base theory $\forall_1\text{–Th}(\mathbb{N})$, the set of all universal sentences that are true in the $\mathcal{L}$-structure $\mathbb{N}$. Suppose that $\theta(\bar{x}, \bar{y})$ is $E_1(\mathcal{L})$ and

$$IOpen(\mathcal{L}) \vdash \forall \bar{x}\, \exists \bar{y}\, \theta(\bar{x}, \bar{y}).$$

Then for some tuple of functions $\bar{F}$ which are computable in deterministic time n^2 using oracles for the functions and relations from $\mathcal{L}$,

$$\mathbb{N} \models \forall \bar{x}\, \theta(\bar{x}, \bar{F}(\bar{x})).$$

Proof: Without loss of generality, we may assume that θ is quantifier-free.

The idea of the proof is to find a Skolemization of $IOpen(\mathcal{L})$ in which the new functions added to the language have a natural interpretation in $\mathbb{N}$ in which they are $DTIME(n^2)$-computable using oracles for $\mathcal{L}$ and are of at most linear growth-rate, i.e., are bounded by some polynomial (just as all $\mathcal{L}$ functions are). It will follow from the Skolemization (using Herbrand's theorem) that there are terms $\bar{F}$ in the Skolem language such that

$$\mathbb{N} \models \forall \bar{x}\, \theta(\bar{x}, \bar{F}(\bar{x})).$$

The key point to note is that this class of oracle-$DTIME(n^2)$-computable linear growth-rate functions is closed under composition, and hence that the terms $\bar{F}$ are in this class. To see this, consider a composite $f(g(x))$. Suppose $g(x)$ is computed in time at most Cn^2, where n is the length of x, $f(y)$ is computed in time at most Dm^2, where m is the length of y, and $g(x) < x^k$ for all but finitely many x, where C, D, k are constants. Then $f(g(x))$ is computed in time at most $D|g(x)|^2 < Dk^2n^2$ for all but finitely many x.

It remains to find the Skolemization. The only axioms of $IOpen(\mathcal{L})$ that are not of $\forall_1$ form are the induction axioms themselves; these can be written as

$$\forall \bar{a}, b\, \exists x < b\, \big(\neg\theta(0, \bar{a}) \vee \theta(b, \bar{a}) \vee (\theta(x, \bar{a}) \wedge \neg\theta(x+1, \bar{a}))\big)$$

where θ is quantifier-free. We shall witness the x in this axiom by an interval-halving algorithm. Note first that, by closure under composition and a straightforward induction on formulas, we can decide the truth value of $\theta(x, \bar{a})$ in $DTIME(n)$ using oracles for $\mathcal{L}$, where n is the length of the input, $\langle x, \bar{a}\rangle$. Thus, given $\bar{a}$ and b, it is possible to verify if $\theta(0, \bar{a})$ and $\neg\theta(b, \bar{a})$ hold. The interval from 0 to b is then halved successively, at each stage retaining an interval $[u, v]$ in which $\theta(u, \bar{a})$ and $\neg\theta(v, \bar{a})$ and checking the truth values of $\theta(u + \lfloor (u+v)/2 \rfloor, \bar{a})$ and $\theta(u + \lceil (u+v)/2 \rceil, \bar{a})$. (It is convenient to check both of these truth values to be sure to eliminate the case when $u+v$ is odd and the truth value of $\theta(x, \bar{a})$ in the range $x \in [u, v]$ happens to change at exactly this point and no other.) After at most $\lfloor \log_2(b) \rfloor$ stages we will have found x such that $(\theta(x, \bar{a}) \wedge \neg\theta(x+1, \bar{a}))$ as required. The total time taken will then be $C\lfloor \log_2(b) \rfloor n$ where n is the length of $\langle b, \bar{a}\rangle$ and C is a constant. ∎

COROLLARY 6 If $\mathcal{L}$ and $I\,Open$ are as in the theorem, but in addition all symbols in $\mathcal{L}$ are computable in $DTIME(n^k)$ for some fixed $k \in \mathbb{N}$, $\theta(\bar{x}, \bar{y})$ is $E_1(\mathcal{L})$ and

$$I\,Open(\mathcal{L}) \vdash \forall \bar{x}\, \exists \bar{y}\, \theta(\bar{x}, \bar{y})$$

then for some tuple of functions $\bar{F}$ in $DTIME(n^{k+1})$,

$$\mathbb{N} \models \forall \bar{x}\, \theta(\bar{x}, \bar{F}(\bar{x})).$$

■

As is usual with this kind of result, there is a version for relations too. Let us say *a formula $\theta(\bar{x})$ is ∇_1 in the theory T (or model M)* iff there are E_1 and U_1 formulas $\theta_E(\bar{x})$ and $\theta_U(\bar{x})$ in the language of T (M) such that T proves (or M satisfies)

$$\forall \bar{x}\, \big((\theta_U(\bar{x}) \leftrightarrow \theta(\bar{x})) \wedge (\theta_E(\bar{x}) \leftrightarrow \theta(\bar{x}))\big).$$

COROLLARY 7 (i) With $\mathcal{L}$ and $I\,Open$ as in the theorem, if $\theta(\bar{x})$ is ∇_1 in $I\,Open(\mathcal{L})$, then $\theta(\bar{x})$ defines a predicate in $\mathbb{N}$ that is $DTIME(n^2)$-computable with oracles for $\mathcal{L}$.

(ii) With $\mathcal{L}$, k, and $I\,Open$ as in the previous corollary, if $\theta(\bar{x})$ is ∇_1 in $I\,Open(\mathcal{L})$, then $\theta(\bar{x})$ defines a $DTIME(n^{k+1})$-computable predicate.

Proof: For (i), consider the theorem applied to

$$I\,Open \vdash \forall \bar{x}\, \exists y < 2\, \big((y = 0 \wedge \neg\theta_U(\bar{x})) \vee (y = 1 \wedge \theta_E(\bar{x}))\big).$$

(ii) is similar. ■

As an application we consider factorization. The problem of factorizing a given integer can be conveniently stated as a decision problem in the following manner:

$$FACT(a, b) \Leftrightarrow_{\text{def}} \exists x < b\, x | a.$$

Notice that $FACT(a, b)$ is naturally written as an E_1 formula. $FACT$ is widely believed to be not polynomial-time computable. Assuming the weaker hypothesis that $FACT$ is not computable in $DTIME(n^3)$ the corollary shows that $FACT(a, b)$ is not equivalent to a U_1 formula in the theory $I\,Open(\lfloor x/y \rfloor)$. Interpreted model theoretically, this says,

COROLLARY 8 Suppose $FACT \notin DTIME(n^3)$. Then there are models $M_1 \subseteq M_2$ of $IOpen(\lfloor x/y \rfloor)$ with $a, b \in M_1$ such that

$$M_1 \vDash \forall x < b \, \neg x | a$$

but

$$M_2 \vDash \exists x < b \, x | a.$$

Proof: By the preservation theorem for $\forall_1$ sentences, it is sufficient to obtain a model M_1 of

$$IOpen(\lfloor x/y \rfloor) \; + \; \forall x < b \, \neg x | a \; + \; \forall_1 - \mathrm{Th}\big(IOpen(\lfloor x/y \rfloor) + \exists x < b \, x | a\big)$$

in the language $\mathcal{L} = \mathcal{L}_\mathsf{A} \cup \{\lfloor x/y \rfloor, a, b\}$ where a, b are new constant symbols. ($\forall_1 - \mathrm{Th}(T)$ denotes the set of $\forall_1$ sentences provable from T.)

To see that this last theory is indeed consistent, suppose at first that it isn't. Then there is a universal sentence $\forall \bar{y} \, \theta(a, b, \bar{y})$ (where θ is quantifier-free in $\mathcal{L}_\mathsf{A} + \lfloor x/y \rfloor$) such that

$$IOpen(\lfloor x/y \rfloor) \; + \; \exists x < b \, x | a \; \vdash \; \forall \bar{y} \, \theta(a, b, \bar{y})$$

and

$$IOpen(\lfloor x/y \rfloor) \; + \; \forall \bar{y} \, \theta(a, b, \bar{y}) \; \vdash \; \exists x < b \, x | a.$$

This last statement can be rewritten as

$$IOpen(\lfloor x/y \rfloor) \; \vdash \; \exists \bar{y} \, (\neg \theta(a, b, \bar{y}) \vee \exists x < b \, x | a)$$

and so by Parikh's theorem there is a term $t(a, b)$ such that

$$IOpen(\lfloor x/y \rfloor) \; \vdash \; \forall \bar{y} < t(a, b) \, \theta(a, b, \bar{y}) \to \exists x < b \, x | a).$$

Clearly, we also have

$$IOpen(\lfloor x/y \rfloor) \; \vdash \; \exists x < b \, x | a \to \forall \bar{y} < t(a, b) \, \theta(a, b, \bar{y}),$$

so $FACT(a, b)$ is ∇_1 in $IOpen(\lfloor x/y \rfloor)$, and hence is $DTIME(n^3)$ computable. ■

It is worth mentioning that Macintyre and Marker have proved results similar to (and indeed, stronger than) the last corollary for the theories *IOpen* and *IOpen* + normality, without making any complexity-theoretic assumptions. Thus their results can be regarded as a kind of lower bound on the computational complexity of primality, and proved using algebraic and model theoretic methods.

Combining results presented above we can deduce several rather curious results, of which the following is a sample.

THEOREM 9' Suppose *FACT* is not computable in $DTIME(n^3)$ then *either*:

1. the U_1 formula '$\forall y < b\, \neg y|a$' is not equivalent to any E_1 formula in the theory IE_1 (and, in particular, IE_1 does not prove the MRDP theorem on the diophantine representation of r.e. predicates); *or*

2. $IOpen(\lfloor x/y \rfloor) \not\vdash LOpen(\lfloor x/y \rfloor)$. ■

Of course, the methods used hold for many other theories of open induction (axiomatized in a similar way but using different languages), other decision problems in place of *FACT*, and other complexity classes in place of $DTIME(n^3)$, but the last result seems to be particularly elegant. Results like these are curious however, because I do not at present know which of the two alternatives is true (although one may suspect that they both are). No witnessing theorems are known for the $\forall E_1$ consequences of IE_1, even though the strongest statement of this form that has been found is only the statement of Euclid's algorithm for highest common factor (Wilmers, 1985), but on the other hand it is not apparently very easy to construct models of $IOpen(\lfloor x/y \rfloor)$.

4 Independence results

Most of the independence results known in this area are obtained by a rather ad hoc examination of simple models, such as Shepherdson's model described above. An exception to this rule is the general method used by Macintyre and Marker (1989) to construct models of Open induction with rather wild behaviour of the notions 'prime', 'irreducible', etc. This is done by using the algebraic criterion for a model to be extendible to a model of *IOpen* or *IOpen* + normality given by Wilkie (1978) and van den Dries (1980) respectively, and then constructing a suitable union of chains, dovetailing in suitable properties to be forced in the final structure. But since it depends so much on this algebraic criterion, the method is not applicable (at least at present) to stronger systems. It does illustrate however the importance of understanding the universal consequences, $\forall_1 - \mathrm{Th}(IOpen(\mathcal{L}))$, of the theory $IOpen(\mathcal{L})$ of open induction in the language $\mathcal{L}$: since *IOpen* is $\forall\exists$ axiomatized two obvious approaches to constructing models are the union of chains method, and model theoretic forcing with quantifier-free conditions, and for both a knowledge of $\forall_1 - \mathrm{Th}(IOpen(\mathcal{L}))$ is vital.

As examples of the more ad hoc methods, I shall give two arguments that establish the following result.

THEOREM 10 $IOpen \nvdash IOpen(\lfloor x/y \rfloor)$

Proof: Take a nonstandard model $M \models \mathsf{PA}$. Choose $\delta > \mathbb{N}$ in M and $a \in M$ such that $M \models a = \delta! \cdot p$ for some large prime $p \in M$ with $p > \delta!^n$ for all standard n. Let Z be the full subring generated by a, R the real closure of Z (taking $R \subseteq \mathcal{RC}(M)$), and let K be the structure with domain consisting of all $k \in M$ that are integral parts of nonnegative elements of R. Then $a \in K$ and K is a model of $IOpen$ by Shepherdson's criterion. But for all $x \in K$, $K \models x|a$ if and only if $M \models x|a$ since M and K are models of $IOpen$ and $x|y$ is ∇_1 in $IOpen$. Thus, for any $n > 0$ from $\mathbb{N}$, $M \models n|a$, so $K \models n|a$ since $n < \delta$. But no nonstandard $y < \delta!$ is in K, for if it were it would be $\lfloor r \rfloor$ for some nonstandard $r \in R$ with $r^n < a$ for all $n \in \mathbb{N}$, contradicting the transcendence degree of R being 1. Thus all other divisors of a in K (other than those in $\mathbb{N}$) are multiples of p. But by construction $p > \lfloor \sqrt{a} \rfloor \in K$. Let $b = \lfloor \sqrt{a} \rfloor$. Then

$$\mathbb{N} = \{x \in K : x = 0 \vee (x < b \wedge x \cdot \lfloor a/x \rfloor = a)\}$$

so K fails to satisfy induction in the language $\mathcal{L}_{\mathsf{A}} \cup \{\lfloor x/y \rfloor\}$. ■

Second proof: I shall show that in fact Shepherdson's model, described in section 2 above, does not satisfy $IOpen(\lfloor x/y \rfloor)$ (although the instance of the induction scheme that I will show fails in this model is slightly more complicated).

Let M be Shepherdson's model and consider the subset

$$S = \{y : y = 0 \vee (y|X \wedge y + 1|X)\}.$$

If $y \in \mathbb{N}$ then $y \in S$ for $1/y$ is (trivially) real and algebraic when $y > 0$. It is easy to check that, if $y > \mathbb{N}$ and $y|X$ then $y = aX^q$ for some real algebraic a and some $q \in \mathbb{Q}$ with $0 < q \leqslant 1$, so $y \notin S$. Thus $S = \mathbb{N}$ and hence fails to satisfy induction. ■

It is also interesting to contemplate upon which of the theories I have been looking at could prove the MRDP theorem. Of course, we suspect that the answer is 'none of them', but the challenge is to prove it!

Using results of the last section we can make the following observation: if $IOpen(\lfloor x/y \rfloor)$ proves the MRDP theorem, then every Δ_0 formula is ∇_1 in $IOpen(\lfloor x/y \rfloor)$, so defines a $DTIME(n^3)$ predicate. This seems particularly unreasonable (but, as I write, I can't see an argument that shows it impossible).

It *is* possible to give a more direct argument why the original theory $IOpen$ cannot prove the MRDP theorem. One way of seeing this is by observing that the last theorem gave a Δ_0 (in fact, quantifier-free) formula ϕ defining $\mathbb{N}$ in the recursive model M. If M satisfied the MRDP theorem every Δ_0 formula would be equivalent in M to both $\exists_1$ and $\forall_1$ formulas. In particular any predicate defined by an arithmetic formula $\theta(\bar{n})$ would then be recursive: $\mathbb{N} \models \theta(\bar{n})$ would be equivalent to $M \models \theta^\phi(\bar{n})$ where θ^ϕ is the formula obtained by relativizing all unbounded quantifiers in θ to ϕ; θ^ϕ is Δ_0 in M (by using a nonstandard parameter to bound all unbounded quantifiers) so is equivalent to both $\exists_1$ and $\forall_1$ forms; but then, since M is recursive, it follows that $\theta^\phi(\bar{n})$ is both r.e. and co-r.e., so is recursive.

The argument in the last paragraph can be given many interesting generalizations, and these suggest a method (for me, by far the most hopeful method) of showing the MRDP theorem to be independent of certain extensions of $IOpen$ or possibly even of IE_1 or—the ultimate goal—of Π_1−Th($\mathbb{N}$) itself. These ideas have already been mentioned in Kaye 1989 and also (briefly) in the hints to exercise 11.12 on p. 157 of Kaye 1991, but I crave the reader's indulgence to let me digress and discuss them in slightly more detail here.

THEOREM 11 A nonstandard arithmetical model M of PA^- + MRDP satisfies Δ_0-overspill.

Proof: The proof has more or less already been given. If Δ_0 overspill fails, i.e., if $\mathbb{N}$ is definable in M by a Δ_0 formula, then the truth of any arithmetical predicate can be reduced to a predicate that is recursive in the operations $+$ and $\cdot$ of M. ∎

There is a variation for a *uniform* version of the MRDP theorem too. For the purposes of the next result, say M satisfies *the uniform MRDP theorem* if there is a $\mathcal{L}_\mathsf{A}$-definable function $f\colon \mathbb{N} \to \mathbb{N}$ such that for all $\theta(\bar{x})$ in Σ_1 there is $\varphi(\bar{x})$ in $\exists_1$ with $\ulcorner\varphi\urcorner = f(\ulcorner\theta\urcorner)$ such that

$$M \models \forall\bar{x}\,\big(\theta(\bar{x}) \leftrightarrow \varphi(\bar{x})\big).$$

Then

THEOREM 12 If $M \models \mathsf{PA}^-$ satisfies the uniform MRDP theorem and fails to satisfy Δ_0 overspill, then the degree of M (i.e., the degree of the sup of M's $+$ and $\cdot$ operations) is greater than or equal to $\mathbf{0}^{(\omega)}$.

Proof: Under the circumstances given, the complete satisfaction relation on $\mathbb{N}$ reduces to $+$ and $\cdot$ on M. ■

From these results, the usual corollaries such as Tennenbaum's theorem etc., follow.

COROLLARY 13 A nonstandard model satisfying PA^- + MRDP is non-recursive. ■

COROLLARY 14 An arithmetical nonstandard model of PA^- + MRDP has standard system that is a Scott set. ■

Thus any construction of models that are not very saturated—i.e., they fail to satisfy Δ_0 overspill—will give independence results for the MRDP theorem, and would be of great interest. Obviously, one can try to apply these ideas directly to theories of open induction. It is not quite so obvious how to apply them to the theories IE_1 or $\Pi_1{-}\mathrm{Th}(\mathbb{N})$ since the former's nonstandard models are rather complicated, by George Wilmers' results, and the latter theory proves $I\Delta_0$ so all its nonstandard models have Δ_0 overspill. For these theories I remind the reader of the following results from Kaye 1990:

RESULT 15 1. If T is a true Π_1-axiomatized $\mathcal{L}_\mathsf{A}$-theory that proves the MRDP theorem, then T has a $\forall E_1$ axiomatization.

2. If $IE_1 \vdash$ MRDP, then IE_1 is equivalent to a finite fragment of its parameter-free version, IE_1^-.

It follows that, if each true $\forall E_1$ sentence σ has an arithmetical model failing to satisfy Δ_0 overspill, then $\Pi_1{-}\mathrm{Th}(M)$ does not prove the MRDP theorem, and hence $E_1^{\mathbb{N}} \neq U_1^{\mathbb{N}}$, and if each finite fragment of IE_1^- has a nonstandard model in which $M{\restriction}+$ is not recursively saturated, then $IE_1 \nvdash$ MRDP.

5 Conclusions

The work above collects together several of my thoughts concerning connections between complexity and arithmetic in the realm of IE_1 and $IOpen$.

The connections are not very tight, but this suggests the use of algebraic and number theoretic methods to tackle problems which are sufficiently similar to the main-line complexity theoretic problems to be of great interest all round. Indeed some of the main questions posed in this area may turn out to be equivalent to main-line complexity theory problems: it is unknown if NP is equal to the set of problems

$$\{\exists \bar{y} < 2^{(|x|^k)} \, p(x, \bar{y}) = 0 : k \in \mathbb{N}, p(x, \bar{y}) \in \mathbb{Z}[x, \bar{y}]\}$$

to be solved over $\mathbb{N}$. Similarly (but presumably more difficult to show) it is plausible that $E_1^{\mathbb{N}} = \Sigma_1^{\mathsf{lin}}$. This would relate the above discussion to complexity theory in a much tighter way. One is also reminded of the non-relativizable proof of the MRDP theorem, and the facts that $P = NP$ and similar problems are known not to relativize.

If the approach advocated above of looking at stronger theories of Open induction, axiomatized by extending the language, is to make sense in the long run, it will be necessary to find sentences that the extensions can prove but the original $IOpen$ cannot. Two examples were given in section 4 above, but these were rather artificial. Unfortunately, I do not know of any $\forall_1$ sentences of $\mathcal{L}_{\mathsf{A}}$ that are provable in (say) $IOpen(\lfloor x/y \rfloor)$ but not in $IOpen$ itself. Nor do I know any such extension of $IOpen$ that proves the irrationality of $\sqrt{2}$, normality, or Fermat's last theorem for exponent 3. Certainly one of the biggest problems left over from the current work is to understand the universal consequences of the theories I consider, just as Wilkie did for $IOpen$. Here is a sample question: are $IOpen$ and $IOpen(\lfloor x/y \rfloor)$ conservative for $\forall_1$ sentences of $\mathcal{L}_{\mathsf{A}}$? A positive answer would show that $IOpen(\lfloor x/y \rfloor)$ and $LOpen(\lfloor x/y \rfloor)$ are not equal, since there is no such conservation result between $IOpen$ and IE_1. As I mentioned in section 3, understanding the universal consequences of these theories is an important prerequisite for constructing models.

6 References

Buss, S. (1986) *Bounded arithmetc.* Bibliopolis, Napoli.

van den Dries, L. (1980) Some model theory and number theory for models of weak systems of arithmetic. In: *Model theory or algebra and arithmetic,* Proceedings, Karpacz 1979 (ed. L. Pacholski et al) *Lecture Notes in Mathematics,* **834**, pp. 346–62, Springer-Verlag, Berlin.

Kaye, R. (1989) Parameter-free universal induction. *Zeitschrift für Math. Logik,* **35**, pp. 443–56.

Kaye, R. (1990) Diophantine induction. *Annals of Pure and Applied Logic*, **46**, pp. 1–40.

Kaye, R. (1991) *Models of Peano arithmetic*. Oxford Logic Guides, 15. Oxford University Press.

Krajíček, J., P. Pudlák, and G. Takeuti (1991) Bounded arithmetic and the polynomial hierarchy. *Annals of Pure and Applied Logic*, **52**, pp. 143–53.

Macintyre, A.J. and D. Marker (1989) Primes and their residue rings in models of open induction. *Annals of Pure and Applied Logic*, **38**, pp. 57–77.

Parikh, R. (1971) Existence and feasibility in arithmetic. *Journal of Symbolic Logic*, **36**, pp. 494-508.

Schmerl, U.R. (1988) Diophantine equations in fragments of arithmetic. *Annals of Pure and Applied Logic*, **38**, pp. 135–70.

Shepherdson, J.C. (1964) A non-standard model for a free variable fragment of number theory. *Bull. Polish Acad. Sci.*, **12**, pp. 79–86.

Shepherdson, J.C. (1965) Non-standard models for fragments of number theory. In: *Symposium on the theory of models* (ed. J.W. Addison et al) pp. 342–358, North Holland, Amsterdam.

Wilkie, A.J. (1978) Some results and problems on weak systems of arithmetic. In: *Logic Colloquium '77* (ed. A. Macintyre et al) pp. 285–296, North Holland, Amsterdam.

Wilmers G.M. (1985) Bounded existential induction *Journal of Symbolic Logic*, **50**, pp. 72–90.

Wrathall, C. (1978) Rudimentary predicates and relative computation. *SIAM Journal of Computing* **7** number 2.

Richard Kaye
Jesus College
Oxford University
Oxford, OX1 3DW
United Kingdom
email: KAYE@vax.oxford.ac.uk

Using Herbrand-type theorems to separate strong fragments of arithmetic

Richard Kaye

1 Introduction

Many authors have emphasized the analogies between the fragments Σ_n^{b}-IND of $I\Delta_0 + \forall x\, x^{\log x}$ *exists* and the fragments $\mathrm{I}\Sigma_n$ of PA. Sometimes this is helpful, but often one feels that the bounded hierarchy of theories is of a rather different nature, and new techniques must be developed to answer the key questions concerning them. But this may not hold the other way round: techniques developed for Σ_n^{b}-IND are often applicable in some form to the stronger theories $\mathrm{I}\Sigma_n$.

A beautiful result, due to Krajíček, Pudlák and Takeuti (1991) separates the fragments Σ_i^{b}-IND and $\Sigma_{i+1}^{\mathrm{b}}$-LIND of bounded arithmetic under the assumption $\Pi_{i+2}^{\mathrm{p}} \neq \Sigma_{i+2}^{\mathrm{p}}$. By the analogy referred to in the last paragraph, a similar argument should separate the strong fragments $\mathrm{I}\Sigma_n$, but no such argument was previously known: to use only the assumption $\Pi_{i+2}^{\mathrm{p}} \neq \Sigma_{i+2}^{\mathrm{p}}$ Krajíček, Pudlák and Takeuti had to bypass the usual approach which attempts to formalize and use truth definitions within a model or a theory. It therefore would seem interesting to find an argument separating, say, $\mathrm{I}\Sigma_n$ and $\mathrm{I}\Sigma_{n+1}$ based on the type of Herbrand witnessing theorem they use. This is what I shall do here. In fact, I shall give a new proof of the well-known result that the true Π_2 theory of $\mathbb{N}$ does not prove Σ_1 induction. The argument clearly relativizes to all higher levels, giving $\Pi_{n+1} - \mathrm{Th}(\mathbb{N}) \nvdash \mathrm{I}\Sigma_n$, and I shall omit most of the details. After this, I shall discuss how to modify the technique to give the results $\Pi_2 - \mathrm{Th}(\mathbb{N}) + \mathrm{B}\Sigma_1 \nvdash \mathrm{I}\Sigma_1$ and $\Pi_{n+1} - \mathrm{Th}(\mathbb{N}) + \mathrm{B}\Sigma_n \nvdash \mathrm{I}\Sigma_n$.

These arguments give new information concerning which induction axioms are provable in $\Pi_2 - \mathrm{Th}(\mathbb{N})$, as we shall see. In fact, some of the arguments in the current paper are similar in style to those in Adamowicz 1988, in which she gives an alternative argument that shows $\Pi_2 - \mathrm{Th}(\mathbb{N}) \nvdash \mathrm{B}\Sigma_1$. As far as I know, Adamowicz was the first to use the more complex

Herbrand-type theorems to separate fragments of arithmetic. She, however, used truth definitions at the last step of her argument rather than the tricky combinatorics of Krajíček, Pudlák and Takeuti, or the direct recursion-theoretic argument I shall use here.

2 Results

I shall work with two languages of arithmetic: the usual language, $\mathcal{L}_{\mathsf{A}} = \{0, 1, +, \cdot, <\}$; and $\mathcal{L}_{\mathsf{rec}}$, the language consisting of (symbols for) all recursive functions and relations on $\mathbb{N}^m$ for some m. The standard model for $\mathcal{L}_{\mathsf{A}}$, $\mathbb{N}$, is expanded to an $\mathcal{L}_{\mathsf{rec}}$-stucture $\mathbb{N}_{\mathsf{rec}}$ in the natural way. The collection, $\forall_1 - \mathrm{Th}(\mathbb{N}_{\mathsf{rec}})$, of true $\forall_1$ sentences of $\mathcal{L}_{\mathsf{rec}}$ can be thought of as a Skolemization of the collection, $\Pi_2 - \mathrm{Th}(\mathbb{N})$, of true Π_2 sentences of $\mathcal{L}_{\mathsf{A}}$.

I shall prove here that $\Pi_2 - \mathrm{Th}(\mathbb{N})$ does not prove certain instances of $\mathrm{I}\Sigma_1$. For convenience, I shall take a different formulation of $\mathrm{I}\Sigma_1$ than is usual, namely the scheme of *strong* Σ_1 *collection* given by

$$\forall a, z\, \exists t\, \forall x \leqslant z\, \big(\, \exists y\, \theta(x, y, a) \to \exists y \leqslant t\, \theta(x, y, a)\, \big)$$

for all θ from Σ_1 (equivalently Δ_0) with the free variables shown. (This scheme is apparently due to Clote who proved it equivalent to $\mathrm{I}\Sigma_1$ in 1982, and subsequently published this observation as theorem 4 on page 40 of Clote 1985. Ressayre coined the term 'strong collection'. A proof that it is equivalent to the usual induction scheme over a finite fragment of $\mathrm{I}\Delta_0$ can also be found in Kaye 1991, lemma 10.6 on p.135 and exercise 10.6 on p.139. In particular, note that the totality of 2^x can be proved by taking t such that

$$\forall z \leqslant x\, \big(\, \exists y\, y = 2^z \to \exists y \leqslant t\, y = 2^z\, \big)$$

and then looking at the least $z \leqslant x$ such that $\neg \exists y \leqslant t\, y = 2^z$ if such z exists.)

The main Herbrand-style lemma concerning this strong collection scheme is the following.

Lemma 1 Suppose $\theta(x, y, a)$ is Δ_0 and

$$\Pi_2 - \mathrm{Th}(\mathbb{N}) \vdash \forall a, z\, \exists t\, \forall x \leqslant z\, \big(\, \exists y\, \theta(x, y, a) \to \exists y \leqslant t\, \theta(x, y, a)\, \big).$$

Then there are $k \in \mathbb{N}$ and recursive functions $f_1, \ldots, f_k$ such that for all $a, z \in \mathbb{N}$

either
(0) $\forall x \leqslant z \, \forall y \, (\theta(x,y,a) \to \exists y \leqslant f_1(a,z) \, \theta(x,y,a))$
or
(1) for all $x_1 \leqslant z$, and y_1 such that $\theta(x_1, y_1, a)$ and $\forall y \leqslant f_1(a,z) \, \neg\theta(x_1, y, a)$,
either
(1.0) $\forall x \leqslant z \, \forall y \, (\theta(x,y,a) \to \exists y \leqslant f_2(a,z,x_1,y_1) \, \theta(x,y,a))$
or
(1.1) for all $x_2 \leqslant z$, and y_2 such that $\theta(x_2, y_2, a)$ and $\forall y \leqslant f_2(a,z,x_1,y_1) \, \neg\theta(x_2,y,a)$, ...
...

$$(1.1\ldots 0) \quad \forall x \leqslant z \, \forall y \left(\begin{array}{l} \theta(x,y,a) \to \\ \exists y \leqslant f_k(a,z,x_1,y_1,\ldots,x_{k-1},y_{k-1}) \, \theta(x,y,a) \end{array} \right).$$

Proof: If the conclusion was false, then for some sequence

$$f_1(a,z), f_2(a,z,x_1,y_1), f_3(a,z,x_1,y_1,x_2,y_2), \ldots$$

of all terms in $\mathcal{L}_{\mathsf{rec}}$ with the variables shown there would be elements $a, z, \bar{x}, \bar{y}$ of $\mathbb{N}$ such that, for all k,

$$\begin{array}{rl} \mathbb{N}_{\mathsf{rec}} \vDash & \theta(x_1,y_1,a) \wedge \forall y \leqslant f_1(a,z) \, \neg\theta(x_1,y,a) \\ & \wedge\theta(x_2,y_2,a) \wedge \forall y \leqslant f_2(a,z,x_1,y_1) \, \neg\theta(x_2,y,a) \\ & \wedge \ldots \\ & \wedge\theta(x_k,y_k,a) \wedge \forall y \leqslant f_k(a,z,x_1,y_1,\ldots,x_{k-1},y_{k-1}) \, \neg\theta(x_k,y,a). \end{array}$$

A simple compactness argument then gives a model M with

$$a, z, x_1, y_1, \ldots, x_k, y_k, \ldots \in M \vDash \mathrm{Th}(\mathbb{N}_{\mathsf{rec}})$$

such that

$$M \vDash \theta(x_k,y_k,a) \wedge \forall y \leqslant f_k(a,z,x_1,y_1,\ldots,x_{k-1},y_{k-1}) \, \neg\theta(x_k,y,a)$$

for each k. Then the substructure K consisting of all elements

$$f_k(a,z,x_1,y_1,\ldots,x_{k-1},y_{k-1})$$

from M satisfies $\forall_1 - \mathrm{Th}(\mathbb{N}_{\mathsf{rec}})$, hence $\Pi_2 - \mathrm{Th}(\mathbb{N})$, and also

$$\theta(x_k,y_k,a) \wedge \forall y \leqslant f_k(a,z,x_1,y_1,\ldots,x_{k-1},y_{k-1}) \, \neg\theta(x_k,y,a)$$

since θ is Δ_0. Hence

$$K \nvDash \exists t \, \forall x \leqslant z \, (\exists y \, \theta(x,y,a) \to \exists y \leqslant t \, \theta(x,y,a))$$

since every $t \in K$ is $f_k(a,z,x_1,y_1,\ldots,x_{k-1},y_{k-1})$ for some k. ∎

We now use this to prove...

THEOREM 2 Suppose $\theta(x,y,a)$ is Δ_0 and

$$\Pi_2 - \mathrm{Th}(\mathbb{N}) \vdash \forall a, z\, \exists t\, \forall x \leqslant z\, \big(\, \exists y\, \theta(x,y,a) \rightarrow \exists y \leqslant t\, \theta(x,y,a)\,\big).$$

Then $\forall x\, \exists y\, \theta(x,y,a)$ is equivalent to

$$\exists c \left\{ \begin{array}{l} \forall x\, \theta(x, f(x,a,c), a) \wedge \\ \forall x, y\, (\theta(x,y,a) \rightarrow \theta(x, f(x,a,c), a)) \end{array} \right\}$$

for some recursive function f.

Proof: Consider the following algorithm $A(a,z,l)$ with inputs a, z and $l \leqslant k$. A runs in l steps:

Step 1: compute $f_1 = f_1(a,z)$;

Step 2: search systematically (in some predetermined fashion, not depending on l) for x_1, y_1 such that $x_1 \leqslant z$, $y_1 > f_1$, $\theta(x_1, y_1, a)$, and $\forall y < y_1\, \neg\theta(x_1, y, a)$; compute $f_2 = f_2(a, z, x_1, y_1)$;

...

Step l: search for x_{l-1}, y_{l-1} such that $x_{l-1} \leqslant z$, $y_{l-1} > f_{l-1}$, $\theta(x_{l-1}, y_{l-1}, a)$, and $\forall y < y_{l-1}\, \neg\theta(x_{l-1}, y, a)$; compute $f_l = f_l(a, z, x_1, y_1, \ldots, x_{l-1}, y_{l-1})$.

Note that, for given a, z, $A(a,z,l)$ only converges (halts) for small enough l. If $f_1, \ldots, f_k$ are given as in the lemma, $A(a,z,l)$ converges, but $A(a,z,l+1)$ does not, then the output f_l of $A(a,z,l)$ satisfies

$$\forall x \leqslant z\, \forall y\, (\theta(x,y,a) \rightarrow \exists y' \leqslant f_l\, \theta(x,y',a)).$$

Define

$$l(a) = \max\{l \leqslant k : l \geqslant 1 \text{ and for cofinally many } z \in \mathbb{N}\, A(a,z,l) \text{ converges}\}$$

and

$$g(a) = \max\{z : A(a,z,l(a)+1) \text{ converges}\},$$

or $g(a) = 0$ if $l(a) = k$. Then the 'advice' c in the statement of the theorem is the pair $c = \langle l(a), g(a)\rangle$. Given this, the recursive function $f(x,a,c)$ in the theorem is computed following the following algorithm:

Given x, a and c as above, search for $z > g(a)$ such that $A(a,z,l(a))$ converges, and compute f_l, the output of $A(a,z,l(a))$. We are guaranteed to find such z by the definition of $g(a)$ and $l(a)$. Then, by the definition of $g(a)$, it is easy to see that $A(a,z,l(a)+1)$ does not converge, so to verify if y exists satisfying $\theta(x,y,a)$ it suffices to search through all $y \leqslant f_{l(a)}$. Do this, and let $f(x,a,c)$ be the first such y found (if any).

This completes the proof of the theorem. ■

The proof easily relativizes, so we get

THEOREM 3 Suppose $A \subseteq \mathbb{N}$, $\theta(x,y,a)$ is Δ_0^A (i.e., is Δ_0 in the language $\mathcal{L}_\mathsf{A}$ expanded by adding a new predicate for A) and the Π_2^A formula $\forall x\, \exists y\, \theta(x,y,a)$ is not equivalent to any Σ_2^A formula. Then

$$\Pi_2 - \mathrm{Th}(\mathbb{N}, A) \nvdash \forall a, z\, \exists t\, \forall x \leqslant z\, \big(\exists y\, \theta(x,y,a) \rightarrow \exists y \leqslant t\, \theta(x,y,a)\big)$$

■

If $n \geqslant 1$ and A is any set with degree $\mathbf{0}^{(n)}$ (e.g., the usual Σ_n truth definition Sat_{Σ_n}—see for example Kaye 1991, chapter 9) then $\Pi_2 - \mathrm{Th}(\mathbb{N}, A)$ is a conservative extension of $\Pi_{n+2} - \mathrm{Th}(\mathbb{N})$ (because the Sat_{Σ_n} is reducible to the degree $\mathbf{0}^{(n)}$ set A and we are working with the full Π_2 theory of $\mathbb{N}$), so the relativized version of the theorem gives $\Pi_{n+2} - \mathrm{Th}(\mathbb{N}) \nvdash \mathrm{I}\Sigma_{n+1}$.

The relativized theorem gives us extra information that can be used to improve the well-known result that $\Pi_2 - \mathrm{Th}(\mathbb{N}) \nvdash \mathrm{I}\Sigma_1$. Recall that an r.e. set A is *high* if the degree of its jump, A', is $\mathbf{0}^{(2)}$, and A is *low* if the degree of A' is $\mathbf{0}^{(1)}$. It is easy to check that if $\Sigma_2 \subseteq \Pi_2^A$ then A is high. Thus we have

COROLLARY 4 Suppose A is r.e. and not high. Then $\Pi_2 - \mathrm{Th}(\mathbb{N}, A) \nvdash \mathrm{I}\Sigma_1$. ■

Using the last corollary, we can recover a version of the well-known fact that $\mathrm{I}\Sigma_1$ is not provable from $\Pi_2 - \mathrm{Th}(\mathbb{N}) + \mathrm{B}\Sigma_1$. (See for example Paris and Kirby 1978.) Recall that $\mathrm{B}\Sigma_1$, the scheme of Σ_1 *collection*, can be axiomatized by

$$\forall a\, \big(\forall x\, \exists y\, \theta(x,y,a) \rightarrow \forall t\, \exists z\, \forall x < t\, \exists y < z\, \theta(x,y,a)\big)$$

for $\theta(x,y,a)$ from Δ_0; indeed, by using a Σ_1-complete formula, $\mathrm{B}\Sigma_1$ can be seen to be equivalent (modulo $\Pi_2 - \mathrm{Th}(\mathbb{N})$) to a single instance of the above scheme.

Lemma 5 There is a function $f : \mathbb{N} \to \mathbb{N}$ of low degree such that $\Pi_2 - \mathrm{Th}(\mathbb{N}, f)$ proves the scheme $\mathrm{B}\Sigma_1$.

Sketch proof: By taking a Σ_1 complete relation for example, it is sufficient to show how to witness

$$\forall a, t \, \exists x < t \, (\forall y \, (\theta(x, y, a) \to \forall x' < t \, \exists y' \leqslant y \, \theta(x', y', a)))$$

by $f(a, t) = x$ of low degree satisfying

$$\forall y \, (\neg\theta(x, y, a) \vee \forall x' < t \, \exists y' \leqslant y \, \theta(x', y', a))$$

for all arguments a, t, where $\theta(x, y, a)$ is a particular Δ_0 formula. (This alternative formulation of the collection scheme is due to Adamowicz, and roughly states that 'there is $x < t$ for which the suitable y satisfying $\theta(x, y, a)$—if such exists—is maximal'.) We do this by constructing an infinite finitely-branching recursive tree T and defining f from an infinite path of the tree, using the low basis theorem.

Intuitively, a node of the tree at height $n = \langle a, t, l \rangle$ contains the following information:

a value $x < t$ and the assertion that

$$\forall y \, (\theta(x, y, a) \to \forall x' < t \, \exists y' \leqslant y \, \theta(x', y', a));$$

A node of the form just described can exist only if it is consistent with all previous assertions on the same path, according to some suitable definition of 'consistent'.

I shall now be a little more precise.

The trees I have in mind are sets of *nodes*, each node being a finite sequence of natural numbers. The length of the sequence corresponding in this way to a node is also called the node's *height*. The *parent* of a node $k_0 k_1 \ldots k_n$ of height $n+1$ is the node $k_0 k_1 \ldots k_{n-1}$. *Trees* are sets of nodes such that the parent of any node of positive height in the tree is also in the tree. The *childeren* of a node $k_0 k_1 \ldots k_{n-1}$ in a tree T are the nodes $k_0 k_1 \ldots k_{n-1} k_n$ in the tree.

I will now define a particular tree with the intuitive meaning sketched out above. The empty node ϵ is in the tree T and, for any $n \in \mathbb{N}$, where $n = \langle a, t, l \rangle$, the children of the node $k_0 k_1 \ldots k_{n-1}$ are among $k_0 k_1 \ldots k_{n-1} k_n$ where $k_n = x$ and x is less than t. Not all possibilities are allowed though: we make the condition that

if $k_n = x$ then $\forall y < n\,(\theta(x,y,a) \to \forall x' < t\,\exists y' \leqslant y\,\theta(x',y',a))$
and for no $n' = \langle a,t,l'\rangle < n$ is $k_{n'} \neq k_n$;

all possibilities not disallowed by these rules are allowed.

Note that the conditions above imply that if $k_n = x$ but $\forall y\,(\theta(x,y,a) \to \forall x' < t\,\exists y' \leqslant y\,\theta(x',y',a))$ is false, then this particular node cannot be extended indefinitely to nodes of arbitrary height that are still in the tree T. (This is the meaning of the part above saying that the 'assertion'

$$\forall y\,(\theta(x,y,a) \to \forall x' < t\,\exists y' \leqslant y\,\theta(x',y',a))$$

is made.

Clearly the tree just defined is recursive. Any infinite path through the tree codes enough information to witness the instance of $\mathrm{B}\Sigma_1$, for, given $\langle a,t\rangle$, one need only look at nodes of height $n = \langle a,t,l\rangle$ for large enough l. If the (unique!) child of such a node $k_0\ldots k_{n-1}$ is $k_0\ldots k_{n-1}x$ then x satisfies $\forall y\,(\theta(x,y,a) \to \forall x' < t\,\exists y' \leqslant y\,\theta(x',y',a))$ since there is an infinite path above this node in the tree.

Certainly this tree is finitely branching, and indeed the amount of branching at a given node of height n is bounded by a predetermined number (given by the coding function). Thus the tree can be recursively encoded as a binary tree, in such a way that any infinite path through this tree determines an infinite path through the original one in an effective way. Thus we can apply the low basis theorem to obtain (an oracle which may be used to compute) our required function f, and this oracle and hence f have low degree, as required. ■

COROLLARY 6 $\Pi_2 - \mathrm{Th}(\mathbb{N}) + \mathrm{B}\Sigma_1 \nvdash \mathrm{I}\Sigma_1$ ■

Notice that the use of the Σ_1 complete formula in the above argument (showing that $\mathrm{B}\Sigma_1$ is finitely axiomatized over $\Pi_2 - \mathrm{Th}(\mathbb{N})$) was for convenience only: the lemma can easily be modified to give witnessing functions for any finite collection of axioms of $\mathrm{B}\Sigma_1$, the join of the degrees of them being low.

Once again, the result stated in the last corollary relativizes, so $\mathrm{I}\Sigma_n$ is not provable in $\Pi_{n+1} - \mathrm{Th}(\mathbb{N}) + \mathrm{B}\Sigma_n$, and this result can be extended further in the style of corollary 4. I omit the details.

3 Closing remarks

The work presented above is only a beginning: there is still much to do in determining precise connections between the structure of Turing degrees and subsystems of arithmetic. (On reading the first draft of this paper, the referee has also commented that understanding the meaning of other Herbrand-type theorems for fragments of PA promises to be at least as interesting: for example, there are ralatively straightforward characterisations of the functions Σ_m-definable in $I\Sigma_n$ for $n < m$, similar in flavour to lemma 1 above.) In addition to this, the work I have presented here may also suggest new areas of study for subsystems of the weak system $I\Delta_0 + \Omega_1$ (there is an extensive literature on degrees under polynomial-time reducibilities with may be useful).

Having made the comments in the previous paragraph I would add that the most interesting fragments of arithmetic are the *natural* fragments (of either strong or weak arithmetic) which are typically interesting because of their elegant axiomatizations and because of their combinatorial and number-theoretic consequences. Some of the suggested lines of investigation in the last paragraph may take researchers out of the realm of these natural theories, and—despite those suggestions—I sincerely hope that the bulk of future work on fragments of arithmetic will concentrate on the 'nice' theories rather than the pathological ones.

4 References

Adamowicz, Z. (1988) A recursion-theoretic characterization of instances of $B\Sigma_n$ provable in $\Pi_{n+1}(N)$. *Fundamenta Mathematicae*, **129**, pp. 231–6.

Clote, P. (1985) Partition relations in arithmetic. In *Methods in Mathematical Logic*, proceedings, Caracas 1983, ed. C.A. DiPrisco, Springer Lecture Notes in Mathematics **1130**, pp. 32–68.

Kaye, R. (1991) *Models of Peano arithmetic.* Oxford University Press.

Krajíček, J., P. Pudlák, and G. Takeuti (1991) Bounded arithmetic and the polynomial hierarchy. *Annals of Pure and Applied Logic*, **52**, pp. 143–53.

Paris, J.B., and L.A.S. Kirby (1978) Σ_n collection schemas in arithmetic. *Logic Colloquium '77* (ed. A.J. Macintyre et al), pp. 199-209, North Holland, Amsterdam.

Richard Kaye
Jesus College
Oxford University
Oxford, OX1 3DW
United Kingdom
email: KAYE@vax.oxford.ac.uk

An Equivalence between Second Order Bounded Domain Bounded Arithmetic and First Order Bounded Arithmetic

Alexander A. Razborov

Abstract

We introduce a bounded domain version $V_2^1(BD)$ of Buss's second order theory V_2^1 of bounded arithmetic and show that this version is equivalent to the first order theory S_3^1. More precisely, we construct two natural interpretations $V_2^1(BD) \rightsquigarrow S_3^1$ and $S_3^1 \rightsquigarrow V_2^1(BD)$ which are inverse to each other and preserve the syntactic structure of bounded formulae. As a corollary, for the bounded domain case we obtain Buss's result concerning $\Sigma_1^{1,b}$-expressibility in V_2^1 as a direct consequence of his main result for first order theories. Using only plain corollaries of the cut elimination theorem, we show that V_2^1 and $V_2^1(BD)$ prove the same $\Sigma^{1,b}$-formulae and the same *closed* $\forall^0\exists^0\Sigma^{1,b}$-formulae where $\forall^0, \exists^0$ stand for first order quantifiers. Combined with the above mentioned result this gives an alternative proof of Buss's characterization of $\Sigma_1^{1,b}$-definable in V_2^1 functions. All this readily extends to the case $V_k^i(BD)$ vs. S_{k+1}^i $(i, k \geq 1)$.

1. Introduction

The study of weak fragments of Peano Arithmetic has received much attention in last years partly because of close connection with important questions studied in the Computational Complexity. Originally the main target of this research was the theory $I\Delta_0$ and its subtheories. However, one considerable disadvantage of $I\Delta_0$ is that we can code in this theory finite sequences only of a fixed length prescribed in advance by the coding scheme. The smallest extension of $I\Delta_0$ where we can code sequences of length comparable with lengths of their elements (which is e.g. necessary for formalizing methamatematics in a natural way) is the theory $I\Delta_0 +$ "$\forall x\,(x^{\log x}$ exists)". This last theory (under the name S_2) and its fragments S_2^i, T_2^i obtained by further restricting the induction scheme were systematically studied in the seminal work of S.R.Buss [Bus86]. The main result of that work was a characterization of levels of the Meyer-Stockmeyer hierarchy in the Complexity theory (see e.g. [GJ79]) in terms of expressibility in certain fragments of S_2 [Bus86, theorems 3.1 and 5.1].

To capture higher levels of the complexity hierarchy such as $PSPACE$ or $EXPTIME$, Buss also introduced in [Bus86] second order theories U_2^1 and V_2^1 respectively. Clote and Takeuti [CT86] extended Buss's result concerning definability of $EXPTIME$ in V_2^1 to n-fold exponential time computable functions by utilizing some many-sorted theories of Bounded Arithmetic.

Theories U_2^1 and V_2^1 are included into hierarchies U_k^i and V_k^i which relate to first order hierarchies S_k^i and T_k^i more or less like the analytical hierarchy relates to the arithmetic hierarchy in the classical case. Hence it is not surprising that many definitions and proofs in second order theories are straightforward analogues of the corresponding definitions and proofs for first order theories.

But Bounded Arithmetic also gives an opportunity to *directly* interpret second order objects in a first order language. The antagonism "second order vs. first order" translates via this interpretation to "arbitrary integers vs. those "small" integers x for which $f(x)$ exists" where f is a rapidly growing function whose existence is not provable in Bounded Arithmetic.

This idea was first explored by Buss at the end of his book [Bus86] where he, based upon techniques of Solovay, Nelson and of Wilkie (see [Pud83]), sketched a proof that a bounded domain version of the second order bounded arithmetic can be interpreted in S_2 and hence (by a result of Wilkie and Nelson) is predicative in the sense of Nelson.

The main (informal) goal of this paper is to develop logical formalism which would reveal that second order theories of Bounded Arithmetic (at least those from the hierarchy V_k^i) not only *look* similar to their first order counterparts but basically *are* the same. Formally we do the following.

First we define[1] bounded domain versions $V_k^i(BD)$ of theories V_k^i. We

[1]The first attempt to do this was undertaken by Buss in [Bus86, §10.8]. This attempt however is not quite satisfactory since the theory $V_2^1(BD)$, as defined in [Bus86], strictly speaking is not a second order theory at all. For example, it contains the axiom $\forall x \forall \phi^a\, (x > a \supset \neg\phi^a(x))$ where ϕ^a is a second order variable of the sort associated with the term a. However, the instance of this axiom obtained by, say, substituting 0 for a is not intended to be a theorem of $V_2^1(BD)$.

will shortly discuss underlying principles of these theories in section 3. Here we only note that with the suggested version of the definition Buss's interpretation of $V_2(BD)$ in S_2 works out smoothly.

Then we step by step establish the main result of this paper. It says that there are two natural interpretations $\flat : V_2^1(BD) \rightsquigarrow S_3^1$ and $\sharp : S_3^1 \rightsquigarrow V_2^1(BD)$ such that $\flat$ maps closed $\Sigma_i^{1,b}$-formulae into closed Σ_i^b-formulae, $\sharp$ maps closed Σ_i^b-formulae into closed $\Sigma_i^{1,b}$-formulae and $\flat, \sharp$ are inverse to each other in the sense that $V_k^i(BD) \vdash A \equiv A^{\flat\sharp}$ and $S_{k+1}^i \vdash B \equiv B^{\sharp\flat}$ for any closed A, B in the appropriate language.

I would like to comment on the somewhat surprising fact that removing second order variables from, say, V_2^1 costs us exactly introducing the new symbol $\#_3$ from a computer scientist's point of view. It is well recognized in Theoretical Computer Science that when we change the representation of integers from binary to unary, the class $EXPTIME$ becomes *not* the class P but the class $TIME\left(2^{(\log n)^{O(1)}}\right)$. So the theory V_2^1 must correspond to a first order theory which captures this latter class. S_3^1 is exactly such a theory.

A straightforward application of this equivalence is a proof of the fact that functions which are $\Sigma_i^{1,b}$-definable in $V_2^i(BD)$ are exactly the functions from the i-th level of the exponential hierarchy. Aside from this equivalence, the proof uses only the corresponding result for S_3^i.

Lastly we compare the power of different bounded and unbounded domain versions of the theories V_2^i and show that all they prove the same $\Sigma^{1,b}$-formulae and the same closed $\forall^0\exists^0\Sigma^{1,b}$-formulae ($\forall^0, \exists^0$ stand for first order quantifiers). Combined with the above mentioned result this gives an alternative proof of Buss's characterization of $\Sigma_1^{1,b}$-definable in V_2^1 functions which uses only a plain consequence of the cut elimination theorem for Second Order Bounded Arithmetic.

The paper is organized as follows. In Section 2 we briefly remind the reader the necessary definitions and facts from [Bus86]. In Section 3 we define and discuss bounded domain counterparts of Buss's theories V_k^i. In two following sections we construct interpretations $\flat : V_2^1(BD) \rightsquigarrow S_3^1$ and $\sharp : S_3^1 \rightsquigarrow V_2^1(BD)$ respectively and establish their main properties (Lemmas 4.2, 5.4). In Section 6 we prove that these interpretations are inverse to each other (Lemmas 6.1, 6.2). In Section 7 we summarize in the Main Theorem (Theorem 7.1) facts established in previous sections and show that this theorem fairly easy allows us to translate results on Σ_i^b-definability in S_{k+1}^i to corresponding results on $\Sigma_i^{1,b}$-definability in $V_k^i(BD)$ (Theorem 7.2). In the next section 8 we show that all (bounded and unbounded domain) variants of V_k^i considered in [Bus86] and this paper prove the same $\Sigma^{1,b}$-formulae and the same closed $\forall^0\exists^0\Sigma^{1,b}$-formulae. We conclude the paper with a couple of remarks in Section 9.

Remark 1.1. *After the original version of this paper was disseminated, I learned that very similar results had been earlier proven by Takeuti* [Tak90, Tak91]. *The latter paper also contains a generalization to the case $U_k^i(BD)$ vs. R_{k+1}^i where R_k^i is the theory introduced in various forms by Allen* [All89], *Clote* [Clo89] *and Takeuti* [Tak91].

2. Preliminaries

This section consists of primary definitions and facts concerning Bounded Arithmetic mostly borrowed from [Bus86]. We do not intend to give a self-contained account so some familiarity with [Bus86] is desirable.

Let the first order language L_k ($k > 1$) with equality consist of the function symbols $0, S, +, \cdot, \lfloor \frac{1}{2}x \rfloor, |x|, x\#_2 y, \ldots, x\#_k y$ and of the predicate symbol $\leq$. Sometimes we will denote $\#_2$ by $\#$ and $\#_1$, for uniformity of notation, will always mean "$\cdot$". For a vector $\vec{x} = (x_1, \ldots, x_n)$ we will write $\bigwedge_{i=1}^n x_i \leq y$ and $\bigwedge_{i=1}^n x_i < y$ in the simplified form $\vec{x} \leq y$ and $\vec{x} < y$ respectively (where of course $x < y \rightleftharpoons x \leq y \wedge x \neq y$). The intended meaning of symbols $x\#_j y$ ($2 \leq j \leq k$) is given recursively by

$$x\#_j y = 2^{|x|\#_{j-1}|y|} (2 \leq j \leq k). \qquad (1)$$

The meaning of all other symbols is obvious.

Quantifiers of the form $(\forall x \leq t)$ and $(\exists x \leq t)$ are called *bounded*, quantifiers of the form $(\forall x \leq |t|)$ and $(\exists x \leq |t|)$ are called *sharply bounded* (here t is any term not involving x). A formula is *bounded* if all quantifiers in it are bounded. The hierarchy Σ_i^b of bounded formulae is defined by counting alternations of bounded quantifiers, ignoring the sharply bounded quantifiers. Also Σ^b stands for the set of all bounded formulae that is $\Sigma^b \rightleftharpoons \bigcup_{i \geq 0} \Sigma_i^b$.

Definition 2.1. The theory S_k^i is the first order theory in the language L_k with the following axioms:

a) 33 open axioms $BASIC_k$. This list coincides with Buss's list [Bus86, §2.2] with the difference that we generalize his axiom (13) $|x\#y| =$

$S(|x| \cdot |y|)$ to

$$|x\#_j y| = S(|x|\#_{j-1}|y|) \; (2 \le j \le k) \tag{2}$$

and add the new axiom

$$z < x\#_j y \equiv |z| < |x\#_j y| \; (2 \le j \le k). \tag{3}$$

b) $\Sigma_i^b - PIND$ that is the scheme

$$A(0) \wedge \forall x \left(A\left(\lfloor \frac{1}{2} x \rfloor \right) \supset A(x) \right) \supset \forall x A(x) \tag{4}$$

where $A \in \Sigma_i^b$.

Note that axioms (2) and (3) uniquely describe $\#_3, \ldots, \#_k$. Since these symbols are not used in bootstrapping S_k^i (see [Bus86, §2.4-2.5]), we do not care about including more their properties into $BASIC_k$.

Any new function symbol defined in S_k^i by a Σ_1^b-formula and any predicate symbol defined by a Δ_1^b-formula can be freely used in the formula A from (4). Among these are, for example, the following symbols:

- for each fixed $r > 0$ a function symbol $\langle x_1, \ldots, x_r \rangle$ which implements a one-to-one mapping $\mathbf{N}^r \longrightarrow \mathbf{N}$ and r unary symbols $\Pi_1^r, \ldots, \Pi_r^r$ representing the inverse mapping $\mathbf{N} \longrightarrow \mathbf{N}^r$. W.l.o.g. we may assume that
$$S_1^1 \vdash \langle x_1, \ldots, x_r \rangle \le y \supset \vec{x} \le y. \tag{5}$$
Also there exists a term $B^r(y)$ such that
$$S_1^1 \vdash \vec{x} \le y \supset \langle x_1, \ldots, x_r \rangle \le B^r(y), \tag{6}$$

- predicate symbol $Power2(x)$ and function symbol $max(x_1, \ldots, x_r)$ with obvious meaning,

- function symbol $BIT(i, x)$ which means "the bit of x in the 2^i position" and
$LSP(x, y)$ which means "the integer presented by y last bits of x".

All symbols above are already defined in S_1^1.

The main result proved by Buss concerns Σ_i^b-definability in the theory S_k^i. We state it here only for the case $i = 1$.

Proposition 2.2 ([Bus86, theorem 3.1 + theorem 5.1]). *Let $k \geq 2$.*

a) *for each function $f(n_1, \ldots, n_r) : \mathbf{N}^r \longrightarrow \mathbf{N}$ computable in time $|t(n_1, \ldots, n_r)|$ where t is a term of L_k (interpreted on $\mathbf{N}$ using (1)) there exists a Σ_1^b-formula $A(x_1, \ldots, x_r)$ in L_k such that:*

a1) $S_k^1 \vdash \exists y \leq t \; A(\vec{x}, y)$,
a2) $S_k^1 \vdash (A(\vec{x}, y) \wedge A(\vec{x}, z) \supset y = z)$,
a3) $\mathbf{N} \models A(\vec{n}, f(\vec{n}))$.

b) *conversely, suppose $S_k^1 \vdash \forall \vec{x} \exists y \; A(\vec{x}, y)$ where $A(\vec{x}, y)$ is a Σ_1^b-formula with all free variables displayed. Then there exists a term $t(\vec{x})$ in the language L_k, a Σ_1^b-formula $B(\vec{x}, y)$ and a function $f(\vec{n})$ computable in time $|t(\vec{n})|$ such that:*

b1) $S_k^1 \vdash B(\vec{x}, y) \supset A(\vec{x}, y)$,
b2) $S_k^1 \vdash B(\vec{x}, y) \wedge B(\vec{x}, z) \supset y = z$,
b3) $S_k^1 \vdash \exists y \leq t \; B(\vec{x}, y)$,
b4) *for all $\vec{n}$,* $\mathbf{N} \models B(\vec{n}, f(\vec{n}))$.

Let $\mathcal{L}_k$ be the second order language corresponding to L_k. It is well known that second order variables for functions can be easily simulated by second order variables for predicates (see e.g. [Bus86, Lemma 9.6]). So we define $\mathcal{L}_k$ to be the language obtained from L_k by augmenting it with second order variables $\{\alpha_i^r \mid i, r \in \mathbf{N};\; r \geq 1\}$ where r denotes the arity of the variable. The superscript r will be dropped whenever it can not create confusion.

Remark 2.3. Since we are going to talk of interpretations, it is important that we treat $\mathcal{L}_k$ as a many-sorted language (see e.g. [Tak75, §1.8]). Formally, this language has sorts $0, 1, 2, \ldots$ where 0 is reserved for the sort of first order variables and $r > 0$ is the sort of r-ary second order variables. It contains special predicate symbols $Value_r$ of type

$$(r+1; r, \underbrace{0, \ldots, 0}_{r \text{ times}})$$

such that $\alpha^r(x_1, \ldots, x_r) \rightleftharpoons Value_r(\alpha^r, x_1, \ldots, x_r)$. In particular, we are going to freely introduce function and predicate symbols containing second order variables as their arguments.

A term in the language L_k is called a *first order term.* A second order formula is *bounded* if all first order quantifiers in it are bounded. The hierarchy $\Sigma_i^{1,b}$ of bounded second order formulae is defined by counting second order quantifiers ignoring bounded first order quantifiers; $\Sigma^{1,b} \rightleftharpoons \bigcup_{i\geq 0} \Sigma_i^{1,b}$. Note that, like in the first order case, function symbols introduced by $\Sigma_1^{1,b}$-formulae and predicate symbols introduced by $\Delta_1^{1,b}$-formulae can be freely used in second order formulae since such usage does not bring bounded formulae out of the classes from the hierarchy $\Sigma_i^{1,b}$ they originally belonged to.

Definition 2.4. For a class Φ of formulae, Φ *comprehension axioms* (or $\Phi - CA$) is the following scheme:

$$\exists\alpha^r \forall\, x_1, \ldots, x_r\, \{\alpha^r(x_1, \ldots, x_r) \equiv A(x_1, \ldots, x_r)\}$$

where A is in Φ and does not contain α.

Definition 2.5. The second order theory $\widetilde{V}_k^i$ in the language $\mathcal{L}_k$ has the following axioms:

a) $BASIC_k$,

b) $\Sigma_i^{1,b} - IND$,

c) $\Sigma_0^{1,b} - CA$.

The theory $\overset{\circ}{V}{}_k^i$ differs from $\widetilde{V}_k^i$ in allowing a slightly stronger comprehension axiom.

Definition 2.6. $\overset{\circ}{V}{}_k^i$ is the second order theory in $\mathcal{L}_k$ with the following axioms:

a) $BASIC_k$,

b) $\Sigma_i^{1,b} - IND$,

c) $\Delta_1^{1,b} - CA$.

3. Bounded Domain Variants

In this section we define two bounded domain versions of the theories $\widetilde{V}_k^i$ and $\overset{\circ}{V}{}_k^i$ and prove their primary properties. Given our general goal, the idea is to describe in the second order language $\mathcal{L}_k$ exactly those properties which S_{k+1}^i can prove about sequences $Bit(0,x), Bit(1,x), \ldots, Bit(y,x), \ldots$ where $x \in \mathbf{N}$. And the first thing we discover is that the general CA-scheme becomes unsound and we should replace it by the following restricted scheme.

Definition 3.1. $\Phi - BCA$, Φ *bounded comprehension axioms* is the following scheme:

$$\exists \alpha^r \forall\, x_1, \ldots, x_r \, \{\alpha^r(x_1, \ldots, x_r) \equiv (A(x_1, \ldots, x_r) \wedge \vec{x} \leq t)\}$$

where A is in Φ and does not contain α and t is a first order term which does not contain variables from $\vec{x}$.

Definition 3.2. $V_k^i(bd)$ is the theory in the language $\mathcal{L}_k$ with the following set of axioms:

a) $BASIC_k$,

b) $\Sigma_i^{1,b} - IND$,

c) $\Sigma_0^{1,b} - BCA$.

$V_k^i(bd)$ is a "neutral" theory which can neither prove nor disprove the existence of sequences with infinitely many ones. However we know that S_{k+1}^i can disprove this. So we must add a new axiom forbidding such sequences. This leads to the following

Definition 3.3. $V_k^i(BD)$ has the following axioms:

a) $BASIC_k$,

b) $\Sigma_i^{1,b} - IND$,

c) $\Sigma_0^{1,b} - BCA$,

d) $\exists y \forall \vec{x} \, (\alpha(\vec{x}) \supset \vec{x} < y)$.

Now we are going to establish some simple properties of $V_k^i(BD)$. First we introduce the *second order equality* by

$$\alpha^r = \beta^r \rightleftharpoons \forall \vec{x} \, (\alpha^r(\vec{x}) \equiv \beta^r(\vec{x})) \tag{7}$$

and *bounded second order equality* by

$$\alpha^r \overset{y}{=} \beta^r \rightleftharpoons \forall \vec{x} \leq y \, (\alpha^r(\vec{x}) \equiv \beta^r(\vec{x})) \, . \tag{8}$$

Note that whereas the unbounded second order equality is *not* allowed in axiom schemes b), c) of Definition 3.3, the bounded second order equality is given by the $\Sigma_0^{1,b}$-formula (8) and hence can be freely used there.

Now apply induction on y to the $\Sigma_0^{1,b}$-formula $A(\alpha, y, z) \rightleftharpoons \exists \vec{x} \leq z \, (\neg \vec{x} < y \wedge \alpha(\vec{x}))$. $V_1^0(BD) \vdash \neg A(\alpha, z+1, z)$ and hence we have

$$\left. \begin{array}{l} V_1^0(BD) \vdash \forall \vec{x} \leq z \, (\neg \alpha(\vec{x})) \;\vee\; \exists y > 0 \\ \qquad \left\{ \begin{array}{l} \exists \vec{x} \leq z \, (\alpha(\vec{x}) \wedge max(x_1, \ldots, x_r) = y \dot{-} 1) \;\wedge \\ \forall \vec{x} \leq z \, (\alpha(\vec{x}) \supset \vec{x} < y) \, . \end{array} \right. \end{array} \right\} \tag{9}$$

Let z be such that $\forall \vec{x} \, (\alpha(\vec{x}) \supset \vec{x} \leq z)$; the existence of such z is assured by Definition 3.3 d). We obtain from (9):

$$\left. \begin{array}{l} V_1^0(BD) \vdash \forall \vec{x} \, (\neg \alpha(\vec{x})) \;\vee \\ \exists y > 0 \;\{\exists \vec{x} \, (\alpha(\vec{x}) \wedge max(x_1, \ldots, x_r) = y \dot{-} 1) \;\wedge \\ \forall \vec{x} \, (\alpha(\vec{x}) \supset \vec{x} < y)\} \, . \end{array} \right\} \tag{10}$$

If we define now

$$\begin{array}{l} B(\alpha, y) \rightleftharpoons \forall \vec{x} \, (\alpha(\vec{x}) \supset \vec{x} < y) \;\wedge \\ \{y > 0 \supset \exists \vec{x} \, (\alpha(\vec{x}) \wedge max(x_1, \ldots, x_r) = y \dot{-} 1)\} \end{array}$$

then (10) implies $V_1^0(BD) \vdash \forall \alpha \exists y B(\alpha, y)$ and an easy logical analysis gives us $V_1^0(BD) \vdash B(\alpha, y) \wedge B(\alpha, y') \supset y = y'$. Hence we may introduce function symbols $\ell^r(\alpha^r)$ $(r \geq 1)$ of type $(1; r, 0)$[2] with defining axioms

$$\forall \vec{x}(\alpha(\vec{x}) \supset \vec{x} < \ell(\alpha)), \tag{11}$$
$$\ell(\alpha) > 0 \supset \exists \vec{x}\,(\alpha(\vec{x}) \wedge max(x_1, \ldots, x_r) = \ell(\alpha) \dot{-} 1)\,. \tag{12}$$

These symbols will be of crucial importance for the following; their intended meaning in the case $r = 1$ is "the length of α".

The next thing to do is to note that w.l.o.g. we may consider only unary second order variables which will highly simplify the notation. For this we implement one-to-one correspondence between $\mathbf{N}^r$ and $\mathbf{N}$ using $\Sigma_0^{1,b}$-definable function symbols

$$\langle x_1, \ldots, x_r \rangle,\ \Pi_1^r, \ldots, \Pi_r^r$$

from previous section. Using $\Sigma_0^{1,b} - BCA$ we prove that for each α^r there exists α^1 such that $\alpha^r(x_1, \ldots, x_r) \equiv \alpha^1(\langle x_1, \ldots, x_r \rangle)$ and vice versa (the bounding term t in Definition 3.1 is changed accordingly to (5), (6)). More formally, this leads to the following:

Theorem 3.4. *For a second order formula A define the formula $A^\star$ by replacing all its atomic subformulae of the form $\alpha_i^r(t_1, \ldots, t_r)$ by*

$$\alpha_{i,r}^1(\langle t_1, \ldots, t_r \rangle)$$

(variables $\alpha_{i,r}$ of sort 1 are assumed pairwise distinct). Then:

a) *for each $i \geq 0$, if A is $\Sigma_i^{1,b}$ then so is $A^\star$,*

b) *$\star$ defines an interpretation of $V_k^i(BD)$ in the theory obtained from $V_k^i(BD)$ by restricting the language $\mathcal{L}_k$ to variables of sorts 0 and 1 only,*

c) *$\star$ is inverse to the identical interpretation in the sense that for each formula A without free second order variables, $V_k^0(BD) \vdash A \equiv A^\star$.*

In view of Theorem 3.4, we will often restrict our considerations to the language which contains variables of types 0 and 1 only. Theorem 3.4 always allows us to extend corresponding results to the general case.

[2]the notation $(1; r, 0)$ reads as "1-ary function symbol with the argument of type r taking values in type 0" (see [Tak75, §1.8])

The reader may have observed that, unlike Section 2, in this section we have not used any special notation to indicate the strength of comprehension axioms involved in our theories. Note that it is open whether $\widetilde{V}_k^i \equiv \overset{\circ}{V}{}_k^i$ [Bus86, §10.8]. The next result shows that in bounded domain the answer to the analogous question is positive.

Theorem 3.5. *$V_k^i(bd)$ proves $\Delta_i^{1,b} - BCA$.*

Proof. Let $A(x)$ be a $\Delta_i^{1,b}$-formula. We prove by induction on y that

$$\exists\alpha\,\{\alpha(x) \equiv (A(x) \wedge x < y)\} \tag{13}$$

(note that this is a $\Sigma_i^{1,b}$-formula). The base $y = 0$ is obvious.

If we already have α_y such that $\alpha_y(x) \equiv (A(x) \wedge x < y)$ then in the case $\neg A(y)$ α_y also satisfies (13) with $y+1$ instead of y. If $A(y)$ then the required α_{y+1} is obtained from α_y by applying $\Sigma_0^{1,b} - BCA$ to the formula $(\alpha(x) \vee x = y) \wedge x \leq y$. This completes the inductive step.

Substituting $t+1$ for y in (13) completes the proof of the theorem.■

Now, by $\Sigma_0^{1,b} - BCA$, $V_k^0(bd) \vdash \exists\beta\forall x\,\{\beta(x) \equiv (\alpha(x) \wedge x \leq y)\}$. It is easy to see that such a β is unique and hence we have

$$V_k^0(bd) \vdash \exists!\beta\forall x\,\{\beta(x) \equiv (\alpha(x) \wedge x \leq y)\}\,.$$

So we may introduce in $V_k^0(bd)$ the function symbol $\alpha|_y$ of type (2;1,0,1) with the defining axiom $\alpha|_y(x) \equiv (\alpha(x) \wedge x \leq y)$. The following properties of $\alpha|_y$ are easy to check:

$$V_1^0(bd) \vdash \alpha|_y \overset{y}{=} \alpha, \tag{14}$$

$$V_1^0(BD) \vdash \ell\,(\alpha|_y) \leq y+1, \tag{15}$$

$$V_1^0(BD) \vdash \ell\,(\alpha) \leq y+1 \supset \alpha|_y = \alpha, \tag{16}$$

$x = \ell(\alpha|_y)$ and $x \leq \ell(\alpha|_y)$ are $\Sigma_0^{1,b}$ – formulae (in $V_1^0(BD)$). (17)

Part a) of the following theorem is a bounded version of [Tak75, Proposition 15.13] (see also [Bus86, Lemma 10.9]).

Theorem 3.6. *For any bounded formula $A(\alpha)$ there exists a first order term t such that:*

a) $V^0_k(bd) \vdash \alpha \stackrel{t}{=} \beta \supset A(\alpha) \equiv A(\beta)$,

b) $V^0_k(BD) \vdash \exists \alpha A(\alpha) \ \equiv \ \exists \alpha \{\ell(\alpha) \leq t+1 \wedge A(\alpha)\}$.

Proof. a) Induction on complexity of A. If $A \equiv \alpha(t)$ then clearly $\alpha \stackrel{t}{=} \beta \supset \alpha(t) \equiv \beta(t)$. If $A \equiv B \circ C$ where $\circ$ is a Boolean connective, we let $t_A \rightleftharpoons t_B + t_C$. In the case $A \equiv \exists x \leq sB(x)$ we set $t_A \rightleftharpoons t_B\,[s/x]$ (note that $S^1_k \vdash x \leq s \supset t \leq t\,[s/x]$ since first order terms are provably monotone). All other cases are trivial.

b) We assume that t is such that a) holds. By (15), (14) and a), $A(\alpha)$ implies $\ell(\alpha|_t) \leq t+1 \wedge A(\alpha|_t)$. This gives us part b) of the theorem.■

To conclude this section we mention the possibility to define in $V^1_k(BD)$ new function symbols (which may in general depend on second order variables) using "bounded" $\Delta^{1,b}_1$-abstracts.

Lemma 3.7. *Let $A(x_1,\ldots,x_r,\vec{y},\vec{\alpha})$ be a $\Delta^{1,b}_1$-formula and $t(\vec{y},\vec{z})$ be a first order term such that*

$$V^i_k(BD) \vdash A(\vec{x},\vec{y},\vec{\alpha}) \supset \vec{x} \leq t(\vec{y},\ell(\vec{\alpha})). \tag{18}$$

Then, provided $i \geq 1$, we can introduce in $V^i_k(BD)$ the function symbol $f(\vec{y},\vec{\alpha})$ taking values of sort r with the defining axiom

$$f(\vec{y},\vec{\alpha})(x_1,\ldots,x_r) \equiv A(x_1,\ldots,x_r,\vec{y},\vec{\alpha}). \tag{19}$$

Moreover, for each $j \leq 1$, any $\Sigma^{1,b}_j$-formula possibly containing f is equivalent in $V^i_k(BD)$ to a $\Sigma^{1,b}_j$-formula without f. Hence f can be freely used in $\Sigma^{1,b}_i - IND$, $\Delta^{1,b}_1 - BCA$ and repeating applications of this lemma.

Proof. By Theorem 3.5, we may apply $\Delta^{1,b}_1 - BCA$ to show

$$V^1_k(BD) \vdash \exists \beta \forall \vec{x}\,\{\beta(\vec{x}) \equiv (A(\vec{x},\vec{y},\vec{\alpha}) \wedge \vec{x} \leq t(\vec{y},\vec{z}))\}.$$

Substituting here $\ell(\vec{\alpha})$ for $\vec{z}$ and applying (18), we get

$$V^i_k(BD) \vdash \exists \beta \forall \vec{x}\,\{\beta(\vec{x}) \equiv A(\vec{x},\vec{y},\vec{\alpha})\}.$$

The uniqueness is obvious hence we may introduce the desired symbol f with the defining axiom (19). The second part of the lemma follows by standard argument (see e.g. [Bus86, theorem 2.4]).■

4. Interpretation $\flat$

In this section we construct an interpretation[3] $\flat : V_k^i(BD) \rightsquigarrow S_{k+1}^i$ and show that it preserves the syntactic structure of bounded formulae. The idea resembles the idea used by Buss in [Bus86, §10.8]. But instead of "pushing" first order variables down to the level defined by existence of the *double* exponent, we stay on the "logarithmic" level and insert instead the next smash function $\#_{k+1}$. It is the novelty which will allow us to show in subsequent sections that our interpretation is exact.

We define $\flat$ as follows. Both first and second order variables are interpreted by first order variables of the language L_{k+1}. First order quantifiers are relativized to the formula $\exists y\,(|y| = x)$. Second order quantifiers are not relativized. Symbols from the language L_k are interpreted by themselves. $\alpha(x)$ is interpreted as $Bit(x, \alpha) = 1$.

Before proving that so defined $\flat$ is really an interpretation with desired properties, let us establish the following useful auxiliary fact.

Lemma 4.1. *For each term $t(\vec{x})$ in the language L_k there exists a Σ_1^b-definable in S_{k+1}^1 function symbol $t^+(\vec{x})$ such that $S_{k+1}^1 \vdash |t^+(x_1, \ldots, x_r)| = t\,(|x_1|, \ldots, |x_r|)$.*

Proof. First we show that there exists a term $\bar{t}^+$ such that

$$S_{k+1}^1 \vdash t\,(|\vec{x}|) \leq |\bar{t}^+(\vec{x})|. \tag{20}$$

This is easily done by induction on complexity of t; for the case of smash functions $\#_j$ $(1 \leq j \leq k)$ use (2). Now the desired symbol t^+ is defined by the following Σ_1^b-formula:

$$t^+(\vec{x}) = y \;\rightleftharpoons\; |y| = t(|\vec{x}|) \wedge (y = 0 \vee Power2(y))\,.$$

The existence condition follows from (20) and the fact

$$S_1^1 \vdash 0 < x \leq |z| \supset \exists y\,(|y| = x \wedge Power2(y))\,;$$

---

[3] In this paper we use the standard notion of interpretability (see e.g. [Sho67, Section 4.7]) extended in an obvious way to the many-sorted case. Bounded quantifiers $\exists x \leq t\, A(x)$ and $\forall x \leq t\, A(x)$ are simply treated as shorthand terms for $\exists x(x \leq t \wedge A(x))$ and $\forall x(x \leq t \supset A(x))$ respectively and do not require special care.

uniqueness easily follows from general properties of *Power*2 provable in S^1_1.■

Now we are ready to establish the main result of this section.

Lemma 4.2. *Let $A(x_1, \ldots, x_n)$ be a formula in $\mathcal{L}_k$ with all free first order variables displayed. Then:*

a) *if A is $\Sigma^{1,b}_0$ then $A^\flat(|x_1|, \ldots, |x_n|)$ is Δ^b_1 with respect to S^1_{k+1},*

b) *if A is $\Sigma^{1,b}_i$ $(i \geq 1)$ then $A^\flat(|x_1|, \ldots, |x_n|)$ is equivalent in S^1_{k+1} to a Σ^b_i-formula,*

c) *if $i \geq 1$ then $\flat$ is an interpretation of $V^i_k(BD)$ in S^i_{k+1}.*

Proof. a). Induction on complexity of A. The only nontrivial case to be considered is $A(\vec{x}) \equiv \exists y \leq t(\vec{x}) B(\vec{x}, y)$. By definition of $\flat$, $S^1_k \vdash A^\flat(\vec{x}) \equiv \exists z \left(|z| \leq t(\vec{x}) \wedge B^\flat(\vec{x}, |z|)\right)$. By Lemma 4.1,

$$S^1_{k+1} \vdash A^\flat(|\vec{x}|) \equiv \exists z \left(|z| \leq |t^+(\vec{x})| \wedge B^\flat(|\vec{x}|, |z|)\right) \equiv$$
$$\exists z \leq 2t^+(\vec{x}) \left(|z| \leq |t^+(\vec{x})| \wedge B^\flat(|\vec{x}|, |z|)\right).$$

By inductive assumption, this gives us a Σ^b_1-representation of $A^\flat(|\vec{x}|)$. To get a Π^b_1-representation, we merely note that

$$S^1_{k+1} \vdash \exists z \leq u\, C(|z|) \equiv \exists y \leq |u| \forall z \leq u\, (|z| = y \supset C(|z|)). \tag{21}$$

a) is proved.

As an intermediate step, we prove now the following weak form of c):

Statement 4.3. *$\flat$ is an interpretation of $V^0_k(BD)$ in S^1_{k+1}.*

Proof of Statement 4.3. $BASIC_k$ axioms are interpreted by themselves. $\Sigma^{1,b}_0 - IND$ translates to

$$A^\flat(0) \wedge \forall z \left\{A^\flat(|z|) \supset A^\flat\left(S(|z|)\right)\right\} \supset \forall z\, A^\flat(|z|)$$

which is equivalent in S^1_{k+1} to

$$A^\flat(0) \wedge \forall z \left(A^\flat\left(\left|\lfloor\frac{1}{2}z\rfloor\right|\right) \supset A^\flat(|z|)\right) \supset \forall z\, A^\flat(|z|).$$

By part a) of Theorem 4.2, this is equivalent to an instance of $\Sigma^b_1 - PIND$.

$\Sigma^{1,b}_0 - BCA$ translates to

$$\exists\alpha\forall z\left(Bit(|z|,\alpha) = 1 \;\equiv\; \left(A^\flat(|z|) \wedge |z| \leq |t^+(\vec{x})|\right)\right)$$

which is easily seen to be equivalent in S^1_{k+1} to

$$\left.\begin{array}{l}\exists\alpha \leq 4t^+(\vec{x}) + 1\; \forall z \leq t^+(\vec{x}) + 1\; Bit(|z|,\alpha) = 1 \;\equiv \\ \left(A^\flat(|z|) \wedge |z| \leq |t^+(\vec{x})|\right).\end{array}\right\} \quad (22)$$

Now, it is easy to prove by $\Sigma^b_1 - PIND$ on u that

$$\exists\alpha \leq 4u + 1\; \forall z \leq u + 1\left(Bit(|z|,\alpha) = 1 \;\equiv\; \left(A^\flat(|z|) \wedge |z| \leq |u|\right)\right)$$

(note this is Σ^b_1 by (21)). (22) follows.

Lastly, Definition 3.3 d) translates to $\exists y \forall z\, (Bit(|z|,\alpha) = 1 \supset |z| < |y|)$ which follows in S^1_1 from $Bit(x,\alpha) = 1 \supset x < |\alpha|$. This completes the proof of Statement 4.3.∎

Now we continue the proof of Lemma 4.2.

b) By Statement 4.3, $\ell^\flat$ is properly defined in S^1_{k+1} and it follows from $(11)^\flat$,$(12)^\flat$ that $S^1_{k+1} \vdash \ell^\flat(\alpha) \equiv |\alpha|$. We apply again induction on complexity of A. The case $A \equiv \exists y \leq tB$ is considered as in the proof of part a). Let $A(\vec{x}) \equiv \exists\alpha B(\alpha,\vec{x})$. By Theorem 3.6 b) and Statement 4.3,

$$S^1_{k+1} \vdash A^\flat(|\vec{x}|) \equiv \exists\alpha\left(|\alpha| \leq t(|\vec{x}|) + 1 \wedge B^\flat(\alpha, |\vec{x}|)\right). \quad (23)$$

for some first order term t. Since, by Lemma 4.1, $S^1_{k+1} \vdash |\alpha| \leq t(|\vec{x}|) + 1 \supset \alpha \leq 4t^+(\vec{x}) + 1$, the quantifier $\exists\alpha$ in (23) is bounded and b) is proved.

c) The $\flat$-image of the last remaining axiom scheme $\Sigma^{1,b}_i - IND$ is provable in S^i_{k+1} by b) and the same argument which was used in the proof of Statement 4.3 to deal with $\Sigma^{1,b}_0 - IND$.∎

5. Interpretation $\sharp$

In this section we construct an interpretation $\sharp : S^i_{k+1} \rightsquigarrow V^i_k(BD)$ $(i, k \geq 1)$ which interprets first order variables of the language L_{k+1} by second order variables of $\mathcal{L}_k$. The universe of $\sharp$ consists of all α that is first order quantifiers become second order unrelativized quantifiers. Recall from Section 4 that bounded quantifiers are simply treated as shorthand terms. In order to complete the definiton of $\sharp$, we are only left to interpret nonlogical symbols of the language L_{k+1}.

We divide them into two groups. The first group consists of

$$\left\{0, \lfloor\frac{1}{2}x\rfloor, S, +\right\}.$$

These symbols will be called *local* and will be $\Delta^{1,b}_1$-defined using standard bit-operating computational algorithms. The remaining symbols

$$\{|x|, x\#_2 y, \ldots, x\#_k y, \leq\}$$

are *global*; their definitions will be based on the symbol ℓ.

We start with local symbols.

Let us apply Lemma 3.7 to $A(x) \rightleftharpoons 0 = 1,\ t \rightleftharpoons 0$. We will get a second order constant 0 with the defining axiom

$$0(x) \;\equiv\; 0 = 1.$$

Let now $A(x, \alpha) \;\rightleftharpoons\; \alpha(x+1)$ and $t(z) \;\rightleftharpoons\; z$. Clearly, (11) implies $V^0_1(BD) \vdash \alpha(x+1) \supset x \leq \ell(\alpha)$ hence we may introduce by Lemma 3.7 the symbol $\lfloor\frac{1}{2}\alpha\rfloor$ of type (1;1,1) with the defining axiom

$$\lfloor\frac{1}{2}\alpha\rfloor(x) \equiv \alpha(x+1).$$

It is easy to see that $V^0_1(BD) \vdash (\alpha(x) \equiv \exists y < x\ \neg\alpha(y)) \supset x \leq \ell(\alpha)$. Hence we may define $S\alpha$ such that

$$S\alpha(x) \equiv \{\alpha(x) \equiv \exists y < x\ \neg\alpha(y)\}.$$

Addition is defined by the "school" algorithm. Namely, what we would like to have is

$$\left.\begin{array}{l} (\alpha_1 + \alpha_2)(x) \equiv \exists\beta \\ \quad\left\{\begin{array}{l} \neg\beta(0) \wedge \\ \forall 0 < y \leq x \\ \{\beta(y) \equiv T_2\left(\alpha_1(y \dot- 1), \alpha_2(y \dot- 1), \beta(y \dot- 1)\right)\} \wedge \\ \alpha_1(x) \oplus \alpha_2(x) \oplus \beta(x). \end{array}\right. \end{array}\right\} \quad (24)$$

Here $T_2(x,y,z) \rightleftharpoons (x \wedge y) \vee (x \wedge z) \vee (y \wedge z)$ is the *threshold* function and $x \oplus y \rightleftharpoons x \not\equiv y$ is *addition mod* 2. The idea is that $\beta(y)$ represents the carry bit to the y's position.

To justify this definition denote

$$\neg\beta(0) \wedge \forall 0 < y \leq x \, \{\beta(y) \equiv T_2(\alpha_1(y \dot- 1), \alpha_2(y \dot- 1), \beta(y \dot- 1))\}$$

by $Carry(\alpha_1, \alpha_2, \beta, x)$. Then $\Sigma_1^{1,b} - IND$ proves

$$\exists\beta \; Carry(\alpha_1, \alpha_2, \beta, x)$$

and

$$\forall\beta, \gamma \left(Carry(\alpha_1, \alpha_2, \beta, x) \wedge Carry(\alpha_1, \alpha_2, \gamma, x) \;\supset\; \beta \overset{x}{=} \gamma \right)$$

which implies that the right-hand side of (24) actually belongs to $\Delta_1^{1,b}$. Moreover, $\neg\alpha_1(y) \wedge \neg\alpha_2(y)$ implies $\neg\beta(y+1)$ hence if in addition $\neg\alpha_1(y+1) \wedge \neg\alpha_2(y+1)$, then $\neg(\alpha_1 + \alpha_2)(x+1)$. All this shows that Lemma 3.7 is applicable (with $t(z_1, z_2) \rightleftharpoons z_1 + z_2$) and the definition (24) of $\alpha_1 + \alpha_2$ is justified.

Before defining the multiplication it is convenient to introduce two auxiliary symbols. The *projection* $\gamma^2|_y$ is the function symbol of the type (2;2,0,1) defined by

$$\gamma^2|_y(x) \equiv \gamma^2(x,y).$$

Define also "the easy case of multiplication" $2^y * \beta$ of type (2;0,1,1) by

$$2^y * \beta(x) \equiv x \geq y \wedge \beta(x \dot- y).$$

Now we define the multiplication by

$$\left.\begin{array}{l} (\alpha \cdot \beta)(x) \equiv \exists\gamma^2 \\ \quad \left\{ \begin{array}{l} \gamma^2|_0 \overset{x}{=} 0 \;\wedge\; \forall 0 < y \leq x+1 \\ \left(\neg\alpha(y) \supset \gamma^2|_y \overset{x}{=} \gamma^2|_{y \dot- 1} \;\wedge \right. \\ \left. \alpha(y) \supset \gamma^2|_y \overset{x}{=} \gamma^2|_{y \dot- 1} + 2^{y \dot- 1} * \beta \right) \;\wedge \\ \gamma^2(x, x+1). \end{array} \right. \end{array} \right\} \quad (25)$$

The intuitive idea behind (25) is that the last $(x+1)$ bits of $\gamma^2|_y$ represent the last $(x+1)$ bits of the product $LSP(\alpha, y) \cdot \beta$.

To justify this definition we, similarly to the case of addition, define

$$Table(\alpha,\beta,\gamma^2,x) \rightleftharpoons \gamma^2|_0 \overset{x}{=} 0 \ \wedge\ \forall 0<y\leq x+1 \left\{ \begin{array}{l} \neg\alpha(y)\supset \gamma^2|_y \overset{x}{=} \gamma^2|_{y\dot{-}1} \ \wedge \\ \alpha(y)\supset \gamma^2|_y \overset{x}{=} \gamma^2|_{y\dot{-}1} + 2^{y\dot{-}1} * \beta \end{array}\right.$$

and prove that

$$V_1^1(BD) \vdash \exists\gamma^2\ Table(\alpha,\beta,\gamma^2,x)$$

and

$$V_1^1(BD) \vdash Table(\alpha,\beta,\gamma^2,x)\wedge Table(\alpha,\beta,\delta^2,x) \supset \forall y\leq x{+}1 \left(\gamma^2|_y \overset{x}{=} \delta^2|_y\right).$$

This gives us that the right-hand side of (25) belongs to $\Delta_1^{1,b}$. Put $t(z_1,z_2) \rightleftharpoons z_1+z_2$. By $\Sigma_0^{1,b}-IND$ on y we prove the following:

$$Table(\alpha,\beta,\gamma^2,x)\wedge\forall u\leq x+1\,(\beta(u)\supset u<z_2)\wedge\gamma^2(x,y) \ \supset\ x\leq y+z_2$$

and then

$$\begin{array}{l} Table(\alpha,\beta,\gamma^2,x)\wedge\forall u\leq x+1\,(\beta(u)\supset u<z_2)\ \wedge \\ \forall u\leq x+1\,(\alpha(u)\supset u<z_1)\wedge\gamma^2(x,y)\ \supset\ x\leq z_1+z_2. \end{array}$$

This clearly implies (18) for multiplication.

Let us $\Delta_1^{1,b}$-define two more symbols $id(y)$ and $exp(y)$, of type (1;0,1) each:

$$id(y)(x) \ \equiv\ Bit(x,y)=1, \tag{26}$$
$$exp(y)(x) \ \equiv\ x=y. \tag{27}$$

Now we are in position to interpret remaining (global) symbols of the language L_{k+1}. We do this as follows:

$$\begin{array}{rcl} |\alpha| & \rightleftharpoons & id(\ell(\alpha)), \\ \alpha\#_j\beta & \rightleftharpoons & exp\left(\ell(\alpha)\#_{j-1}\ell(\beta)\right),\ (2\leq j\leq k+1) \\ \alpha\leq\beta & \rightleftharpoons & \alpha=\beta\vee\exists x\,(\neg\alpha(x)\wedge\beta(x)\wedge\forall y>x\,\{\alpha(y)\equiv\beta(y)\}). \end{array} \tag{28}$$

This completes the definition of a translation from the set of all (first order) formulae in the language L_{k+1} to the set of (second order) formulae in the language $\mathcal{L}_k$. We denote this translation by $\sharp$. The rest of the section is devoted to proving that $\sharp$ is actually an interpretation.

First we indicate several easy properties of $\sharp$.

Lemma 5.1 (soundness of id**).** *For each first order term* $t(\vec{x})$ *in* L_k (!) *we have*

$$V_k^1(BD) \vdash id(t(\vec{x})) = t^\sharp(id(\vec{x})).$$

Proof. It suffices to check this for function symbols. This is done straightforwardly. For example, for the case of multiplication we prove by induction on y that

$$Table\left(id(x_1), id(x_2), \gamma^2, j\right) \wedge y \leq j+1 \;\supset\; id\left(LSP(x_1, y) \cdot x_2\right) \stackrel{j}{=} \gamma^2|_y.$$

(2),(3) imply $S_k^1 \vdash Bit\,(z, x\#_j y) = 1 \equiv z = |x|\#_{j-1}|y|$ which, along with (26),(27) gives us

$$id(x\#_j y) = exp\left(|x|_{j-1}|y|\right) = exp\left(\ell(id(x))\#_{j-1}\ell(id(y))\right) = id(x)\#_j id(y)$$

etc.■

The following lemma is in a sense dual to Lemma 4.1.

Lemma 5.2. *For each term* $t(\vec{x})$ *in the language* L_{k+1} *there exists a term* $t^-(\vec{x})$ *in the language* L_k *such that* $V_k^1(BD) \vdash \ell\left(t^\sharp(\vec{\alpha})\right) \leq t^-(\ell(\vec{\alpha}))$.

Proof. Straightforward (note that for local function symbols f the desired upper bound on $\ell\left(f^\sharp(\vec{\alpha})\right)$ was established in the process of justifying their definitions).■

Lemma 5.3. *For each term* $t(x_1, \ldots, x_r)$ *in the language* L_{k+1} *there exists a* $\Delta_1^{1,b}$*-formula* $A_f(\alpha_1, \ldots, \alpha_r, z_1, \ldots, z_r, y)$ *such that*

$$V_k^1(BD) \vdash t^\sharp(\alpha_1, \ldots, \alpha_r)(y) \equiv A_t(\alpha_1, \ldots, \alpha_r, \ell(\alpha_1), \ldots, \ell(\alpha_r), y).$$

Proof. First check this for function symbols.

The claim of our lemma is obvious for local symbols; we merely take their defining axioms as corresponding A_f's (z-variables are dummy). Also let $A_{|x|}(\alpha, z, y) \rightleftharpoons Bit(y, z) = 1$ and $A_{x_1\#_j x_2}(\alpha_1, \alpha_2, z_1, z_2, y) \rightleftharpoons y = z_1\#_{j-1}z_2$.

Now if $t(\vec{x}) \equiv f(\vec{s}(\vec{x}))$ then

$$f^\sharp\left(\vec{s}^\sharp(\vec{\alpha})\right)(y) \equiv A_f\left(\vec{s}^\sharp(\vec{\alpha}), \ell\left(\vec{s}^\sharp(\vec{\alpha})\right), y\right) \equiv \text{ (by Lemma 5.2 and (16))}$$
$$\exists \vec{u} \leq \vec{s}^-(\ell(\vec{\alpha}))\left(A_f\left(\vec{s}^\sharp(\vec{\alpha}), \vec{u}, y\right) \wedge \vec{u} = \ell\left(\vec{s}^\sharp(\vec{\alpha})\big|_{\vec{s}^-(\ell(\vec{\alpha}))}\right)\right).$$

This is $\Delta_1^{1,b}$ in $\vec{\alpha}, \ell(\vec{\alpha}), y$ by (17) and inductive assumption.■

Now we are in position to establish the main result of this section.

Lemma 5.4. **a)** *Let $A(x_1, \ldots, x_r)$ be a formula in the language L_{k+1} with all its free variables displayed. Then there exists a formula $\bar{A}^\sharp(\alpha_1, \ldots, \alpha_r, z_1, \ldots, z_r)$ such that $V_k^1(BD) \vdash A^\sharp(\vec{\alpha}) \equiv \bar{A}^\sharp(\vec{\alpha}, \ell(\vec{\alpha}))$ and:*

a1) *if A is Σ_0^b then $\bar{A}^\sharp$ is $\Delta_1^{1,b}$ with respect to $V_k^1(BD)$,*

a2) *if A is Σ_i^b $(i \geq 1)$ then $\bar{A}^\sharp$ is $\Sigma_i^{1,b}$.*

b) *$\sharp$ is an interpretation of S_{k+1}^i in $V_k^i(BD)$ provided $i \geq 1$.*

Proof. a) We construct $\bar{A}^\sharp$ by induction on complexity of A. First we set $\overline{t = s}^\sharp \rightleftharpoons t^\sharp(\vec{\alpha}) \stackrel{z}{=} s^\sharp(\vec{\alpha})$ where $z \rightleftharpoons t^-(\ell(\vec{\alpha})) + s^-(\ell(\vec{\alpha}))$. We have $\overline{t = s}^\sharp \equiv (t^\sharp = s^\sharp)$ by Lemma 5.2, (16), (14) and it is in $\Delta_1^{1,b}$ by Lemma 5.3. $\overline{t \leq s}^\sharp$ is defined similarly. If $A \equiv B \circ C$ where $\circ$ is a Boolean connective then $\bar{A}^\sharp \rightleftharpoons \bar{B}^\sharp \circ \bar{C}^\sharp$.

Consider the case $A \equiv \exists x \leq |t(\vec{y})|\, B(x, \vec{y})$. We have

$$A^\sharp \equiv \exists \alpha \leq id\left(\ell(t^\sharp(\vec{\beta}))\right) B^\sharp(\alpha, \vec{\beta}) \equiv \exists \alpha \leq id\left(\ell(t^\sharp(\vec{\beta}))\right) \bar{B}^\sharp(\alpha, \vec{\beta}, \ell(\alpha), \ell(\vec{\beta})).$$

Let $\bar{A}^\sharp \rightleftharpoons \exists y \leq \ell(t^\sharp(\vec{\beta}))\, \bar{B}^\sharp(id(y), \vec{\beta}, |y|, \ell(\vec{\beta}))$. Like in the proof of Lemma 5.3 note that this is equivalent to

$$\exists y \leq t^-(\ell(\vec{\beta}))\left(y \leq \ell\left(t^\sharp(\vec{\beta})\big|_{t^-(\ell(\vec{\beta}))}\right) \wedge \bar{B}^\sharp(id(y), \vec{\beta}, |y|, \ell(\vec{\beta}))\right)$$

which has the same complexity as $\bar{B}^\sharp$. The equivalence $\bar{A}^\sharp \equiv A^\sharp$ follows from the fact

$$V_1^1(BD) \vdash \alpha \leq id(x) \equiv \exists y \leq x(id(y) = \alpha). \tag{29}$$

To prove (29), we show by induction on x that $\exists y(|y| \leq |x|+1)\left(id(y) \overset{|x|}{=} \alpha\right)$ and note that $V_1^0(BD) \vdash \alpha \leq id(x) \wedge |y| \leq |x|+1 \wedge id(y) \overset{|x|}{=} \alpha \supset id(y) = \alpha$.

Consider the last remaining case $A \equiv \exists x \leq t(\vec{y})\ B(x, \vec{y})$. We have $A^\sharp \equiv \exists \alpha \leq t^\sharp(\vec{\beta})\ \bar{B}^\sharp(\alpha, \vec{\beta}, \ell(\alpha), \ell(\vec{\beta}))$. The problem is to eliminate $\ell(\alpha)$. Since ℓ is monotone, we can use the same trick as before and let

$$\bar{A}^\sharp \rightleftharpoons \exists \alpha \exists z \leq t^-(\ell(\vec{\beta}))\left(\overline{\alpha \leq t^\sharp(\vec{\beta})} \wedge z = \ell\left(\alpha\Big|_{t^-(\ell(\vec{\beta}))}\right) \wedge \bar{B}^\sharp(\alpha, \vec{\beta}, z, \ell(\vec{\beta}))\right).$$

The first conjunctive term is in $\Delta_1^{1,b}$ and the second is in $\Sigma_0^{1,b}$ by (17).

The definition of the formula $\bar{A}^\sharp$ is completed. The syntactic analysis of its structure presented in the course of construction proves part a) of Lemma 5.4.

b) First we will show that $V_k^1(BD)$ is strong enough to prove primary properties $(BASIC_k)^\sharp$ of function symbols and relations introduced in this section. This is straightforward for most axioms so we give only proof sketches for a few illustrating examples. We refer to the $\sharp$-image of the i-th axiom in the $BASIC$ list [Bus86, §2.2] as to $(\mathbf{Bi})^\sharp$.

$(\mathbf{B1})^\sharp$ $(\beta \leq \alpha \supset \beta \leq S\alpha)$. Applying $\Sigma_0^{1,b} - IND$ on x to the formula $\alpha(x)$, define the function symbol $\ell^*(\alpha)$ with the defining axiom $\neg\alpha\left(\ell^*(\alpha)\right) \wedge \forall x < \ell^*(\alpha)\ \alpha(x)$. Then $S\alpha$ is alternatively described as

$$S\alpha(x) \equiv \begin{cases} 0 = 1, & x < \ell^*(\alpha) \\ 0 = 0, & x = \ell^*(\alpha) \\ \alpha(x), & x > \ell^*(\alpha). \end{cases}$$

Now $(\mathbf{B1})^\sharp$ easily follows from the definition (28) using case analysis.

$(\mathbf{B5})^\sharp$ $(\alpha \neq 0 \supset 2 \cdot \alpha \neq 0)$. First we prove

$$V_1^1(BD) \vdash 2^y * \beta = exp(y) \cdot \beta. \tag{30}$$

This is done by applying $\Sigma_1^{1,b} - IND$ on u to the formula

$$Table\left(exp(y), \beta, \gamma^2, x\right) \wedge 0 < u \leq x+1 \supset \gamma^2|_u \overset{x}{=} \begin{cases} 0, & u \leq y \\ 2^y * \beta, & y < u \leq x+1. \end{cases}$$

Now $(\mathbf{B5})^\sharp$ is implied by (30) with $y = 1$.

$(\mathbf{B6})^\sharp$ $(\alpha \leq \beta \vee \beta \leq \alpha)$. $\Delta_1^{1,b}$-define the function symbol $\alpha \oplus \beta$ such that

$$(\alpha \oplus \beta)(x) \equiv (\alpha(x) \oplus \beta(x)).$$

$\alpha \neq \beta$ implies $\alpha \oplus \beta \neq 0$ and $\ell(\alpha \oplus \beta) > 0$. Now, $\alpha(x)$ and $\beta(x)$ coincide for $x \geq \ell(\alpha \oplus \beta)$ and differ at $x = \ell(\alpha \oplus \beta) \dot{-} 1$. $(\mathbf{B6})^\sharp$ follows.

$(\mathbf{B10})^\sharp$ $(\alpha \neq 0 \supset |2 \cdot \alpha| = S(|\alpha|))$. By (30), $V_1^1(BD) \vdash \alpha \neq 0 \supset \ell(2 \cdot \alpha) = S(\ell(\alpha))$. Applying id and using Lemma 5.1, we get $V_1^1(BD) \vdash \alpha \neq 0 \supset |2 \cdot \alpha| = id(S(\ell(\alpha))) = S(id(\ell(\alpha))) = S(|\alpha|)$.

$(\mathbf{B12})^\sharp$ $(\alpha \leq \beta \supset |\alpha| \leq |\beta|)$. $V_1^1(BD)$ proves that both ℓ and id are monotonic.

$(2)^\sharp$ $V_k^1(BD) \vdash |\alpha \#_j \beta| = id\left(\ell\left(exp\left(\ell(\alpha) \#_{j-1} \ell(\beta)\right)\right)\right) =$
$id\left(S\left(\ell(\alpha) \#_{j-1} \ell(\beta)\right)\right) =$ (by Lemma 5.1) $S\left(id(\ell(\alpha)) \#_{j-1} id(\ell(\beta))\right) =$
$S\left(|\alpha|_{j-1} |\beta|\right)$.

$(\mathbf{B17})^\sharp$ $(|\alpha| = |\beta| \supset \alpha \# \gamma = \beta \# \gamma)$. V_2^1 readily proves $id(x) = id(y) \supset x = y$ and hence $|\alpha| = |\beta| \supset \ell(\alpha) = \ell(\beta)$. But the definition of $\alpha \# \gamma$ depends only on $\ell(\alpha), \ell(\gamma)$.

$(\mathbf{B19})^\sharp$ $(\alpha_1 \leq \alpha_1 + \alpha_2)$. Prove by induction on y that

$$Carry(\alpha_1, \alpha_2, \beta, x) \wedge 0 < y \leq x \wedge \neg\beta(y) \ \supset\ \alpha_1\big|_{y \dot{-} 1} \leq (\alpha_1 + \alpha_2)\big|_{y \dot{-} 1}$$

and substitute $x = y = \ell(\alpha_1) + \ell(\alpha_2)$.

$(\mathbf{B21})^\sharp$ $(\alpha + \beta = \beta + \alpha)$. The definition of addition is symmetric.

$(\mathbf{B24})^\sharp$ $((\alpha + \beta) + \gamma = \alpha + (\beta + \gamma))$. Define the ternary *symmetric* function symbol $(\alpha + \beta + \gamma)$ in the obvious way. Then prove separately $(\alpha + \beta) + \gamma = (\alpha + \beta + \gamma)$ and $\alpha + (\beta + \gamma) = (\alpha + \beta + \gamma)$ by joint induction on all natural relations between $\alpha(x), \beta(x), \gamma(x), (\alpha + \beta)(x), (\alpha + \beta + \gamma)(x)$ etc. and all carry bits involved in the computations.

$(\mathbf{B29})^\sharp$ $(\alpha \cdot (\beta + \gamma) = (\alpha \cdot \beta) + (\alpha \cdot \gamma))$. First we prove $2^y * (\beta + \gamma) = (2^y * \beta) + (2^y * \gamma)$. Then we get $(\mathbf{B29})^\sharp$ from this and $(\mathbf{B21})^\sharp$,$(\mathbf{B24})^\sharp$ by applying induction on y to the formula

$$\begin{aligned} &Table(\alpha, \beta + \gamma, \delta^2, x) \wedge Table(\alpha, \beta, \delta_1^2, x) \wedge \\ &Table(\alpha, \gamma, \delta_2^2, x) \wedge 0 < y \leq x + 1 \ \supset \\ &\delta^2|_y \stackrel{x}{=} \delta_1^2|_y + \delta_2^2|_y. \end{aligned}$$

$(\mathbf{B30})^\sharp$ $(\alpha \geq 1 \ \supset\ \alpha \cdot \beta \leq \alpha\gamma \equiv \beta \leq \gamma)$. In view of $(\mathbf{B6})^\sharp$,$(\mathbf{B7})^\sharp$ we only have to check $V_1^1(BD) \vdash \alpha \neq 0 \wedge \beta < \gamma \ \supset\ \alpha \cdot \beta < \alpha \cdot \gamma$. This is done by induction like in $(\mathbf{B5})^\sharp$, $(\mathbf{B29})^\sharp$.

Now we check $V_k^i(BD) \vdash \left(\Sigma_i^b - PIND\right)^\sharp$. By already proven part a), it suffices to show that

$$\left.\begin{array}{l} V_k^i(BD) \vdash \\ \left[B(0,0) \wedge \forall\alpha \left(B\left(\lfloor\frac{1}{2}\alpha\rfloor, \ell\left(\lfloor\frac{1}{2}\alpha\rfloor\right)\right) \supset B(\alpha,\ell(\alpha))\right)\right] \supset \\ \forall\alpha\ B(\alpha,\ell(\alpha)) \end{array}\right\} \quad (31)$$

where $B(\alpha, z)$ is in $\Sigma_i^{1,b}$. Let us introduce the function symbol $2^{-y}\alpha$ by

$$2^{-y}\alpha(x) \equiv \alpha(x+y).$$

Now induction on y applied to the $\Sigma_i^{1,b}$-formula

$$C(\alpha, z, y) \rightleftharpoons B\left(2^{-(z \dot{-} y)}\alpha, y\right)$$

implies (31) after substituting $z = y = \ell(\alpha)$.

The proof of Lemma 5.4 is completed.■

6. $\flat$ and $\sharp$ are inverse to each other

The primary purpose of this section is to justify its title by proving the following two statements:

Lemma 6.1. *Let $A(\vec{x})$ be any formula in $\mathcal{L}_k$ with all free first order variables displayed. Then $V_k^1(BD) \vdash A^{\flat\sharp}(id(\vec{x})) \equiv A(\vec{x})$. In particular, if A does not contain free first order variables, $V_k^1(BD) \vdash A^{\flat\sharp} \equiv A$.*

Lemma 6.2. *Let A be any formula in L_{k+1}. Then $S_{k+1}^1 \vdash A^{\sharp\flat} \equiv A$.*

Proof of Lemma 6.1. Induction on complexity of A. If $A \equiv t(\vec{x}) \circ s(\vec{x})$ where $\circ$ is $=$ or $\leq$ then $A^{\flat\sharp} \equiv t^\sharp(\vec{\alpha}) \circ s^\sharp(\vec{\alpha})$. By Lemma 5.1, $V_k^1(BD) \vdash A^{\flat\sharp}(id(\vec{x})) \equiv id(t(\vec{x})) \circ id(s(\vec{x})) \equiv A(\vec{x})$ since $V_1^1(BD)$ readily proves $x \circ y \equiv id(x) \circ id(y)$.

Assume now that $A \equiv \alpha(t(\vec{x}))$. Then $A^{\flat\sharp} \equiv Bit^\sharp\left(t^\sharp(\vec{x}), \alpha\right) = 1$ and $A^{\flat\sharp}(id(\vec{x})) \equiv Bit^\sharp\left(id(t(\vec{x})), \alpha\right) = 1$. So we only have to understand that

$$V_1^1(BD) \vdash Bit^\sharp(id(y), \alpha) = 1 \equiv \alpha(y). \quad (32)$$

The easiest way of doing this is to translate via $\sharp$ identities $Bit(0, 2x) = 0$, $Bit(0, 2x + 1) = 1$, $Bit(y + 1, x) = Bit(y, \lfloor \frac{x}{2} \rfloor)$ provable in S_1^1 to identities in $V_1^1(BD)$. Having these we easily establish by induction on z that $V_1^1(BD) \vdash Bit^\sharp \left(id(z), 2^{-(y \dot{-} z)}\alpha\right) = 1 \equiv \alpha(y)$ which (with $z = y$) gives us (32).

Suppose $A \equiv \exists x\ B(x)$. Then $A^\flat \equiv \exists z\ B^\flat(|z|)$ and $A^{\flat\sharp} \equiv \exists \alpha\ B^{\flat\sharp}(|\alpha|)$. To apply the inductive hypothesis we only have to show that $V_k^1(BD) \vdash \exists \alpha\ C(|\alpha|) \equiv \exists x\ C(id(x))$. It is clear in one direction since $|\alpha| \equiv id(\ell(\alpha))$. In another direction this follows from the identity $V_1^1(BD) \vdash id(x) = |exp(x)|$.

All remaining cases are trivial.■

The following immediate corollary is interesting in its own right.

Corollary 6.3. *For each $i \geq 1$ the interpretation $\flat$ is exact. In other words, for each closed formula A in $\mathcal{L}_k$, $V_k^1(BD) \vdash A$ iff $S_{k+1}^1 \vdash A^\flat$.*

Proof of Lemma 6.2. Induction on complexity of A. The only nontrivial case is when A is atomic and it suffices to show for that case that

$$S_{k+1}^1 \vdash t^{\sharp\flat}(\vec{x}) = t(\vec{x}) \tag{33}$$

for any term t in L_{k+1} and $S_{k+1}^1 \vdash x \leq^{\sharp\flat} y \equiv x \leq y$. It is sufficient to establish (33) only for the case when t is a function symbol of the language L_{k+1}. The easiest way to do this is to note that the system $\left\{ f^{\sharp\flat}, \leq^{\sharp\flat} \mid f \in L_{k+1} \right\}$ satisfies all $BASIC_k$ axioms. Moreover, $f^{\sharp\flat}$ are Σ_1^b-defined and $\leq^{\sharp\flat}$ is Δ_1^b-defined so we may freely apply $\Sigma_1^b - PIND$ to formulas containing those symbols. But S_{k+1}^1 readily proves that axioms from the list $BASIC_k$ uniquely determine symbols from the language L_{k+1}.■

7. Summary

We summarize results established in the previous sections in the following theorem:

Theorem 7.1 (Main). *For each $i, k \geq 1$ there exist interpretations $\flat$: $V_k^i(BD) \rightsquigarrow S_{k+1}^i$ and $\sharp : S_{k+1}^i \rightsquigarrow V_k^i(BD)$ such that:*

a) *for any* $\Sigma_j^{1,b}$*-formula* $A(x_1, \ldots, x_n)$ *in* $\mathcal{L}_k$ *(where* $j \geq 1$ *and all free first order variables in* A *are displayed),* $A^\flat(|x_1|, \ldots, |x_n|)$ *is equivalent in* S^1_{k+1} *to a* Σ_j^b*-formula,*

b) *for any* Σ_j^b*-formula* A *in* L_{k+1}*,* $A^\sharp$ *is equivalent in* $V_k^1(BD)$ *to a* $\Sigma_j^{1,b}$*-formula (again, provided* $j \geq 1$*),*

c) *for any formula* $A(x_1, \ldots, x_n)$ *in* $\mathcal{L}_k$ *(all free first order variables are displayed),*

$$V_k^1(BD) \vdash A^{\flat\sharp}(id(x_1), \ldots, id(x_n)) \equiv A(x_1, \ldots, x_n),$$

d) *for any formula* A *in* L_{k+1}*,*

$$S^1_{k+1} \vdash A^{\sharp\flat} \equiv A.$$

Proof. This is the content of Lemmas 4.2, 5.4, 6.1, 6.2.■

As an application we obtain the analog of Buss's main result for the theory $V_2^1(BD)$ as a *direct* consequence of Proposition 2.2:

Theorem 7.2. **a)** *For each function* $f(\vec{x}) \in EXPTIME$ *of polynomial growth rate there exist a* $\Sigma_1^{1,b}$*-formula* $A(\vec{x}, y)$ *and a first order term* $t(\vec{x})$*, both in* $\mathcal{L}_2$ *such that:*

a1) $V_2^1(BD) \vdash \exists y \leq t(\vec{x})\ A(\vec{x}, y)$,

a2) $V_2^1(BD) \vdash A(\vec{x}, y) \wedge A(\vec{x}, z) \supset y = z$,

a3) *for all* $\vec{n}$, $\mathbf{N} \models_{BD} A(\vec{n}, f(\vec{n}))$ *where* $\models_{BD}$ *corresponds to the model where second order variables range over all* α *with finitely many ones.*

b) *Conversely, suppose* $V_2^1(BD) \vdash \forall\vec{x}\exists y\ A(\vec{x}, y)$ *where* $A(\vec{x}, y)$ *is a* $\Sigma_1^{1,b}$*-formula with all free variables displayed. Then there exist a* $\Delta_1^{1,b}$*-formula* $B(\vec{x}, y)$*, a first order term* $t(\vec{x})$ *(in* $\mathcal{L}_2$*) and an* $EXPTIME$ *function* $f(\vec{x})$ *of polynomial growth rate such that:*

b1) $V_2^1(BD) \vdash B(\vec{x}, y) \supset A(\vec{x}, y)$,

b2) $V_2^1(BD) \vdash \exists y \leq t(\vec{x})\ B(\vec{x}, y)$,

b3) $V_2^1(BD) \vdash B(\vec{x}, y) \wedge B(\vec{x}, z) \supset y = z$,

b4) *for all* $\vec{n}$, $\mathbf{N} \models_{BD} B(\vec{n}, f(\vec{n}))$.

Proof. We start with the part b).

By Theorem 7.1 a), $A^\flat(|\vec{x}|, |y|)$ is equivalent in S_3^1 to a Σ_1^b-formula. Apply Proposition 2.2 b) with $k = 3$ to the formula $A^\flat(|\vec{x}|, |y|)$ and find corresponding triple (B, t, f) which we redenote as $(B'(\vec{x}, y), t'(\vec{x}), f'(\vec{x}))$. We define (B, t, f) by

$$\begin{aligned} t(\vec{x}) &\rightleftharpoons (t')^-(\vec{x}), \\ B(\vec{x}, y) &\rightleftharpoons \exists\beta \left(B'^\sharp\left(exp(\vec{x}), \beta|_{t(\vec{x})}\right) \wedge \ell\left(\beta|_{t(\vec{x})}\right) = y\right), \\ f(\vec{x}) &\rightleftharpoons \left|f'\left(2^{\vec{x}}\right)\right|. \end{aligned}$$

Let us check that these B, t, f possess all required properties.

Since $f'(\vec{n})$ is computable in time $\exp\left(\log^{O(1)}|\vec{n}|\right)$, $f'\left(2^{\vec{n}}\right)$ and hence $f(\vec{n})$ are computable in time $\exp\left(|\vec{n}|^{O(1)}\right)$. Now $|f'(\vec{n})| \leq \exp\left(\log^{O(1)}|\vec{n}|\right)$ implies $f(\vec{n}) = \left|f'\left(2^{\vec{n}}\right)\right| \leq \exp\left(|\vec{n}|^{O(1)}\right)$ that is $|f(\vec{n})| \leq |\vec{n}|^{O(1)}$. Which means f has polynomial growth rate.

Proposition 2.2 b1) says that $S_3^1 \vdash B'(\vec{x}, y) \supset A^\flat(|\vec{x}|, |y|)$. Applying $\sharp$ gives us

$$V_2^1(BD) \vdash B'^\sharp(\vec{\alpha}, \beta) \supset A^{\flat\sharp}(|\vec{\alpha}|, |\beta|).$$

Substituting here $exp(\vec{x})$ for $\vec{\alpha}$ and $\beta|_{t(\vec{x})}$ for β, we will have $V_2^1(BD) \vdash B(\vec{x}, y) \supset A^{\flat\sharp}(id(\vec{x}), id(y))$. Which, along with Theorem 7.1 c), gives us b1).

Similarly, applying $\sharp$ to Proposition 2.2 b3), we have

$$V_2^1(BD) \vdash \exists\beta \leq t'^\sharp(\vec{\alpha})\ B'^\sharp(\vec{\alpha}, \beta).$$

If β is such that $\beta \leq t'^\sharp(exp(\vec{x}))$ and $B'^\sharp(exp(\vec{x}), \beta)$ then

$$\ell(\beta) \leq \ell\left(t'^\sharp(exp(\vec{x}))\right) \leq \text{(by Lemma 5.2) } (t')^-(\vec{x}) = t(\vec{x})$$

hence $\beta|_{t(\vec{x})} = \beta$ and $B(\vec{x}, \ell(\beta))$. This proves b2).

b3) is proved similarly.

B is equivalent to a $\Sigma_1^{1,b}$-formula by Theorem 7.1 b). Already proven b2) and b3) imply its $\Pi_1^{1,b}$-representation $V_2^1(BD) \vdash B(\vec{x}, y) \equiv \forall z \leq t(\vec{x})\,(B(\vec{x}, z) \supset z = y)$. Hence B is in $\Delta_1^{1,b}$.

Lastly, $\mathbf{N} \models B'(\vec{n}, f'(\vec{n}))$ implies $\mathbf{N} \models_{BD} B'^\sharp(id(\vec{n}), id(f'(\vec{n})))$ since $\sharp$ is sound with respect to $\models_{BD}$. Substitute here $2^{\vec{n}}$ for $\vec{n}$ and let $\eta \rightleftharpoons$

$id\left(f'\left(2^{\vec{n}}\right)\right)$. Then, as in the proof of b2), $\mathbf{N} \models_{BD} \eta|_{t(\vec{n})} = \eta$ and hence $\mathbf{N} \models_{BD} B\left(\vec{n}, \ell(\eta)\right)$. We only have to note now that $\ell(\eta) = \left|f'\left(2^{\vec{n}}\right)\right| = f(\vec{n})$.

The proof of part b) is completed.

a) Let $f'(\vec{n}) \rightleftharpoons 2^{f(|\vec{n}| \dot{-} 1)} \dot{-} 1$. $f(|\vec{n}|)$ is computable in time

$$\exp\left(\log^{O(1)}|\vec{n}|\right)$$

and $f(|\vec{n}|) \leq \exp\left(\log^{O(1)}|\vec{n}|\right)$. Hence $f'(\vec{n})$ is computable in time

$$\exp\left(\log^{O(1)}|\vec{n}|\right) \leq |t'(\vec{n})|$$

for some term t' of the language L_3. Apply Proposition 2.2 a) with $k = 3$ to find corresponding A'. Now the same proof allows us to lift f', t', A' to some $\tilde{f}, t, A$ such that a1), a2) and $\mathbf{N} \models_{BD} A(\vec{n}, \tilde{f}(\vec{n}))$. But $\tilde{f}(\vec{n}) = \left|f'\left(2^{\vec{n}}\right)\right| = f\left(\left|2^{\vec{n}}\right| \dot{-} 1\right) = f(\vec{n})$. The proof of part a) is completed.■

Remark 7.3. The same argument can be applied to the case $i = 1$ i.e. $V_1^1(BD)$ vs. S_2^1. This will imply that $V_1^1(BD)$ can $\Sigma_1^{1,b}$-define exactly those functions $f(\vec{n})$ which are computable in time $2^{O(|\vec{n}|)}$ and have *linear* growth rate. Similarly, $V_1(BD)$ (which can be viewed as a weak monadic extension of $I\Delta_0$) $\Sigma^{1,b}$-defines exactly those functions which belong to the linear exponential time hierarchy $E \cup E^{NE} \cup E^{NE^{NE}} \cup \ldots$ and have linear growth rate (here $E \rightleftharpoons DTIME\left[2^{O(n)}\right]$) etc.

8. Conservativity results

In this section we will show that the theories $\widetilde{V}_k^i, \overset{\circ}{V}_k^i, V_k^i(bd)$ and $V_k^i(BD)$ prove the same $\Sigma^{1,b}$-formulae and the same closed $\forall^0\exists^0\Sigma^{1,b}$-formulae (the superscript "0" indicates that the quantifiers are first order). This implies that in the part concerning such formulae, the equivalence between $V_k^i(BD)$ and S_{k+1}^i established in previous sections is extended to $\widetilde{V}_k^i, \overset{\circ}{V}_k^i, V_k^i(bd)$.

Lemma 8.1. *$\overset{\circ}{V}_k^i$ is $\Sigma^{1,b}$-conservative over $V_k^i(bd)$.*

Proof. In the course of this proof it will be convenient for us to change the framework and consider $\overset{\circ}{V}{}^i_k$ and $V^i_k(bd)$ as Gentzen style second order theories rather than merely many sorted theories.

Assume $\overset{\circ}{V}{}^i_k \vdash A$ where A is bounded. Consider the theory $\overset{\circ}{V}{}^i_k(\delta)$ which is obtained from $\overset{\circ}{V}{}^i_k$ by adding to the language new predicate symbols δ for all $\Delta^{1,b}_1$ predicates and replacing the scheme $\Delta^{1,b}_1 - CA$ by $\Sigma^{1,b}_0(\delta) - CA$ (see [Bus86, §9.7]). By [Bus86, Corollary 9.21] there exists a Gentzen style $\overset{\circ}{V}{}^i_k(\delta)$-proof of the sequent $\rightarrow A$ such that all formulae in this proof are bounded. We can eliminate now in this proof symbols δ to get a proof *in* $\overset{\circ}{V}{}^i_k$ with the same property (compare [Bus86, Proposition 9.18]). So we only have to check that the comprehension rules

$$\frac{\Gamma \rightarrow \Delta, F(V)}{\Gamma \rightarrow \Delta, \exists\phi F(\phi)} \qquad\qquad \frac{F(V), \Gamma \rightarrow \Delta}{\forall\phi F(\phi), \Gamma \rightarrow \Delta}$$

where V is a $\Delta^{1,b}_1$-abstract *and F is bounded* are admissible in $V^i_k(bd)$.

To see this, note that the proofs of theorem 3.6 a) and (14) generalize to showing that for F, V as above there exists a first order term t such that $V^0_k(bd) \vdash F(V) \equiv F(V|_t)$ where $V|_t$ is defined in the obvious way. But the rules

$$\frac{\Gamma \rightarrow \Delta, F(V|_t)}{\Gamma \rightarrow \Delta, \exists\phi F(\phi)} \qquad\qquad \frac{F(V|_t), \Gamma \rightarrow \Delta}{\forall\phi F(\phi), \Gamma \rightarrow \Delta}$$

are equivalent to $\Delta^{1,b}_1 - BCA$ which is provable in $V^1_k(bd)$ by Theorem 3.5.■

Define the translation $*$ from the set of formulae in the language $\mathcal{L}_k$ to itself by relativizing all second order quantifiers $\exists\alpha, \forall\alpha$ to the domain $\exists x \forall y\,(\alpha(y) \supset y < x)$.

Lemma 8.2. *For each bounded formula A, $V^0_k(bd) \vdash A^* \equiv A$.*

Proof. Induction on complexity of A. The only nontrivial case $A \equiv \exists\alpha\ B(\alpha)$ is taken care of by Theorem 3.6 a) and (14) like in the proof of previous lemma.■

Corollary 8.3. $*$ *is an interpretation of $V^i_k(BD)$ in $V^i_k(bd)$.*

Proof. Obviously follows from Lemma 8.2.∎

Corollary 8.4. *$V^i_k(BD)$ is $\Sigma^{1,b}$-conservative over $V^i_k(bd)$.*

Theorem 8.5. *$\widetilde{V}^i_k, \mathring{V}^i_k, V^i_k(bd)$ and $V^i_k(BD)$ prove the same bounded formulae.*

Proof. Immediate from Lemma 8.1 and Corollary 8.4.∎

Theorem 8.6. *All four theories $\widetilde{V}^i_k, \mathring{V}^i_k, V^i_k(bd)$ and $V^i_k(BD)$ prove the same closed $\forall^0\exists^0\Sigma^{1,b}$-formulae.*

Proof. Let V^i_k be one of the two theories $\mathring{V}^i_k, V^i_k(BD)$ which proves $\forall\vec{x}\exists y_1, \ldots, y_r\ A(\vec{x}, \vec{y})$ where A is a bounded formula with all free variables displayed. Decoding the vector $\vec{y}$ by $\langle y_1, \ldots, y_r\rangle$, we may assume that $r = 1$. As before, it suffices to check that

$$V^i_k(bd) \vdash \exists y\ A(\vec{x}, y). \tag{34}$$

By the extension of Parikh's theorem [Par71] to second order Bounded Arithmetic in the case $V^i_k \equiv \mathring{V}^i_k$ and by Theorem 7.2 b) in the case $V^i_k \equiv V^i_k(BD)$, we have a first order term $t(\vec{x})$ such that $V^i_k \vdash \exists y \leq t(\vec{x})\ A(\vec{x}, y)$. Applying Theorem 8.5 gives us $V^i_k(bd) \vdash \exists y \leq t(\vec{x})\ A(\vec{x}, y)$ which implies (34).∎

9. Concluding remarks

In this paper we considered bounded domain versions $V^i_k(BD)$ of Buss's theories V^i_k, developed a logical formalism allowing one to understand their power and showed that these versions talk of essentially the same domain and have exactly the same power as their first order counterparts S^i_{k+1} (Theorem 7.1). The bounded domain versions do not differ much from original Buss's (unbounded domain) theories while we are concerned with bounded (or "almost bounded") formulae (Theorems 8.5, 8.6). Also they are robust with respect to the choice of the bounded comprehension axiom scheme (Theorem 3.5). The analogue of the latter property is not expected in the case of unbounded domain.

Probably this equivalence can be without difficulties extended to higher order theories. A typical result might say that the "bounded domain" theory obtained from S^i_k by allowing functionals of type $\leq l$ is equivalent to S^i_{k+l}. The equivalence should interpret functionals of type $j \leq l$ by those integers x for which the j-th fold exponent exists (compare with [CT86]). However I did not try to develop this systematically.

10. Acknowledgment

I am grateful to Lev Beklemishev and anonymous referee for extensive comments made on an earlier version of this paper. I also wish to express my sincere gratitude to Samuel Buss and Gaisi Takeuti for pointing the sources [All89, Clo89, Tak90, Tak91] out to me.

References

[All89] B. Allen. *Arithmetizing uniform NC*. PhD thesis, University of Hawaii, 1989.

[Bus86] S. R. Buss. *Bounded Arithmetic*. Bibliopolis, Napoli, 1986.

[Clo89] P. Clote. A first-order theory for the parallel complexity class NC. Technical Report BCCS-89-01, Boston College, January 1989. Published in expanded form jointly with G. Takeuti in "Arithmetics for NC, $ALOGTIME$, L and NL", to appear in *Annals of Pure and Applied Logic*.

[CT86] P. Clote and G. Takeuti. Exponential time and bounded arithmetic. In *Proceedings of the 1st Structure in Complexity Theory Annual Conference*, pages 125–143, 1986.

[GJ79] M. R. Garey and D. S. Johnson. *Computers and Intractability. A guide to the theory of NP-completeness.* W. H. Freeman, 1979. Рус. пер.: М. Гэри, Д. Джонсон, *Вычислительные машины и труднорешаемые задачи*, М., Мир, 1982.

[Par71] R. J. Parikh. Existence and feasibility in arithmetic. *Journal of Symbolic Logic*, 36:494–508, 1971.

[Pud83] P. Pudlák. Some prime elements in the lattice of interpretability types. *Transactions of the AMS*, 280:255–275, 1983.

[Sho67] J. Shoenfield. *Mathematical logic.* Reading, Mass., Addison-Wesley Pub. Co., 1967. Рус. пер.: Дж. Шенфилд, *Математическая логика*, М., Наука, 1975.

[Tak75] G. Takeuti. *Proof Theory.* North-Holland, 1975. Рус. пер.: Г. Такеути, *Теория доказательств*, М., Мир, 1978.

[Tak90] G. Takeuti. S_3^i and $\overset{\circ}{V}{}_2^i(BD)$. *Archive for Math. Logic*, 29:149–169, 1990.

[Tak91] G. Takeuti. *RSUV* isomorphisms. This volume, 1991.

Alexander A. Razborov
Steklov Mathematical Institute
Vavilova 42
117966, GSP–1,
Moscow, RUSSIA
email: razb@log.mian.su

Integer Parts of Real Closed Exponential Fields

J.-P. Ressayre

This paper is an extended abstract: the detailed proofs are being prepared for future publication. We define real exponential fields as ordered fields with a function 2^x satisfying a set E of four basic axioms. We show that every real closed exponential field in this sense possesses an "integer part", that is a discrete subring Z such that Z^+ is closed under 2^x, and every element x of the field has an "integer part" $\lfloor x \rfloor \in Z$, such that $\lfloor x \rfloor \leq x < \lfloor x \rfloor + 1$. By using such integer parts, we show that $Th(\mathbb{R}, 2^x)$ is equivalent to $E + Th(\mathbb{R}, 2^{1/(1+x^2)})$; and we deduce the model completeness of $Th(\mathbb{R}, 2^x)$ from the model completeness of $Th(\mathbb{R}, 2^{1/(1+x^2)})$ by an argument which differs from Wilkie's original proof.

§I. Survey of Main Results

§I.A Integer parts and the theory of $(\mathbb{R}, 2^x)$

Notations: Throughout "field" stands for *ordered* field. An *exponential* field is a field with a function 2^x satisfying the set E of axioms:

E_1: $2^1 = 2$, $2^{x+y} = 2^x \cdot 2^y$
E_2: $x < y \Leftrightarrow 2^x < 2^y$
E_3: $2^x > x^n$ for all x somewhat larger than n
E_4: $\forall x > 0, \exists y$ such that $2^y = x$ (this y is denoted $\log x$)

RC denotes the theory of real closed fields, $\mathbb{R}$ denotes the field of the reals, RCE denotes $RC + E$, the theory of *real closed exponential fields*. Z is an *integer part* (denoted IP) of the field K iff Z is a discrete subring of K, and $\forall x \in K, \exists \lfloor x \rfloor \in Z, \lfloor x \rfloor \leq x < \lfloor x \rfloor + 1$. Z is an IP of the *exponential* field $(K, 2^x)$ if in addition Z^+ is closed under 2^x.

Theorem 1 Every RCE field has an IP.

Theorem 2 Let T_i, $i = 0, 1$ be a pair of theories extending RCE;

a) there exist models $(M_i, 2^x)$ of T_i with IPs Z_i such that

$$(M_0, 2^x|Z_0^+) \simeq (M_1, 2^x|Z_1^+).$$

b) If the T_i's contain $Th(\mathbb{R}, 2^x|[0,1])$, then in (a) the conclusion can be strengthened to

$$(M_0, 2^x|Z_0^+, 2^x|[0,1]) \simeq (M_1, 2^x|Z_1^+, 2^x|[0,1])$$

hence to $(M_0, 2^x) \simeq (M_1, 2^x)$ (by the fact below).

Fact 1 Assume that ϕ is an isomorphism

$$\phi : (M_0, 2^x|Z^+, 2^x|[0,1]) \simeq (M_1, 2^x|Z_1^+, 2^x|[0,1])$$

where $(M_1, 2^x)$ are E fields, and Z_i is IP of M_i; then actually

$$\phi : (M_0, 2^x) \simeq (M_1, 2^x).$$

Proof

$$\phi(2^x) = \phi(2^{\lfloor x \rfloor}) \cdot \phi(2^{x-\lfloor x \rfloor}) = 2^{\phi(\lfloor x \rfloor)} \cdot 2^{\phi(x-\lfloor x \rfloor)} = 2^{\phi(x)} \Box$$

Thus we get:

Theorem 3 $Th(\mathbb{R}, 2^{1/1+x^2}) + E$ is complete, hence equivalent to $Th(\mathbb{R}, 2^x)$.

Proof $Th(\mathbb{R}, 2^{1/1+x^2})$ is bi-interpretable with $Th(\mathbb{R}, 2^x \lceil [0,1])$, so Th 2 applies to show that any two complete theories T_i containing $Th(\mathbb{R}, 2^{1/1+x^2})+ E$ have isomorphic models, hence are equivalent. $\Box$

The proof of Theorem 3 is easy to extend to the case where the language of T_i contains extra constants and both T_i's include the simple diagram DA of a given RCE field $(A, 2^{1/1+x^2})$. Thus it shows the completeness of

$$Th(\mathbb{R}, 2^{1/1+x^2}) + E + DA,$$

hence the model completeness of $Th(\mathbb{R}, 2^x)$.

§I.B Dyadic representation in RCE fields

We also prove the following suggestive strengthening of Th. 1:

THEOREM (Dyadic Representation of RCE fields)
If $(K, 2^K)$ is RCE, then K has a subring Z s.t.

i)$Z = IP$ of $(K, 2^K)$

ii) each $y \in K$ has a "dyadic development" of the form $y = \sum_{i<\alpha} \epsilon_i 2^{z_i}$, where α is an ordinal, $\epsilon_i = \pm 1$, and:

$$z_i \searrow \text{ in } Z \text{ (that is: } z_i \text{ is strictly decreasing in } Z).$$

iii) Z happens to be the ring of all elements $y \in K$, whose development has only *positive* exponents $z_i \in Z^+$.

Here (ii) is a bit imprecise; below we shall give a precise form, in which the dyadic series will have coefficients $a_i \in \mathbf{R}$, not only $\epsilon_i = \pm 1$. But replacing each a_i by its dyadic representation in $\mathbb{R}$ then yields the above form.

Definition Let G be an ordered abelian group, then

$$\mathbb{R}((2^G)) =_{\text{def}} \{\sum_{i<\alpha} a_i 2^{g_i} : \alpha \text{ ordinal}, a_i \in \mathbb{R}, g_i \searrow \text{ in } G\}.$$

Two elements of $\mathbb{R}((2^G))$ are said **equal** if they differ only by terms whose coefficient is 0.

Proposition

a) $\mathbb{R}((2^G))$ is a field, with the natural addition and multiplication of series.

b) This field has a unique field order $<$ extending $<_{\mathbb{R}}$, such that

$$\mathbb{R} < 2^g \Leftrightarrow g > 0.$$

This order satisfies:

$$g_0 < g \Rightarrow 2^{g_0} = o(1) \cdot 2^g;$$

where $o(1)$ denotes some element smaller than $1/n, \forall n \in \mathbb{N}$.
In fact, for every element $\sum_{i<\alpha} a_i 2^{g_i}$ of $\mathbb{R}((2^G))$, we even have:

$$g_0 < g \Rightarrow \sum_{i<\alpha} a_i 2^{g_i} = o(1) \cdot 2^g.$$

c) Let $\sum 2^G$ be a subfield of $\mathbb{R}((2^G))$, *closed under truncation:*

$$\sum_{i<\alpha} a_i 2^{g_i} \in \sum 2^G \Rightarrow \sum_{i<\beta} a_i 2^{g_i} \in \sum 2^G, \forall \beta < \alpha.$$

Then $\mathbb{Z} + \sum^+ 2^G$ is IP of $\sum 2^G$; where

$\sum^+ 2^G =_{\text{def}}$ the set of all series $\sum_{i<\alpha} a_i 2^{g_i}$, such that $g_i > 0, \forall i < \alpha$.

Proof of (a) + (b) – Malcev and B. von Neumann, see [K].

Proof of (c)

$$y = \sum_{i<\alpha} a_i 2^{g_i} \in \sum 2^G \Rightarrow y = \sum_{g_i>0} a_i 2^{g_i} + b + \sum_{g_i<0} a_i 2^{g_i} \ ;$$

the first term belongs to $\sum^+ 2^G$, the second to $\mathbb{R}$, and the third is $o(1)$. Put $[y] = \sum_{g_i>0} a_i 2^{g_i} + \lfloor b \rfloor_{\mathbb{Z}}$; then $[y] \in \mathbb{Z} + \sum^+ 2^G$ and:

$$|y - [y]| = |b - \lfloor b \rfloor| + o(1) < 1.$$

Dyadic representation of RCE fields (precise form).

Theorem 4 If $(K, 2^x)$ is RCE, then there is G, subgroup of $(K, +, <)$, such that

i) $\exists \phi : K \overset{\simeq}{\rightarrow} \sum 2^G \subseteq \mathbb{R}((2^G))$ with $\sum 2^G$ *closed under truncation.*
ii) $\phi(G) = \sum^+ 2^G (= \{\sum a_i 2^{g_i} \in \sum 2^G : g_i > 0, \forall i\})$
iii) $Z = \mathbb{Z} + G$ is IP of $(K, 2^x)$

Th. 1 only asserts (iii) of Th. 4; but in practice, one needs to prove Th. 4, just in order to obtain Th. 1. Moreover, this strengthening is also needed in the proof of Th. 2 and 3. We end the introduction with a survey of former work on IPs.

§II. IPs of ordered fields

In this section we forget about the exponential function, and consider only (ordered) fields.

Question: which fields possess an IP?

Boughattas, Mourgues, Ressayre have attempted a systematic survey of this question, and their results are beginning to give a general idea of the answer. Before recalling them, we give simple examples of IPs:

a) $\mathbb{Z}$ is the only IP of $\mathbb{R}$ and of any subfield of $\mathbf{R}$; hence of any *archimedian* field (that is, any field without infinite elements).

b) So we consider only *non-archimedian* fields. The first example of an IP in such a field comes from "non-standard models of analysis"; namely, consider $({}^*\mathbb{R}, {}^*\mathbb{Z}) \succ (\mathbb{R}, \mathbb{Z})$, then clearly ${}^*\mathbb{Z}$ is an IP of the field ${}^*\mathbb{R}$ (where "$\succ$" means "strict elementary extension").

But this IP ${}^*\mathbb{Z}$ is far from being unique: in fact, ${}^*\mathbb{R}$ is RC, and every RC (non-archimedian) field has an *infinity* of IPs. On the other hand IPs do not always exist.

Theorem A [M-R].
Every RC field has an IP.

This is optimal, in the sense that:

Theorem B [B].
For every $p \in \mathbb{N}$, there is a p-RC field which does not have *any* IP.

However,

Theorem C [B].
For every field K, there is an ultra-power of K which possesses an IP. (Thus no theory of an ordered field can forbid the existence of an IP.)

Remark: *arithmetic and algebraic Integer Parts*
Our first examples $\mathbb{Z}$, ${}^*\mathbb{Z}$ of IPs were models of the strongest possible theory of arithmetic, which is very undecidable. This contrasts with the theory of RC fields; and it suggests that the whole subject of IPs has to do with Arithmetic, but little to do with Algebra. Algebraists may even think that a systematic mixing of IPs with Algebra betrays a lack of understanding of the spirit of Algebra.

But among the IPs of RC fields, some have rather odd properties from the point of view of Arithmetic (for instance: their fraction field contains all algebraic numbers); and quite reasonable ones from the point of view of Algebra. Such *algebraic IPs* are the main tool of this paper; below they are constructed only in RCE fields, but [M-R] shows their existence in all RC fields.

We return to exponential fields.

§III. Sections of RCE fields

Definition 1
Let K be a field, R a maximal archimedian subfield; identify R with its unique image in $\mathbb{R}$, so that $R = K \cap \mathbb{R}$.

For every $x \in K$ we say that x is

infinite iff $|x| > R$
finite iff it is not infinite
infinitesimal iff $|y| < \frac{1}{n}, \forall n \in \mathbb{N}^+$.

$O(1) =_{\text{def}} \{x : x \text{ is finite}\}$; $o(1) =_{\text{def}} \{x : x \text{ is infinitely small}\}$. (We also write $x = O(1)$, $x = o(1)$ instead of $x \in O(1), x \in o(1)$.)

Proposition 1 $O(1) = R + o(1)$, that is, u finite $\Leftrightarrow \exists a \in R$ s.t. $u = a + o(1)$.

Definition 2

We say that x and y are *comparable* iff $x = O(1) \cdot y$ <u>and</u> $y = O(1) \cdot x$. y is comparable to $\{x_i, i \in I\} \Leftrightarrow y$ is comparable to x_i for some $i \in I$.

Definition 3

Let $(K, 2^x)$ be an E field; we say that a family $2^G = \{2^g : g \in G\}$ of powers of 2 by elements g of K is a *section* iff

I) G is a subgroup of $(K, +)$ (so 2^G is a subgroup of $(K^*, \cdot)$)

II) $Rg \subseteq G, \forall g \in G$ (so G is a vector space over R)

III) $\forall y \in K^*, \exists$ at most one g in G s.t. y is comparable to 2^g

The section 2^G is *complete* iff

IV) $\forall y \in K^*, y$ is comparable to 2^G.

Remarks

a) An easy application of Zorn's Lemma shows that complete sections always exist.

b) (Let v denote the valuation of K over R; then y is comparable to x iff x and y have the same valuation; and when 2^G is a complete section, then G is anti-isomorphic to $v(K)$. This remark can be ignored by people not knowing valuation theory.)

c) Also, y is comparable to x iff y has a "development at order 1": $y = a \cdot x + o(1) \cdot x$, with $a \in R^*$. (This follows from the Proposition above Def. 2). Thus if 2^G is a complete section, then every $y \in K^*$ has exactly one development at order one w.r. to 2^G. Therefore by iteration y has one development at order n, for each n:

$$y = \sum_{i \leq n} a_i 2^{g_i} + o(1) \cdot 2^{g_n} a, \text{ with } g_i \searrow \text{ in } G.$$

Roughly speaking, this process can be iterated transfinitely, allowing the representation of K by transfinite series. This is what we examine below; but for us it will be useful to examine as well the case where the section 2^G is not complete in K.

Let 2^G be a section, complete or not; we next study how to (partially) embed K into $R((2))^G$. To this end, we assume that a partial embedding

$\phi : K \to R((2))^G$ is already constructed, with domain a field A containing 2^G. Without loss of generality we can assume that ϕ is id_A and

$$A = K \cap R((2))^G,$$

just because it shortens notations. We can also assume that A is RC, since otherwise ϕ has a unique extension to the real closure of A. And finally we assume that

A is closed under truncation in $R((2))^G$,

that is:

$$\sum_{i<\alpha} a2^{g_i} \in A, \beta < \alpha \Rightarrow \sum_{i<\beta} a_i 2^{g_i} \in A.$$

Now given any $y \in K \setminus A$, we examine the problem of how to extend ϕ to y. We next define a series called the *development of y over A*, which in the good case will provide the image of y under the extension of ϕ.

Definition 4

The development of any y at order 0 is 0. $\sum_{i<\alpha} a_i 2^{g_i}$ is **the development of y over A at order $\alpha > 0$** iff

i) $\forall \beta < \alpha, \sum_{i<\beta} a_i 2^{g_i} \in A$
ii) $\forall \beta \leq \alpha, \forall j < \beta, y - \sum_{i<\beta} a_i 2^{g_i} = o(1) \cdot 2^{g_j}$

Lemma 1

a) If this development $\sum_{i<\alpha} a_i 2^{g_i}$ exists then it is unique and $\forall \beta < \alpha, \sum_{i<\beta} a_i 2^{g_i}$ is the development at order β.

b) There is a unique development of y over A at *maximum* order α (this is called *the* development of y over A: if the order is not indicated then it is maximum).

c) Let $\sum_{i<\alpha} a_i 2^{g_i}$ be the development of y over A.

Case 1 $\sum_{i<\alpha} a_i 2^{g_i} \notin A$. Then ϕ can be extended to an embedding ψ of the real closure of $A(y)$ such that $\psi(y) = \sum_{i<\alpha} a_i 2^{g_i}$ and the image of ψ is closed under truncation.

Case 2 $\sum_{i<\alpha} a_i 2^{g_i} \in A$. Then if $y' = y - \sum_{i<\alpha} a_i 2^{g_i}$,

i) y' is incomparable with 2^G and $y' = o(1) \cdot 2^{g_i}$ for each $i < \alpha$.
ii) 2^G can be extended to a section 2^H containing $|y'|$.
iii) ϕ has an extension to an embedding ψ of the real closure of $A(y)$ into $R((2^H))$, such that $\psi(y) = \sum_{i<\alpha} a_i 2^{g_i} + \epsilon 2^h$, where $2^h = |y'|$ and ϵ is the sign of y', and the image of ψ is closed under truncation.

If 2^G is a complete section, then only Case 1 arises; and by transfinite iteration of this case one gets an embedding of all of K into $R((2^G))$.

If $\sum 2^G$ is a partial section, both Case and 2 arise, depending on y. But if we iterate Case 1 as long as we find elements to which it applies, we get a maximal partial embedding into $R((2^G))$; this partial embedding is also obtainable by use of (b), below:

Proposition 2 – Suppose $(K, 2^x)$ RCE, 2^G a section.

a) 2^G complete $\Rightarrow K \simeq \sum 2^G \subseteq R((2^G))$. One can take $\sum 2^G$ closed under truncation.

b) If 2^G is incomplete, $\exists$ maximal $L \subseteq K$ such that 2^G is complete in L.

c) Hence $\exists \phi : L \overset{\simeq}{\rightarrow} \sum 2^G \subset R((2^G))$ (by (a) applied to L)

d) If $2^{G_0} \subset 2^G$ is given as well as $L_0 \subset K$ and $\phi : L_0 \overset{\simeq}{\rightarrow} \sum 2^{G_0} \subset R((2^{G_0}))$, then one can choose ϕ so as to extend ϕ_0. And if $\sum 2^{G_0}$ is closed under truncation, one can take $\sum 2^G$ closed too.

e) In what follows we identify L and $\sum 2^G$:

$$\sum 2^G = K \cap R((2^G)), \phi = id_{\Sigma 2^G}.$$

Then $\forall y \in K \setminus \sum 2^G$, y can be written in a unique way as $y = \sigma + y_1$, with $\sigma \in \sum 2^G$, y_1 incomparable to 2^G, and $y_1 = o(1) \cdot 2^{g_i}$, $\forall g_i$ exponent of σ.

Notation Henceforth whenever we fix a section 2^G of K, we (often tacitly) also fix a maximal subfield of K embedded in $R((2^G))$; and we denote it $\sum 2^G$. ($\sum 2^G$ exists by Lemma 1.e) Usually our choice of $\sum 2^G$ is not arbitrary, it is one which fits our purpose. And if 2^G extends a previous section 2^{G_0}, then we require: $\sum 2^{G_0} \subset \sum 2^G$, and the fixed embedding $\sum 2^G \hookrightarrow R((2^G))$ extends the previous embedding $\sum 2^{G_0} \hookrightarrow R((2^{G_0}))$. Usually we identify $\sum 2^G$ with its image under the fixed embedding; so that $\sum 2^G = K \cap R((2^G))$, and the fixed embedding is $id_{\Sigma 2^G}$.

We always assume that $\sum 2^G$ is closed under truncation in $R((2^G))$; then by a former proposition, the sum of rings $Z + \sum^+ 2^G$ is an IP of $\sum 2^G = K$; where:

$$\textstyle\sum^+ 2^G = \{\sum_{i<\alpha} a_i 2^{g_i} \in \sum 2^G : g_i > 0, \forall i < \alpha\}.$$

With the somewhat abusive but convenient identification of K with $\sum 2^G$, proving the dyadic representation theorem I.4 amounts to prove:

Theorem 1
Every RCE field $(K, 2^x)$ has a complete section 2^G such that $G = \sum^+ 2^G$.

(We shall call **dyadic sections** the sections with this property.)

Proof that Th. 1 implies Th. I.4 Let 2^G be a dyadic section of K; then condition (i) + (ii) of Th I.4 is satisfied, since with our identification of K and $\sum 2^G$, it simply reads: $K = \sum 2^G$, and $G = \sum^+ 2^G$. Moreover we just recalled that $Z = \mathbb{Z} + \sum^+ 2^G = \mathbb{Z} + G$ is an IP of K, which is part of (iii). There only remains to fulfill (iii), by showing that Z^+ is closed under 2^x. So let $n + \sigma$ (n in $\mathbb{Z}$ and $\sigma > 0$ in $\sum^+ 2^G$) be an element of Z^+; $2^{n+\sigma} = b \cdot 2^h$ with $b = 2^n$ in $\mathbb{R}$, and $h > 0$ in G, since h is σ. So $2^{n+\sigma}$ is in $\sum^+ 2^G$. □

§IV. Dyadic sections

In order to construct a dyadic section of the RCE field $(K, 2^x)$, we take the following:
Inductive assumption — 2^G is a (partial) section such that $G = \sum^+ 2^G$.
Given any element y of $K \setminus \sum 2^G$, we want to extend 2^G to a section 2^H in such a way that still $H = \sum^+ 2^H$, but now $y \in \sum 2^H$.
We can assume that

y is *incomparable* to 2^G

(if not, replace y by y', where $y = \sigma + y'$ is the development of y w.r. to 2^G, given by the above Prop. 2.e; hence y' is incomparable to 2^G)

y is *infinite*

(if not, $y \to 1/y$); and

$y > 0$ (otherwise $y \to -y$);

MAIN LEMMA

a) Let

$$\log y = \sigma + \epsilon y_1$$

be the development of $\log y$ over $\sum 2^G$ (Prop. 2.e); where $\sigma \in \sum 2^G$, y_1 is incomparable to 2^G, and $\epsilon = \pm 1$ is chosen so that y_1 is positive; then y_1 is infinite, just like y.
So by repeating this, we get $(y_n)_{n \in \mathbb{N}}$ such that $y_0 = y$, $\log y_n$ has development $\sigma_n + \epsilon_n y_{n+1}$ with $\sigma_n, \epsilon_n, y_{n+1}$ as σ, ϵ, y_1 above.

b) Then $2^G \cup \{y_n : n \in \mathbb{N}\}$ *generates* a section $2^{G'}$ such that $G' \subset \sum^+ 2^{G'}$.

c) Assume G' as in (b), and $\sigma \in \sum^+ 2^{G'}$; then $2^{G'(R\cdot\sigma)}$ is a section 2^H such that $\sigma \in H \subset \Sigma^+ 2^H$. Moreover $2^{H(R\cdot 2^\sigma)}$ is a section $2^{H'}$ such that $2^\sigma \in H' \subset \sum^+ 2^{H'}$.

d) Assume $(K_1, 2^x)$ is another RCE, with $K \cap K_1 \supset \mathbb{R}$, $K \cap \mathbb{R}((2^G)) = \sum 2^G \subset K_1$, $(K, 2^x)\restriction G^+ = (K_1, 2^x)\restriction G^+$ and $G = \sum^+ 2^G$. Then the order type of y is finitely satisfiable in K_1; if z satisfies this order type in K_1 (or in $K_1', 2^x) \succ (K_1, 2^x)$), then there is $(z_n)_{n\in\mathbb{N}}$ in K_1 satisfying with respect to $z = z_0$ the same properties as (y_n) with respect to y in K:

- $\log z_n = \sigma_n + \epsilon_n z_{n+1}$, where σ_n, ϵ_n come from (a).
- order type of z_n over 2^G = order type of y_n over 2^G.

e) Let $2^{G_1'}$, 2^{H_1}, $2^{H_1'}$ be the sections generated in K_1' from 2^G, $\{z_n : n \in \mathbb{N}\}$, σ and 2^σ in the same way as $2^{G'}$, 2^{H_1}, $2^{H_1'}$ were generated from 2^G, $\{y_n : n \in \mathbb{N}\}$, σ, and 2^σ in K; and let ϕ be the identity map over $\sum 2^G$. Then ϕ has an extension $\psi : \sum 2^{H'} \to \sigma 2^{H_1'} \subset K_1''$, such that

$\psi(y_n) = z_n$,
$\psi(2^g) = 2^{\psi(g)}$ for each element $2^g \in \sum^+ 2^{H'}$, and
$(K_1', 2^x) \prec (K_1'', 2^x)$.

e′) Assume $(\mathbb{R}, 2^x) = (K, 2^x) \cap (K_1, 2^x)$ (not only $\mathbb{R} = K \cap K_1$). For any partial map ϕ between structures $(K_i, 2^x)$ $(i = 0, 1)$ we say that ϕ is elementary iff it is elementary with respect to $(K_i, 2^{1/1+x^2})$ and the domain and image of ϕ are elementary submodels of $(K_i, 2^{1/1+x^2})$. Then if the application ϕ of (e) was elementary in this sense, we can ensure that its extension ψ is elementary too.

f) Assume that 2^H is any section of $(K_1, 2^x)$, $S \subset \sum 2^H \subset (K_1, 2^x)$, S is a field closed under truncation, $H \subset S \cap \sum^+ 2^H$ and we have an embedding $\Phi : S \to K$ (for instance, Φ is the inverse of the function ψ of (e) and (e′)). Then there exists an embedding $\Psi \supset \Phi$ such that $\Psi : \sum 2^H \to K'$, where $(K', 2^x) \succ (K, 2^x)$. If $(\mathbb{R}, 2^x) = (K, 2^x) \cap (K_1, 2^x)$ and Φ is elementary, then Ψ can be taken elementary too.

Remarks

a) All the theorems of §*I* follow by transfinite iteration of the Main Lemma.

b) The proof of model completeness of $Th(\mathbb{R}, 2^x)$ outlined in §*I.A* actually shows the following:

Let L be any definable expansion of the language of $(\mathbf{R}, 2^{1/1+x^2})$, such that $(\mathbb{R}, 2^{1/1+x^2})$ admits elimination of quantifiers in L; then the structure $(\mathbb{R}, 2^{1/1+x^2}, 2^x, \log x)$ admits elimination of quantifiers in its language expanded by L.

c) The proof of (e′) and (f) in the Main Lemma rests on the property of $Th(\mathbb{R}, 2^{1/1+x^2})$ which is proved in [W2]: namely "valdim $\leq$ dim". Thus our "new" proof of model completeness of $(\mathbb{R}, 2^x)$ relies for a large part on Wilkie's original proof; nevertheless, there are substantial differences, of interest for the future.

References

[B] S. Boughattas, Optimal results on the existence of Integer Parts in ordered fields; to appear in *Journal of Symbolic Logic*. See also: Résultats positifs et négatifs sur l'existence d'une Partie Entière dans les corps ordonnés, *C.R.A.S.* (1991).

[M-R] M.H. Mourgues, J.-P. Ressayre, Every real closed field has an Integer Part, submitted to *Journal of Symbolic Logic*. See also: Tout corps réel clos possède une Partie Entière, *C.R.A.S.* (1992).

[K] A. Kaplansky, *Duke Journal of Maths* **9** (1942), pp. 303–321.

[W1] A. Wilkie, Some model completeness results for expansions of the field of real numbers by Pfaffian functions, preprint.

[W2] A. Wilkie, Smooth *o*-minimal theories and the model completeness of the real exponential field, manuscript.

Jean-Pierre Ressayre
Equipe de Logique UA 753
CNRS et Université Paris 7
Tour 45-55, 5-ième Etage
2 Place Jussieu
75251 Paris Cedex 05
France
email: ressayre@logique.jussieu.fr

Making infinite structures finite in models of second order Bounded Arithmetic

Søren Riis *

Abstract

We state two finitisation principles and prove by forcing that they hold for the theory $S_2^2(\alpha)$.

We prove that both finitisation principles do not work if $S_2^2(\alpha)$ is replaced by $T_2^2(\alpha)$.

As a corollary we get new proofs of the proper inclusions $T_2^1(\alpha) \subset S_2^2(\alpha) \subset T_2^2(\alpha)$.

1 Introduction

In this paper we focus on certain second order versions of S_2^i and T_2^i. In the preliminaries it is shown that these second order versions are first order conservative over their corresponding first order versions. We show that S.Buss' conservation result [3] (which is known to relativise) holds for the second order versions of T_2^1 and S_2^2.

The main aim is to show that there is a fundamental difference between $S_2^2(\alpha)$ and levels in the hierarchy which are as least as strong as $T_2^2(\alpha)$. There is a general feeling amongst those working in bounded arithmetic that a proof of the nonfinite axiomatizability of the first order theory S_2 ($= T_2$) would be of great importance. An important step in that direction would be to separate T_2^1, S_2^2 and T_2^2. These theories are only known to be different under additional hypotheses from complexity theory.

In [12] T_2^i and S_2^{i+1} were conditionally separated under the assumption that the polynomial hierarchy does not collapse on level $i+2$. In [7] S_2^2

*Supported by The University of Copenhagen and the Danish Research Academy

and T_2^2 were conditionally separated under the assumption $L^{\Sigma_2^p} \neq \Delta_3^p$. The relativised cases $S_2^2(\alpha)$ and $T_2^2(\alpha)$ were first separated in [6] and in [13].

We present new proofs for these relativised cases. Indeed we prove that two finitisation principles are consistent with $S_2^2(\alpha)$ and that they are inconsistent with $T_2^2(\alpha)$. Since an understanding of the relativised cases seems to precede an understanding of the unrelativised cases our results could be useful in separating S_2^2 from T_2^2 unconditionally.

Essentially, the first finitisation principle says that if a finite set Δ of formulas in some first order language has an infinite model in the real world, there is a S_2^2-world where Δ has a finite model. It follows from the first finitisation principle that unless further second order principles are added, the second order version of S_2^2 does not prove statements such as, for example "A finite set cannot be ordered as a dense linear ordering" or "If A is finite and contains more than one point there is no bijection f from $A \times A$ to A".

It follows from the second finitisation principle that the second order version of S_2^2 does not prove, for example "Every finite vector space of n point has a basis of size $\leq \log_2(n)$". All these three statements are provable in our second order version of T_2^2.

The results in this paper are "low-level" results in the sense that they deal mainly with low levels in the hierarchy T_2^i, S_2^{i+1}. Results in Ajtai [5], for example, are "high-level" which deal with $I\Delta_0$ or even S_2 etc.

High-level results such as those in [5] generally speaking involve more combinatorics than the low-level results. This is because we need a sufficiently strong "switching lemma" (which does not always seem to be possible). Such a switching lemma makes it possible to "collapse" high-level formulas into low-level formulas. Because we do not need such a collapse we can use forcing conditions with only sharply-bounded domains rather than $n - n^\epsilon$ as in [5]. Using forcing conditions with sharply-bounded domains makes it possible to deal with the sharply-bounded quantifiers which appear in the induction axioms.

2 Preliminaries

2.1 Basic Definitions

In this section we recall some basic notations and facts, essentially all from [2]. Let **BASIC** denote a finite set of quantifier free formulas relating constants, functions and relations in the first order language

$$L = L(0, 1, +, \cdot, |\cdot|, \sharp, \lfloor \frac{x}{2} \rfloor, \leq, =).$$

Here $\sharp$ denotes the function given by $a\sharp b = 2^{|a|\cdot|b|}$ where $|a| = \lceil \log_2(a+1) \rceil$. An example of a proper choice of **BASIC** (without coding functions) can be found in [2]. It is convenient to add other functions to the language. We will assume that a function $(w)_x$ which takes the value of the xth element in the sequence coded by w is part of the language. As long as additional functions are polynomially time computable, the results in these notes can be stated with no change.

In the first order case atomic formulas are of the form $t = s$ or $t \leq s$ where s and t are terms in L, while in the second order case, atomic formulas can also be of the form $t \in X$ or $X =_2 Y$ (where "$=_2$" denotes equality between second order variables).

A first (second) order formula is bounded if all its quantifiers are of the form ....$\forall x \leq t$... or ....$\exists x \leq t$..... Second order quantifiers are not allowed in bounded formulas. Atomic formulas $X =_2 Y$ are not allowed because they smuggle in an unbounded first order quantifier (see extensionality axiom below).

A first (second) order formula is sharply bounded if it is bounded and all quantifiers are of the form ....$\forall x \leq \mid t \mid$ ... or ....$\exists x \leq \mid t \mid$ ....

The class of bounded formulas can be stratified as follows: Let $\Sigma_0^b = \Pi_0^b$ be the set of sharply bounded formulas (first or second order formulas depending on the context).

Let Σ_{i+1}^b [Π_{i+1}^b] ; $i \geq 0$ be the smallest class of formulas which contains Π_i^b [Σ_i^b] and is closed under $\wedge$, $\vee$, sharply bounded quantification, and bounded existential quantification [bounded universal quantification]. Notice that modulo minor syntactical changes any bounded formula belongs to some Σ_i^b and to some Π_i^b.

Finally, let strict-$\Sigma_{i+1}^b \subseteq \Sigma_{i+1}^b$ denote the set of Σ_{i+1}^b-formulas which are of the form $\exists y_1 \leq t_1 \exists y_2 \leq t_2 ... \exists y_r \leq t_r\ \phi$ where $\phi \in \Pi_i^b$. Similarly let strict-$\Pi_{i+1}^b \subseteq \Pi_{i+1}^b$ denote the set of Π_{i+1}^b-formulas which are of the form $\forall y_1 \leq t_1 \forall y_2 \leq t_2 ... \forall y_r \leq t_r\ \phi$ where $\phi \in \Sigma_i^b$.

2.2 The first order theories

Let S_2^i denote the first order theory consisting of **BASIC**, together with the following "polynomial time" induction scheme Σ_i^b-PIND:

$$\varphi(0) \wedge \forall x(\varphi(\lfloor \frac{x}{2} \rfloor) \Rightarrow \varphi(x)) \Rightarrow \forall x \varphi(x)$$

where $\varphi \in \Sigma_i^b$.

Let T_2^i denote the first order theory consisting of **BASIC** together with the Σ_i^b induction scheme Σ_i^b-IND:

$$\varphi(0) \wedge \forall x(\varphi(x) \Rightarrow \varphi(x+1)) \Rightarrow \forall x \varphi(x)$$

where $\varphi \in \Sigma_i^b$.

2.3 The second order theories

The second order versions of these theories $S_2^i(\alpha)$ $[T_2^i(\alpha)]$ consist of **BASIC**, Σ_i^b-PIND (Σ_i^b-IND) together with the extensionality axiom

$$\text{EXT}: \forall X, Y(X =_2 Y \Leftrightarrow \forall x(x \in X \Leftrightarrow x \in Y)).$$

We do not allow the full comprehension axiom, but follow [2] and equip $S_2^i(\alpha)$ $[T_2^i(\alpha)]$ with the following "NP $\cap$ co-NP" comprehension axiom-scheme Δ_1^b-comprehension:

$$\forall x(\varphi(x, \vec{z}, \vec{Z}) \Leftrightarrow \neg\eta(x, \vec{z}, \vec{Z}))$$

$$\Rightarrow \exists X \forall x\ (x \in X \Leftrightarrow \varphi(x, \vec{z}, \vec{Z}))$$

where $\varphi, \eta \in \Sigma_1^b$.

The underlying logic of these theories is second order predicate logic with second order equality $=_2$. Later it is proved that no deductive strength is lost if $X =_2 Y$ is taken to be shorthand for $\forall z(z \in X \leftrightarrow z \in Y)$, and if EXT and the equality axioms in the underlying logic are dropped.

Our results are also true for theories $S_2^{i,fin}(\alpha)$ and $T_2^{i,fin}(\alpha)$, which are defined as their counterparts $S_2^i(\alpha)$ and $T_2^i(\alpha)$ except that the axiom FIN: $\forall X \exists x \forall y\ (y \in X \Rightarrow y < x)$ is added and the Δ_1^b-comprehension axiom scheme is replaced by the following modified scheme: $\Delta_1^{b,fin}$-comprehension:

$$\forall a[\forall x \leq a\ (\varphi(x, \vec{z}, \vec{Z}) \Leftrightarrow \neg\eta(x, \vec{z}, \vec{Z}))$$

$$\Rightarrow \exists X \forall x \leq a\ (x \in X \Leftrightarrow \varphi(x, \vec{z}, \vec{Z}))]$$

where $\varphi, \eta \in \Sigma_1^b$.

2.4 Models of second order theories

A model of a second order theory T is a pair $(M, \hat{R})$, where $\hat{R} \subseteq P(M)$, the power set of M, and where M is a model for the first order part of T. The satisfaction relation $\models$ is defined inductively such that second order variables are taken to be the subsets of M which are in $\hat{R}$. The well known main advantage of using this notion of a model, and not requiring that $\hat{R} = P(M)$, is that the Compactness Theorem, the Completeness Theorem and Skolem-Löwenheim's Theorems hold with minor changes.

These facts follow easily from the natural isomorphism:

Observation 1 *(Transitive collapse)*

Let $M^{\triangle} = M^{\triangle}_{num} \cup M^{\triangle}_{set}$ be a first order model which contains two kinds of elements. One kind(x,y,z,...) denotes numbers and belongs to $M^{\triangle}_{num}$, the other kind(X,Y,Z,...) denotes "sets" and belongs to $M^{\triangle}_{set}$. If $\tilde{\in}$ is a binary relation on $M^{\triangle}$, with domain $M^{\triangle}_{num}$ and range $M^{\triangle}_{set}$ such that

$$M^{\triangle} \models (X = Y \Leftrightarrow \forall x(x\tilde{\in}X \Leftrightarrow x\tilde{\in}Y))$$

the map

$$\psi_{collapse} : X \longrightarrow \{x \in M^{\triangle}_{num} \mid M^{\triangle} \models "x\tilde{\in}X"\}$$

maps $M^{\triangle}_{set}$ bijectively onto a class $\hat{S}$ of subsets of $M^{\triangle}_{num}$.

Furthermore $(M, \hat{S}) :\equiv (M^{\triangle}_{num}, \psi_{collapse}{}^{"}M^{\triangle}_{set})$ is isomorphic to $M^{\triangle}$.

From this we get the following version of the Completeness Theorem.

Proposition 2 $T \cup \mathrm{EXT}$ *is consistent* $\Leftrightarrow$ T *has a model.*

$$T \cup \mathrm{EXT} \vdash \phi \text{ if and only if}$$

$$(M, \hat{S}) \models T \Rightarrow (M, \hat{S}) \models \phi \text{ for all models } (M, \hat{S}).$$

2.5 The expansion theorem

Let $T(i)$ be either S^i_2 or T^i_2 $i \geq 1$, and let $T(i, \alpha)$ be the corresponding second order theory defined above. Let $T(i; U_1, ..., U_r)$ be $T(i)$ in the language $L(U_1, ..., U_r)$.

The second order theory $T(i, \alpha)$ is conservative over $T(i)$ with respect to first order consequenses. Actually each model $M \models T(i)$ has an expansion to a model $(M, \tilde{S}) \models T(i, \alpha)$.

Theorem 3 *(Expansion theorem) If*

$$(M, \tilde{R}) \models T(i)$$

there is an <u>*expansion*</u> $\tilde{S} \supseteq \tilde{R}$ *such that*

$$(M, \tilde{S}) \models T(i, \alpha) + \Delta^b_1\text{-comprehension}.$$

Corollary 4 *Every model of T has an expansion to a model of $T(i,\alpha)$. Let $U_1, U_2, ..., U_r$ be unary relation symbols added to L.*

Every model of $T(i; U_1, ..., U_r)$ has an expansion to a model of $T(i,\alpha)$.

In both cases the proof is straightforward.

2.6 The well ordering axiom in $S_2^1(\alpha)$.

The well ordering axiom (WOA) is the principle:

$$\text{WOA}: \quad \forall X(X \neq \emptyset \Rightarrow \exists y(y \in X \wedge \forall z < y \quad z \notin X)).$$

WOA says that "<" well orders any set X, and should not be confused with the stronger principle WOA* stating that <u>any</u> linear ordering of a bounded set is a well ordering.

Proposition 5 WOA *is equivalent to $T_2^1(\alpha)$ over $S_2^1(\alpha)$.*

2.7 Some conservation results

Let Π_1^1 denote the class of formulas where all second order universal quantifiers appear positively, and where all second order existential quantifiers appear negatively. There are no restrictions on the first order quantifiers.

Lemma 6 *If $\psi \in \Pi_1^1$ then*

$$T(i,\alpha) \vdash \psi \text{ iff}$$

$$T(i,\alpha) + \Delta_1^b\text{−comprehension} \vdash \psi.$$

Proof: From the Expansion Theorem. □

Let $X < x$ be shorthand notation for $\forall y\ y \in X \Rightarrow y < x$. By a bounded second order universal quantifier we understand a quantifier which appears in the context $\forall X(X < t \Rightarrow ...)$. Let $\Pi_1^{1,b}(BD) \subseteq \Pi_1^1$ denote the class of Π_1^1-formulas of the form

$$\forall \vec{x} \forall \vec{X} < \vec{t}(\vec{x})\ \theta(\vec{x}, \vec{X})$$

where θ can be any first order formula. The proofs of the next two lemmas are straight forward.

Lemma 7 *Let* $\psi \in \Pi_1^{1,b}(BD)$.

$$T^{fin}(i) + (\Delta_1^{b,fin}\text{–comprehension}) \vdash \psi \textit{ iff}$$

$$T^{fin}(i)\text{–FIN} + (\Delta_1^{b,fin}\text{–comprehension}) \vdash \psi.$$

Lemma 8 *Let* ψ *be a formula. Suppose* ψ^* *is obtained from* ψ *by replacing each appearance of* $X =_2 Y$ *with* $\forall x(x \in X \Leftrightarrow x \in Y)$. *Let* T *be any set of formulas which does not contain* $=_2$.

$$T + \text{EXT} \vdash \psi \text{ iff } T \vdash \psi^*.$$

Notice that had the second order language L_2 contained relations of a higher type, say a unary predicate R^3, $X =_2 Y \Rightarrow (R^3(X) \Leftrightarrow R^3(Y))$ would be an equality axiom, which would not have been provable without EXT.
This also explains why in set-theory (unlike our case), the axiom of extensionality is not conservative over the other axioms.

So summarily if $S_2^i(R)$ is the first order system S_2^i with a unary predicate $R(x)$ $[x \in R]$ then

Proposition 9

$$S_2^i(\alpha) \vdash \forall x \forall X < x \ \psi(x, X)$$

iff

$$S_2^i(R) \vdash \forall x \ (R < x \Rightarrow \psi(x, R))$$

iff

$$S^{i,fin}(\alpha) \vdash \forall x \forall X < x \ \psi(x, X)$$

where ψ *can be any first order formula.*

2.8 S.Buss' result for the theory $S_2^1(\alpha)$

The following theorem is essentially due to S.Buss [2].

Theorem 10 *Suppose*

$$S_2^1(\alpha) \vdash \forall x \forall X < x \exists y \ \psi(x, X, y).$$

There is a poly-time oracle Turing machine f and a polynomial P such that
(1) $N \models \forall x \forall X f^X(x)$ *stops in* $\leq P(|x|)$ *steps.*
(2) There is a code e for f such that

$$S_2^1(\alpha) \vdash \forall x \forall R \subseteq x \exists y\ (\{e\}^R(x) = y \wedge \psi(x, R, y)).$$

In particular
(3) $N \models \psi(x, R, f^R(x))$.

Proof: By relativization of Buss' witnessing Theorem [2]. □

2.9 Some relations between $T_2^i(\alpha)$ and $S_2^{i+1}(\alpha)$.

Proposition 11 *[2]*

$$S_2^{i+1}(\alpha) \vdash T_2^i(\alpha)$$

for all $i \geq 0$.

Proof: By the method of shortening cuts. □

Theorem 12 *[3]* $S_2^{i+1}(\alpha)$ *is* $\forall\Sigma_{i+1}^b$-conservative over $T_2^i(\alpha)$.

Proof: By relativising one of the proofs of S.Buss' theorem stating that for any $i \geq 0$ S_2^{i+1} is $\forall\Sigma_{i+1}^b$-conservative over T_2^i, [8] [3]. □

2.10 Further definitions and assumptions:

Let $L_2(0, 1, +, \cdot, f_{fast}, f_{slow}, =, \leq)$ be a second order language where through some basic axioms f_{fast} and f_{slow} are ensured to define functions such that f_{fast} is "fast"-growing, and that f_{slow} is "slow" growing. Further it is assumed that f_{slow} is slower than f_{fast} is fast (!). More precisely assume:
(1) f_{fast}, f_{slow} are increasing.
(2) $f_{fast} \geq x^2$.
(3) For any fixed k, for all sufficiently large n:

$$f_{slow}(f_{fast}^{(k)}(n)) \leq n$$

where

$$f^{(k)} = \underbrace{f \circ f \circ \ldots \circ f}_{k}.$$

Definition 1 *A formula $\psi \in L_2$ is* <u>*sharply bounded*</u> *if each quantifier appears in the context $\exists x \leq f_{slow}(t)$ or $\forall x \leq f_{slow}(t)$, where $t \in \mathrm{Term}_L$.*

A formula $\psi \in L_2$ is <u>*bounded*</u> *if each quantifier appears in a context $\exists x \leq t$ or $\forall x \leq t$, where $t \in \mathrm{Term}_L$.*

Define Σ_i^b-formulas, Π_i^b-formulas similar to the earlier definition.

Definition 2 *$T^1(\alpha)$ denotes the second order theory consisting of a proper basic theory* **BASIC** *together with*

$$\mathrm{EXT} + \Delta_1^b-\text{comprehension} + \Sigma_1^b-\mathrm{IND}.$$

Given some additional relations $\vec{R}$ we define the first order theories $T^1(\vec{R})$ and $T^{1,fin}(\vec{R})$ in obvious way. <u>Notice</u> that if $f_{fast}(< x, y >) := x\sharp y$ and $f_{slow}(x) := |x|$, for a proper choice of **BASIC** $T^1(\alpha)$ $[T^{1,fin}(\alpha)]$ becomes $T_2^1(\alpha)$ $[T_2^{1,fin}(\alpha)]$.

Furthermore notice that in this case

$$f_{fast}(n) \leq f_{fast}(< n, n >) = 2^{|n|^2}$$

and therefore

$$|\, (f_{fast}^{(k)}(n))\, | \leq |\, (f_{fast}^{(k)}(< n, n >)\, | = |\, n\, |^{2^k}\ .$$

Clearly for any standard k and non-standard n

$$|\, n\, |^{2^k} < n$$

so the consideration below applies to the theory $T_2^1(\alpha)$.

Definition 3 *Let $< x_1, x_2, ..., x_k >$ be a natural code of the k-tuple. For each k we introduce quantifiers $\forall^k$ and $\exists^k$ such that $Q^k x\psi(x)$ is shorthand notation for $Qx_1Qx_2...Qx_k\ \psi(< x_1, ..., x_k >)$ where $Q \equiv \forall$ or $Q \equiv \exists$.*

Alternatively we could have used a many-sorted underlying logic, with one sort of second order quantifiers for each k, ranging over k-ary relations.

Definition 4 *Let $Q^k x < t...$ be shorthand notation for*

$$Qx_1 < tQx_2 < t...Qx_k < t \quad < x_1, x_2, ..., x_k >= x...$$

Let $\theta(\vec{x})$ be a formula in some relational language $L(R_1, R_2, ..., R_l)$. For a relation symbol S, the formula $\theta_S(\vec{x})$ denotes the formula which appears if each quantifier in θ is restricted to S.

By $\theta_{<a}(\vec{x})$ we understand the formula which appears by restricting each quantifer in θ to $[0, a)$.

3 How to make infinite objects finite in models of $T_2^1(\alpha)$

3.1 An impossible version of the completeness theorem

In the real mathematical universe it is not true that any finite consistent set of first order sentences has a finite model. But there are $T_2^1(\alpha)$-universes where such a strong form of the completeness theorem holds. Essentially "the first finitisation principle" says that if a sentence has an infinite model in the real universe, it is consistent with $T_2^1(\alpha)$ [or $S_2^2(\alpha)$] that it has a finite model!

Main Theorem 13 *(First finitisation principle)*
Let $\theta(\vec{R})$ be a first order property expressed in some relational language $\tilde{L} = L(\vec{R}, =)$. Suppose $\eta(n)$ is an arithmetical first order property expressible in the language of arithmetic and suppose $\eta(n)$ holds for arbitrarily large n.

If $\theta(\vec{R})$ has an infinite model then

$$T^1(\vec{R}) + \exists n \; \eta(n) \wedge \theta_{<n}(\vec{R})$$

is consistent.

Firstly we shall make some preparations for the proof. Let $\tilde{L}$ and $\theta(\vec{R})$ be given as in the theorem. We can assume that all relations are r-ary. $\theta(\vec{R})$ is assumed to have an infinite model, so by Skolem-Löwenheim's Theorem, we can assume $\theta(\vec{R})$ has a countable model S_{st} on a subset of the natural numbers. Furthermore, we can assume S_{st} is a model with an underlying co-countable set, and if convenient, that an extra unary relation symbol denoting membership of S_{st} is added to the language.

Let (M, S) be a countable non-standard model for the language $\tilde{L}(R) := L \cup \tilde{L}$ which is elementarily equivalent to the standard model (N, S_{st}). Use overspill to pick a non-standard number n such that $M \models \eta(n)$.

Fix a non-standard number $b_0 < n$ such that $b_0^k < n$ for each standard number k, and such that $f_{slow}(f_{fast}^{(k)}(n)) < b_0$ for every standard number k. Use overspill to pick $c \in M$ non-standard such that $f_{slow}(c) < b_0$ and such that $c > f_{fast}^{(k)}(n)$ for all standard numbers k.

Definition 5 *Let P_k, $k \in \omega$, be the set of all partial (1-1)-maps α from $[0, c]$ to M such that $| \; dom(\alpha) \; | \leq b_0^k$, such that α maps $[0, n)$ to S, and maps $[0, c] \setminus [0, n)$ to $M \setminus S$.*

Let $P = \cup_{k \in \omega} P_k$ and let $(P, \subseteq)$ be P ordered under inclusion.

3.2 Sketch of proof

Now let me sketch the proof. We are given a non-standard model $(M, S) \equiv (N, S_{st})$ in which the "structure" S we want to miniaturize is a part. We can assume that the language contains the language of arithmetic together with extra relations, denoting the relations in S.

Firstly, it is shown that some (actually any) generic $\tilde{\alpha}$ is a bijection from $[0, c]$ to M, mapping $[0, n)$ onto S. At this stage we also have to show certain lemmas about the forcing relation in order to ensure it behaves well.

Secondly, it is shown that for some (actually any) generic map $\tilde{\alpha}$

$$(M, \tilde{\alpha}) \models \exists^* - \text{LNP},$$

where $\exists^*-$LPN denotes the least number principle for formulas in which all quantifiers are either existential (positive appearance) or are restricted to $[0, b_0)$ (i.e. essentially sharply bounded). This part of the argument is based on the same idea as the proof of theorem 21 in [9].

Thirdly, it is noticed that the constants, relations and functions in the miniaturized structure $S_{mini} := \tilde{\alpha}^{-1}(S)$ are $\exists \cap \forall$-definable in the generic model $(M, \tilde{\alpha})$.

Fourthly, it is shown that each formula expressing a Σ_1^b- property about the miniaturized model and numbers in $[0, c)$, can be translated into an $\exists^*$-formula. This ensures that LNP holds for Σ_1^b-formulas with parameters in $[0, c)$.

Finally, a concrete model M^* is constructed as the smallest initial segment of $[0, c)$ which contains $[0, n)$ and is closed under f_{fast}.

3.3 Generic maps

Let us now make a few basic definitions:

Definition 6 *$D \subseteq P$ is called dense if $\forall \alpha \in P \exists \beta \in D$ such that $\beta \supseteq \alpha$*

A subset $S \subseteq M$ is quasidefinable in M if there is $\theta(x) \in L(R_\omega)$ such that

$$S = \{x \in M \mid \ (M, \omega) \models \theta(x)\}$$

where R_w is a predicate denoting the standard numbers. θ might contain parameters from M.

Notice that in [5] the similar notion (just called definable) is $S = \{x \in M \mid \exists n \in R_w \ \theta(x, n)\}$, which would not work in our case because we can not force formulas in general to be equivalent to existential formulas.

It should also be noticed that any extension of the notion of quasi-definability which does not produce uncountably many dense subsets of P, would work.

Example 1 *(1) P is quasi-definable.*

(2) $[0, b_0^\omega)$ is quasi-definable.

Definition 7 *$G \subseteq P$ is generic if*

$$\forall \alpha, \beta \in G \ \exists \gamma \in G \ \gamma \supseteq \alpha \wedge \gamma \supseteq \beta$$

$$\forall \alpha \in G \ \forall \beta \in P \ \alpha \supseteq \beta \Rightarrow \beta \in G$$

$$G \cap D \neq \emptyset \quad \textit{for each quasidefinable dense set} \quad D \subseteq P.$$

Lemma 14 *For every $\alpha_0 \in P \ \exists G \subseteq P$ generic such that $\alpha_0 \in G$.*

Proof: M is assumed to be countable so that there are at most countably many quasi-definable sets $D \subseteq P$. List those as $D_1, D_2, D_3, \ldots$.

Pick $\alpha_1 \supseteq \alpha_0$ such that $\alpha_1 \in D_1$, pick $\alpha_2 \supseteq \alpha_1$ such that $\alpha_2 \in D_2$, etc. and let

$$G := \{\beta \in P \mid \exists j \in \omega \quad \beta \subseteq \alpha_j\}$$

It is straightforward to check that G is generic.

Definition 8 *$\tilde{\alpha} \subseteq M$ is* <u>*generic*</u> *if there is a generic $G \subseteq P$ such that $\tilde{\alpha} = \cup_{\alpha \in G} \alpha$.*

3.4 The forcing relation

Before we define the forcing relation let us reduce the number of logical constants. So suppose that $\forall :\equiv \neg\exists\neg$ and $\wedge :\equiv \neg \vee \neg$.

Extend the language $\tilde{L}(R)$ by an extra binary relation symbol $\bar{\alpha}$.

For formulas in this language $\tilde{L}(\vec{R}, \bar{\alpha})$ we define the forcing relation inductively as follows:

$\alpha \Vdash \psi$ iff ψ does not contain $\bar{\alpha}$, is atomic and true.

$\alpha \Vdash \bar{\alpha}(a, b)$ iff $\alpha(a)$ is defined and equals b.

$\alpha \Vdash \neg\bar{\alpha}(a, b)$ iff $\alpha(a)$ is defined, is different from b or $\alpha(c) = b$ for some $c \neq a$.

$\alpha \Vdash \psi \vee \eta$ iff $\alpha \Vdash \psi$ or $\alpha \Vdash \eta$.

$\alpha \Vdash \exists x\ \psi(x)$ iff for some $x \quad \alpha \Vdash \psi(x)$.

The forcing relation for negation satisfies:

$$\alpha \Vdash \neg\psi \text{ iff for no } \beta \supseteq \alpha,\ \ \beta \in P\ \beta \Vdash \psi.$$

3.5 Soundness of the forcing relation

We have to make sure that the forcing relation satisfies certain key properties. Except for lemma 16 below, the reader who is familiar with forcing techniques could skip this section.

Definition 9 *Let $\Sigma(b_0)$ be the set of formulas where all quantifiers are restricted to $[0, b_0]$.*

Notice that $f_{slow}(c) < b_0$ and therefore in the final model all sharply bounded quantifiers are restricted to $[0, b_0]$.

Lemma 15 *(Extension lemma) If $\alpha \Vdash \psi$ and $\beta \supseteq \alpha$ then $\beta \Vdash \psi$.*

Proof: First notice that the extension lemma holds for atomic formulas. Clearly the extension lemma holds if $\psi \equiv \psi_0 \vee \psi_1$, or if $\psi \equiv \exists y \psi_0(y)$. Suppose $\psi \equiv \neg\psi_0$, $\alpha \Vdash \psi$, and $\beta \supseteq \alpha$ is given. By definition for no $\alpha' \supseteq \alpha$,

$\alpha' \vdash \psi_0$. The ordering $\supseteq$ of the forcing conditions P is transitive so for no $\alpha' \supseteq \beta$, $\alpha' \Vdash \psi_0$. By definition $\beta \Vdash \psi$. □

Lemma 16 *For each $\psi(\vec{x}) \in \Sigma(b_0)$ there is $k \in \omega$ which does not depend on the parameters in $\psi(\vec{x})$ and M-definable maps $\vec{x} \longrightarrow V^D_{\vec{x}}$ and $\vec{x} \longrightarrow V^R_{\vec{x}}$ such that*

$$1) \quad \mathrm{Card}(V^D_{\vec{x}}) \leq b_0^k, \quad \mathrm{Card}(V^R_{\vec{x}}) \leq b_0^k.$$

$$2) \quad \text{For all } \alpha \in P \text{ with } \mathrm{Dom}(\alpha) \supseteq V^D_{\vec{x}} \text{ and } \mathrm{Ran}(\alpha) \supseteq V^R_{\vec{x}}$$

$$\alpha \Vdash \psi(\vec{x}) \text{ or } \alpha \Vdash \neg\psi(\vec{x}).$$

$$3) \quad \text{For all } \alpha \in P,\ \alpha_{|V^D_{\vec{x}} \cup V^R_{\vec{x}}} \Vdash \psi(\vec{x}) \text{ if } \alpha \Vdash \psi(\vec{x}).$$

Proof: If $\psi(x)$ is atomic and is of the form $\alpha(u,v), \alpha(x,v), \alpha(u,x)$ or $\alpha(x_1,x_2)$ let $V^D_x := \{u\}, V^D_x := \{u\}, V^D_x := \{v\}$ or $V^D_{x_1,x_2} := \{x_1\}$, and similarly let $V^R_x := \{v\}$, $V^R_x := \{v\}$, $V^R_x := \{v\}$ and $V^R_{x_1,x_2} := \{x_2\}$.

If $\psi \equiv \psi_0 \vee \psi_1$ let $V^D_{\vec{x}} := V^{D,0}_{\vec{x}} \cup V^{D,1}_{\vec{x}}$, and let $V^R_{\vec{x}} := V^{R,0}_{\vec{x}} \cup V^{R,1}_{\vec{x}}$.

If $\psi \equiv \exists u \leq b_0\ \psi_0(u,\vec{x})$ let

$$V^D_{\vec{x}} := \bigcup_{u \leq b_0} V^{D,0}_{u,\vec{x}} \text{ and } V^R_{\vec{x}} := \bigcup_{u \leq b_0} V^{R,0}_{u,\vec{x}}$$

and notice that

$$\mathrm{Card}(V^{D/R}_{\vec{x}}) \leq b_0 \cdot b_0^k = b_0^{k+1}$$

for some $k \in \omega$.

If $\psi \equiv \neg\psi_0$ let $V^{D/R}_{\vec{x}} := V^{D/R,0}_{\vec{x}}$.

Now we prove (2) and (3) by induction on the number of logical constants in $\psi(\vec{x})$. If $\psi(\vec{x})$ is atomic it is easy to check(2) and (3). Suppose $\alpha \not\Vdash \exists\vec{x} \leq b_0\ \psi(\vec{x})$ where $\psi \in \Sigma(b_0)$.

We need to show that $\alpha_{|V} \Vdash \neg\exists x \leq b_0\ \psi(x)$. Conversely suppose that for some $\beta \supseteq \alpha_{|V}$, $\beta \Vdash \exists x \leq b_0\ \psi(x)$. By definition for some $a \leq b_0$, $\beta \Vdash \psi(a)$,and by induction

$$\beta_{|V^D_a \cup V^R_a} \Vdash \psi(a).$$

Now as $\alpha_{|V^D_a \cup V^R_a} = \beta_{|V^D_a \cup V^R_a}$

$$\alpha_{|V^D_a \cup V^R_a} \Vdash \exists x \leq b\ \psi(x)$$

and then by the extension lemma $\alpha_{|V} \Vdash \exists x \leq b_0\ \psi(x)$, which is in contradiction with the assumption $\alpha \not\Vdash \exists x \leq b_0\ \psi(x)$. □

Lemma 17 *(Consistency) For no $\alpha \in P$ and for no ψ, both $\alpha \Vdash \psi$ and $\alpha \Vdash \neg\psi$ hold.*

Proof: Suppose both $\alpha \Vdash \psi$ and $\alpha \Vdash \neg\psi$ and that ψ is chosen of minimal complexity with that property.

The only case where the induction could fail is in the case of atomic formulas! In our case it is straightforward to check this situation. □

Lemma 18 *For any ψ, $D_1 := \{\alpha \in P \mid \alpha \Vdash \psi\}$ is quasi-definable.*
For any ψ, $D_2 := \{\alpha \in P \mid \alpha \Vdash \psi \vee \neg\psi\}$ is quasi-definable and dense.

Proof: The quasi-definability is shown by induction after the logical complexity of ψ and by using the fact that P is quasi-definable. That D_2 is dense follows from the definition of the forcing relation. □

Lemma 19 *(Completeness) For each generic set $G \subseteq P$, and for each ψ there is $\alpha \in G$ such that $\alpha \Vdash \psi$ or $\alpha \Vdash \neg\psi$.*

Proof: Let $D := \{\alpha \mid \alpha \Vdash \psi \vee \neg\psi\}$. By lemma 18 D is quasi-definable and dense so there is $\alpha_0 \in G \cap D$.
By definition either $\alpha_0 \Vdash \psi$ or $\alpha_0 \Vdash \neg\psi$.□
Finally we have

Lemma 20 *(Soundness) If for a generic map $\tilde{\alpha}$, $(M, \tilde{\alpha}) \models \psi$, there are $\alpha \in G$ such that $\alpha \Vdash \psi$.*
Conversely if $\alpha \Vdash \psi$ then $(M, \tilde{\alpha}) \models \psi$ for <u>any</u> generic $\tilde{\alpha} \supseteq \alpha$.

Proof: Both claims are proved simultaneously using induction on the number of logical constants in ψ. The case where ψ is atomic is straightforward, and so is the case where $\psi \equiv \psi_0 \vee \psi_1$ or $\psi \equiv \exists y\ \psi_0(y)$.
If $(M, \tilde{\alpha}) \models \neg\psi$, by induction there cannot be $\alpha \in G$ such that $\alpha \Vdash \psi$. By lemma 19 (completeness) there is $\alpha \in G$ such that $\alpha \Vdash \neg\psi$.
If $\alpha \Vdash \neg\psi$ but $(M, \tilde{\alpha}) \models \psi$ for some generic map $\tilde{\alpha} \supseteq \alpha$, there is $\beta \in G$ such that $\beta \Vdash \psi$. By definition α and β have a common extension in

P. By lemma 15 (extension lemma) and lemma 17 (consistency) we get a contradiction. □

Corollary 21 *Any generic* $\tilde{\alpha}$ *is a bijection from* $[0, c]$ *to* M*,which maps* $[0, n)$ *onto* S

Proof: Let $a \in [0, c]$ be an arbitrary element. Notice that

$$D_a := \{\alpha \mid \alpha(a) \text{ is defined}\}$$

is both dense and quasi-definable. By definition there is $\alpha_0 \in D \cap G$.

As $\alpha_0 \subseteq \tilde{\alpha}$ this shows that $\tilde{\alpha}$ has domain $[0, c]$. The other properties are proved by a similar argument.□

3.6 How much induction does a generic object satisfy?

We have already defined $\exists^*$ to be the set of formulas in which all quantifiers are either restricted to $[0, b_0]$ (sharply bounded quantifiers) or are existential quantifiers which appear positively.

According to our plan, in order to prove the main theorem we have to prove that some (any) generic $\tilde{\alpha}$ satisfies the $\exists^*$ -LNP scheme.

Let $\exists^*_{Strict}$ be the set of formulas $\exists \vec{x} \psi(\vec{x})$, where $\psi \in \Sigma(b_0)$ and where there are no restrictions on the parameters in ψ. First we prove that:

Lemma 22 *For any generic map* $\tilde{\alpha}$, $(M, \tilde{\alpha})$ *satisfies the* $\exists^*_{Strict}$*-LNP scheme.*

Proof: Let $\psi(z) \equiv \exists \vec{x} \psi_1(\vec{x}, z)$ be given. ($\psi_1 \in \Sigma(b_0)$). Let $\alpha_0 \in P_k$ be given such that $\alpha_0 \Vdash \psi(a)$ for some $a \in M$. Let $a_0 \in M$ be the smallest element such that for some $\beta \supseteq \alpha_0$ with $\beta \in P_{k+r}$, and some $\vec{x}_0$,

$$\beta \Vdash \psi_1(\vec{x}_0, a_0).$$

This definition makes sense because the forcing relation is definable for $\Sigma(b_0)$-formulas (by lemma 16). Let D be the set of $\beta' \in P$ which are either incompatible to α_0 or extensions of a β with the property just mentioned above. D is quasi-definable and dense so there is $\alpha \in D \cap G$. Clearly $\alpha \Vdash \psi(a_0)$. All that remains is to check that if r is chosen properly (lemma 16) there is no $a_1 < a_0$ and $\alpha' \supset \alpha$ with $\alpha' \Vdash \psi(a_1)$.□

3.7 A minor problem

Now we want to prove that for some (any) generic $\tilde{\alpha}$, $(M, \tilde{\alpha})$ actually satisfies the $\exists^*$-LNP scheme. It should be noticed that because of the presence of the generic object $\tilde{\alpha}$ it is not clear why any $\exists^*$-formula should be equivalent to an $\exists^*_{Strict}$ formula.

Let $\psi(z)$ be a given $\exists^*$-formula. Let us try to follow the same strategy as in the proof of lemma 22 above. Without loss of generality we can assume

$$\psi(z) \equiv \exists x_0 \forall u_0 \leq b_0 \exists x_1 \forall u_1 \leq b_0 ... \exists x_k \theta(\vec{x}, \vec{u}, z)$$

where $\theta \in \Sigma(b_0)$. Let $b = b_0$. Find smallest z_0 such that there are a_0, $a_{1,0}, a_{1,1}, ..., a_{1,b}$, and $a_{2,0,0}, a_{2,0,1}, ..., a_{2,0,b}$ and $a_{2,1,0}, a_{2,1,1}, a_{2,1,2}, ..., a_{2,1,b}$ $..., a_{2,b,b}$ ...

$$... \underbrace{a_{k,b,b,...b}}_{k\ b's}$$

and $\beta \supseteq \alpha_0$ in P_{k+r} such that for any choice $u_0 \leq b$, $u_1 \leq b$, ... $u_k \leq b$

$$\beta \Vdash \theta(a_0, a_{1,u_0}, a_{2,u_0,u_1}, ..., a_{k,u_0,...,u_{k-1}}, u_0, u_1, ...u_{k-1}, z_0).$$

Again if there is any z_0 satisfying the condition, there is also a smallest such z_0. This holds for each choice of $r \in \omega$.

As a minor technical problem we need to show (what might be obvious to the reader) that if β is chosen as above, for no $\beta' \supseteq \beta$ can we have $\beta' \Vdash \psi(z_1)$ for some $z_1 < z_0$.

Now in general $(M, \tilde{\alpha})$ has <u>more</u> definable functions on $[0, b_0]$ than M. If namely for each formula $\theta(x, \alpha)$ and each generic map $\tilde{\alpha}$, α could be eliminated, i.e. there was a formula $\psi(x)$ (without α) such that

$$\forall x \in [0, b_0]\ \theta(x, \alpha) \Leftrightarrow \psi(x)$$

in the case where $f_{slow}(x) := log(x)$, then induction would hold up to $b_0 \geq log(c)$ and therefore up to c, which would be a contradiction.

Essentially we have to check that $\psi(z_1)$ can not have a sequence of witnesses which was not definable in M.

In the case where $\psi(x)$ was a $\exists^*_{Strict}$-formula there was no such problem because the search in M was only a search for standard finitely many witnesses. We have to show that there cannot be such a z_1. This is essentially done by showing that in the case of $\exists^*$-formulas for any generic $\tilde{\alpha}$ there is an M definable sequence of witnesses.

For $\theta \in \exists^*$ we make the following definition of "θ is true at α":

Definition 10 $\alpha \Vdash^D \ \forall x \leq b_0\ \theta(x)$ *iff* $\forall x \leq b_0\ \alpha \Vdash^D \ \theta(x)$.

$\alpha \Vdash^D \exists x\theta(x)$ *iff* $\exists x\ \alpha \Vdash^D \theta(x)$.
$\alpha \Vdash^D \neg\neg\theta$ *iff* $\alpha \Vdash^D \theta$.

Notice that for $\theta \in \exists^*$ if

$$\alpha \Vdash^D \theta$$

for each extension (not nessesary a generic extension!) $\tilde{\alpha}$ of α

$$(M, \tilde{\alpha}) \models \theta.$$

Observation 23 *For each $\alpha_0 \in P$, each $k \in \omega$ and each $\psi \in \exists^*$*

$$\{<\alpha, x>| \ \alpha \supseteq \alpha_0 \wedge \alpha \in P_k \wedge \alpha \Vdash^D \psi(x)\}$$

is definable in M.

The key problem at this stage is whether it is possible for given α_0 and $\theta \in \Sigma(b_0)$ to have $\alpha \supseteq \alpha_0$ with $\alpha \Vdash \forall x \leq b_0 \exists y\ \theta(x,y)$ but for no $\beta \supseteq \alpha_0,\ \beta \in P_r:\ \beta \Vdash^D \forall x \leq b_0 \exists y\ \theta(x,y)$?

Assume there is $\alpha \supseteq \alpha_0$ such that

$$\alpha \Vdash \forall x \leq b_0 \exists y\ \theta(x,y).$$

Claim: There is an extension $\alpha_1 \supseteq \alpha$ such that $\forall x \leq b_0 \exists y\ \alpha_1 \Vdash \theta(x,y)$.
Pick for each $j \leq b_0$, α^j such that

$$\alpha \subseteq \alpha^0 \subseteq \alpha^1 \subseteq ...\alpha^{b_0} = \alpha_1$$

and such that

$$\alpha_j \Vdash \exists y \theta(j,y).$$

Furthermore by lemma 16 each extension can be chosen to be of size $\leq b_0^l$ for some fixed $l \in \omega$ (l can be chosen to be the number of "$\forall x \leq b_0$" quantifiers in θ). So α_1 is a $\leq b_0^{l+1}$ extension of α. By lemma 16 for each $x \leq b_0$ and witness $y(x)$ there is a set $V_{x,y(x)}$ with cardinality $\leq b_0^l$ for some fixed standard l.

Let $V := \bigcup_{x \leq b_0} V_{x,y(x)}$. Notice that Card$(V) \leq b_0^{l+1}$ and that

$$\forall x \leq b_0 \exists y\ \alpha_{1,|V} \Vdash \psi(x,y).$$

Let $\beta = (\alpha_1)_{|V}$ and notice that

$$\beta \Vdash^D \forall x \leq b_0 \exists y\ \theta(x,y).$$

We have just proved the first part of the next lemma:

Lemma 24 *Let $\psi(z) = \forall x \leq b_0 \exists y\ \theta(x, y, z)$ where $\theta \in \Sigma(b_0)$. There is $r \in \omega$ such that if $\beta \supseteq \alpha_0$ is the b_0^r-extension of α_0 with the smallest z_0 such that*

$$\beta \Vdash^D \ \forall x \leq b_0 \exists y\ \theta(x, y, z_0)$$

then for <u>any</u> generic model $(M, \tilde{\alpha})$ with $\tilde{\alpha} \supseteq \alpha_0$

$$(M, \tilde{\alpha}) \models \psi(z_0) \wedge \forall z < z_0\ \neg\psi(z).$$

More generally:

$$\text{Let } \psi(z) = \exists x_0 \forall u_0 \leq b_0 \exists x_1 \forall u_1 \leq b_0 .... \exists x_k\ \theta(\vec{u}, \vec{x}, z) \quad ; \theta \in \Sigma(b_0).$$

There is a standard number r which does not depend on the parameters in ψ such that if $\beta \supseteq \alpha_0$ is the b_0^r-extension of α_0 with the smallest z_0 satisfying

$$\beta \Vdash^D \ \psi(z_0)$$

then for <u>any</u> generic model $(M, \tilde{\alpha})$ with $\tilde{\alpha} \supseteq \alpha_0$

$$(M, \tilde{\alpha}) \models \psi(z_0) \wedge \forall z < z_0\ \neg\psi(z).$$

Proof: Let α_0 be given. <u>Assume</u> that for some $\alpha \supseteq \alpha_0$

$$\alpha \Vdash \forall x_1 \leq b_0 \exists y_1 \forall x_2 \leq b_0 ... \exists x_k\ \theta(\vec{x}, \vec{y})$$

but that for <u>no</u> $\beta \supseteq \alpha_0$, $\beta \in P_r$

$$\beta \Vdash^D \ \forall x_1 \leq b_0 \exists y_1 ... \theta(\vec{x}, \vec{y}).$$

There is $\alpha' \supseteq \alpha$ such that

$$\forall x_1 \leq b \exists y_1\ \alpha' \Vdash \forall x_2 \leq b_0 \exists y_2 .... \theta(\vec{x}, \vec{y}).$$

By induction on k, we can assume we have already proved that if for some fixed x_1 and y_1, $\beta \supseteq \alpha$, $\beta \Vdash \forall x_2 \leq b_0 \exists y_2 ... \theta(\vec{x}, \vec{y}), \beta$ can be chosen to be an b^{l+k-1}-extension, where l is the number of $\forall$-quantifiers in θ. There is $\alpha'' \supseteq \alpha'$ (making $b_0^2\ \ b_0^{k+l-2}$-extensions of α'), such that

$$\forall x_1 \leq b_0 \exists y_1 \forall x_2 \leq b_0 \exists y_2\ \alpha'' \Vdash \forall x_3 \leq b_0 ... \theta(\vec{x}, \vec{y}).$$

Continuing like this, after k steps we obtain $\alpha^* \supseteq \alpha$ such that

$$\alpha^* \Vdash^D \ \psi.$$

We can not use α^* as β in the lemma, because α (and therefore α^*) could be too large. But by lemma 16, there are b_0^k sets

$$V_{x_1,y(x_1),x_2,y(x_2),\ldots,x_k,y(x_k)}$$

each with $\leq b_0^l$ elements where l can be taken to be the number of $\forall$-quantifiers in θ. Let

$$V := \bigcup_{x_1 \leq b_0} \bigcup_{x_2 \leq b_0} \cdots \bigcup_{x_k \leq b_0} V_{x_1,y(x_1),\ldots,x_k,y(x_k)}.$$

Notice that $\mathrm{Card}(V) \leq b_0^{k+l}$. Let $\beta := \alpha^*_{|V}$ and notice that $\beta \Vdash^D \psi$. This contradicts the assumption. □

Corollary 25 *For <u>any</u> generic map* $\tilde{\alpha}$

$$(M, \tilde{\alpha}) \models \exists^*\text{–LNP}.$$

Proof: Let $\tilde{\alpha}$ be an arbitrary generic map. We need to show that $(M, \tilde{\alpha}) \models \exists x \psi(x) \Rightarrow \exists x_0 \leq x \forall z < x_0\ \psi(x_0) \wedge \neg\psi(z)$. Suppose $(M, \tilde{\alpha}) \models \psi(a)$ for some $a \in M$ (otherwise there is nothing to prove). According to lemma 19 (completeness) there is $\alpha_0 \subseteq \tilde{\alpha}$ such that $\alpha_0 \in P$ and $\alpha_0 \Vdash \psi(a)$. Assume $\alpha_0 \in P_k$
Consider $D \subseteq P$ defined by:

$$D := \{\alpha \mid \alpha \not\supseteq \alpha_0 \ \vee \ (\alpha \supseteq \alpha_0 \wedge \exists x\ \alpha \Vdash^D \ \psi(x)$$
$$\wedge\ \forall y < x(\forall \alpha' \supseteq \alpha\ \alpha' \nVdash^D \ \psi(y)))\}.$$

From what has already been proved it follows that D is well defined, dense and quasi-definable. For <u>any</u> generic $G \subseteq P$ there are $\alpha_1 \in D \cap G$. By lemma 24 for any $\tilde{\beta} \supseteq \alpha$ (in particular $\tilde{\alpha}$)

$$(M, \tilde{\beta}) \models \exists x \psi(x) \Rightarrow \exists x_0 \leq x \forall z < x_0\ \psi(x_0) \wedge \neg\psi(z).$$

As $\alpha_1 \subseteq \tilde{\alpha}$ we are done. □

3.8 Proof of the finitisation principle

The previous section has given us a "generic model" $(M, \tilde{\alpha})$ which satisfies the $\exists^*$-LNP scheme. Clearly $\tilde{\alpha}$ induces a miniaturized version of S on $[0, n)$. Constants, relations and functions on S correspond to constants,

relations and functions on $[0, n)$. Now we prove the important fact that all the miniaturized relations etc. are $\exists \cap \forall$-definable in $(M, \tilde{\alpha})$.
For each relation $R \subseteq S^r$ (with quantifier-free definition in M), we define the corresponding miniaturized relation R_{mini} "existentially" by:

$$\{(x_1, x_2, ..., x_r) \mid \exists z_1, z_2, ..., z_r \ \wedge_{i=0}^{i=r} \ \tilde{\alpha}(x_i, z_i) \wedge R(z_1, z_2, ..., z_r)\}.$$

Notice that R_{mini} also has a "universal" definition:

$$\{(x_1, x_2, ..., x_r) \mid \forall z_1, z_2, ..., z_r \ \wedge_{i=0}^{i=r} \ \tilde{\alpha}(x_i, z_i) \Rightarrow R(z_1, z_2, ..., z_r)\}.$$

Add the miniaturized relations to the language. Consider the sublanguage L_{mini} which contains L and names for the miniaturized constants, relations and functions.

Lemma 26 *If $\psi(x) \in L_{mini}$ is a Σ_1^b-formula with all quantifiers restrictet to $[0, c)$, there is a $\exists^*$-formula $\psi^*(x)$ such that:*

$$\{x \leq c \mid (M, \tilde{\alpha}) \models \psi(x)\} = \{x \leq c \mid (M, \tilde{\alpha}) \models \psi^*(x)\}.$$

Proof: Let $\psi(x) \in L_{mini}$ be a given Σ_1^b-property. Replace each appearance of a "miniaturized" relation R_{mini} by either the $\exists$-definition or the $\forall$-definition according to whether the R_{mini} appears positively or negatively in $\psi(x)$. Notice that this gives an $\exists^*$-formula $\psi^*(x)$ which satisfies the lemma. □

Now we are ready to construct a model of $T_2^1(\alpha)$ in which there is a bounded substructure S_{mini} on $[0, n)$ where $\eta(n)$ such that $S_{mini} \models \theta_{<n}(\vec{R})$. We have already got a countable model M of true arithmetic in which a possibly unbounded model of $\theta(\vec{R})$ is coded. We are given $n, b_0, c \in M$ as above. Let $\tilde{\alpha}$ be an arbitrary generic map, as described earlier. Let

$$M^*_{\tilde{\alpha}} = \{x \leq c \mid \exists k \in \omega \quad x < f_{fast}^{(k)}(n)\}.$$

Let $\hat{R}$ be the set of all L_{mini}-definable relations on $[0, n)$. It follows from what has already been proved that $(M, \hat{R}) \models \Sigma_1^b$-IND scheme, so because of the expansion theorem there is an expansion of $\hat{R}$ to $\hat{S} \supseteq \hat{R}$ such that $(M, \hat{S}) \models T_2^1(\alpha)$. As second order existence statements are preserved with respect to expansions, we have proved the main theorem in the T^1-case.

Corollary 27 *Let S be a countable mathematical structure. There is a model of $T_2^1(\alpha)$ in which S appears as a bounded set.*

4 Generalization to $S_2^2(\alpha)$

4.1 Krajiček's trick

Corollary 28 *The first finitisation principle holds if $T^1(\alpha)$ is replaced by $S_2^2(\alpha)$.*

Proof: To obtain the result for $S_2^2(\alpha)$ we uses Krajiček's trick (personal communication). Assume

$$S_2^2(\alpha) \vdash \forall n[\eta(n) \Rightarrow \forall \vec{R} < n \ \neg\theta_{<n}(\vec{R})]$$

where

$$\theta_{<n}(\vec{R}) \equiv \forall x_1 \exists y_1 ... \forall x_k \exists y_k \ \tilde{\theta}(\vec{x}, \vec{y}, n; \vec{R})$$

and where $\tilde{\theta}$ is quantifier free. Now by the first finitisation principle (the $T_2^1(\alpha)$-case) used on the skolemization of the sentence

$$T_2^1(\alpha) + \exists n \ [\eta(n) \wedge \exists \vec{R} < n \exists F_1, ..., F_k < n$$

$$\forall \vec{x} < n \ \tilde{\theta}(x_1, F_1(x_1), x_2, F_2(x_1, x_2), ..., n, \vec{R})]$$

is consistent. If an extra constant n°, extra function symbols $F_1, ..., F_k$ and extra relations symbols $\vec{R}^\circ$ are added to the basic language such that we obtain a language $\tilde{L}_2$, the theory

$$T_2^1(\tilde{\alpha}) + \eta(n^\circ) \wedge \forall \vec{x} < n^\circ \ \tilde{\theta}(x_1, F_1^\circ(x_1), x_2, F_2^\circ, ..., n^\circ, \vec{R}^\circ)$$

is consistent.

Now by theorem 12, (S.Buss [3]), (which relativises)

$$S_2^2(\tilde{\alpha}) + \eta(n^\circ) \wedge \forall \vec{x} < n^\circ \ \tilde{\theta}(x_1, F_1^\circ(x_1), x_2, F_2^\circ, ..., n^\circ, \vec{R}^\circ)$$

is consistent. As the skolemization of θ implies θ

$$S_2^2(\tilde{\alpha}) + \exists n(\eta(n) \wedge \exists \vec{R} < n \ \theta_{<n}(\vec{R})$$

is consistent. This contradicts the assumption.□

The following version of the finitisation principle follows as a corollary:

Corollary 29 *(First finitisation principle) Let $\theta(\vec{R})$ be any first order property in some relational language $\tilde{L} = L(\vec{R}, =)$. Then*

$$\text{Predicate logic} \vdash \theta(\vec{R})$$

iff

$$T^1(\alpha) \vdash \forall n \theta_{<n}(\vec{R})$$

iff

$$S_2^2(\alpha) \vdash \forall n \theta_{<n}(\vec{R}).$$

Proof: "$\Rightarrow$": Let T be $T^1(\alpha)$ or $S_2^2(\alpha)$. Assume $\theta(\vec{R})$ holds in all models S. If T does not prove $\theta_{<n}(\vec{R})$ there is a non-standard model M^* of T in which $\neg\theta_{<n}(\vec{R})$ hold in some initial segment $[0, n)$. Clearly $([0, n), \vec{R}) \models \neg\theta(\vec{R})$ contradicting the assumption.

"$\Leftarrow$": If $\not\vdash \theta(\vec{R})$ there is a model S^- of $\neg\theta(\vec{R})$. If S^- is finite the implication is easy. If S^- is infinite the implication follows from the previous version of the finitisation principle. □

Notice that our results hold in the special case where $T^1(\alpha)$ is the second order theory which consists of

1) Induction for existential formulas.
2) The $\forall \cap \exists$-comprehension axiom scheme.

4.2 Separating $T_2^1(\alpha)$ and $S_2^2(\alpha)$

The next theorem shows that we also obtain a separation result between $T_2^1(\alpha)$ and $S_2^2(\alpha)$. For any generic map $\tilde{\alpha}$ we have constructed a model $(M_{\tilde{\alpha}}^*, \tilde{S}_{\tilde{\alpha}})$.

Theorem 30 *For any generic map* $\tilde{\alpha}$

$$(M_{\tilde{\alpha}}^*, \tilde{S}_{\tilde{\alpha}}) \models T_2^1(\alpha).$$

If S contains relations which make it possible by using only the relations, functions and parameters in S, to divide S into two parts, each of size $\geq b_0^\omega$, the for any generic map $\tilde{\alpha}$

$$(M_{\tilde{\alpha}}^*, \tilde{S}_{\tilde{\alpha}}) \not\models S_2^2(\alpha).$$

Proof: The first part of the theorem has already been proved. It follows from the examples below that the second part holds at least for some constructions (i.e. for some S).

To prove the second part let $U(x)$ be a new unary predicate symbol which holds exactly in one component of S. Let U_{mini} denote the corresponding predicate in L_{mini}.

Put $f_{slow} :=\mid x \mid$ and $f_{fast} := x^{|x|}$.
Consider the formula

$$A(z) := \exists u_1 < u_2 \leq n \; u_2 - u_1 \geq b^z \; \forall x \in [u_1, u_2) \; U_{mini}(x)).$$

Clearly $A(k)$ is valid for all standard k and hence, by overspill, $A(z)$ holds for some non-standard z (and $z <\mid n \mid$). But this is a contradiction as for any given interval of length $< b^\omega$ the set of $\alpha \in P$ such that α maps an element of $S \setminus U$ into it, is dense and quasidefinable.

According to [2] Σ_2^b-PIND is sufficient to ensure the validity of the Σ_2^b-LMAX principle which in turn implies the validity of the Σ_2^b-overspill just used.

I thank one of the referees for improving the exposition of the proof. □

Notice that if S is an infinite (co-countable) set with no additional structure

$$(M^*_{\tilde{\alpha}}, \tilde{S}_{\tilde{\alpha}}) \models T_2^\infty(\alpha).$$

This shows that the second part of the theorem becomes false if there is no condition on S.

Corollary 31

$$T_2^1(\alpha) \neq S_2^2(\alpha)$$

This is a new proof of the result which was first proved in [12].

5 Some applications of our first finitisation principle

5.1 The weak pigeon hole principle

Definition 11 *A set B satisfies the weak pigeon hole principle (*WPHP*) if there is no injective map*

$$f : B \times B \to B$$

More formally:

$$\forall x, y \in B \exists ! z \in B \; R(x, y, z) \Rightarrow$$
$$[\exists u_1, u_2, v_1, v_2, z \in B \quad R(u_1, u_2, z) \wedge R(v_1, v_2, z)$$
$$\wedge < u_1, u_2 > \neq < v_1, v_2 >].$$

Example 2 *Consider the sentence:*

$$\psi(R) \equiv \exists x, y\ x \neq y \wedge [\forall x, y \exists z R(x, y, z)] \wedge$$

$$\neg[\exists x_1, y_1, x_2, y_2, z\ R(x_1, y_1, z) \wedge R(x_2, y_2, z) \wedge (x_1 \neq x_2 \vee y_1 \neq y_2)].$$

The statement ψ states that the universe has at least two elements and that some 3-ary relation R violates the weak pigeon hole principle. Clearly ψ has an infinite model A, and then according to the first finitisation principle it is consistent with $S_2^2(\alpha)$ that ψ has a finite model. More formally

$$S_2^2(\alpha) + \exists n \exists R \subseteq [0, n)\ \psi_{<n}(R)$$

are consistent.

5.2 More examples

Example 3 *There is an infinite model A where "R defines a partition of A into pairs". Let $\eta(n) \equiv$ "n is odd". According to the first principle:*
$T^1(\alpha) \nvdash$ "If [0,n) can be divided into pairs, n is even".

Example 4 *According to the first principle:*
(1) $T^1(\alpha) \nvdash$ "every linear ordering R (of a finite set) has an isolated point".
(2) $T^1(\alpha) \nvdash$ "every linear ordering R (of a finite set) is discrete".
(3) $T^1(\alpha) \nvdash$ "every linear ordering R (of a finite set) is a well ordering".
In all cases $T^1(\alpha)$ can be replaced by $S_2^2(\alpha)$.

Notice that example 4 shows (using $T_2^1(\alpha) \equiv$ WOA, Proposition 5) that the well ordering axiom for arbitrary linear orderings (WOA*) R does not follow from WOA.

5.3 Separating $S_2^2(\alpha)$ and $T_2^2(\alpha)$

Theorem 32 *The statement "Every lin.ordering of an interval has a smallest element" is provable in $T_2^2(\alpha)$, but not provable in $S_2^2(\alpha)$.*

The statement "There is no dense linear ordering" is provable in $T_2^2(\alpha)$, but not provable in $S_2^2(\alpha)$.

Proof: We have already proved the second part of the claims. To prove the first part let $(M, \hat{S}) \models T_2^2(\alpha)$ be an arbitrary model. Assume R is a M-definable linear ordering of a subset $X \subseteq [0, n)$ for some $n \in M$. Let

$$Y = \{b \in [0, n) \mid \exists a_0 \leq b \forall x \leq b \quad xRa_0 \Rightarrow x = a_0\}$$

Notice that Y is Σ_2^b-definable and that $0 \in Y$ and $b \in Y \Rightarrow b + 1 \in Y$ for $b < n$. As $(M, \hat{S}) \models \Sigma_2^b$-IND, $(n - 1) \in Y$. By definition of Y, R has a smallest element. The second part follows from the first part.□

The following were proved in [6]. Above we have obtained a new proof of the second part of the theorem.

Theorem 33 *[6] The weak pigeon hole principle (*WPHP*) is provable in $T_2^2(\alpha)$, but <u>not</u> provable in $S_2^2(\alpha)$.*

Proof: The second part of the theorem have already been proved. The first part follows from the fact (Theorem 12) that $S_2^3(\alpha)$ is $\forall\Sigma_3^b$-conservative over $T_2^2(\alpha)$, that WPHP is a $\forall\Sigma_2^b$-formula and the following subclaim:

Lemma 34 WPHP *is provable in $S_2^3(\alpha)$.*

Proof: A careful analysis of the S_2-proof in [10] of WPHP shows that the proof is actually a $S_2^3(R)$-proof.□

6 Another finitisation principle

Our second finitisation principle says that for any given $r \in N$ if some second order existential property $P \equiv \exists X P(\vec{R}, X)$ is only witnessed by infinite sets (in the real universe), it is consistent with $S_2^2(\alpha)$ that there is an n and relatons $\vec{R} \subseteq [0, n)$ such that no subset X of $[0, n)$ with size $\leq log_2^r(n)$ witnessed $P(\vec{R}, X)$.

Main Theorem 35 *(Finitisation principle)*

Let $\theta \equiv \exists X \psi(\vec{R}, X)$, be a second order existential formula where ψ can be any first order formula in the relational language $L(\vec{R}, X, =)$. Let $k \in N$ be a given natural number.

In general (1) implies (2):

(1) There is a <u>*countable*</u> *model S of the language $L(\vec{R}, =)$, such that for* <u>*no finite*</u> *set $X \subseteq S$, $S \models \theta(\vec{R}, X)$*

(2) $T +$ " $\exists n \forall X \subseteq [0,n)$ $\text{Card}(X) \leq (\log(n))^k \rightarrow \neg\theta_{<n}(\vec{R}, X)$" is consistent.

If in the real universe

$$\forall n \exists X \subseteq [0,n) \ \text{Card}(X) \leq (\log(n)^k \theta_{<n}(\vec{R}, X)$$

(2) implies (1).

This finitisation principle states that if a second order existential property has (in the real universe) a <u>countable</u> model where the existential quantifier is not <u>finitely</u> witnessed then it is consistent with $S_2^2(\alpha)$ that there is a <u>finite</u> model where the existential quantifier is not <u>$(\log(n))^k$</u>-witnessed.

This shows we have the translations: countable $\longrightarrow$ finite, finite $\longrightarrow$ "polynomial" when we pass from the real universe of sets to a universe which only contains "feasible" sets. Perhaps the principle is best explained by some examples.

In the real world there is a binary tree with no finite branch. By the second finitisation principle $S_2^2(\alpha)$ + " there is n and a tree T $\subseteq [0,n)$, with no branch (coded by a number) of length $\leq log_2^k(n)$" is consistent.

In the real world there is a vector space over Z_2 with no finite basis so by the second principle $S_2^2(\alpha)$ + "there is n and a vectorspace $V \subseteq [0,n)$ with no basic (coded by a number) of size $\leq log_2^k(n)$" is consistent.

Proof: (1) $\Rightarrow$ (2): Is proved by a similar construction as the proof of the first finitisation principle. Pick S according to (1). Choise S as a countable model in the standard model on say the even numbers. If suitable, extra relations can be added to the language of arithmetic at this stage. Extend the standard model to an elementary equivalent countable non-standard model. Pick non-standard numbers $b_0 < n < c$ as above. Consider the same set of forcing conditions (P,$\subseteq$) as above. The forcing conditions ensure that each generic $\tilde{\alpha}$ maps "small" (i.e $\leq b_0^k$ points) M-definable sets to "small" M-definable sets. The map $\tilde{\alpha}$ induces a miniaturized structure S_{mini} with underlying set $[0,n)$. For each M-definable subset $B \subseteq [0,n)$, $A := \tilde{\alpha}^{-1}(B)$ can <u>not</u> witness $\theta(\vec{R}, A)$ because the set A has cardinality $\leq b_0^k$ which is finite <u>in</u> the model M. As in the proof of the first principle without any complications we construct a model of $T_2^1(\alpha)$ where $\theta_{<n}(\vec{R}, X)$ are not witnessed of any set A of size $\leq b_0^k$ for any standard k.

To lift the result to the $S_2^2(\alpha)$-case we use Krajiček's trick. Suppose

$$S_2^2(\alpha) \vdash \forall n \forall \vec{R} \subseteq [0,n) \exists b \ \psi_{<n}(\vec{R}, Seq(b))$$

where

$$\psi_{<n} \equiv \forall x_1 < n \exists x_2 < n ... Q x_{2r+1} < n \; \tilde{\theta}(\vec{x}, n, \vec{R}, Seq(b))$$

with $\tilde{\theta}$ quantifier free.

We have already proved the $T_2^1(\alpha)$-case, so

$$T_2^1(\alpha) + \exists n \exists \vec{R} \subseteq [0, n) \exists F_1, F_3, ..., F_r \forall b \forall \vec{x} < n$$

$$\neg\tilde{\theta}(x_1, F_1(x_1), x_3, F_3(x_1, x_3), ..., n, \vec{R}, Seq(b))$$

is consistent. But then

$$T_2^1(\tilde{\alpha}) + \forall \vec{x} \forall b \; \tilde{\theta}(x_1, F_1^{\circ}(x_1), x_3, F_3^{\circ} ..., n^{\circ}, \vec{R}^{\circ}, Seq(b))$$

is consistent, where the language is extended by extra relation symbols $n^0, \vec{R}^{\circ}, F_1^{\circ}, F_3^{\circ}, ...etc...$ By theorem 12, which also works in this case where the underlying language is $\tilde{L}(R)$,

$$S_2^2(\alpha) + \forall \vec{x} < n \forall b \neg\tilde{\theta}(x_1, F_1^{\circ}(x_1), x_3, ..., n^{\circ}, \vec{R}^{\circ}, Seq(b))$$

is consistent. Clearly this contradicts our initial assumption.

(2) $\Rightarrow$ (1): Pick a non standard model

$$(M^*, \vec{R}) \models T + \exists n \forall X \subseteq [0, n)[\text{Card}(X) \leq (log(n))^k \Rightarrow \neg\theta_n(\vec{R}, X)$$

where n can be assumed (additional assumption) to be non-standard. Clearly $([0, n), \vec{R}) \models \text{Card}(X) \in \omega \Rightarrow \neg\theta_{<n}(\vec{R}, X)$. □

Notice that we have the following "gap" principle as a corollary:

Corollary 36 *If*

$$\exists X \subseteq [0, n) \; \text{Card}(X) \leq (\log(n))^{k_1} \wedge \theta_{<n}(\vec{R}, X)$$

holds in the real world for some constant k_1 but does not hold for some other constant $k_2 < k_1$, $S_2^2(\alpha)$ does not prove the formula for any constant k.

7 Some applications of our second finitisation principle

By use of the second finitisation principle we get the following theorem.

Theorem 37 *$S_2^2(\alpha)$ does not prove any of the following statements:*

(1) "Every vector space $V \subseteq [0,n)$ over Z_2 has a basis of size $\leq log^k(n)$."

(2) "Every binary tree $T \subseteq [0,n)$ has a branch of length $\leq log^k(n)$."

(3) "Every total irreflectively oriented graph on $[0,n)$ has a dominating set of size $\leq log(n)$."

In (1) and (2), k is an an arbitrary fixed standard number. On the other hand $T_2^2(\alpha)$ proves (1)-(2).

Proof: The first part of the Theorem follows from the second finitisation principle (In case (1) from a slightly modified version of the principle) and the facts:

There is an (infinite) vector space with no finite basic.

There is an (infinite) binary tree with no finite branch.

There is an (infinite) irreflexively oriented graph with no finite dominating set.

To prove the last part of the Theorem, assume V is a vector space on $[0,n)$ with no coded basic. Pick independent vectors $v_1, v_2, ..., v_{\lceil log 2n \rceil + 1}$ and get a Δ_1^b-definable injective map from $[0, 2n)$ to $[0, n)$. This violates WPHP which are provable in $T_2^2(\alpha)$ (Theorem 33).

Finally $T_2^2(\alpha)$ proves that every tree $T \subseteq [0, n)$ has a coded branch. Let

$$T^* := \{x \in T \mid \exists b \; b \text{ codes a path of length} \leq \lceil log(n) \rceil + 1 \text{ from the root to x}\}.$$

Notice that T^* also has the Π_1^b-definition:

$$T^* := \{x \in T \mid \forall b \; b \text{ codes a path from the root to x and the length of the parth is} \leq \lceil log(n) \rceil + 1\}.$$

Define a Δ_1^b-relation by F(x)=y iff $\forall b$ b codes a path $x_1, x_2, ..., x_r$ from the root to x, and bit(y,j)=1 $\leftrightarrow x_j$ is a right son.

Again we get a contradiction to the WPHP (Theorem 33). □

At present it is not known whether $T_2^2(\alpha) \vdash (3)$. A proof of even $T_2^\infty(\alpha) \vdash (3)$ would have interesting consequences. (See [11] and [12] for more details.)

8 Conclusion and acknowledgements

The second order theories $S_2^2(\alpha)$ and $T_2^2(\alpha)$ allow induction on formulas of the same complexity. The only syntactical difference between the theories is that induction in $S_2^2(\alpha)$ is "polynomial" while it is normal in $T_2^2(\alpha)$.

The first main result says that $S_2^2(\alpha)$ *is never able to conclude that an object is infinite on the basis of the object's first order properties, while for some infinite objects this can be done by the theory* $T_2^2(\alpha)$.

The second main theorem can also be phrased in philosophical terms: *Contary to* $T_2^2(\alpha)$, *the theory* $S_2^2(\alpha)$ *can never distinguish "infinite and not finitely witnessed" from "finite and not poly-logaritmic (in Card(P)) witnessed".*

To end, I would like to thank the participants in SanDiego and in Prague for many fruitful discussions and for enabling me to present this paper. I would also like to thank A.Visser and R.Verbrugge for inviting me to present the paper at a well-organised conference in Utrecht. Finally I would like to thank A.Wilkie for his encouragement and support and to thank the referees for many helpful suggestions improving the readability of the paper.

References

[1] F.Ferreira; Polynomial Time Computable Arithmetic and Conservative extensions. Ph.D dissertation, Pennsylvania State University, 1988.

[2] S.Buss; Bounded Arithmetic. Ph.D. dissertation,Princeton University, (June 85) As book, Bibliopolis, Napoli 1986.

[3] S.Buss; Axiomatization and conservation results for fragments of bounded arithmetic, Proc.Workshop in Logic and Computation (1987), Contemporary Mathematics AMS, vol. 106 (1990), pp 57-84.

[4] A.Wilkie; A model-theoretic proof of Buss' characterisation of the polynomial time computable functions. Notes. 1985.

[5] M.Ataij; On the complexity of the pigeonhole principle. 29th Annual symp. on Found. Comp. Sci.(1988),pp 340-355. appear.

[6] J.Krajiček; No counter-example interpretation and interactive computation. To appear in: Logic in Computer science, ed. Y.N.Moschovakis, RSRI Publ.21,(1992), pp 287-293.

[7] J.Krajiček; Fragments of bounded arithmetic and bounded query classes. To appear in the Transactions of AMS.

[8] P.Hajek,P.Pudlák; Metamathematics of first order arithmetic, to app. in Springer-Verlag.

[9] J.Paris, A.Wilkie; Counting problems in bounded Arithmetic, in: Methods in Mathematical Logic,LNM 1130, Springer 1985, pp 317-340.

[10] J.Paris, A.Wilkie, A.Woods; Provability of the pigeonhole principle and the existence of infinitely many primes. Journal of Symbolic Logic 53. 1988, pp 1235-1244.

[11] Pudlák. Ramsey's theorem in Bounded Arithmetic. To appear.

[12] J.Krajiček, P.Pudlák, G.Takeuti; Bounded Aritmetic and the Polynomial Hierarchy, Annals of pure and applied Logic 52, (1991), pp 143-153.

[13] P.Pudlák; Some relations between subsystems of arithmetic and complexity of computations. In: Logic in Comp. Science, ed. Y.N.Moschovakis, RSRI Pub;. 21, (1992), pp. 499-519.

Søren Riis
University of Oxford
Mathematical Institute
24-29 St. Giles
SMRIIS@vax.oxford.ac.uk

Ordinal Arithmetic in $I\Delta_0$

Richard Sommer

Abstract

We show that much of the basic theory of the ordinal functions and relations that are involved in proof-theoretic ordinal analysis, can be carried out in a very weak fragment of arithmetic, namely $I\Delta_0$, on an arbitrary recursive ordinal. General conditions are given for the Δ_0-representability of ordinal relations and functions, and $I\Delta_0$-provability of their basic properties.

1 Introduction and Informal Discussion of the Problem

The recursive ordinals are the primary measuring stick for measuring the proof-theoretic strength of subsystems of arithmetic and analysis. Not only are ordinals applied in the study of theories, but they have an important role in the study of hierarchies of sets and functions such as the Grzegorczyk, Wainer, and hyperarithmetic hierarchies. However, ordinals (beyond ω) have had practically no role in the proof-theoretic study of weak systems such as subsystems of primitive recursive arithmetic (**PRA**). Similarly, they have had practically no role in the study of complexity classes lower than primitive recursive.

On the other hand, properties of weak subsystems and low complexity classes have had no role in the ordinal analysis and consistency programs for strong systems. This paper aims to investigate some of the interplay of recursive ordinals and weak fragments of arithmetic. The following are natural questions along these lines.

1. Can "ordinal analysis" be useful in the study of weak fragments of arithmetic and low complexity classes?

2. How much of the theory of ordinal numbers can be carried out in weak fragments of arithmetic?

3. What are the computational complexities of various ordinal operations and relations? Can such measures be used to show that the proof-theoretic ordinals of strong subsystems of analysis or set theory are, in some sense, more complex than those of the weaker subsystems?

4. Can the ordinal analysis of subsystems of analysis be done over a feasible base theory?

These questions are broad and vague, and this paper will provide only partial answers. We focus on the narrower problem of determining how much of the "basic theory" of ordinals can be dealt with in weak fragments of arithmetic, with the result that a significant amount for arbitrary recursive ordinals can be handled in such systems. In this section we describe these results somewhat informally, and then discuss how they can be applied to the general questions stated above. Here terminology is used without being defined; terminology that will be needed later in this paper is defined in Sections 2 and 3; other definitions appear in the cited literature. Section 2 will provide some preliminary material, then in Section 3 we set up a framework for describing and proving the results of Sections 4–6.

Suppose U is a recursive set of ordinals, R and F are sets of relations and functions on those ordinals, and B is a set of basic axioms regarding the relations and functions in $R \cup F$. By "basic axioms" we mean only algebraic and defining axioms for the relations and functions in $R \cup F$; it will not include any formulation of well foundedness for $\leq$. Informally, we seek $sys(U, R, F, B)$, the "least" pair (C, T) such that C is a complexity class and T a theory with the properties that the elements of $\{U\} \cup R \cup F$ all have C-representations, and for these representations the functions in F are all T-provably total and T proves the formulas in B. In this case we say that (U, R, F, B) **is** (C, T)**-representable**. A more precise definition of this appears in Section 2.

Of course there are many options of how one might make the above notion of "least" precise. Typically, we would have a hierarchy of complexity classes and a hierarchy of theories in mind and we would either fix a complexity class and ask for the weakest theory, or we would fix a theory and ask for the lowest complexity class in our hierarchies, or we might consider the complexity classes and theories together in pairs. As an example of the first possibility we may ask for the weakest theory, T, of the form **ERA** + "F_n is total" where **ERA** is a version of "elementary recursive arithmetic" (e.g. **ERA** could be a conservative extension of $I\Delta_0 + exp$),

F_n is the nth function in the Ackermann (or fast growing) hierarchy, and C is taken to be the elementary recursive predicates (cf. [Ros84]). As an example of the third possibility we may consider (C, T) to be of the form (Σ_i^b, S_2^i), where Σ_i^b and S_2^i are from [Bus86] of S. Buss.

Another formulation of the problem is to suppose T, C, R, F, and B are given and ask for $ord(T, C, R, F, B)$, the least ordinal α such that (α, R, F, B) is not (C, T)-representable. For example, for $B = \{$"$\leq$ is a linear order", "+ is associative"$\}$, it is folklore that,

$$ord(\mathbf{PRA}, PR, \{\leq\}, \{+\}, B\}) \geq \epsilon_0,$$

where PR is the set of primitive recursive predicates and ϵ_0 is the least solution to $\alpha = \omega^\alpha$. The results of this paper are significantly stronger.

The theory T should include a base theory, T_0, sufficient for handling coding of sequences. Natural choices, amongst fragments of arithmetic, for such a base theory are $T_0 = I\Delta_0$ and $T_0 = S_2^1$. Complexity classes, C_0, associated with these are $C_0 = \Delta_0$ and $C_0 = \Sigma_1^b$, respectively. (Σ_1^b is the same as NP, and the class of Δ_0-definable sets is exactly the class of sets in the linear time hierarchy—i.e. sets recognized by alternating Turing machines operating in linear time (cf. [Woo86]). Definitions of Δ_0 and $I\Delta_0$ are given in the following section.)

We will only consider the case $T_0 = I\Delta_0$ and $C_0 = \Delta_0$. We are confident that similar results can be obtained using $T_0 = S_2^1$ and $C_0 = \Sigma_1^b$; however, the proofs would be sufficiently different that it would be too much work to include both cases here. We believe that the results for the one case are enough to provide satisfactory answers to the questions at hand. So $(\Delta_0, I\Delta_0)$ is the "least pair" to be considered below. $I\Delta_0$ and Δ_0 have been a part of many interesting results and there continue to be significant open questions in the area of their study; we refer the reader to [HP92] for a presentation of such results and questions, and as the source of a good bibliography for this area.

The main results of this paper are of the form

(1) $\qquad (U, R, F, B)$ can be $(\Delta_0, I\Delta_0)$-represented,

where R, F, and B are natural or frequently occurring in the literature, and U is an arbitrary recursive ordinal closed under the functions in F. Since Δ_0 and $I\Delta_0$ are respectively the lowest complexity class and weakest theory we are considering, (1) leads to,

$$sys(U, F, R, B) = (\Delta_0, I\Delta_0).$$

Also by (1),

$$ord(I\Delta_0, \Delta_0, F, R, B) = \omega_1^{CK},$$

where ω_1^{CK} is the least nonrecursive ordinal. Section 4 proves (1) for the case U is an arbitrary recursive ordinal, $R = \{U, \leq\}$, $F = \emptyset$, and $B = \{$ "$\leq$ is a linear order"$\}$. Section 5 handles extending this to include $+$ in F and many basic properties of addition in B. In Section 6 we provide a more general theorem that gives simple conditions on U, R, F, and B for the Δ_0-representabilty of (U, R, F, B) in $I\Delta_0$. This general theorem is capable of handling cases where F includes the Veblen φ-function, and it was designed to handle versions of the Feferman-Aczel θ-functions.

It would be too much for this paper to cover such examples in any kind of detail. Much will be left out, with the idea that the reader can apply the methods outlined to particular examples of interest. Also certain important functions on ordinals are not dealt with in this paper (and are not handled by the methods of Section 6); examples include the fundamental sequence function, for which the author does not know of $(\Delta_0, I\Delta_0)$-representations. In [Som90], the author shows the $(\Delta_0, I\Delta_0 + \Omega_1)$-representability of the fundamental sequence function. In Section 7 we further discuss such avenues for future research.

Now we address what the results of the form (1) say in regard to questions 1–4 above. First we consider question 1. By weak fragment of arithmetic we mean a subsystem of **ERA**; Buss' theories S_2^i are examples of weak fragments. The central open question regarding the S_2^i's is the question of whether these theories form a true hierarchy, i.e. whether $S_2^i = S_2^{i+1}$ for any $i \geq 1$. (The general belief is that the S_2^i's do form a true hierarchy.) One attempt to use ordinal analysis to solve this problem would be to somehow assign ordinals to theories with the hope that different S_2^i's would be assigned different ordinals, and theories that are assigned different ordinals are distinct.

One possible move in this direction would be to assign ordinals according to how much transfinite induction the theory can prove; however, there are strong indications that this will not distinguish the theories. It has been shown that, over $I\Delta_0$, induction up to ω^2 for Δ_0-formulas proves the axioms of $I\Sigma_1$ ($I\Sigma_1$ extends **PRA** and is thus very much stronger than the weak fragments of arithmetic), while induction for ordinals $< \omega^2$ for Δ_0-formulas is already provable in $I\Delta_0$ (cf. [Som90]). Similar arguments can be applied to show that S_2^1, and hence any of Buss' theories S_2^i or T_2^i for $i \geq 1$, together with transfinite induction up to ω^2 for Σ_1^b-formulas, proves $I\Sigma_1$—so none of Buss' systems prove transfinite induction up to ω^2. Also, the system T_2^1, which is at least as strong as S_2^1, but no stronger than S_2^2, proves transfinite induction up to all ordinals less than ω^2. So the only possibility of separating theories in Buss' hierarchy in this way, would be to separate S_2^1 from the rest of the hierarchy, and this is equivalent to the question of whether $S_2^1 = T_2^1$; ordinals, in this case, do not seem to provide anything new in the direction of solving the problem.

The author has considered other variations of well ordering principles (and plans to expand on this in a future paper). Such considerations all lead to the same sort of consequences; namely the principle, up to a small ordinal, is implied by theories at the bottom of the hierarchy, and for all larger ordinals the principle implies theories that have strength at least that of **PRA**, and hence go far beyond the hierarchy. Thus, so far, transfinite induction does not appear to be useful for separating hierarchies of weak fragments.

An alternative approach is the assignment of ordinals according to the provability of basic properties of ordinals, but the results of the form (1), as given in Sections 4–6, indicate strongly that theories cannot be distinguished in this manner either. We cannot rule out the possibility of inventing some new method of assigning ordinals to weak theories that will yield separation results, but so far the prospects do not look good.

One way to separate complexity classes using ordinals would be to work with a similar approach to the one applied above to theories; that is, one can assign ordinals to complexity classes by considering functions and relations on ordinals and search for the least ordinal for which those functions and relations cannot be represented in the complexity class. Results of the form (1) indicate that the complexity of ordinal functions and relations on arbitrary recursive ordinals is very low (within the linear time hierarchy), and hence this will not yield a method of distinguishing complexity classes with ordinals. Furthermore, the author does not know of any enlightening uses of ordinals to distinguish low level complexity classes. (An example of an unenlightening use is the natural assignment of ordinals $< \omega^\omega$ to polynomials and then to the sets recognized in polynomial time, according to the polynomials that measure the time complexity of the sets.) Overall, the indications are that the answer to question 1 is negative.

The above remarks concerning provability of well foundedness of ordinals, together with results of the form (1), suggest a partial answer to question 2. In particular, the weak fragments prove only trivial cases of well foundedness, but, on the other hand, the basic theory of ordinals is provable in weak fragments.

The fact already mentioned, that the computational complexity of the basic structure of a recursive ordinal is very low, provides an answer to part of question 3; results of this form where only the less-than relation is considered are well known. In [Gri90], Grigorieff obtains upper bounds within linear time for the complexity of recursive linear ordering relations. Descriptions of the ordinals showing up in the ordinal analysis being carried out by proof theorists become seemingly more complicated as one goes to stronger and stronger systems (e.g. ϵ_0 has a much simpler description than, say, the Howard ordinal). In light of the above discussion, this complexity is not a facet of the basic structure of the ordinal; formally we can only describe the complexity of the ordinal in terms of the difficulty in recognizing the well foundedness of particular representations of the ordinal.

In regard to question 4, we note that many proof theorists are completely satisfied using **PRA** as a base theory for their work, so to them this question may not be of interest. But a proof theorist who has philosophical motivations such as those suggested by a modified Hilbert's program, might find question 4 very relevant. In particular, one might be concerned with a version of a modified Hilbert's program with "finitistic" replaced by "feasible". By no means does the author want to suggest that proof theorists do all their work over a feasible base theory, but he does feel that question 4 and its solution are both relevant and interesting. Much of question 4 is answered directly by the results of the form (1), since much of the adequacy of a choice of base theory is determined by the base theory's ability to deal with the ordinals and ordinal functions and relations involved. But there are other constructions that are dealt with in proof theoretic ordinal analysis, such as infinitary proofs and cut elimination procedures, which we do not deal with in this paper (this is addressed with respect to **PRA** in [Buc91]).

2 Preliminaries

2.1 Δ_0 and $I\Delta_0$

We will use $\mathcal{L}(X)$ to denote the language with nonlogical symbols from X, and if $L = \mathcal{L}(X)$ and Y is a set of nonlogical symbols then $L(Y)$ will denote the language obtained by adjoining the symbols in Y to L. L_0 will denote the usual language of arithmetic i.e. the language $\mathcal{L}(\{0, S, +, \cdot, \leq\})$.

A **Δ_0-formula** is any formula built up from the atomic formulas of L_0 using $\rightarrow$, $\leftrightarrow$, $\wedge$, $\vee$, $\neg$, $(\exists x \leq \tau)$, and $(\forall x \leq \tau)$ (where $(\exists x \leq \tau)\varphi$ and $(\forall x \leq \tau)\varphi$ are defined, for τ an L_0-term not containing x, by $\exists x(x \leq \tau \wedge \varphi)$ and $\forall x(x \leq \tau \rightarrow \varphi)$ respectively).

We let P^- be the set of axioms for a discretely ordered commutative semi-ring. P^- then is a version of the "usual" set of basic axioms of arithmetic that is commonly used together with an axiom schema to axiomatize theories such as Peano arithmetic. Let $I\mathcal{C}$ denote the theory axiomatized by P^- together with the axiom schema:

$$\forall \vec{z}(\varphi(0, \vec{z}) \wedge \forall y(\varphi(y, \vec{z}) \rightarrow \varphi(y+1, \vec{z})) \rightarrow \forall x \varphi(x, \vec{z})) : \ \varphi \in \mathcal{C}.$$

$I\Sigma_n$, $I\Pi_n$, and $I\Delta_0$ are commonly studied theories of this form. We note that in these cases, i.e. for $\mathcal{C} = \Sigma_n$, Π_n, or Δ_0, $I\mathcal{C}$ is equivalent to the theory P^- together with the strong induction axioms:

$$\forall \vec{z}(\forall x((\forall y < x)\varphi(y, \vec{z}) \to \varphi(x, \vec{z})) \to \forall x \varphi(x, \vec{z})) : \ \varphi \in \mathcal{C}.$$

2.2 Sequences and Coding

In this section we introduce a Gödel numbering of sequences that has nice properties in $I\Delta_0$. We use the terms "Gödel number" and "code" interchangeably. We use ω to denote the set $\{0, 1, 2, \ldots\}$. For any set A, we use A^* to denote $\bigcup_{n \in \omega} A^n$, where A^n denotes the set of all sequences of length n. I.e. $A^n \stackrel{\text{def}}{=} \{(a_1, \ldots, a_n) : \text{for } i = 1, \ldots, n, \ a_i \in A\}$. We will use $*$ to denote concatenation of elements of A^*, i.e.,

$$(a_1, \ldots, a_n) * (b_1, \ldots, b_m) \stackrel{\text{def}}{=} (a_1, \ldots, a_n, b_1, \ldots, b_m).$$

As long as we are in a situation where we are not considering sequences of sequences, there will be no ambiguity if we use $a_0 * (a_1, \ldots, a_n)$ to denote $(a_0)*(a_1, \ldots, a_n)$, and if we denote the sequence $(a_1, \ldots a_n)$ by $a_1 * \cdots * a_n$. If we have enough symbols on hand so that each element of A can be denoted by a unique symbol then we will leave off the $*$'s. E.g. for $1*1*2*1 \in \{1,2\}^*$ we may write 1121; however, when $1*1*2*1$ is regarded as an element of ω^*, and we are using strings of digits not separated by $*$'s to denote numerals in base ten, we will always include the $*$'s to avoid ambiguity. The **length** of $(a_1, \ldots, a_n)$ is n. We use $()$ to denote the **empty sequence**, the unique sequence of length 0. For any sequences s and t we say s is an **initial segment** of t, or t is an **end extension** of s, if for some sequence u, $t = s * u$, and we write $s \subseteq_{\mathrm{e}} t$. And we say s is a **part of** t if for some sequences u and v, $t = u * s * v$; we write $s \subseteq_{\mathrm{p}} t$. We use the convention that if $j < i$, then $a_i * \cdots * a_j = ()$.

For $n \geq 2$, we define the **Smullyan base n representation** (cf. [Smu61]) of an integer x, denoted $\overline{x}^n$, by:

i. $\overline{0}^n \stackrel{\text{def}}{=} ()$,

ii. $\overline{n \cdot q + r}^n \stackrel{\text{def}}{=} \overline{q}^n * r$, where $0 < r \leq n$.

If $\overline{x}^n = a_0 * a_1 * \ldots * a_k$ then for all $i \leq k$, we have $1 \leq a_i \leq n$, and

$$x = a_0 \cdot n^k + a_1 \cdot n^{k-1} + \cdots + a_{k-1} \cdot n + a_k.$$

For all $n \geq 2$, $\overline{}^n$ is a bijection from ω to $\{1, \ldots, n\}^*$; so for any $n \geq 2$, each $x \in \omega$ may be construed as an element of $\{1, \ldots, n\}^*$. We use $x *_n y$

to denote the natural number z such that $\overline{z}^n = \overline{x}^n * \overline{y}^n$, and we use $|x|_n$ to denote the length of $\overline{x}^n$. If $\overline{x}^n = a_0 * \cdots * a_k$ then $|x|_n = k+1$ and we let,

$$(x)_{n,i} \stackrel{\text{def}}{=} \begin{cases} a_i & \text{if } i \leq k; \\ 0 & \text{otherwise.} \end{cases}$$

For each $n \geq 2$, the functions $x \mapsto |x|_n$, $*_n$, and $(x,i) \mapsto (x)_{n,i}$ are Δ_0-definable. This follows from the fact that there is a Δ_0-formula expressing the graph of the exponential function (cf., for example, [HP92]), and the following:

$$\begin{aligned} |x|_n = l \quad &\leftrightarrow \quad n^l - 1 \leq x \cdot (n-1) \leq n \cdot (n^l - 1); \\ x *_n y = z \quad &\leftrightarrow \quad (\exists l \leq y)(l = |y|_n \wedge z = x \cdot n^l + y); \\ (x)_{n,i} = z \quad &\leftrightarrow \quad (\exists u, v \leq x)(x = (u *_n z) *_n v \wedge |u|_n = i \wedge 1 \leq z \leq n) \\ &\qquad \vee (i \geq |x|_n \wedge z = 0). \end{aligned}$$

We write $x \subseteq_n y$ if $\overline{x}^n \subseteq_e \overline{y}^n$. Since for each $n \geq 2$, $x \subseteq_n y \leftrightarrow (\exists z \leq y)(x *_n z = y)$, it is clear that the relation $\subseteq_n$ is Δ_0-definable. We state some properties of these functions and relations that hold in $I\Delta_0$.

Proposition 2.1 *For all $n \geq 2$, the following hold in $I\Delta_0$:*

i. " $|\ |_n$, $*_n$ *and* $(x,i) \mapsto (x)_{n,i}$ *are total functions".*

ii. "$*_n$ *is associative".*

iii. The Δ_0-definable relations are closed under the quantifiers $\forall x \subseteq_n \tau$ and $\exists x \subseteq_n \tau$, where τ is any L_0-term.

iv. $n^{|x|_n} - 1 \leq x \cdot (n-1)$, *hence* $2^{|x|_2} \leq x + 1$.

v. $x \cdot (n-1) \leq n \cdot (n^{|x|_n} - 1)$; *also* $x + 1 < n^{|x|_n + 1}$.

vi. $|x+1|_n \leq |x|_n + 1$.

vii. $x \leq y \rightarrow |x|_n \leq |y|_n$.

viii. $|x|_n - 1 < \log_n(x) < |x|_n + 1$.

By assuming a familiarity of $I\Delta_0$, we are assuming a familiarity with results similar to those listed in the above proposition. We only sketch the key ideas in the proofs. i follows from a straightforward application of Δ_0-induction, noting that if $z = |x|_n$, then $z \leq x$, if $z = x *_n y$ then

$z = x \cdot n^{|y|_n} + y \leq nxy + x + y$ (cf. iv), and if $z = (x)_{n,i}$ then $z \leq x$. ii uses the definition of $*_n$ and the properties of the Δ_0-definition of exponentiation. iii uses the fact that $I\Delta_0$ proves $x \overset{\subseteq}{n} y \to x \leq y$ for all x and y. iv, v, vi, vii, and viii follow fairly easily from the definition of $|\ |_n$ and properties of the Δ_0-definition of exponentiation.

Next we show how elements of ω^* are coded. We use $\langle s_0, \ldots, s_k \rangle$ to denote the code of the sequence $(s_0, \ldots, s_k)$. We obtain $\overline{\langle \vec{s} \rangle}^3$ from $(\vec{s})$, i.e. the Smullyan base 3 notation for the code of the sequence $\vec{s}$, by replacing each s_i by $3 * \overline{s_i}^2$, i.e. a "3" followed by the Smullyan base 2 notation for s_i. This is given by

a. $\langle () \rangle \overset{\text{def}}{=} 0$.

b. $\overline{\langle (\vec{s}) * a \rangle}^3 \overset{\text{def}}{=} \overline{\langle (\vec{s}) \rangle}^3 * 3 * \overline{a}^2$.

If $x = \langle a_0 * \cdots * a_k \rangle$ then let $\text{len}(x) \overset{\text{def}}{=} k + 1$, and

$$(x)_i \overset{\text{def}}{=} \begin{cases} a_i & \text{if } i \leq k; \\ 0 & \text{otherwise.} \end{cases}$$

We note that $\langle\ \rangle$ is a bijection from ω^* to a Δ_0-definable subset of ω; we let Sq denote this subset. The following can easily be transformed into a Δ_0-definition of Sq.

$$\text{Sq}(x) \overset{\text{def}}{\longleftrightarrow} x = 0 \vee (x)_{3,0} = 3.$$

To obtain Δ_0-definitions for some of the above functions we use the following:

Proposition 2.2 *There is a Δ_0-formula $\nu(x, y, i, k)$ expressing "there are exactly y occurrences of the digit i in the Smullyan base k representation of x", such that the following are provable in $I\Delta_0$.*

i. $\nu(x, y, i, k) \to 1 \leq i \leq k$.

ii. $\forall x, i, k(1 \leq i \leq k \to \exists! y \nu(x, y, i, k))$.

iii. $\nu(x, y, i, k) \to y \leq |x|_k$.

iv. $(0 \leq x \leq k) \wedge (1 \leq i \leq k) \wedge \nu(x, y, i, k) \to (y \leq 1 \wedge (y = 1 \leftrightarrow x = i))$.

v. $\nu(x_1, y_1, i, k) \wedge \nu(x_2, y_2, i, k) \to \nu(x_1 *_k x_2, y_1 + y_2, i, k)$.

For an example of such a formula, ν, see Lemma 2.2 of [Woo86]; in this example "Smullyan base k representation" is replaced by "k-ary expansion of"; the above proposition can be obtained with only slight modifications of that formula. We will not prove this proposition.

Using the formula $\nu(x, y, i, k)$ from the above proposition, the relations $\overset{\subset}{\mathrm{e}}$ and $\overset{\subset}{\mathrm{p}}$, as well as the functions $x \mapsto \langle x \rangle$, len, $*$, and $(x, i) \mapsto (x)_i$ are Δ_0-definable, noting:

$$
\begin{aligned}
z = \langle x \rangle \quad &\leftrightarrow \quad (\exists l \leq z)(|z|_3 = l \wedge |x|_2 = l - 1 \wedge (z)_{3,0} = 3 \\
&\qquad\qquad \wedge (\forall i \leq l - 2)((z)_{3,i+1} = (x)_{2,i})) \\
\mathrm{len}(x) = l \quad &\leftrightarrow \quad \nu(x, l, 3, 3) \\
x * y = z \quad &\leftrightarrow \quad x *_3 y = z \\
(x)_i = y \quad &\leftrightarrow \quad (\exists u, v \leq x)(\mathrm{Sq}(u) \wedge \mathrm{Sq}(v) \wedge x = (u * \langle y \rangle) * v \wedge \mathrm{len}(u) = i) \\
&\qquad \vee ((i \geq \mathrm{len}(x) \vee \neg \mathrm{Sq}(x)) \wedge y = 0) \\
x \overset{\subset}{\mathrm{e}} y \quad &\leftrightarrow \quad \mathrm{Sq}(x) \wedge \mathrm{Sq}(y) \wedge (\exists z \leq y)(\mathrm{Sq}(z) \wedge x * z = y) \\
x \overset{\subset}{\mathrm{p}} y \quad &\leftrightarrow \quad \mathrm{Sq}(x) \wedge \mathrm{Sq}(y) \wedge (\exists w, z \leq y)(\mathrm{Sq}(w) \wedge \mathrm{Sq}(z) \\
&\qquad \wedge w * x * z = y)
\end{aligned}
$$

We will use obvious conventions such as writing $(x)_i < (y)_j$ in place of

$$
\begin{aligned}
&\mathrm{Sq}(x) \wedge \mathrm{Sq}(y) \wedge \mathrm{len}(x) > i \wedge \mathrm{len}(y) > j \wedge \\
&\qquad\qquad\qquad \exists u, v((x)_i = u \wedge (y)_j = v \wedge u < v).
\end{aligned}
$$

The following is a variation of well known results:

Proposition 2.3

i. $\langle\ \rangle$ *is a bijection from* ω^* *to* Sq.

ii. For all $k \in \omega$, $I\Delta_0 \vdash \forall x_1, \ldots, x_k \exists z(z = \langle x_0, \ldots, x_k \rangle)$.

iii. $I\Delta_0 \vdash$ " len, $*$, *and* $(x, i) \mapsto (x)_i$ *are total functions*".

iv. The Δ_0*-definable relations are closed under the quantifiers* $\exists x \overset{\subset}{\mathrm{e}} \tau$, $\exists x \overset{\subset}{\mathrm{p}} \tau$, $\forall x \overset{\subset}{\mathrm{e}} \tau$ *and* $\forall x \overset{\subset}{\mathrm{p}} \tau$.

v. $I\Delta_0 \vdash \forall x, y((x = y) \vee (\mathrm{len}(x) \leq \mathrm{len}(y) \wedge (\forall i < \mathrm{len}(x))((x)_i = (y)_i)) \leftrightarrow x \overset{\subset}{\mathrm{e}} y)$.

2.3 Notational Conventions

We will allow notation to be misused rather than become cumbersome as long as the meaning is clear from the context. Symbols denoting relations and functions on ordinals and natural numbers in the metatheory will also be used as formal symbols. For example "$\leq$" will be used both in the metatheory and as a formal symbol, and in both cases it will be used to denote the less-than relation, sometimes on natural numbers, sometimes on arbitrary ordinals, and sometimes on subsets of ordinals; also "$\leq$" will be used to denote the relation induced by the less-than relation onto formal terms involving function symbols denoting functions on ordinals. In Section 6 we introduce some notation for formal symbols when such abuses of notation as just described would make our meaning unclear.

We will frequently use corner quote marks, $\ulcorner\;\urcorner$, with primes and subscripts to denote codes of objects. If we have codes for the elements of A then,

$$\ulcorner A \urcorner \stackrel{\text{def}}{=} \{\ulcorner a \urcorner : a \in A\}.$$

Similarly, $\ulcorner x_1, \ldots, x_n \urcorner \stackrel{\text{def}}{=} \ulcorner x_1 \urcorner, \ldots, \ulcorner x_n \urcorner$.

We will not be specific about our choice of metatheory; certainly any reasonable set theory will do. Statements such as "$\theta(n)$ holds" where $\theta \in L_0$ and $n \in \omega$, mean that $\theta(n)$ is satisfied in the standard model of the natural numbers.

3 Representability

Definition 3.1 Suppose we are given the following:

a. A set of ordinals U and sets R and F of relations and functions on the ordinals in U. We require that $U \in R$.

b. A set B of basic properties of the ordinal functions and relations in $R \cup F$, expressed in the language $\mathcal{L}(R \cup F)$.

c. A theory, T, in the language $L_0 \stackrel{\text{def}}{=} \mathcal{L}(\{0, S, +, \cdot, \leq\})$, for proving the basic properties, B, about representations of $R \cup F$.

d. A subset C of L_0, to serve as a complexity class for representing the elements of $R \cup F$.

e. A subset D of C of the form $\{\theta_s : s \in R \cup F\}$, which will be used for defining representations of the elements of $R \cup F$.

We say that D, C-**represents** (U, R, F, B) **in** T if there is a Gödel numbering $\alpha \mapsto \ulcorner\alpha\urcorner$ of U, i.e. an injection, $\ulcorner\ \urcorner$, from U into ω, such that

i. For all $r \in R$, $\{(\ulcorner\vec{\alpha}\urcorner) : r\vec{\alpha}\} = \{(\vec{n}) : \theta_r(\vec{n})\}$.

ii. For all $f \in F$, $\{(\ulcorner\vec{\alpha}, \beta\urcorner) : f(\vec{\alpha}) = \beta\} = \{(\vec{n}, m) : \theta_f(\vec{n}, m)\}$.

iii. For all $f \in F$, $T \vdash \forall\vec{x}\exists! y\theta_f(\vec{x}, y)$.

iv. $T + D' \vdash B^U$, where B^U is the set of sentences obtained from the sentences in B by restricting all quantifiers to U, and

$$D' \stackrel{\text{def}}{=} \{(\theta_r(\vec{x}) \leftrightarrow r\vec{x}) : r \in R\} \cup \{(\theta_f(\vec{x}, y) \leftrightarrow f(\vec{x}) = y) : f \in F\}.$$

If for a given U, R, F, B, C, and T such a D exists we say (U, R, F, B) **is** (C, T)-**representable**. We may write (U, R, F) in place of (U, R, F, B) in the case $B = \emptyset$.

As mentioned in Section 1, our main results take $C = \Delta_0$ and $T = I\Delta_0$. The following is straightforward so we omit the proof.

Proposition 3.2 *The conjunction of clauses i and ii in definition 3.1 is equivalent to the conjunction of the following three clauses.*

i$'$. *For all $r \in R$ and $\vec{\alpha} \in U$, $(r\vec{\alpha} \iff \theta_r(\ulcorner\vec{\alpha}\urcorner))$.*

ii$'$. *For all $f \in F$ and $\vec{\alpha}, \beta \in U$, $(f(\vec{\alpha}) = \beta \iff \theta_f(\ulcorner\vec{\alpha}\urcorner, \ulcorner\beta\urcorner))$.*

iii$'$. *For all $s \in R \cup F$ and $\vec{n} \in \omega$, if $\theta_s(\vec{n})$ then for some $\vec{\alpha} \in U$ we have $\vec{n} = \ulcorner\vec{\alpha}\urcorner$.*

For the case that U is an ordinal and $\{U, \leq\} \subseteq R$, the Gödel numbering $\alpha \mapsto \ulcorner\alpha\urcorner$ is completely determined by the formulas θ_U and $\theta_{\leq}$. Letting,

$$\mathcal{U} \stackrel{\text{def}}{=} \{n \in \omega : \theta_U(n)\}, \tag{2}$$

$$n \preceq m \stackrel{\text{def}}{\iff} \theta_{\leq}(n, m), \text{ and} \tag{3}$$

$$n \prec m \stackrel{\text{def}}{\iff} n \preceq m \text{ and } n \neq m, \tag{4}$$

then $\ulcorner\ \urcorner$ is given by the following transfinite recursion:

$$\ulcorner\beta\urcorner = \text{the } \preceq\text{-least element } b \text{ of } \mathcal{U} \text{ such that for all } \gamma < \beta,\ \ulcorner\gamma\urcorner \prec b. \tag{5}$$

In Sections 4 and 5 we will be defining formulas θ_s^i for various ordinal relations and functions s, and for various $i \in \omega$. We will use the following notational conventions.

(6) $$s_i \stackrel{\text{def}}{=} \{(\vec{n}) : \theta_s^i(\vec{n})\},$$

and if U is an ordinal and $\{U, \leq\} \subseteq R$, we let $\ulcorner\ \urcorner^i$ be given by (5), i.e. we let,

(7) $$\ulcorner\beta\urcorner^i = \text{the } \leq_i\text{-least element } b \text{ of } U_i \text{ such that for all } \gamma < \beta,\ \ulcorner\gamma\urcorner^i <_i b.$$

We will only apply this definition in the event $\leq_i$ is a well ordering relation.

4 The Case $R = \{U, \leq\}$, $B = \{$ "$\leq$ is a linear order" $\}$, and U is an Arbitrary Recursive Ordinal

We begin with a lemma that asserts that the $(\Delta_0, I\Delta_0)$-representability of $(\alpha, \{\alpha, \leq\}, \emptyset, \emptyset)$ follows from the existence of a Δ_0-definable well ordering of order type α. For any binary relation R we write $\|R\| = \alpha$ in the event α is an ordinal and there is an isomorphism from $(fld(R), R)$ to $(\alpha, \leq)$, where $fld(R) \stackrel{\text{def}}{=} \{x : \exists y(Rxy \vee Ryx)\}$. This section, as well as the following sections will make heavy use of the following lemma.

Lemma 4.1 *Suppose θ_U^i and $\theta_\leq^i$ are Δ_0-formulas such that, using (6),*

$$fld(\leq_i) = U_i, \text{ and } \| \leq_i \| = \alpha,$$

for some ordinal α; then $\{\theta_U^i, \theta_\leq^i\}$, Δ_0-represents $(\alpha, \{\alpha, \leq\}, \emptyset, \emptyset)$ in $I\Delta_0$.

Proof:

Using (7), the clauses of Definition 3.1 are all seen to be satisfied with $\ulcorner\ \urcorner^i$ in place of $\ulcorner\ \urcorner$ (of course only clause i is relevant here).

q.e.d.

Let α be a fixed recursive ordinal; below we will establish the $(\Delta_0, I\Delta_0)$-representability of $(\alpha, \{\alpha, \leq\}, \emptyset, \{$"$\leq$ is a linear order"$\})$ by successively constructing formulas θ_U^i and $\theta_\leq^i$ for $i = 0$, 1, and 2, each satisfying more conditions of Definition 3.1 than the previous one. Let $\theta_U^0(x)$ be the formula

$x = x$, and suppose $\theta^0_\leq \in \Sigma_1$ is such that, using (6), $\| \leq_0 \| = \alpha$. Since there is a Δ_0-formula representing the Kleene T-predicate we may suppose $\theta^0_\leq$ is of the form $\exists w \varphi_0$ where $\varphi_0 \in \Delta_0$. Note that for $x, y, \in \omega$,

$$(8) \qquad [x \neq y \rightarrow (\exists w \varphi_0(w, x, y) \leftrightarrow \forall w \neg \varphi_0(w, y, x))] \wedge$$

$$(9) \qquad \exists w(\varphi_0(w, x, y) \vee \varphi_0(w, y, x)) \wedge$$

$$(10) \qquad (\exists w \varphi_0(w, x, y) \wedge \exists w \varphi_0(w, y, x) \leftrightarrow x = y).$$

We cannot apply Lemma 4.1 since $\leq_0$ may not be Δ_0; also note that $I\Delta_0$ may not prove that $\leq_0$ has the properties of a linear order. As a first step around these problems we present the following formulas θ^1_U and $\theta^1_\leq$, which are easily seen to be equivalent to Δ_0-formulas. Following their definitions we give an intuitive explanation of these formulas.

$$(11)\ \theta^1_U(\langle x, m \rangle) \stackrel{\text{def}}{\Longleftrightarrow} (\forall y \leq x)(\exists w \leq m)(\varphi_0(w, x, y) \vee \varphi_0(w, y, x)) \\ \wedge(\exists y \leq x)(\forall w < m)(\neg \varphi_0(w, x, y) \wedge \neg \varphi_0(w, y, x))$$

$$(12) \qquad \theta^1_\leq(\langle x, m_x \rangle, \langle y, m_y \rangle) \stackrel{\text{def}}{\Longleftrightarrow} \theta^1_U(\langle x, m_x \rangle) \wedge \theta^1_U(\langle y, m_y \rangle) \wedge \\ (\exists w \leq m_x + m_y)\varphi_0(w, x, y)$$

Letting $\ulcorner\ \urcorner^i$ be given by (7), and letting m_x denote the unique m such that $\langle x, m \rangle \in U_1$, the motivation for the definition of θ^1_U can be explained as follows. $\ulcorner\alpha\urcorner^1$ is the result of adjoining to $\ulcorner\alpha\urcorner^0$ the least upper bound, m_x, for least witnesses to "$z \leq_0 \ulcorner\alpha\urcorner^0$" and "$\ulcorner\alpha\urcorner^0 \leq_0 z$" for all $z \leq \ulcorner\alpha\urcorner^0$. So every pair of elements of U_1 codes a bound on the size of the least witness, w, to $\exists w \varphi_0$ that determines their $\leq_0$ relationship.

Lemma 4.2 $\{\theta^1_U, \theta^1_\leq\}$ *Δ_0-represents* $(\alpha, \{\alpha, \leq\}, \emptyset, \emptyset)$ *in* $I\Delta_0$.

Proof:

We will show that the function $f : \omega \rightarrow U_1$, given by

$$(13) \qquad f(x) \stackrel{\text{def}}{=} \langle x, m_x \rangle$$

is an isomorphism from $(U_0, \leq_0)$ to $(U_1, \leq_1)$. Then by Lemma 4.1, this lemma is easily obtained. It is straightforward to see that f is a bijection. We need for $x, y \in \omega$,

$$(14) \qquad x \leq_0 y \Longleftrightarrow f(x) \leq_1 f(y).$$

For the $\Rightarrow$-direction, first suppose $x \geq y$, then

$$(\exists w \leq m_x)(\varphi_0(w,x,y) \vee \varphi_0(w,y,x)).$$

Since $x \leq_0 y$ we have $(\exists w \leq m_x)\varphi_0(w,x,y)$, so clearly

$$(\exists w \leq m_x + m_y)\varphi_0(w,x,y).$$

I.e. $\theta^1_{\leq}(\langle x, m_x\rangle, \langle y, m_y\rangle)$, and so $\langle x, m_x\rangle \leq_1 \langle y, m_y\rangle$; thus $f(x) \leq_1 f(y)$. The argument is similar in the case $y \geq x$.

Now if $f(x) \leq_1 f(y)$, i.e. if $\langle x, m_x\rangle \leq_1 \langle y, m_y\rangle$, we have

$$(\exists w \leq m_x + m_y)\varphi_0(w,x,y).$$

If $x \geq y$ then, by (9), (11), and the fact that $\theta^1_U(\langle x, m_x\rangle)$, we have $(\exists w \leq m_x)(\varphi_0(w,x,y) \vee \varphi_0(w,y,x))$. By (10), we must have $(\exists w \leq m_x)\varphi_0(w,x,y)$, i.e. $x \leq_0 y$. A similar argument works for the $y \geq x$ case.
q.e.d.

Although $\leq_1$ is Δ_0-definable, $I\Delta_0$ may fail to prove the linear ordering properties of $\leq_1$; we next present Δ_0-formulas θ^2_U and $\theta^2_{\leq}$ such that, using (6),

1. $\| \leq_2 \| = \alpha$, $fld(\leq_2) = U_2$, and
2. $I\Delta_0 \vdash$ "$\leq_2$ linearly orders U_2".

We note that $I\Delta_0$ does not necessarily prove that for all x there is an m_x such that $\langle x, m_x\rangle \in U_1$.

U_2 will consist of triples $\langle x, \overline{m_x}, s_x\rangle$ where $\overline{m_x} = \max\{m_y : y \leq x\}$ is a bound for witnesses to the formula $\exists w \varphi_0(w,u,v)$ for all $u, v \leq x$. Let $\hat{x}$ denote $\langle x, \overline{m_x}, s_x\rangle$. s_x is to be construed as a binary sequence (of 1's and 2's) that codes the relation $<_2$ for all $\hat{v}$ and $\hat{w}$ such that $v < w$ and $\langle v, w\rangle < \langle x, x\rangle$ (we will use 2 for "true"). Thus affixed to each element of U_2 is a record of its $<_2$ relationships to elements of U_2 that have smaller corresponding first element in their triple; this makes the computation of $\leq_2$ quite simple. First priority in the determination of $\hat{v} <_2 \hat{w}$, for $v < w$, goes to making sure transitivity will hold for all triples $\hat{u}, \hat{v}, \hat{w}$, where $u < \min\{v, w\}$. For example if $u < v < w$ and we have decided that $\neg(\hat{u} <_2 \hat{v})$ and $\hat{u} <_2 \hat{w}$ then we will make $\hat{v} <_2 \hat{w}$. Intuitively, we may think of going through an enumeration of pairs (v, w) where $v < w$ and deciding the relation $<_2$. In $I\Delta_0$ it can be shown that there will be no ambiguity using this method; i.e. it will not be the case that considering two different values of u, transitivity has contradictory implications. We set things up so that such contradictory implications will rule out the existence of $\hat{x}$

for x too large (nonstandard), but this will have no effect on our results. If the relation $\hat{v} <_2 \hat{w}$ is not determined by transitivity then it will be set to coincide with $<_1$. This is expressed more formally as follows. Let $\varphi_2(x, m, s)$ express the following.

1. For each $y \leq x$, $m_y \leq m$, and

2. a. If u, v, and w are such that $u < v < w$, $\langle v, w \rangle < |s|_2$, and $(s)_{2,\langle u,v\rangle} \neq (s)_{2,\langle u,w\rangle}$ then $(s)_{2,\langle v,w\rangle} = (s)_{2,\langle u,w\rangle}$ (i.e. $\hat{v} <_2 \hat{w}$ is determined by transitivity).

   b. If v and w are such that $v < w$ and $\langle v, w \rangle < |s|_2$, and for all $u < v$ we have $(s)_{2,\langle u,v\rangle} = (s)_{2,\langle u,w\rangle}$, then $(s)_{\langle v,w\rangle} = 2$ or 1 according to whether or not $\langle v, m_v \rangle <_1 \langle w, m_w \rangle$.

   c. $|s|_2 \geq \langle x, x \rangle$.

3. $\langle m, s \rangle$ is least, i.e. 1 and 2 above fail if m and s are replaced by m' and s' where $\langle m', s' \rangle < \langle m, s \rangle$.

And let θ^2_U and $\theta^2_\leq$ be given by,

$$\theta^2_U(z) \stackrel{\text{def}}{\iff} \exists x, m, s \leq z(z = \langle x, m, s \rangle \wedge \varphi_2(x, m, s)) \tag{15}$$

$$\begin{aligned} \theta^2_\leq(\hat{x}, \hat{y}) \stackrel{\text{def}}{\iff} \; & \theta^2_U(\hat{x}) \wedge \theta^2_U(\hat{y}) \wedge \\ & [(x < y \wedge (s_y)_{2,\langle x,y\rangle} = 2) \vee \\ & (y < x \wedge (s_x)_{2,\langle y,x\rangle} = 1) \vee x = y] \end{aligned} \tag{16}$$

$$\theta^2_<(\hat{x}, \hat{y}) \stackrel{\text{def}}{\iff} \hat{x} \leq_2 \hat{y} \wedge x \neq y \tag{17}$$

Lemma 4.3 *$\{\theta^2_U, \theta^2_\leq\}$ Δ_0-represents $(\alpha, \{\alpha, \leq\}, \emptyset, \emptyset)$ in $I\Delta_0$.*

Proof:

We will show that the function f defined by $f(x) \stackrel{\text{def}}{=} \hat{x}$ is an isomorphism from $(U_0, \leq_0)$ to $(U_2, \leq_2)$. As in the proof of Lemma 4.2, this yields the result. To see that $\text{dom}(f) = U_0 = \omega$, we note that $\leq_1$ is transitive (since $\leq_0$ is transitive) so if we pick s_x least such that $|s_x|_2 = \langle x, x \rangle$, and for all $u < v$ such that $\langle u, v \rangle < \langle x, x \rangle$,

$$(s_x)_{2,\langle u,v\rangle} = \begin{cases} 2 & \text{if } \langle u, m_u \rangle <_1 \langle v, m_v \rangle, \\ 1 & \text{if } \langle v, m_v \rangle <_1 \langle u, m_u \rangle, \end{cases} \tag{18}$$

and if we take $\overline{m_x} \overset{\text{def}}{=} \max\{m_y : y \leq x\}$ then clearly $\varphi_2(x, \overline{m_x}, s_x)$. Thus f is defined on ω. It is straightforward to see that f is a bijection onto U_2.

We need to show for all $x, y \in \omega$,

$$x \leq_0 y \iff \hat{x} \leq_2 \hat{y}. \tag{19}$$

Using Lemma 4.2 it is enough to show that for $x, y \in \omega$,

$$\langle x, m_x \rangle \leq_1 \langle y, m_y \rangle \iff \langle x, \overline{m_x}, s_x \rangle \leq_2 \langle y, \overline{m_y}, s_y \rangle. \tag{20}$$

(20) is easy to obtain noting that $\leq_1$ is transitive (so clause 2a in the definition of φ_2 yields the same results as clause 2b).

q.e.d.

Lemma 4.4 *$I\Delta_0 \vdash$ "$\leq_2$ linearly orders U_2".*

The proof of this lemma will rely on the following:

Lemma 4.5 *$I\Delta_0$ proves the following.*

1. *If $x < y$ and s_x and s_y exist then $s_x \subset_e s_y$.*
2. *If $x < y$ then $\hat{x} <_2 \hat{y}$ if and only if $(s_y)_{2,\langle x,y \rangle} = 2$, and $\hat{y} <_2 \hat{x}$ if and only if $(s_y)_{2,\langle x,y \rangle} = 1$.*

Proof of Lemma 4.5:

We argue, informally, in $I\Delta_0$.

1. Assume $x < y$ and s_x and s_y exist; we use induction on i to show,

$$(i < |s_x|_2 \rightarrow (s_x)_{2,i} = (s_y)_{2,i}).$$

This is a Δ_0-induction. If clauses 2a or 2b in the definition of φ_2 determine the value of $(s_x)_{2,i}$ then by the induction hypothesis they will force the same value on $(s_y)_{2,i}$ (since it will depend on values of $(s_x)_{2,j}$ for $j < i$). On the other hand, if 2a and 2b are not relevant then $(s_x)_{2,i} = (s_y)_{2,i} = 1$ for satisfying the leastness condition given by clause 3 of the definition of φ_2.

The proof of clause 2 of the lemma is straightforward.

q.e.d.

Proof of Lemma 4.4:

Throughout this proof we work in $I\Delta_0$. First we show that the trichotomy property holds. Suppose $x \neq x'$, say $x < x'$. If $\langle x, \overline{m_x}, s_x\rangle$, $\langle x', \overline{m_{x'}}, s_{x'}\rangle \in U_2$ then we need to show that exactly one of the following holds:

a. $\langle x, \overline{m_x}, s_x\rangle <_2 \langle x', \overline{m_{x'}}, s_{x'}\rangle$, or

b. $\langle x', \overline{m_{x'}}, s_{x'}\rangle <_2 \langle x, \overline{m_x}, s_x\rangle$.

First note that $(s_{x'})_{2,\langle x,x'\rangle} = 1$ or $(s_{x'})_{2,\langle x,x'\rangle} = 2$, so, by Lemma 4.5, a or b holds. If both a and b held then, using Lemma 4.5, we would have $(s_{x'})_{2,\langle x,x'\rangle} = 1$ and $(s_{x'})_{2,\langle x,x'\rangle} = 2$, a clear contradiction, so the trichotomy property holds.

It remains to show that transitivity holds. Suppose not. Working in $I\Delta_0$, suppose a, b, and c are such that,

$$\hat{a} <_2 \hat{b} <_2 \hat{c} <_2 \hat{a}. \tag{21}$$

We may suppose $b, c < a$. Using Lemma 4.5 we obtain,

$$(s_a)_{2,\langle c,a\rangle} = 2, \text{ and} \tag{22}$$

$$(s_a)_{2,\langle b,a\rangle} = 1. \tag{23}$$

Suppose $b < c$ then,

$$(s_a)_{2,\langle b,c\rangle} = 2. \tag{24}$$

By clause 2a in the definition of φ_2, using (23) and (24), we would have

$$(s_a)_{2,\langle c,a\rangle} = (s_a)_{2,\langle b,a\rangle},$$

contradicting (22) and (23). On the other hand, if $c < b$ then, from (21),

$$(s_a)_{2,\langle c,b\rangle} = 1. \tag{25}$$

From (22) and (25) we have, using clause 2a in the definition of φ_2,

$$(s_a)_{2,\langle b,a\rangle} = (s_a)_{2,\langle c,a\rangle},$$

contradicting (22) and (23). So (21) fails. We have proven transitivity completing the proof that $<_2$ is a linear order.

q.e.d.

Lemmas 4.3 and 4.4 easily yield,

Theorem 4.6 $\{\theta^2_U, \theta^2_\leq\}$ *Δ_0-represents* $(\alpha, \{\alpha, \leq\}, \emptyset, \{$'$\leq$ *is a linear order*'$\})$ *in* $I\Delta_0$.

5 The Case $R = \{U, \leq\}$, $F = \{+\}$, and U is an Arbitrary Recursive Ordinal

In this section we deal with a very simple function on ordinals, namely addition. The technique used here to code the ordinals, which relies on the Cantor normal form theorem, is well known. This section shows that such coding techniques work well when dealing with low complexity classes and weak theories, but more importantly, this section provides us with a simple prototype for the kind of methods that will be applied in the following section to more complicated notation systems.

In terms of the Gödel numberings $\ulcorner\ \urcorner^i$ given by (7), the Gödel numbers of ordinals increase as i goes from 0 to 2. Put another way, the cost of reducing the complexity of the ordinal less-than relation, and of forcing basic properties to be easily proven, was taken up by an increase in the size of codes of ordinals. This has not yet been a problem; however, this could be a problem when one considers low complexity definitions for ordinal functions; for example the functions $\ulcorner\alpha\urcorner^i \mapsto \ulcorner\alpha+1\urcorner^i$ for $i = 0, 1$, and 2, may not even be recursive. In fact,

Theorem 5.1 *There are Δ_0-formulas θ_U and $\theta_\leq$ such that for $\mathcal{U}$, $\preceq$ and $\ulcorner\ \urcorner$ given by (2), (3) and (5) from Section 3, we have,*

i. $\| \preceq \| = \omega$, *and*

ii. the function $\ulcorner\alpha\urcorner \mapsto \ulcorner\alpha + 1\urcorner$ *is not recursive.*

Proof:

Suppose T is a version of Kleene's T-predicate, i.e. $T(e, x, y)$ is a formalization of "y is a computation of the eth program on input x". Let,

$$(26) \qquad \theta_U(y) \overset{\text{def}}{\iff} (\exists\langle e, x\rangle \leq y)T(e, x, y).$$

Since there are Δ_0-formulas representing T we may suppose that θ_U is a Δ_0-formula. Now define $\theta_\leq$ as follows.

$$\begin{aligned}(27)\quad \theta_\leq(y_0, y_1) \overset{\text{def}}{\iff}\ & \theta_U(y_0) \wedge \theta_U(y_1) \wedge \\ & (\exists\langle e_0, x_0\rangle \leq y_0)(\exists\langle e_1, x_1\rangle \leq y_1)(T(e_0, x_0, y_0) \\ & \wedge T(e_1, x_1, y_1) \wedge \langle e_0, x_0\rangle \leq \langle e_1, x_1\rangle).\end{aligned}$$

So $\preceq$ orders computations, y, of $\{e\}(x)$, for all $\langle e, x\rangle$ such that $\{e\}(x)\downarrow$, in the order of increasing $\langle e, x\rangle$. Clearly this is a linear order of order type ω. (We assume there is at most one halting computation for any given program and input.)

We may also suppose our machine is such that there is a computation, y_0, of $\{0\}(0)$, so y_0 is the least element of our ordering, i.e. $y_0 = \ulcorner 0 \urcorner$. Suppose, for a contradiction, that $\ulcorner \alpha \urcorner \mapsto \ulcorner \alpha + 1 \urcorner$ is recursive; we could then determine whether $\{e\}(x)$ converges by successively finding $\ulcorner 0 \urcorner$, $\ulcorner 1 \urcorner$, $\ulcorner 2 \urcorner$, etc., and reading off for what program and input each is a computation. Eventually we will reach a computation of $\{e'\}(x')$ for $\langle e', x'\rangle \geq \langle e, x\rangle$; whether or not the least such $\langle e', x'\rangle$ equals $\langle e, x\rangle$ corresponds to whether or not $\{e\}(x)$ converges. This would entail a solution to the halting problem, a contradiction.

q.e.d.

We avoid the difficulty described above by using the Gödel numbering, $\ulcorner\ \urcorner^2$, from Section 4 for coding only the closure ordinals of certain functions; codes for the remaining ordinals will be obtained from normal form properties of the ordinals in terms of these functions. This idea was suggested by S. Feferman, and is applied to similar problems in [Fef68]. In this section we handle the case where the only function considered is addition; the method will be the prototype for the next section where we deal with a wide variety of functions.

In the case of addition the closure ordinals are ordinals of the form ω^α, i.e. the additive principal ordinals, and the normal form is given by the Cantor normal form theorem.

Proposition 5.2 (Cantor) *For α an ordinal such that $\alpha > 0$, α has a unique representation:*

$$\alpha = \omega^{\alpha_1} + \cdots + \omega^{\alpha_k}$$

where $k \in \omega - \{0\}$, and $\alpha_1, \ldots, \alpha_k$ are ordinals such that $\alpha_1 \geq \cdots \geq \alpha_k$.

We refer the reader to [End77] for a proof of this Proposition; this will also serve as a reference for some of the "well known facts about ordinals" that we sight below without proof. We call the representation, $\omega^{\alpha_1} + \cdots + \omega^{\alpha_k}$, given in the above proposition, the **normal form** of α.

Below we take α to be a fixed recursive ordinal. In light of the above theorem we identify each ordinal $\beta < \omega^\alpha$ with the sequence $(\beta_1, \ldots \beta_k)$, where $\omega^{\beta_1} + \cdots + \omega^{\beta_k}$ is the normal form of β. We identify 0 with the empty sequence. This motivates defining Δ_0-formulas θ^3_U and $\theta^3_\leq$ such that the Gödel numbering obtained by (7), $\ulcorner\urcorner^3$, will code the ordinal with normal form $\omega^{\beta_1}+\cdots+\omega^{\beta_k}$ by the code of the sequence $(\ulcorner\beta_1\urcorner^2,\ldots,\ulcorner\beta_k\urcorner^2)$, and will code 0 by 0. This is accomplished by the following.

(28) $\theta^3_U(s) \overset{\text{def}}{\iff} s \in \mathrm{Sq} \wedge (\forall i < \mathrm{len}(s))(U_2((s)_i) \wedge (\forall j < i)((s)_i \leq_2 (s)_j))$

(29) $\theta^3_{\leq}(s,t) \overset{\text{def}}{\iff} \theta^3_U(s) \wedge \theta^3_U(t) \wedge (s \subseteq_{\mathrm{e}} t \vee (\exists i < \mathrm{len}(s))((\forall j < i)((s)_j = (t)_j) \wedge (s)_i <_2 (t)_i))$

We will apply the following basic fact about ordinals to define the formula θ^3_+.

Proposition 5.3 *If β and γ have normal forms $\omega^{\beta_1}+\cdots+\omega^{\beta_k}$ and $\omega^{\gamma_1}+\cdots+\omega^{\gamma_l}$, respectively, then, if $\beta_1 < \gamma_1$, the normal form of $\beta+\gamma$ is the same as the normal form of γ; otherwise the normal form of $\beta+\gamma$ is given by,*

$$\omega^{\beta_1}+\cdots+\omega^{\beta_m}+\omega^{\gamma_1}+\cdots+\omega^{\gamma_l},$$

where m is maximum such that $\beta_m \geq \gamma_1$. Also, for an arbitrary β, the normal forms of $\beta+0$ and $0+\beta$ are both the same as the normal form of β.

In light of this proposition, we define $\theta^3_+(s,t,r)$, to represent $s+t=r$, as follows.

(30) $$\begin{aligned}&\theta^3_+(s,t,r) \overset{\text{def}}{\iff}\\ &\quad \theta^3_U(s) \wedge \theta^3_U(t) \wedge \theta^3_U(r) \wedge\\ &\quad (\exists s' \subseteq_{\mathrm{e}} s)(r = s' * t \wedge [(t = 0 \wedge s' = s) \vee (t \neq 0 \wedge\\ &\quad (\forall i < \mathrm{len}(s))(\mathrm{len}(s') \leq i \leftrightarrow (s)_i <_2 (t)_0))])\end{aligned}$$

The following proposition lists some of the basic properties of $\leq$ and $+$ that are provable in $I\Delta_0$ with respect to the formulas θ^3_U, $\theta^3_{\leq}$ and θ^3_+. The proposition uses the following definitions.

(31) $\alpha = 1 \overset{\text{def}}{\iff} (\alpha > 0) \wedge (\forall \beta < \alpha)(\beta = 0)$,

(32) $\mathrm{SUC}(\alpha) \overset{\text{def}}{\iff} \exists \beta(\alpha = \beta + 1)$,

(33) $\mathrm{LIM}(\alpha) \overset{\text{def}}{\iff} \alpha \neq 0 \wedge \neg\mathrm{SUC}(\alpha)$.

Proposition 5.4 *Let α be an arbitrary recursive ordinal.* $\{\theta^3_U, \theta^3_{\leq}, \theta^3_+\}$, Δ_0*-represents* $(\omega^\alpha, \{\omega^\alpha, \leq\}, \{+\}, B_+)$, *where* B_+ *consists of the following formulas.*
Trichotomy:

i. $((\alpha = 0) \vee SUC(\alpha) \vee LIM(\alpha)) \wedge$
$(\neg(\alpha = 0 \wedge \mathrm{SUC}(\alpha)) \wedge \neg(\alpha = 0 \wedge \mathrm{LIM}(\alpha)) \wedge \neg(\mathrm{SUC}(\alpha) \wedge \mathrm{LIM}(\alpha)))$

$<$ *is a linear order with least element and successor function:*

ii. $((\alpha < \beta) \vee (\alpha = \beta) \vee (\beta < \alpha)) \wedge$
$(\neg(\alpha < \beta \wedge \alpha = \beta) \wedge \neg(\alpha < \beta \wedge \beta < \alpha) \wedge \neg(\alpha = \beta \wedge \beta < \alpha))$

iii. $\alpha < \beta \wedge \beta < \gamma \rightarrow \alpha < \gamma$

iv. $0 \leq \alpha$

v. $\alpha < \alpha + 1 \wedge (\beta \leq \alpha \vee \alpha + 1 \leq \beta)$

Monotonicity:

vi. $\alpha < \beta \rightarrow (\gamma + \alpha < \gamma + \beta \wedge \alpha + \gamma \leq \beta + \gamma)$

Defining formula for $\leq$*:*

vii. $\alpha \leq \beta \leftrightarrow \exists! \gamma(\alpha + \gamma = \beta)$

Defining formulas for addition:

viii. $\alpha + 0 = \alpha$

ix. $\alpha + (\beta + 1) = (\alpha + \beta) + 1$

x. $\mathrm{LIM}(\lambda) \rightarrow (\alpha + \lambda = \sup_{\delta < \lambda}(\alpha + \delta))$

Associative law:

xi. $(\alpha + \beta) + \gamma = \alpha + (\beta + \gamma)$

Proof:
We provide only a partial proof sketch (most of these are proved in detail for a similar coding in [Som90]). Clause i in Definition 3.1 is handled similarly to the proofs of Lemmas 4.2 and 4.3; however, here we rely on the Cantor normal form theorem. Since in this case $F \neq \emptyset$, we must also deal with clauses ii and iii of Definition 3.1. Clause ii is easy to prove

using Proposition 5.3; we omit the details. Clause iii relies on the fact that $\ulcorner\alpha + \beta\urcorner \leq \ulcorner\alpha\urcorner * \ulcorner\beta\urcorner$; in fact there is a polynomial, p, such that,

$$(34) \qquad I\Delta_0 \vdash \forall s, t, r(\theta^3_+(s,t,r) \rightarrow r \leq p(s,t)).$$

The bound on r allows us to use Δ_0-induction in obtaining clause iii of Definition 3.1; we omit the details.

Below we present a proof of formula ii of B_+, omitting proofs of the other formulas in B_+. We note that the proofs rely on properties of concatenation presented in Section 2, and use Δ_0-induction. Also Proposition 5.3 is used.

To prove formula ii of B_+, we assume $\theta^3_U(x)$, $\theta^3_U(y)$, and $x \neq y$, and we wish to show that either $x \leq_3 y$ or $y \leq_3 x$. Now if $x \subsetneq_e y$ or $y \subsetneq_e x$ we have the result from the definition of $\theta^3_\leq$. Suppose neither $x \subsetneq_e y$ nor $y \subsetneq_e x$. Then there must be an $i < \mathrm{len}(x), \mathrm{len}(y)$, such that $(x)_i \neq (y)_i$. Using Δ_0-induction, there is a least such i. In which case $(x)_j = (y)_j$ for all $j < i$. Using the definition of θ^3_U we have $U_2((x)_i)$ and $U_2((y)_i)$, so by Lemma 4.4, either $(x)_i <_2 (y)_i$ or $(y)_i <_2 (x)_i$, and hence $\theta^3_U(x,y)$ or $\theta^3_U(y,x)$, but not both, completing the proof of formula ii of B_+.

q.e.d.

6 Generalizing to a Variety of Functions

Mainly, this section provides a generalization of the method that took us from $(U_2, \leq_2)$ to $(U_3, \leq_3, +)$. Let U be a set of ordinals generated from G using the functions in F, and suppose we want to Δ_0-represent (U, R, F, B) in $I\Delta_0$. In this section we give very general sufficient conditions on U, R, F, B, and G for this to take place. We denote by $CL_F(G)$ the smallest set of ordinals that includes the ordinals in G and is closed under the functions in F. Throughout this section we will have particular examples of systems of ordinal notations in mind such as the standard notations for the ordinals less than Γ_0 and notations involving the Feferman-Aczel θ-functions. The examples will be discussed in greater detail below.

The above mentioned sufficient conditions for representing U, when $U = CL_F(G)$, require that U has nice normal form properties in terms of F and G; in particular, we require that certain conditions be satisfied on a one to one mapping, $\alpha \mapsto \tau_\alpha$, from U into $TM_{\overline{F}}(\overline{G})$, the set of terms built up from constant symbols from $\overline{G}$, a set of symbols denoting the elements of G, using function symbols from $\overline{F}$, a set of symbols denoting the elements of F. Let NF, for normal form, denote the range of this map; actually NF is a set of canonical closed term representations for the elements of U. For $\sigma \in TM_{\overline{F}}(\overline{G})$, we denote by $\|\sigma\|$, the ordinal that is the intended interpretation of the term σ; we note that the notation "$\| \ \|$" has been used for a different purpose in previous sections.

Definition 6.1 We will abuse notation by using the same symbol for a relation or function $s \in R \cup F$, on ordinals, as for the relation or function on NF induced by the mapping $\alpha \mapsto \tau_\alpha$; however, we will not use U to denote NF. For each relation s, of arity a, and each X such that $X \subseteq U$ or $X \subseteq NF$, we let,

$$s^X \stackrel{\text{def}}{=} s \cap X^a.$$

We will use NF in place of U^{NF}, and more generally, we will use X in place of U^X. Also, for a function f of arity a,

$$f^X \stackrel{\text{def}}{=} f|X^a,$$

where $f|X^a$ is the result of restricting the domain of the function f, to the set X^a. For a set of relations and functions S, we let,

$$S^X \stackrel{\text{def}}{=} \{s^X : s \in S\}. \tag{35}$$

We will often be careless and drop the superscript X when the restriction is clear from the context, as was done throughout the previous sections in our usage of $\leq$ and $+$.

Our "sufficient conditions" require that there be simultaneous "rudimentary" monotone inductive definitions of the relations in R^{NF} as well as the graphs of the functions in F^{NF}, using the relations in $R^{\overline{G}}$ and the graphs of the functions in $F^{\overline{G}}$; we define this formally below. And we require that the normal form of the ordinal denoted by $\overline{f}\vec{\tau}$, where $\vec{\tau} \in NF$, $f \in F$, and $\overline{f}$ is a symbol denoting f, is obtained by permuting and deleting symbols in $\overline{f}\vec{\tau}$. Finally, we require that the totality of the functions in F and the properties in B are provable from properties, B', of relations and functions on the ordinals in G, together with basic properties of terms and subterm induction on a restricted class of formulas. Such conditions imply that if (G, R^G, F^G, B') can be Δ_0-represented in $I\Delta_0$, then so can (U, R, F, B). Below we give a more formal statement of these conditions.

Typically, the elements of G will be inaccessibles for the functions in F (i.e. for $\gamma \in G$, γ is closed under the functions in F). The idea is that we have a way to code G, and we code the ordinals coming in between these inaccessibles using codes of terms involving the symbols in $\overline{F}$ together with the elements of $\overline{G}$. Then when computing the value of applying one of the functions in F to an ordinal, the representation of the result will not involve any new elements of $\overline{G}$, and so we have control over the size of the code of the result.

We now define the notion "rudimentary" used above. Suppose $L_{TM(A)}$ is a formal language for the theory of closed terms built up from symbols in A; i.e. A is a set of function and constant symbols and $L_{TM(A)}$ has constant symbols, $\overline{A}$, intended to denote the elements of A, and $L_{TM(A)}$ has a function symbol, $*$, for concatenation, and a relation symbol, TM, denoting the set of sequences that denote well formed closed terms. The variables of $L_{TM(A)}$ are intended to range over sequences of elements of $\overline{A}$. For the next three definitions we suppose the set A and a set Z of relation and function symbols, are fixed. Let L_{TM} denote the language obtained by adding the symbols in Z to the language L_{TM}, i.e.,

$$L_{TM} \stackrel{\text{def}}{=} L_{TM(A)}(Z). \tag{36}$$

Definition 6.2 A formula φ is L_{TM}-**rudimentary** if φ is built up from the atomic formulas of L_{TM} of the forms

a. $s\vec{x}$, where $s \in Z$ and $\vec{x}$ is a sequence of *variables* of the language $L_{TM(A)}$, and

b. $\sigma = \tau$, where σ and τ are terms of the language L_{TM},

using

c. Boolean connectives, and

d. subterm quantification of the form $(\exists x \subseteq_{\text{st}} y)$ and $(\forall x \subseteq_{\text{st}} y)$, where x and y are variables in the language. (We note that subterm quantification can be defined in the language using TM and $*$.)

Let L_{TM}-Rud denote the set of all L_{TM}-rudimentary formulas. The notion of rudimentary formula is similar to notions used by Smullyan in [Smu61].

From conditions a and d in Definition 6.2, it is clear that if $\varphi(\vec{x})$ is a L_{TM}-Rud formula and $\varphi(\vec{\tau})$ is a sentence (no free variables), then the truth value of $\varphi(\vec{\tau})$ is completely determined by the truth values of the relations $s\vec{\sigma}$, where $s \in Z$ and $\vec{\sigma}$ ranges only over the subterms of the terms $\vec{\tau}$. I.e., if $St(\vec{\tau})$ denotes the set of all subterms of the terms denoted by $\vec{\tau}$, then for any set X such that $St(\vec{\tau}) \subseteq X$,

$$\varphi(\vec{\tau}, Z) \Longleftrightarrow \varphi(\vec{\tau}, Z^X), \tag{37}$$

where Z^X is given by (35). (37) will be applied later; we omit the straightforward proof by induction on the construction of φ.

Definition 6.3 Let T_{TM} be the theory in the language L_{TM} that includes axioms expressing the following.

i. "Every term is either a constant or of the form $\overline{f} * \tau_1 * \cdots * \tau_k$, where f is a k-ary function symbol for some $k > 0$, and $\tau_1, \ldots, \tau_k$ are terms," and

ii. "concatenation is associative".

Note: in all of our applications there are only finitely many function symbols, so it is easy to express i in the language L_{TM}. Otherwise, it would be necessary to do something such as introduce a predicate symbol denoting the set of k-ary function symbols in A for $k > 0$, and here we would need to impose a bound on k.

Definition 6.4 For $\varphi \in L_{TM}$, let $STI(\varphi)$, for **subterm induction on** φ, denote the formula,

(38) $$\forall\tau((\forall\sigma_{\mathrm{st}}^{\subset} \neq \tau)\varphi(\sigma) \rightarrow \varphi(\tau)) \rightarrow \forall\tau\varphi(\tau).$$

For $\mathcal{C}$ a set of L_{TM}-formulas, let,

(39) $$STI(\mathcal{C}) \stackrel{\text{def}}{=} \{STI(\varphi) : \varphi \in \mathcal{C}\}.$$

Definition 6.5 Suppose for Z in (36), $Z = S \cup S'$, where $S = \{s_0, \ldots, s_n\}$ and $S' = \{s'_0, \ldots, s'_{n'}\}$ are sets of relations and functions on terms. We say S is **definable from S' by simultaneous rudimentary monotone inductive definitions** if for each $s_i \in S$ there is an L_{TM}-Rud formula $\chi_{s_i}(\vec{x}, \vec{X}, \vec{X'})$, involving the relation and function symbols $\vec{X} = X_0, \ldots, X_n$ and $\vec{X'} = X'_0, \ldots, X'_{n'}$, of the appropriate arities, that is monotone in the sense that for all $\vec{X}$ and $\vec{Y}$, if $X_i \subseteq Y_i$ for $i = 0, \ldots, n$, then

(40) $$\{(\vec{z}) : \chi_{s_i}(\vec{z}, \vec{X}, \vec{X'})\} \subseteq \{(\vec{z}) : \chi_{s_i}(\vec{z}, \vec{Y}, \vec{X'})\}.$$

And χ_{s_i} has the property that by defining,

(41) $$(s_i)_0 \stackrel{\text{def}}{=} \emptyset, \text{ and}$$

(42) $(s_i)_{m+1} \stackrel{\text{def}}{=} \{(\vec{x}) : \chi_{s_i}(\vec{x}, (s_0)_m, \ldots, (s_n)_m, s'_0, \ldots, s'_{n'})\} \cup (s_i)_m,$

we get, for $i = 0, \ldots, n$,

(43) $$s_i = \bigcup_{m \in \omega} (s_i)_m.$$

If S and S' contain only relation symbols, we can replace the sets S and S' by single relations s and s', by letting,

(44) $$s \stackrel{\text{def}}{=} \{(\bar{i}, \vec{x}, \underbrace{\bar{0}, \ldots, \bar{0}}_{a-a_i\ \bar{0}'s}) : 0 \leq i \leq n \wedge s_i\vec{x}\}, \text{ and}$$

(45) $$s' \stackrel{\text{def}}{=} \{(\bar{i}, \vec{x}, \underbrace{\bar{0}, \ldots, \bar{0}}_{a-a'_i\ \bar{0}'s}) : 0 \leq i \leq n' \wedge s'_i\vec{x}\},$$

where a_i and a'_i are the arities of s_i and s'_i respectively, and a is the maximum of all the a_i's and a'_i's. We note that new symbols $\bar{0}$, $\bar{1}$, ..., $\bar{a}$ are added to A for these definitions. Along the lines of the above, from the χ_{s_i}'s, we can obtain a single formula, χ_s, such that by letting,

(46) $$(s)_0 \stackrel{\text{def}}{=} \emptyset, \text{ and}$$

(47) $$(s)_{m+1} \stackrel{\text{def}}{=} \{(\vec{y}) : \chi_s(\vec{y}, (s)_m, s')\} \cup (s)_m,$$

we have,

(48) $$s = \bigcup_{m \in \omega} (s)_m.$$

χ_s is defined as follows. Let $\tilde{\chi}_{r_i}$ be obtained from χ_{r_i} by replacing each atomic formula of the form $X_j(\vec{y})$ and $X'_j(\vec{y})$ by

$$X(\bar{j}, \vec{y}, \underbrace{\bar{0}, \ldots, \bar{0}}_{a-a_j\ \bar{0}'s}) \text{ and } X'(\bar{j}, \vec{y}, \underbrace{\bar{0}, \ldots, \bar{0}}_{a-a'_j\ \bar{0}'s}),$$

respectively. And let,

(49) $$\chi_s(y_o, \ldots, y_a) \stackrel{\text{def}}{\Longleftrightarrow} \bigvee_{0 \leq i \leq n} [y_0 = \bar{i} \wedge \tilde{\chi}_{r_i}(y_1, \ldots, y_{a_i}) \wedge \bigwedge_{a_i < j \leq a} (y_j = \bar{0})].$$

Although it will not pose any problem in our forthcoming applications, it is worthwhile to note that χ_s not a rudimentary formula since atomic formulas of the form $X\vec{z}$ and $X'\vec{z}$ may occur where $\vec{z}$ contains non-variable symbols.

Using arguments from the theory of monotone inductive definitions and (37), we obtain:

Proposition 6.6 *s, given by (48),is the least fixed point of the operator*

$$X \mapsto X \cup \{\vec{y} : \chi_s(\vec{y}, X, s')\}, \tag{50}$$

i.e.,

$$\forall\vec{y}(\chi_s(\vec{y}, s, s') \rightarrow s(\vec{y})), \tag{51}$$

and if (51) holds for some set Z replacing s then $s \subseteq Z$.

We omit the details of the proof of the above proposition. The key idea is to use (37) to show that if $\chi_s(\vec{\sigma}, s, s')$ holds, then there is an $n \in \omega$ such that $s^{St(\vec{\sigma})} \subseteq (s)_n$, so $\chi_s(\vec{\sigma}, (s)_n, s')$ holds.

Definition 6.7 Suppose U and G are countable sets of ordinals and R and F are finite sets of relations and functions on U such that $F_{\mathrm{rel}} \subseteq R$, where F_{rel} is the set of graphs of the functions in F; further suppose, $U \in R$, $S \subseteq R$, $U = CL_F(G)$, and for Z in (36) we have $Z = \overline{R}$, a set of symbols intended to denote the relations in R. We say that (U, R, F, B) has **nice normal form properties with respect to** (G, S, B'), if for some $NF \subseteq TM_{\overline{F}}(\overline{G})$ and $\alpha \mapsto \tau_\alpha$, a mapping from U into $TM_{\overline{F}}(\overline{G})$, the following hold:

1. $NF = \{\tau_\alpha : \alpha \in U\}$.

2. For each $\alpha \in U$, $\|\tau_\alpha\| = \alpha$.

3. For each $\alpha \in G$ we have $\tau_\alpha \in \overline{G}$; we write $\tau_\alpha = \overline{\alpha}$.

4. R^{NF} is definable from $S^{\overline{G}}$ by simultaneous rudimentary monotone inductive definitions. For each $r \in R$, let χ_r denote the rudimentary formula defining r.

5. For $\sigma = \overline{f}\vec{\tau}$, where $\vec{\tau} \in NF$, if $\rho = \tau_{\|\sigma\|}$ then ρ is obtained from $\overline{f}\vec{\tau}$ by permuting and deleting symbols.

6. $T_{TM} + STI(\{\exists\tau\chi_f(\vec{\sigma},\tau) : f \in F\} \cup L_{TM}\text{-Rud}) + D_R$
$\vdash (\forall\vec{x}\exists!y\chi_f(\vec{x},y))^{NF}$,

   where,

$$D_R \stackrel{\text{def}}{=} \{(\chi_r(\vec{x}) \leftrightarrow r\vec{x}) : r \in R^{NF}\},$$

   and the superscript NF on the formula $\forall\vec{x}\exists!y\chi_f(\vec{x},y)$ indicates that the quantifiers in the formula are restricted to NF. (Recall: NF denotes U^{NF}.)

7. $$T_{TM} + STI(L_{TM}\text{-Rud}) + (B')^G + D \vdash B^{\overline{U}},$$
   where $D = D_R \cup \{f_{\text{rel}}(\vec{x},y) \leftrightarrow f(\vec{x}) = y : f \in F\}$ and f_{rel} is the relation in R that is the graph of f.

Note that clauses 1 and 2 imply $\alpha \mapsto \tau_\alpha$ is a bijection from U to NF. Also clauses 3 and 5 ensure that for each $g \in G$, $g \notin CL_F(G - \{g\})$. To see this suppose $||\sigma|| = g \in G$, then by clause 3, $\tau_{||\sigma||} = \overline{g}$. Applying clause 5 recursively we get that $\overline{g}$ is a subterm of σ, and this implies $||\sigma|| \notin CL_F(G - \{g\})$.

Now we state the main theorem.

Theorem 6.8 *Suppose U, R, F, G, and S have the properties listed in the first sentence of Definition 6.7. If (U,R,F,B) has nice normal form properties with respect to (G,S,B'), and $(G,S^G,\emptyset,B')$ can be Δ_0-represented in $I\Delta_0$ then (U,R,F,B) can be Δ_0-represented in $I\Delta_0$.*

Before proceeding to the proof of Theorem 6.8 we show how the theorem implies that many examples of ordinal functions on recursive sets of ordinals can be Δ_0-represented in $I\Delta_0$.

Example 6.9 Let $U = \omega^\alpha$, $G = \{\omega^\beta : \beta < \alpha\}$, and $F = \{+\}$. This is just the example of the previous section; however, in the previous section, $+$ was essentially treated as an arbitrary arity function. Here, to apply our general theorem, we need to be more careful, and we will treat $+$ as a binary function with the convention of "association of parentheses to the left". In Polish notation this amounts to pushing all the plus signs to the left; for example, given $\alpha > \beta_1 \geq \beta_2 \geq \beta_3$, the term $+ + \omega^{\beta_1}\,\omega^{\beta_2}\,\omega^{\beta_3}$ is in normal form, while the term $+\,\omega^{\beta_1} + \omega^{\beta_2}\,\omega^{\beta_3}$ is not. As in Section 5 we wish to represent $(\omega^\alpha, \{\omega^\alpha, \leq\}, \{+\}, B_+)$ in $(\Delta_0, I\Delta_0)$, where B_+ contains basic properties of ordinal addition and less-than. Below we give explicitly the rudimentary formulas needed to show that $(\omega^\alpha, \{\omega^\alpha, \leq, +\}, \{+\}, B_+)$ has nice normal form properties with respect to $(G, \{G, \leq\}, B_\leq)$, where $G = \{\omega^\beta : \beta < \alpha\}$, and $B_\leq$ contains the single formula expressing that $\leq$ is a linear order. These formulas are just versions of the inductive definitions used to define the corresponding relations.

The meaning of "simultaneous inductive definitions" is that the formulas χ_{NF}, $\chi_\leq$, and χ_+ will work together to define NF, $\leq$, and $+$. For example χ_{NF} uses $+$ in its definition—the idea is that χ_+ will be defined in such a way that the range of $+$ is always in NF. Below, the third disjunct of χ_{NF} ensures that NF will include the range of $+$. To see how these formulas work, following their definitions, we give explicitly $(NF)_m$, $(\leq^{NF})_m$, and $(+^{NF})_m$, according to (41) and (42), for $m = 0$, 1, and 2. We use $+$ to denote the graph of the function $+^{NF}$, thus $+(\sigma, \tau, \rho)$ has intended interpretation $\|\sigma\| + \|\tau\| = \|\rho\|$; $\overline{+}$ is a formal symbol denoting $+$. $\chi_{NF}(\tau)$ defines $NF(\tau)$, $\chi_\leq(\sigma, \tau)$ defines $\sigma \leq \tau$, and $\chi_+(\sigma, \tau, \rho)$ defines $+(\sigma, \tau, \rho)$.

$$
\begin{aligned}
(52)\ &\chi_{NF}(\tau) \overset{\text{def}}{\Longleftrightarrow} \\
&\quad \tau = \overline{0} \vee G(\tau) \vee \\
&\quad (\exists \tau_1, \tau_2 \overset{\subset}{_\text{st}} \tau)(\tau = \overline{+} * \tau_1 * \tau_2 \wedge NF(\tau_1) \wedge NF(\tau_2) \wedge +(\tau_1, \tau_2, \tau))
\end{aligned}
$$

$$
\begin{aligned}
(53)\ \ &\chi_\leq(\sigma, \tau) \overset{\text{def}}{\Longleftrightarrow} NF(\sigma) \wedge NF(\tau) \wedge \\
&\quad [\sigma = \overline{0} \vee \sigma \leq^G \tau \vee \\
&\quad (\exists \sigma_1, \sigma_2 \overset{\subset}{_\text{st}} \sigma)(\sigma = \overline{+} * \sigma_1 * \sigma_2 \wedge G(\tau) \wedge \sigma_1 < \tau) \vee \\
&\quad (\exists \tau_1, \tau_2 \overset{\subset}{_\text{st}} \tau)(\tau = \overline{+} * \tau_1 * \tau_2 \wedge \sigma \leq \tau_1) \vee \\
&\quad (\exists \sigma_1, \sigma_2 \overset{\subset}{_\text{st}} \sigma)(\exists \tau_1, \tau_2 \overset{\subset}{_\text{st}} \tau)(\sigma = \overline{+} * \sigma_1 * \sigma_2 \wedge \tau = \overline{+} * \tau_1 * \tau_2 \wedge \\
&\quad\quad ((\sigma_1 = \tau_1 \wedge \sigma_2 \leq \tau_2) \vee (\sigma_1 < \tau \wedge \sigma_2 < \tau_2)))]
\end{aligned}
$$

$$
\begin{aligned}
(54)\ \ &\chi_+(\sigma, \tau, \rho) \overset{\text{def}}{\Longleftrightarrow} NF(\sigma) \wedge NF(\tau) \wedge \\
&\quad [(\sigma = \overline{0} \wedge \rho = \tau) \vee \\
&\quad (\tau = \overline{0} \wedge \rho = \sigma) \vee \\
&\quad (G(\sigma) \wedge G(\tau) \wedge \\
&\quad\quad [(\tau \leq^G \sigma \wedge \rho = \overline{+} * \sigma * \tau) \vee (\sigma <^G \tau \wedge \rho = \tau)]) \vee \\
&\quad (\exists \sigma_1, \sigma_2 \overset{\subset}{_\text{st}} \sigma)(\sigma = \overline{+} * \sigma_1 * \sigma_2 \wedge G(\tau) \wedge
\end{aligned}
$$

$$[(\tau \leq^G \sigma_2 \wedge \rho = \mp * \sigma * \tau) \vee (\sigma_2 <^G \tau \wedge +(\sigma_1, \tau, \rho))]) \vee$$
$$(\exists \tau_1, \tau_2 \overset{\subset}{_{\mathrm{st}}} \tau)(\tau = \mp * \tau_1 * \tau_2 \wedge$$
$$(\exists \rho_1 \overset{\subset}{_{\mathrm{st}}} \rho)(+(\sigma, \tau_1, \rho_1) \wedge \rho = \mp * \rho_1 * \tau_2))]$$

Applying (41) and (42), we have $(NF)_0 = (\leq^{NF})_0 = (+^{NF})_0 = \emptyset$. $(NF)_1 = \overline{G} \cup \{\overline{0}\}$, but since $(NF)_0 = \emptyset$, we have $(\leq^{NF})_1 = (+^{NF})_1 = \emptyset$. Then since $(+)_1 = \emptyset$, we do not get anything new in $(NF)_2$, i.e. $(NF)_2 = (NF)_1$, but

$$\begin{aligned} (\leq^{NF})_2 &= \leq^{\overline{G}} \cup \{\langle \overline{0}, x\rangle : x \in \overline{G}\}, \text{ and} \\ (+^{NF})_2 &= \{\langle \overline{0}, x, x\rangle : x \in \overline{G}\} \cup \{\langle x, \overline{0}, x\rangle : x \in \overline{G}\} \\ &\quad \cup \{\langle x, y, \mp * x * y\rangle : y \leq^G x\}. \end{aligned}$$

It is straightforward to show that (43) gives,

$$(55)\, NF = \{\overline{0}\} \cup \{\underbrace{\mp \cdots \mp}_{n} \overline{\omega^{\alpha_0}} \cdots \overline{\omega^{\alpha_n}} : n \in \omega \text{ and } \alpha > \alpha_0 \geq \cdots \geq \alpha_n\}.$$

Similarly we could give explicitly $\leq^{NF}$ and $+^{NF}$ as given by (43). Using this it is routine to check clauses 1–5 of Definition 6.7. Clause 6 is obtained by a straightforward subterm induction on τ for the formula $\exists \rho \chi_+(\sigma, \tau, \rho)$. Clause 7 is routine and we only sketch part of the case where the formula in B^+ is the assertion that $\leq^{NF}$ is transitive. We want to show for all σ, τ, and ρ,

$$(56) \qquad (\chi_\leq(\sigma, \tau) \wedge \chi_\leq(\tau, \rho)) \to \chi_\leq(\sigma, \rho).$$

We use subterm induction on σ, τ, and ρ, respectively, for the formula given by (56). There are many cases to consider according to the disjuncts in $\chi_\leq(\sigma,\tau)$ and $\chi_\leq(\tau, \rho)$. For a simple case suppose $G(\sigma)$, $G(\tau)$, $G(\rho)$, $\sigma \leq^G \tau$, and $\tau \leq^G \rho$. Here (56) is immediate using the assumption that $\leq^G$ is transitive (since "$\leq$ is a linear order" is in $B_\leq$). For a more complicated case suppose $\chi_\leq(\sigma, \tau)$ and $\chi_\leq(\tau, \rho)$ both hold according to the last two lines in (54); say $\sigma = \mp\sigma_1\sigma_2$, $\tau = \mp\tau_1\tau_2$, $\rho = \mp\rho_1\rho_2$, $\sigma_1 = \tau_1 \wedge \sigma_2 \leq \tau_2$ and $\tau_1 < \rho \wedge \tau_2 \leq \rho_2$. Then by the induction hypothesis $\sigma_1 < \rho$ and $\sigma_2 < \rho_2$, hence $\chi_\leq(\sigma, \rho)$. We omit the remaining cases.

Now to apply Theorem 6.8, it is enough to show that $(G, \{G, \leq, +\}, \emptyset, B_\leq)$ can be represented in $(\Delta_0, I\Delta_0)$, where $+$ is treated as a relation. Of course G is not closed under $+$, but this is not required; we do require the representability of $+^{NF}$ restricted to G. This is easily accomplished. That $\leq^G$ can be represented follows from Lemma 4.2. Thus Theorem 6.8 can be applied to yield the representability of $(\omega^\alpha, \{\omega^\alpha, \leq\}, \{+\}, B_+)$ in $(\Delta_0, I\Delta_0)$.

Example 6.10 Here we consider the situation where (U, R, F, B) has nice normal form properties with respect to $(G, \{G, \leq\}, B_{\leq})$, where $B_{\leq}$ consists of the single formula expressing "$\leq$ is a linear order", and we wish to Δ_0-represent (U, R, F', B') in $I\Delta_0$ where F' is a certain extension of F, $U = CL_{F'}(U)$, and B' contains B together with properties of the functions in $F' - F$. In particular, we consider a situation similar to situations considered in [Fef68]. Suppose the functions in $F = \{f_1, \ldots, f_m\}$ are increasing functions defined on Ω, the countable ordinals. Let $g : \Omega \to \Omega$ be such that $g(\alpha)$ is the αth closure ordinal of $f_1, \ldots, f_m$, i.e.

(57) $$g(\alpha) = \mu\beta > 0[(\forall\gamma < \alpha)(g(\gamma) < \beta) \wedge (CL_F(\beta) = \beta)],$$

(where μ is the least ordinal operator) and let $f : \Omega^2 \to \Omega$ be obtained by transfinite recursion as follows,

(58) $$f(0, \beta) = g(\beta), \text{ and if } \alpha > 0 \text{ then}$$
(59) $$f(\alpha, \beta) = \mu\gamma[(\forall\beta' < \beta)(f(\alpha, \beta') < \gamma) \wedge (\forall\alpha' < \alpha)(f(\alpha', \gamma) = \gamma)].$$

Let $F' \stackrel{\text{def}}{=} F \cup \{f\}$. So if $F = \{+\}$ then g is the function $\alpha \mapsto \omega^\alpha$, f is the Veblen φ-function, and $F' = \{+, \varphi\}$. We have the following general theorem.

Theorem 6.11 *Suppose F' is defined from $F = \{f_1, \ldots, f_m\}$ as above, $(U$, R, F, $B)$ has nice normal form properties with respect to $(G, \{G, \leq\}, B_{\leq})$, where B includes all formulas of the form $\vec{\alpha} \leq f_i(\vec{\alpha})$ for $i = 1, \ldots, m$, $U = CL_F(G) = CL_{F'}(G)$, and $G = \{\gamma \in U : \gamma = CL_F(\gamma)\}$. Then (U, R, F', B') has nice normal form properties with respect to $(G', \{G', \leq\}, B_{\leq})$, where $G' \stackrel{\text{def}}{=} \{\gamma \in G : \gamma = CL_{F'}(\gamma)\}$, and B' includes B as well as the following formulas,*

1. $f(\alpha_1, \beta_1) \leq f(\alpha_2, \beta_2) \leftrightarrow [\ (\alpha_1 < \alpha_2 \wedge \beta_1 \leq f(\alpha_2, \beta_2)) \vee$
   $(\alpha_1 = \alpha_2 \wedge \beta_1 \leq \beta_2) \vee$
   $(\alpha_2 < \alpha_1 \wedge f(\alpha_1, \beta_1) \leq \beta_2)]$.

2. *For* $i = 1, \ldots, m$, $\vec{\beta} < f(\alpha_1, \alpha_2) \leftrightarrow f_i(\vec{\beta}) < f(\alpha_1, \alpha_2)$.

3. $\beta < \gamma \to f(\alpha, \beta) < f(\alpha, \gamma)$.

4. $\beta \leq f(\alpha, \beta)$.

5. $\lambda \in \mathrm{LIM} \to f(\alpha, \lambda) = \sup_{\beta < \lambda} f(\alpha, \beta)$.

6. $\alpha_1 < \alpha_2 \rightarrow f(\alpha_1, f(\alpha_2, \gamma)) = f(\alpha_2, \gamma)$.

Before sketching the proof of Theorem 6.11 we note the following corollary, which also uses Example 6.9 and Theorem 6.8; we will not prove the corollary.

Corollary 6.12 *$(\gamma,\ \{\gamma, \leq\},\ \{+, \varphi\},\ B_\varphi)$ can be Δ_0-represented in $I\Delta_0$ where γ is any recursive gamma number (i.e. $0 < \gamma < \omega_1^{\mathrm{CK}}$ and γ is closed under φ) and $B_\varphi = B_+ \cup \{$ properties 1–6 from the previous theorem with f replaced by $\varphi\}$, where B_+ is from Example 6.9.*

Proof of Theorem 6.11:

This will be only a sketch. Mainly we modify the formulas χ_s, for $s \in R$, given by the hypothesis of the theorem (i.e., the hypothesis that (U, R, F, B) has nice normal form properties with respect to $(G, \{G, \leq\}, B_{\leq})$) using the usual inductive clauses involving Veblen type functions. To $\chi_{NF}(\tau)$ we add the clause,

$$(60) \vee (\exists \tau_1, \tau_2 \overset{\subset}{_{\mathrm{st}}} \tau)(NF(\tau_1) \wedge NF(\tau_2) \wedge \tau = \overline{f} * \tau_1 * \tau_2 \wedge f_{\mathrm{rel}}(\tau_1, \tau_2, \tau)),$$

where f_{rel} in this formula denotes the graph of the function f^{NF} and f is the function given by (58) and (59). To $\chi_{\leq}(\sigma, \tau)$ we add disjuncts corresponding to clauses 1 and 2 in the statement of the theorem, together with a disjunct expressing that for any $h \in F'$ and $\tau \in G'$, $\vec{\sigma} < \tau \leftrightarrow h(\vec{\sigma}) < \tau$. Then for each χ_r, we replace each occurrence of $G(\sigma)$ by $G'(\sigma) \vee [NF(\sigma) \wedge (\exists \sigma_1, \sigma_2 \overset{\subset}{_{\mathrm{st}}} \sigma)(\sigma = \overline{f} * \sigma_1 * \sigma_2)]$, and we replace each occurrence of $\sigma \leq^G \tau$ by a formula expressing $\sigma \leq \tau$ and each of σ and τ is either in G' or is of the form $\overline{f} * \rho_1 * \rho_2$. Finally we let the formula $\chi_f(\sigma, \tau, \rho)$, expressing $f(\sigma, \tau) = \rho$, be given by,

$$\begin{aligned}(61) \quad \chi_f(\sigma, \tau, \rho) &\overset{\mathrm{def}}{\Longleftrightarrow} NF(\sigma) \wedge NF(\tau) \wedge \\ &[(G'(\tau) \wedge \sigma < \tau \wedge \rho = \tau) \vee \\ &(\exists \tau_1, \tau_2 \overset{\subset}{_{\mathrm{st}}} \tau)(\tau = \overline{f} * \tau_1 * \tau_2 \wedge \sigma < \tau_1 \wedge \rho = \tau) \vee \\ &[(\neg G'(\tau) \vee \tau \leq \sigma) \wedge (\forall \tau_1, \tau_2 \overset{\subset}{_{\mathrm{st}}} \tau)((\tau \neq \overline{f} * \tau_1 * \tau_2 \vee \tau_1 \leq \sigma) \\ &\quad \wedge \rho = \overline{f} * \sigma * \tau)]]\end{aligned}$$

The elements of $R \cup F'_{\mathrm{rel}}$ are easily seen to be the sets inductively defined by the formulas χ_s for $s \in R \cup F'_{\mathrm{rel}}$. So clauses 1–4 of the definition of nice

normal form are clear. Clause 5 easily follows from the hypotheses of the theorem together with the definition of χ_f given in (61). Clause 6 is easy using (61). Clause 7 relies on the fact that $B_{\leq}^{G}$ can be obtained from $B_{\leq}^{G'}$ using T_{TM} and $STI(L_{TM}\text{-Rud})$, noting how $\chi_{\leq}$ was modified above.
q.e.d.

Example 6.13 Clause 5 of the definition of nice normal form was designed to handle notation systems based on versions of the Feferman-Aczel θ-functions (cf. [Sch77], [BS88], [Fef87], [Jag84], [Poh87], and [Rat90]). The author conjectures that the notion of nice normal form is general enough to include such examples.

We now return to the
Proof of 6.8:

Of course we will use a representation of NF to represent U. That is, once we have a Gödel numbering of arbitrary terms, U will be represented by $\ulcorner NF \urcorner$. Likewise we will have Gödel numberings of all elements of $R \cup F$.

Our coding of terms is based on a reverse polish notation. We are assuming that F and R are finite and G is countable, say $F = \{f_i : i = 0, \ldots, m-1\}$ and $G = \{g_i : i \in \omega\}$. Also assume that, for some $a \in \omega$, the arities of the functions in F are all strictly less than a and the arities of the relations in R are less than or equal to a. (There are modifications of our methods that will allow us to consider the case where F is countable and the arities are not bounded but we will not do this here.)

In order to provide the bounds for certain bounded formulas, the Gödel numbering, $\ulcorner\ \urcorner$, for a term τ, will be obtained by padding a preliminary Gödel numbering $\ulcorner\ \urcorner'$. $\ulcorner\ \urcorner'$ is defined on $\overline{F}$ and $\overline{G}$ as follows; for $f_i \in F$ and $g_j \in G$, let,

$$\overline{\ulcorner \overline{f_i} \urcorner'}^3 = 3 * \bar{i}^2 \text{ and } \overline{\ulcorner \overline{g_j} \urcorner'}^3 = 3 * \overline{m+j}^2. \tag{62}$$

Then define $\ulcorner\ \urcorner'$ inductively by,

$$\ulcorner \overline{f} \tau_1, \ldots, \tau_k \urcorner' \stackrel{\text{def}}{=} \ulcorner \overline{f} \urcorner' * \ulcorner \tau_1 \urcorner' * \cdots * \ulcorner \tau_k \urcorner'. \tag{63}$$

$\ulcorner\ \urcorner$ is given by,

$$\ulcorner \tau \urcorner = \underbrace{1 *_3 \cdots *_3 1}_{l \text{ ones}} *_3 \ulcorner \tau \urcorner', \tag{64}$$

where l is least such that,

$$|\ulcorner \tau \urcorner|_2 \geq (\text{number of symbols in } \tau)^{2a}. \tag{65}$$

The proof of the following claim, which we do not present, is straightforward using facts mentioned in Section 2.

Claim 6.14

i. There is a Δ_0*-definable,* $I\Delta_0$*-provably total function that can extract* $\ulcorner\tau\urcorner'$ *from* $\ulcorner\tau\urcorner$*.*

ii. There are Δ_0*-definitions of the sets* $\ulcorner TM_{\overline{F}}(\overline{G})\urcorner$ *and* $\ulcorner TM_{\overline{F}}(\overline{G})\urcorner'$*.*

We call $\overline{f}$ the **main symbol** of the term $\overline{f}\vec{\tau}$. The subterms of a term τ are in one-to-one correspondence with the function and constant symbols appearing in τ. Hence we associate to each occurrence of a 3 in $\overline{\ulcorner\tau\urcorner}^3$ a subterm of τ. This works also for sequences of terms $\vec{\tau}$. In particular, we will refer to the ith subterm of the sequence of terms $\tau_1, \ldots, \tau_k$, by which we will mean the term whose code begins with the ith 3 in the sequence,

$$\overline{\ulcorner\tau_1\urcorner}^3 * \cdots * \overline{\ulcorner\tau_k\urcorner}^3.$$

The following claim introduces functions that allow us to extract the ith subterm of a sequence, $\vec{\tau}$, of terms, as well as the main symbol of that subterm.

Claim 6.15 *There are* Δ_0*-definable and* $I\Delta_0$*-provably total functions* st, st', $\#st$, *and* ms *such that* $st(\langle\vec{x}\rangle, i)$ *is the* $\ulcorner\ \urcorner$*-code of the* i^{th} *subterm of the sequence of terms coded by* $\vec{x}$, $st'(\langle\vec{x}\rangle, i)$ *is the* $\ulcorner\ \urcorner'$*-code of this term,* $\#st(\langle\vec{x}\rangle, i)$ *is the number of subterms, counting multiplicities, of this term, and* $ms(\langle\vec{x}\rangle, i)$ *is the code of the main symbol of this term. Also,* $\#st(\langle\vec{x}\rangle)$ *denotes the number of subterms of the terms coded by* $\vec{x}$ *(counting multiplicities).*

The proof, omitted here, is straightforward using facts mentioned in Section 2. Theorem 6.8 will rely on the following lemma.

Lemma 6.16 *Suppose* U, R, F, G, *and* S *have the properties listed in the first sentence of Definition 6.7. If* (U, R, F, B) *has nice normal form properties with respect to* (G, S, B') *then there are* $\Delta_0(\ulcorner S^{\overline{G}}\urcorner)$*-representations of* $\ulcorner R\urcorner$*.*

Proof:

Assume $R = \{r_i : i = 0, \ldots, n-1\}$. We combine the r_i's and the $r_i^{\overline{G}}$'s that occur in $S^{\overline{G}}$, into single relations, s and s', respectively, as in (44) and (45). Further, we suppose we have a single formula χ_s, given by (49), that provides a monotone inductive definition of s from s'. So we assume s and s' both have arity $a+1$. For this to work we adhere to the following convention. If $\vec{x} = \ulcorner\overline{k}\urcorner, x_1, \ldots, x_{a_k}, \ulcorner\overline{0}\urcorner, \ldots, \ulcorner\overline{0}\urcorner$ ($(a - a_k)$ 0's), then

$$(66) \qquad st(\langle\vec{x}\rangle, i) \text{ denotes } st(\langle x_1, \ldots, x_{a_k}\rangle, i),$$

$$(67) \qquad st'(\langle\vec{x}\rangle, i) \text{ denotes } st'(\langle x_1, \ldots, x_{a_k}\rangle, i), \text{ and}$$

$$(68) \qquad \#st(\langle\vec{x}\rangle) \text{ denotes } \#st(\langle x_1, \ldots, x_{a_k}\rangle, i).$$

I.e., our functions ignore the k and the 0's wherever appropriate. Also we assume $\ulcorner\ \urcorner$ is adjusted so that $\ulcorner\overline{m}\urcorner = m$ for $m = 0, \ldots, a$.

Below we describe a $\Delta_0(\ulcorner s'\urcorner)$-formula, ψ_s, defining $\ulcorner s\urcorner$; i.e. we obtain,

$$(69) \qquad \ulcorner s\urcorner = \{\vec{x} : \psi_s(\vec{x})\}.$$

In order to prevent obscuring the main ideas in technical details, our description of ψ_s will be at times informal. ψ_s will be of the form,

$$(70) \qquad (\exists v \le p(\vec{x}))(|v|_2 = l(\vec{x}) \wedge \chi_s'(\vec{x}, v) \wedge true(\vec{x}, v) \wedge (\forall w < v)(|w|_2 = l(\vec{x}) \rightarrow \neg\chi_s'(\vec{x}, w)),$$

where $l(\vec{x}) \stackrel{\text{def}}{=} n^2 \cdot 3^{a+2} \cdot (\#st(\langle\vec{x}\rangle))^{2a}$. For $\vec{x} = k, x_1, \ldots, x_{a_k}, 0, \ldots, 0$ ($a - a_k$ 0's), v is interpreted as a sequence of 1's and 2's (recall "2 for true") coding the truth value of each atomic formula of the form $\ulcorner r_i\urcorner y_1 \ldots y_{a_i}$, where each y_j codes a subterm of a term coded by one of the x_m's (for $m = 1, \ldots, a_k$). This will include all relations that are used to show that $\ulcorner r_k\urcorner x_1, \ldots, x_{a_k}$ holds. v also serves as a finite approximation of the set $\ulcorner s\urcorner$. The clause $|v|_2 = l(\vec{x})$ ensures that v is large enough to do this; we return to this shortly. χ_s' checks that v is correct according to χ_s (i.e. according to the χ_{r_i}'s), and according to s' (i.e. according to $S^{\overline{G}}$). Essentially this amounts to checking that v is a fixed point of a translation of χ_s. *true* checks that v asserts "$\ulcorner r_k\urcorner x_1, \ldots, x_{a_k}$ is true". The conjunct $(\forall w < v)(|w|_2 = l(\vec{x}) \rightarrow \neg\chi_s'(\vec{x}, w))$ expresses that v is least. The very last part of the proof of the current lemma will show that there is a polynomial, p, that provides a bound on the size of v.

$\chi_s'(\vec{x}, v)$ is the formula,

$$(71) \qquad (\forall k < n)(\forall \vec{i} < \#st(\langle\vec{x}\rangle))(\chi_s^\star(k, \vec{i}, \vec{x}, v) \rightarrow (v)_{2,\langle k,\vec{i}\rangle} = 2),$$

where $\vec{i} = i_1, \ldots, i_a$. $\chi_s^\star$ is obtained by translating χ_s; we describe this shortly. The idea behind χ_s' is that $(v)_{2,\langle j,i_1,\ldots,i_{a_j},0,\ldots,0\rangle} = 2$ or 1 according to whether or not $r_j\tau_1 \ldots \tau_{a_j}$ holds where $\ulcorner\tau_l\urcorner = st(\langle\vec{x}\rangle, i_l)$. Below we will discuss further how χ_s' is just the assertion that v is a kind of fixed point, but it is worthwhile now to compare (71) with (51).

Before describing the formula $\chi^\star$, we show how the requirement $|v|_2 = l(\vec{x})$ ensures that v is large enough to code everything it is supposed to. It is enough to show,

(72) $$\max\{\langle k,\vec{i}\rangle : k < n \text{ and } \vec{i} < \#st(\langle\vec{x}\rangle)\} < |v|_2$$

Now, by the coding of sequences described in Section 2,

$$\overline{\langle k, i_1, \ldots, i_a\rangle}^3 = 3 * \overline{k}^2 * 3 * \overline{i_1}^2 * \cdots * 3 * \overline{i_a}^2.$$

Thus,

$$\begin{aligned} |\langle k, i_1, \ldots, i_a\rangle|_3 &\leq |k|_2 + |i_1|_2 + \cdots + |i_a|_2 + a + 1 \\ &\leq |n-1|_2 + a \cdot \max_{1\leq j\leq a} |i_j|_2 + a + 1 \\ &\leq |n-1|_2 + a \cdot |\#st(\langle\vec{x}\rangle) - 1|_2 + a + 1. \end{aligned}$$

So, by Proposition 2.1 clauses iv and v,

$$\begin{aligned} \langle k, i_1, \ldots, i_a\rangle &< 3^{|n-1|_2 + a\cdot|\#st(\langle\vec{x}\rangle)-1|_2+a+2} \\ &\leq 2^{2\cdot|n-1|_2} \cdot 2^{2a\cdot|\#st(\langle\vec{x}\rangle)-1|_2} 3^{a+2} \\ &\leq n^2 \cdot (\#st(\langle\vec{x}\rangle))^{2a} \cdot 3^{a+2} \\ &= l(\vec{x}). \end{aligned}$$

Hence (72) holds.

Now we describe how $\chi_s^\star$ is obtained from χ_s. The variables in χ_{r_i} were intended to range over terms; in $\chi_s^\star$ we are dealing with the *codes* of certain terms; in fact we primarily deal with the codes of subterms of terms coded by $\vec{x}$. Thus bound variables in χ_s' will be interpreted as denoting *locations* of these subterms. For example, the variables $i_1, \ldots, i_{a_k}$ denote the terms with codes $st(\langle\vec{x}\rangle, i_1), \ldots, st(\langle\vec{x}\rangle, i_{a_k})$. The variables $k, \vec{i}$ in $\chi_s^\star$ are just the free variables $\vec{y}$ in $\chi_s(\vec{y})$, but now they will be interpreted as ranging over integers rather than terms; $\vec{x}$ and v will be introduced into the formula $\chi_s^\star$ as described below.

Atomic formulas will be dealt with in two ways; for atomic formulas involving s, $\chi_{r_i}^\star$ checks the value of a bit in v. Recall from the definition of rudimentary formula, that if the atomic formula $r\vec{x}$ occurs in $\chi_{r_i}(\vec{z})$, where $r \in R$, then the x_j's are variables, and because all quantifications are of the forms $\forall x \subseteq_{\text{st}} y$ and $\exists x \subseteq_{\text{st}} y$, any such atomic formula will always be interpreted as an assertion about subterms of the parameters denoted by $\vec{z}$; all such assertions are coded in the variable v. For atomic formulas involving "=" or s', $\chi_s^\star$ makes use of a $\star$-translation of the terms involved that is a result of replacing variables with expressions of the form $st(\langle\vec{x}\rangle, z)$. This $\star$-translation, for both formulas and terms, is described inductively as follows.

(73) $$z^\star \stackrel{\text{def}}{=} st'(\langle\vec{x}\rangle, z), \text{ for } z \text{ a variable},$$

(74) $$(\overline{f}\vec{\tau})^\star \stackrel{\text{def}}{=} \ulcorner \overline{f} \urcorner *_3 \tau_1^\star *_3 \cdots *_3 \tau_k^\star,$$

(75) $$(s\vec{z})^\star \stackrel{\text{def}}{=} (v)_{2,\langle\vec{z}\rangle} = 2,$$

(76) $$(s'\vec{\tau})^\star \stackrel{\text{def}}{=} \ulcorner s \urcorner' \vec{\tau}^\star,$$

(77) $$(\sigma = \tau)^\star \stackrel{\text{def}}{=} \sigma^\star = \tau^\star,$$

(78) $$(\neg\varphi)^\star \stackrel{\text{def}}{=} \neg(\varphi^\star),$$

(79) $$(\varphi_0 \wedge \varphi_1)^\star \stackrel{\text{def}}{=} \varphi_0^\star \wedge \varphi_1^\star,$$

(80) $$(\exists y \subseteq_{\text{st}} z\varphi)^\star \stackrel{\text{def}}{=} \exists y([z \le y < z + \#st(\langle\vec{x}\rangle, z)] \wedge \varphi^\star),$$

(81) $$(\forall y \subseteq_{\text{st}} z\varphi)^\star \stackrel{\text{def}}{=} \forall y([z \le y < z + \#st(\langle\vec{x}\rangle, z)] \rightarrow \varphi^\star),$$

So for example, $(y = \overline{+}{*}z{*}\overline{0})^\star$ is the formula $st'(\langle\vec{x}\rangle, y) = \ulcorner\overline{+}\urcorner *_3 st'(\langle\vec{x}\rangle, z) *_3 \ulcorner\overline{0}\urcorner$. The idea of the $\star$-translation is that the meaning of a rudimentary formula is preserved as we go from terms to codes for terms; more precisely, we have the following.

Claim 6.17 *Suppose φ is any L_{TM}-Rud formula and $\vec{\tau} = \tau_1, \ldots, \tau_{a_k}$ is a sequence of terms that are all subterms of the terms coded by $\vec{x} = x_1, \ldots, x_{a_k}$, say, for $j = 1, \ldots, a_k$,*

$$\ulcorner\tau_j\urcorner = st(\langle\vec{x}\rangle, i_j),$$

and suppose v accurately codes the relation s on all subterms of the terms coded by $\vec{x}$, i.e.

$$(\forall\vec{i} < \#st(\langle\vec{x}\rangle))((v)_{2,\langle\vec{i}\rangle} = 2 \leftrightarrow \ulcorner s\urcorner(k, st(\langle\vec{x}\rangle, i_0), \ldots, st(\langle\vec{x}\rangle, i_{a_k}), 0, \ldots, 0)),$$

then,

$$\varphi^\star(k, \vec{i}, 0, \ldots, 0, \vec{x}, v) \iff \varphi(\overline{k}, \vec{\tau}, \overline{0}, \ldots, \overline{0}, s).$$

The proof of the claim, which we omit, goes by induction on the construction of φ, and is straightforward.

In order to establish (69), we will show that the formula $\chi^\star$ can be used to inductively define integers $v_{\vec{x}}$ that correspond to finite approximations of the set $\ulcorner s \urcorner$. We will use set-theoretic notation when working with integers, such as v, that are construed as coding finite sets; for example we will write $(\vec{z}) \in v$ in place of $(v)_{2,\langle\vec{z}\rangle} = 2$, etc. We next give an inductive definition of $v_{\vec{x}}$. For each $m \in \omega$ we define $v_{\vec{x},m}$ such that,

(82) $|v_{\vec{x},m}|_2 = l(\vec{x})$, for all $m \in \omega$,

(83) $v_{\vec{x},0} \stackrel{\text{def}}{=} \emptyset$,

(84) $v_{\vec{x},m+1} \stackrel{\text{def}}{=} \{(k,\vec{i}) : k < n \wedge \vec{i} < \#st(\langle\vec{x}\rangle) \wedge \chi_s^\star(k,\vec{i},\vec{x},v_{\vec{x},m})\} \cup v_{\vec{x},m}$,

and let,

(85) $v_{\vec{x}} \stackrel{\text{def}}{=} \bigcup_{m\in\omega} v_{\vec{x},m}$.

Since the $v_{\vec{x},m}$'s are bounded in cardinality, and are increasing (in the sense of $\subseteq$, i.e. $v_{\vec{x},m} \subseteq v_{\vec{x},m+1}$, for all $m \in \omega$), it is clear that for some $m_0 \in \omega$,

(86) $$v_{\vec{x}} = v_{\vec{x},m_0}.$$

We next show that $v_{\vec{x}}$ is a kind of "least fixed point" for χ_s'; more precisely, we show that $v_{\vec{x}}$ is the unique witness, v, to the formula,

(87) $$(|v|_2 = l(\vec{x})) \wedge \chi_s'(\vec{x},v) \wedge (\forall w < v)(|w|_2 = l(\vec{x}) \rightarrow \neg\chi_s'(\vec{x},w)).$$

Clearly $|v_{\vec{x}}|_2 = l(\vec{x})$. To see that $\chi_s'(\vec{x},v_{\vec{x}})$ holds, suppose $k < n$, $\vec{i} < \#st(\langle\vec{x}\rangle)$, and $\chi_s^\star(k,\vec{i},\vec{x},v_{\vec{x}})$. By (86), $\chi_s^\star(k,\vec{i},\vec{x},v_{\vec{x},m_0})$, so by (84), $(k,\vec{i}) \in v_{\vec{x},m_0+1}$, and hence by (85), $(k,\vec{i}) \in v_{\vec{x}}$. This shows, by (71), $\chi_s'(\vec{x},v_{\vec{x}})$.

To establish the leastness of v, suppose $w < v$ and $\chi_s'(\vec{x},w)$. Using the monotonicity of the χ_{r_i}'s it is easy to show that χ_s, and then $\chi_s^\star$, are monotonic. Thus letting $w_{\vec{x},m}$ be given by,

(88) $$w_{\vec{x},0} \stackrel{\text{def}}{=} w,$$

(89) $w_{\vec{x},m+1} \stackrel{\text{def}}{=} \{(k,\vec{i}) : k < n \wedge \vec{i} < \#st(\langle\vec{x}\rangle) \wedge \chi_s^\star(k,\vec{i},\vec{x},w_{\vec{x},m})\} \cup w_{\vec{x},m}$,

we get, by induction and monotonicity, $v_{\vec{x},m} \subseteq w_{\vec{x},m}$ for all $m \in \omega$. But since $\chi_s'(\vec{x},w_{\vec{x},0})$, we have, by (71), $w_{\vec{x},m} = w_{\vec{x},0}$ for all $m \in \omega$. This implies

$$v_{\vec{x}} = v_{\vec{x},m_0} \subseteq w_{\vec{x},m_0} = w_{\vec{x},0} = w,$$

for all $m \in \omega$, and so, by our coding of finite sets, $v_{\vec{x}} \leq w$, a contradiction. Taking $v_{\vec{x}}$ for v, this completes the proof of (87).

From (70) and (87), it is clear that,

$$\psi_s(\vec{x}) \iff (\vec{i}(\vec{x})) \in v_{\vec{x}}, \tag{90}$$

where, for $\vec{x} = x_0, \ldots, x_a$, $\vec{i}(\vec{x}) = i_0, \ldots, i_a$ satisfies $x_k = st(\langle\vec{x}\rangle, i_k)$. More precisely, $\vec{i}(\vec{x})$ is given by,

$$i_0 \stackrel{\text{def}}{=} 0, \text{ and} \tag{91}$$

$$i_{k+1} \stackrel{\text{def}}{=} i_k + \#st(\langle\vec{x}\rangle, i_k), \text{ for } k = 1, \ldots, a-1. \tag{92}$$

Next we argue that to prove (69) it is enough to show,

$$\{\vec{st}(\vec{x}, \vec{i}) : (\vec{i}) \in v_{\vec{x}}\} = (\ulcorner s \urcorner)^{St(\vec{x})}, \tag{93}$$

where

$$St(\vec{x}) \stackrel{\text{def}}{=} \{st(\langle\vec{x}\rangle, i) : i < \#st(\vec{x})\},$$

and for $\vec{i}(\vec{x}) = i_0, \ldots, i_a$,

$$\vec{st}(\langle\vec{x}\rangle, \vec{i}(\vec{x})) \stackrel{\text{def}}{=} st(\langle\vec{x}\rangle, i_0), \ldots, st(\langle\vec{x}\rangle, i_a).$$

(93) implies $\vec{x} \in \ulcorner s \urcorner \Leftrightarrow (\vec{i}(\vec{x})) \in v_{\vec{x}}$, hence using (90), $(\vec{x}) \in \ulcorner s \urcorner \Longleftrightarrow \psi_s(\vec{x})$. Thus (69) will be established once we show (93), which follows from,

$$\text{for all } m \in \omega,\ \{\vec{st}(\langle\vec{x}\rangle, \vec{i}) : (\vec{i}) \in v_{\vec{x},m}\} = (\ulcorner (s)_m \urcorner)^{St(\vec{x})}, \tag{94}$$

where $(s)_m$ is given by (41) and (42). We prove (94) by induction on m. For $m = 0$, (94) is clear. For the induction step we write $\vec{\tau}(\vec{x}, \vec{i})$ to denote the sequence $\tau_0, \ldots, \tau_a$, such that, $\ulcorner \tau_j \urcorner = st(\langle\vec{x}\rangle, i_j)$, and we let $ST(\vec{x})$ denote $\{\tau : \text{for some } i < \#st(\langle\vec{x}\rangle), \ulcorner \tau \urcorner = st(\langle\vec{x}\rangle, i)\}$. We have,

$$\begin{aligned}
&\{\vec{st}(\langle\vec{x}\rangle, \vec{i}) : (\vec{i}) \in v_{\vec{x},m+1}\} = \\
&\quad = \{\vec{st}(\langle\vec{x}\rangle, \vec{i}) : i \in v_{\vec{x},m}\} \cup \{\vec{st}(\langle\vec{x}\rangle, \vec{i}) : \vec{i} < \#st(\langle\vec{x}\rangle) \wedge \chi_s^\star(\vec{i}, \vec{x}, v_{\vec{x},m})\} \\
&\quad\quad \text{(By (84).)} \\
&\quad = (\ulcorner (s)_m \urcorner)^{St(\vec{x})} \cup \{\ulcorner \vec{\tau}(\vec{x}, \vec{i}) \urcorner : \vec{i} < \#st(\langle\vec{x}\rangle) \wedge \chi_s(\vec{\tau}(\vec{x}, \vec{i}), ((s)_m)^{ST(\vec{x})})\} \\
&\quad\quad \text{(By the induction hypothesis and Claim 6.17.)} \\
&\quad = (\ulcorner (s)_m \urcorner)^{St(\vec{x})} \cup \{\ulcorner \vec{\tau}(\vec{x}, \vec{i}) \urcorner : \vec{i} < \#st(\langle\vec{x}\rangle) \wedge \chi_s(\vec{\tau}(\vec{x}, \vec{i}), (s)_m)\} \\
&\quad\quad \text{(By (37).)} \\
&\quad = (\ulcorner (s)_m \urcorner)^{St(\vec{x})} \cup \ulcorner \{\vec{y} : \chi_s(\vec{y}, (s)_m)\}^{St(\vec{x})} \urcorner \\
&\quad = (\ulcorner (s)_{m+1} \urcorner)^{St(\vec{x})}
\end{aligned}$$

Thus completing the proof of (94), and so also the proof of (69). The proof of Lemma 6.16 will be complete once we show that ψ_s is equivalent to a Δ_0-formula. It is clear from the definition of ψ_s, given by (70), (71), (87), and (73)–(81), that it is enough to show,

$$v_{\vec{x}} \leq p(\vec{x}),$$

for some polynomial p.

We have, by (70),

$$|v_{\vec{x}}|_2 = l(\vec{x}) = n^2 \cdot 3^{a+2} \cdot (\#st(\vec{x}))^{2a}.$$

Note,

$$\begin{aligned}(\#st(\vec{x}))^{2a} &\leq (a+1)^{2a} \cdot \max_{0\leq i\leq a} (\#st(x_i))^{2a} \\ &\leq (a+1)^{2a} \cdot \max_{0\leq i\leq a} |x_i|_2, \text{ (By (65))}.\end{aligned}$$

So $|v_{\vec{x}}|_2 \leq k \cdot \max_{0\leq i\leq a} |x_i|_2$, where k is a constant that only depends on a and n. By Proposition 2.1,

$$v_{\vec{x}} < 2^{|v_{\vec{x}}|_2+1} \leq 2 \cdot \max_{0\leq i\leq a} (2^{|x_i|_2})^k \leq 2 \cdot \max_{0\leq i\leq a} (x_i+1)^k,$$

and thus clearly $v_{\vec{x}}$ is bounded by a polynomial in $\vec{x}$. This completes the proof of the lemma.

q.e.d. Lemma 6.16

To complete the proof of Theorem 6.8 we have left to show clauses iii and iv of Definition 3.1. That the representations of the functions in F are $I\Delta_0$-provably total will follow from clauses 5 and 6 of Definition 6.7. Using 5 we easily obtain a polynomial bound, p_f, for the code of the value of a function, $f \in F$, in terms of its arguments. We could, in fact, get by with a much looser restriction than that given by 5, but this is sufficient for all of the examples considered. Using this, together with Lemma 6.16, we have Δ_0-formulas ψ_f and polynomials p_f, for each $f \in F$, such that clause ii of Definition 3.1 is satisfied by the formulas given by,

$$\theta_f(\vec{x}, y) \stackrel{\text{def}}{\Longleftrightarrow} \psi_f(\vec{x}, y) \wedge y \leq p_f(\vec{x}).$$

To obtain clause iii of Definition 3.1, we first note that the appropriate translations of the axioms of T_{TM} are easy to prove in $I\Delta_0$ using results from Section 2. Also $I\Delta_0$ proves induction on $\star$-translations of L_{TM}-Rud formulas. So it is clear that the proofs asserted to exist in clause 6 of Definition 6.7 can be translated into $I\Delta_0$-proofs yielding the totality of the functions defined by the θ_f's.

Similar arguments imply that from clause 7 of Definition 6.7 together with the hypotheses of the theorem we get clause iv in Definition 3.1. The details are straightforward. We note that $I\Delta_0$ in fact proves stronger induction principles than the translations of $STI(L_{TM}\text{-Rud})$.
q.e.d. Theorem 6.8

7 Concluding Remarks

The previous section shows that a significant amount of the basic theory of the ordinals can be represented in weak fragments of arithmetic, but as we mentioned in Section 1 the methods of Section 6 are insufficient for handling some important functions on ordinals. We mention some examples here, and very briefly describe methods that could lead to the representability of these functions.

In many applications one needs a function embedding the natural numbers into the ordinals. For example in carrying out proof-theoretic ordinal analysis one will refer to "the ordinal x" where "x" is a variable ranging over the natural numbers. For all of the examples of Gödel numberings given in the previous two sections, the function taking a natural number onto the corresponding ordinal is not provably total in $I\Delta_0$. In each case this is because the size of the code of the ordinal corresponding to the natural number n is exponential in n. For example in Section 5, the ordinal corresponding to n is given by the code of the sequence $(0, \ldots, 0)$ (n 0's); this code has Smullyan base 3 representation,

$$\underbrace{3 * \cdots * 3}_{n\ 3\text{'s}},$$

which is exponential in n.

This problem can be circumvented by coding up repeated applications of functions. This is essentially what is done in [Som90] to show that the function embedding the natural numbers into the ordinals is $I\Delta_0$-provably total for a particular representation of the ordinals less than ε_0. Roughly, the idea is to code a term of the form $\tau + \cdots + \tau$ (n τ's) by something of the form $\tau \cdot n$ (i.e. we code up n).

We note that not only does the above technique reduce the complexity of the embedding function, but it also reduces the complexity of representations of multiplication and the fundamental sequence function. However, the representations of these functions given by this technique are not $I\Delta_0$-provably total. In [Som90], the author shows that, for a particular representation of ε_0, these functions both have Δ_0-graphs and are provably total in $I\Delta_0 + \Omega_1$, where Ω_1 is the assertion that the function $x \mapsto x^{\log_2(x)}$

is total (this theory is equivalent to Buss' S_2). In order to answer the general question of whether fundamental sequences can be represented on representations of arbitrary recursive ordinals, it seems one would have to exhibit canonical assignments of fundamental sequences to arbitrary recursive ordinals. The author believes that such investigations could further the understanding of the ordinals currently involved in many aspects of proof-theoretic research.

References

[BS88] Wilfried Buchholz and Kurt Schütte. *Proof Theory of Impredicative Subsystems of Analysis. Studies in Proof Theory*, Bibliopolis, 1988.

[Buc91] Wilfried Buchholz. Notation systems for infinitary derivations. *Archive for Mathematical Logic*, 30, 1991.

[Bus86] Samuel R. Buss. *Bounded Arithmetic. Studies in Proof Theory*, Bibliopolis, 1986.

[End77] Herbert Enderton. *Elements of Set Theory.* Academic Press, 1977.

[Fef68] Solomon Feferman. Systems of predicative analysis II: representations of ordinals. *The Journal of Symbolic Logic*, 33(2):193–220, June 1968.

[Fef87] Solomon Feferman. Proof theory: a personal report. In Gaisi Takeuti, *Proof Theory*, pages 447–485, North-Holland, 1987.

[Gri90] Serge Grigorieff. Every recursive linear ordering has a copy in DTIME-SPACE(n, $\log n$). *The Journal of Symbolic Logic*, 55(1):260–276, March 1990.

[HP92] P. Hájek and Pavel Pudlák. Metamathematics of first order arithmetic. 1992. to appear.

[Jag84] Gerhard Jäger. ρ-inaccessible ordinals, collapsing functions and a recursive notation system. *Math. Logik und Grundlagenforschung*, 24:49–62, 1984.

[Poh87] Wolfram Pohlers. Contributions to the Schütte school. In Gaisi Takeuti, *Proof Theory*, pages 406–431, North-Holland, 1987.

[Rat90] M. Rathjen. Ordinal notations based on a weakly Mahlo cardinal. *Archive for Mathematical Logic*, 29(4):249–263, 1990.

[Ros84] H. E. Rose. *Subrecursion: Functions and Hierarchies.* Clarendon Press, 1984.

[Sch77] Kurt Schütte. *Proof Theory.* Springer Verlag, 1977.

[Smu61] Raymond Smullyan. Theory of formal systems. *Annals of Mathematics Studies*, (47), 1961.

[Som90] Richard Sommer. *Transfinite Induction and Hierarchies Generated by Transfinite Recursion within Peano Arithmetic.* PhD thesis, U. C. Berkeley, 1990.

[Woo86] Alan Woods. Bounded arithmetic formulas and Turing machines of constant alternation. In *Logic Colloquium 1984*, pages 355–377, North Holland, 1986.

Richard Sommer
Department of Mathematics
Stanford University
Stanford, CA 94305
sommer@cauchy.Stanford.EDU

$RSUV$ ISOMORPHISMS

Gaisi Takeuti

In [12], we discussed an interpretation of $V_2^i(BD)$ in S_3^i and an inverse interpretation of S_3^i in $V_2^i(BD)$. With a little more care, we shall establish an isomorphism between $V_2^i(BD)$ and S_3^i. Is there any subsystem of S_3^i which is isomorphic to $U_2^i(BD)$? We shall define a first order system R_3^i and prove that R_3^i and $U_2^i(BD)$ are isomorphic. R_3^i is obtained from S_3^i as follows.

(1) Two binary functions $a \dot{-} b$ and $MSP(a, i)$ and axioms on them are added, where the defining axioms of MSP are

$$MSP(a, 0) = a \quad \text{and}$$
$$MSP(a, i+1) = \lfloor \frac{1}{2} MSP(a, i) \rfloor .$$

(2) Σ_i^b-PIND is replaced by the following Σ_i^b-LBIND

$$\frac{F(\lfloor \frac{1}{2} a \rfloor), \Gamma \to \Delta, F(a)}{F(0), \Gamma \to \Delta, F(|t|)},$$

where $F(a)$ is in Σ_i^b in the language of R_3^i and a satisfies the eigenvariable condition, i.e. a does not appear in the lower sequent.

(3) The extensionality axiom of the following form is added

$$|a| = |b|, \forall i < |a| (Bit(i, a) = Bit(i, b)) \to a = b,$$

where Bit is defined by the following equations

$$\text{mod } 2(a) = a - 2 \cdot \lfloor \frac{1}{2} a \rfloor$$

and

$$Bit(i, a) = \text{mod } 2(MSP(a, i)).$$

In [12], it was also shown that V_2^i, $\hat{V}_2^i$, $V_2^i(BD)$, and $\hat{V}_2^i(BD)$ are closely related, where $\hat{V}_2^i$ and $\hat{V}_2^i(BD)$ are obtained from V_2^i and $V_2^i(BD)$ respectively by adding $\Sigma_i^{1,b}$–comprehension schema, namely the following results were shown
(1) $V_2^i(BD)$ and $\hat{V}_2^i(BD)$ are equivalent
(2) $\hat{V}_2^i$ and V_2^i are equivalent with respect to bounded formulas.
(3) V_2^i is a conservative extension of $V_2^i(BD)$ with respect to bounded formulas.
Corresponding to these, we shall show that U_2^i, $\bar{U}_2^i$, $U_2^i(BD)$, and $\bar{U}_2^i(BD)$ are closely related, where $\bar{U}_2^i$ and $\bar{U}_2^i(BD)$ are obtained from U_2^i and $U_2^i(BD)$ respectively by adding $\Delta_i^{1,b}$–comprehension schema. Namely the following results shall be shown.
(1) $U_2^i(BD)$ and $\bar{U}_2^i(BD)$ are equivalent.
(2) $\bar{U}_2^i$ and U_2^i are equivalent with respect to bounded formulas.
(3) U_2^i is a conservative extension of $U_2^i(BD)$ with respect to bounded formulas. Here by bounded formulas we mean the formulas without any first order unbounded quantifiers. Notice that a bounded formula may have many second order quantifiers.

We shall also show that V_2^{i+1} proves

$$\forall x \le a \exists y (A(x,y) \land |y| \le b) \to \exists t \forall x \le a \exists y \le t A(x,y),$$

where $A(x,y)$ is a Σ_i^b–formula. We obtain R_k^i, S_k^i, U_k^i, $V_k^i (k > 2)$ from R_2^i, S_2^i, U_2^i, V_2^i respectively by introducing $\#_k$, where $x \#_{i+1} y = 2^{|x| \#_i |y|}$ and $\#_2$ is $\#$. We also obtain R_1^i, S_1^i, U_1^i, V_1^i from R_2^i, S_2^i, U_2^i, V_2^i respectively by deleting $\#$. Then we shall actually generalize the relations among R_3^i, S_3^i, $U_2^i(BD)$, and $V_2^i(BD)$ to the relations among R_{k+1}^i, S_{k+1}^i, $U_k^i(BD)$ and $V_k^i(BD)$. Namely we shall establish the isomorphism between R_{k+1}^i and $U_k^i(BD)$ and the isomorphism between S_{k+1}^i and $V_k^i(BD)$. The isomorphism theorem between S_{k+1}^i and $V_k^i(BD)$ was also independently proved by Razborov [9] and also by Fernando Ferreira using a model theoretic method. These isomorphism theorems have many corollaries. E.g. R_{k+1}^i and S_{k+1}^i are equivalent iff $U_k^i(BD)$ and $V_k^i(BD)$ are equivalent and R_{k+1}^{i+1} and S_{k+1}^i are equivalent iff $U_k^{i+1}(BD)$ and $V_k^i(BD)$ are equivalent.

B. Allen [1] and P. Clote [4] independently introduced a theory corresponding to the class NC. In this paper, we use the theory TNC in Clote and Takeuti [5] as a theory corresponding to NC. Then we shall show that R_2^1 is equivalent to TNC. Therefore our isomorphism implies that TNC is equivalent to S_2^1 iff $U_1^1(BD)$ is equivalent to $V_1^1(BD)$ and that if $V_1^1(BD)$ and $U_1^1(BD)$ are equivalent with respect to $\Sigma_1^{1,b} \cup \Pi_1^{1,b}$, then $NC = P$. In our notation, we often use Σ_ℓ^b or $\Sigma_\ell^{1,b}$ in the place of $\Sigma_\ell^b(\#_k)$ or $\Sigma_\ell^{1,b}(\#_k)$ respectively when we discuss R_k^i, S_k^i, $U_k^i(BD)$, or $V_k^i(BD)$.

We follow the terminology of [2] and [12] and assume familiarity with them.

§1. The system R^i_k.

In this paper we assume $i > 0$ and $k > 0$. First we discuss a rather trivial relation between R^i_k and S^{i-1}_k.

Theorem 1. *S^{i-1}_k is a subsystem of R^i_k.*

Proof. Let $A(x)$ be in Σ^b_{i-1} and $A(0) \wedge \forall x(A(x) \supset A(x+1))$. It suffices to show $A(|a|)$ in R^i_k. Let $B(a,b)$ be

$$\forall x \le |a| \forall y \le |a|(y \le \lfloor \frac{1}{2} b \rfloor + 1 \wedge A(x \dot{-} y) \supset A(x)).$$

Then

$$B(a, \lfloor \frac{1}{2} b \rfloor) \longrightarrow B(a,b).$$

Therefore by Σ^b_i–LBIND, we have

$$\begin{aligned}&\forall x \le |a| \forall y \le |a|(y \le 1 \wedge A(x \dot{-} y) \supset A(y))\\ &\longrightarrow \forall x \le |a| \forall y \le |a|(y \le |a| + 1 \wedge A(x \dot{-} y) \supset A(x)).\end{aligned}$$

The left–hand side of the sequent follows from $\forall x(A(x) \supset A(x+1))$ and the right–hand side of the sequent implies $A(0) \supset A(|a|)$. □

In [3], S. Buss proved

$$S^i_2 \vdash \exists w \forall x \le |t|(A(x) \longleftrightarrow Bit(x,w) = 1)$$

where $A(x)$ is in Σ^b_i.
In the same way, one can show for $k \ge 2$

$$R^i_k \vdash \exists w \forall x \le ||t||(A(x) \longleftrightarrow Bit(x,w)) = 1),$$

where $A(x)$ is in Σ^b_i. We do not know whether we can improve this result by replacing $\forall x \le ||t||$ by $\forall x \le |t|$. But if $A(x)$ is in Δ^b_i, then we can improve it by replacing $\forall x \le |t|$ to this form as follows. In Clote and Takeuti [5], Π^b_i–separation axioms were introduced in the following form

$$\begin{aligned}\forall i(\neg A(i) \vee \neg B(i)) \longrightarrow \forall x \exists y < 2^{|x|} \forall i < |x|((A(i) \supset Bit(i,y) = 0)\\ \wedge (B(i) \supset Bit(i,y) = 1)),\end{aligned}$$

where $A(i)$ and $B(i)$ are in Π^b_i.

Theorem 2. *R^i_k proves Π^b_i–separation axioms for $k \geq 2$.*

Proof. Let $C(b)$ be

$$\forall u < |a| \exists y < 2^{|a|} \forall i < |a| (u \leq i \leq u + b \supset ((A(i) \supset Bit(i,y) = 0) \\ \wedge (B(i) \supset Bit(i,y) = 1)).$$

Then $\forall i(\neg A(i) \vee \neg(Bi)), C(\lfloor \frac{b}{2} \rfloor) \rightarrow C(b)$. Therefore R^i_3 proves

$$\forall i(\neg A(i) \vee \neg B(i)), \quad C(0) \longrightarrow C(|a|).$$

Obviously $\forall i(\neg A(i) \vee \neg B(i)) \rightarrow C(0)$ and $C(|a|)$ implies

$$\exists y < 2^{|a|} \forall i < |a| ((A(i) \supset Bit(i,y) = 0) \wedge (B(i) \supset Bit(i,y) = 1)). \quad \square$$

Corollary 3. *For $k \geq 2$ R^i_k proves*

$$\forall i(A(i) \longleftrightarrow \neg B(i)) \longrightarrow \exists y < 2^{|a|} \forall i < |a| (A(i) \longleftrightarrow Bit(i,y) = 1)$$

where $A(i)$ and $B(i)$ are in Σ^b_i.

Let $\dot{R}^i_k$ be the theory obtained from R^i_k by replacing Σ^b_i–LBIND by following Σ^b_i–BLIND

$$\frac{F(a), \Gamma \rightarrow \Delta, F(a+1)}{F(0), \Gamma \rightarrow \Delta, F(\|t\|)},$$

where $F(a)$ is in Σ^b_i and a satisfies the eigenvariable condition.

Theorem 4. *R^i_k and $\dot{R}^i_k$ are equivalent.*

Proof. In [2], S. Buss proved an equivalence between Σ^b_i-PIND and Σ^b_i-LIND. We follow his argument. First we show that R^i_k proves

$$A(0) \wedge \forall x(A(x) \supset A(x+1)) \rightarrow A(\|t\|), \tag{1}$$

where $A(a)$ is in Σ^b_i. Let $B(a)$ be $A(|a|)$. Then (1) is easily proved by Σ^b_i-LBIND on $B(a)$.

Next we show that $\dot{R}^i_k$ proves

$$A(0) \wedge \forall x(A(\lfloor \frac{1}{2} x \rfloor) \supset A(x)) \rightarrow A(|a|), \tag{2}$$

where $A(a)$ is in Σ^b_i. Let $B(a,u)$ be $A(MSP(|a|, \|a\| \dot{-} u))$. Then (2) is easily proved by Σ^b_i-BLIND on $B(a,u)$. $\square$

Our proof goes through even if $i = 0$. So we have

Corollary 5. *$\dot{R}^0_k$ and R^0_k are equivalent.*

Let $\bar{S}^0_k$ be the system obtained from S^0_k be adding functions $a \dot{-} b$ and $MSP\ (a, i)$ and axioms on them along with the extensionality axiom. $\dot{S}^0_k$ is obtained from $\bar{S}^0_k$ by replacing Σ^b_0-PIND by Σ^b_0-LIND.

Theorem 6. *R^0_k, $\dot{S}^0_k$ and $\bar{S}^0_k$ are equivalent.*

Proof. Obviously R^0_k is a subsystem of S^0_k. Now we are going to show that Σ^b_0-LIND is provable in R^0_k. Let $A(0) \wedge \forall x(A(x) \supset A(x+1))$ hold, where A(0) is in Σ^b_0. Then R^0_k proves

$$\forall x \le |a| \forall y \le |a| (y \le \lfloor \frac{1}{2} b \rfloor + 1 \wedge A(x \dot{-} y) \supset A(x))$$
$$\to \forall x \le |a| \forall y \le |a| (y \le b + 1 \wedge A(x \dot{-} y) \supset A(x)).$$

Therefore R^0_k proves

$$\forall x \le |a| \forall y \le |a| (y \le 1 \wedge A(x \dot{-} y) \supset A(x))$$
$$\to \forall x \le |a| \forall y \le |a| (y \le |a| + 1 \wedge A(x \dot{-} y) \supset A(x)).$$

Since $\forall x(A(x) \supset A(x+1))$ implies

$$\forall x \le |a| \forall y \le |a| (y \le 1 \wedge A(x \dot{-} y) \supset A(x)) \quad \text{and}$$
$$\forall x \le |a| \forall y \le |a| (y \le |a| + 1 \wedge A(x \dot{-} y) \supset A(x)) \quad \text{implies} \quad A(0) \supset A(|a|),$$

R^0_k proves $A(|a|)$. Now it suffices to show that $\dot{S}^0_k$ proves Σ^b_0-PIND. This is easily done by using MSP as in the proof of Theorem 4. □

Theorem 7. *R^1_2 is equivalent to TNC, where TNC is a theory for NC introduced in Clote and Takeuti [5].*

Proof. Obviously $\dot{R}^1_2$ is a subsystem of TNC. In the same way as in Corollary 5, $\dot{R}^1_2$ and R^1_2 are equivalent. Therefore R^1_2 is a subsystem of TNC. On the other hand, Π^b_1–separation axioms are provable in R^1_2. Therefore TNC is a subsystem of R^1_2. □

§2 The systems U^i_k, $\bar{U}^i_k$, $U^i_k(BD)$ and $\bar{U}^i_k(BD)$.

U^i_k and $\overset{\circ}{U}{}^i_k$ are defined in [2]. $\bar{U}^i_k$ is obtained from U^i_k by adding the following inferences

$$\frac{F(\{x\}A(x)), \Gamma \to \Delta}{\forall \varphi F(\varphi), \Gamma \to \Delta} \quad \text{and} \quad \frac{\Gamma \to \Delta, F(\{x\}A(x)}{\Gamma \to \Delta, \exists \varphi F(\varphi)}$$

where $A(x)$ is in $\Delta_i^{1,b}$ with respect to $\bar{U}_k^i$.

$U_k^i(BD)$ and $\mathring{U}_k^i(BD)$ are obtained from $V_k^i(BD)$ and $\mathring{V}_k^i(BD)$ respectively by replacing $\Sigma_i^{1,b}$–IND by the following $\Sigma_i^{1,b}$–PIND

$$\frac{A(\lfloor \frac{1}{2}a \rfloor),\ \Gamma \to \Delta, A(a)}{A(0),\ \Gamma \to \Delta,\ A(t)},$$

where $A(a)$ is $\Sigma_i^{1,b}$ and a does not occur in the lower sequent of the inference. $\bar{U}_k^i(BD)$ is obtained from $U_k^i(BD)$ by adding the following inferences

$$\frac{F(\{x\}(x \leq t \wedge A(x)),\ \Gamma \to \Delta}{\forall \varphi^t F(\varphi^t),\ \Gamma \to \Delta} \quad \text{and} \quad \frac{\Gamma \to \Delta,\ F(\{x\}(x \leq t \wedge A(x))}{\Gamma \to \Delta,\ \exists \varphi^t F(\varphi^t)}$$

where $A(x)$ is in $\Delta_i^{1,b}$ with respect to $\bar{U}_k^i(BD)$.

In [12], it was shown that U_2^i proves $\exists \varphi \forall x \leq |t|(A(v,x) \longleftrightarrow \varphi(x))$ for $\Sigma_i^{1,b}$–formula $A(v,x)$. In Krajíček and Takeuti [8], it was shown that $U_2^i(BD)$ proves bounded $\Pi_i^{1,b}$–separation axioms. In the same way, we show the following theorem.

Theorem 8. *U_k^i proves the following $\Pi_i^{1,b}$–separation axioms.*

$$\forall t(\neg A(t) \vee \neg B(t)) \to \forall x \exists \varphi \forall t < x((A(t) \supset \varphi(t)) \wedge (B(t) \supset \neg \varphi(t))),$$

where $A(t)$ and $B(t)$ are in $\Pi_i^{1,b}$.

Proof. Let $C(a,b)$ be

$$\forall s \leq b \exists \varphi \forall t \leq b(s \leq t \leq s + a \supset ((A(t) \supset \varphi(t)) \wedge (B(t) \supset \neg \varphi(t)))).$$

Obviously U_k^i proves

$$\forall t(\neg A(t) \vee \neg B(t)), C(\lfloor \frac{a}{2} \rfloor, b) \to C(a,b).$$

By $\Sigma_i^{1,b}$–PIND it follows that

$$\forall t(\neg A(t) \vee \neg B(t)), \quad C(0,b) \to C(b,b).$$

Since $C(0,b)$ is trivially provable, the $\Pi_i^{1,b}$–separation axiom follows. □

Corollary 9. *U_k^i proves the following $\Delta_i^{1,b}$ comprehension axioms*

$$\forall t(A(t) \longleftrightarrow \neg B(t)) \rightarrow \exists\varphi\forall x \le t(A(x) \longleftrightarrow \varphi(x))$$

where $A(t)$ and $B(t)$ are in $\Sigma_i^{1,b}$. $\square$

In the same way, one can show the following theorem.

Theorem 10. *$U_k^i(BD)$ proves*

$$\exists\varphi^t\forall x \le t(A(x) \longleftrightarrow \varphi^t(x)),$$

where x does not occur in t and $A(x)$ is in $\Sigma_i^{1,b}$. Therefore $\bar{U}_k^i(BD)$ and $U_k^i(BD)$ are equivalent.

A formula is said to be bounded if it does not have any first order unbounded quantifiers. It may have many second order quantifiers. In [12] it was shown that $\hat{V}_2^i$ and V_2^i are equivalent with respect to the bounded formulas and $\hat{V}_2^i$ and V_2^i are conservative extensions of $V_2^i(BD)$ with respect to the bounded formulas. In the same way, the following theorem can be proved.

Theorem 11. *$\bar{U}_k^i$ and U_k^i are equivalent with respect to the bounded formulas and $\bar{U}_k^i$ and U_k^i are conservative extensions of $U_k^i(BD)$ with respect to the bounded formulas.*

§3. An interpretation of R_{k+1}^i in $U_k^i(BD)$..

As is discussed in the last section, $U_k^i(BD) = \bar{U}_k^i(BD)$. Since $U_k^i(BD) \subseteq \mathring{U}_k^i(BD) \subseteq \bar{U}_k^i(BD)$, it follows that $U_k^i(BD)$, $\mathring{U}_k^i(BD)$, and $\bar{U}_k^i(BD)$ are all equivalent. We develop an interpretation of R_{k+1}^i in $\mathring{U}_k^i(BD)$.

In [12], we used (a, α^t) to represent $\sum_{i\le a} \epsilon_i 2^i$ where $\epsilon_i = 1$ if $\alpha^t(i)$ and $\epsilon_i = 0$ otherwise. We call $\sum_{i\le a} \epsilon_i 2^i$ the value of (a, α^t) and denote it by $v(a, \alpha^t)$. Then for function f, where f is one of S, t, $\cdot$, $\lfloor\frac{1}{2}a\rfloor$, $|a|$, $a \dot{-} b$, $\#_k$ and $\#_{k+1}$, we constructed a term $T^f(a_1, \ldots, a_n)$ and an abstract

$$V^f = \{x\}(x < T^f(a_1, \ldots, a_n) \wedge A^f(x, a_1, \alpha_1^{a_1}, \ldots, a_n, \alpha_n^{a_n}))$$

with the following properties

(1) $(T^f(a_1, \ldots, a_n), V^f)$ represents f with respect to $(a_1, \alpha_1^{a_1}), \ldots, (a_n, \alpha_n^{a_n})$ in the sense that the value of $(T^f(a_1, \ldots, a_n), V^f)$ is equal to $f(v(a_1, \alpha_1^{a_1}), \ldots, v(a_n, \alpha_n^{a_n}))$.

(2) $A^f(x, a_1, \alpha_1^{a_1}, \ldots, a_n, \alpha_n^{a_n})$ is $\Delta_i^{1,b}$ with respect to $\mathring{U}_k^i(BD)$.

(3) The defining formula of f can be translated with respect to $T^f(a_1, \ldots, a_n)$ and V^f and the result can be proved in $\overset{\circ}{U}{}^i_k(BD)$.

We describe an explicit form of T^f and A^f only for the case where $f(a)$ is $|a|$. In this case we denote T^f and A^f by $T^{|*|}$ and $A^{|*|}$ respectively.

$T^{|*|}(a)$ is $|a|$ and $A^{|*|}(x, a, \alpha^a)$ is

$$x \leq |a| \wedge \exists y \leq a(\alpha^a(y) \wedge \forall z \leq a(y < z \supset \neg\alpha^a(z)) \wedge Bit(x, y+1) = 1).$$

We also defined

$$\{x\}(x \leq t \wedge A(x)) \overset{H}{\leq} \{x\}(x \leq s \wedge B(x))$$

to be

$$\begin{aligned} &\forall x \leq t(A(x) \supset x \leq s \wedge B(x)) \vee \exists x \leq s(B(x) \wedge \neg(x \leq t \wedge A(x)) \\ &\wedge (\forall y \leq t(x < y \wedge A(y)) \supset y \leq s \wedge B(y))) \end{aligned}$$

and $\{x\}(x \leq t \wedge A(x)) \overset{H}{=} \{x\}(x \leq s \wedge B(x))$ to be

$$\begin{aligned} &\{x\}(x \leq t \wedge A(x)) \overset{H}{\leq} \{x\}(x \leq s \wedge B(x)) \\ &\wedge \{x\}(x \leq s \wedge B(x)) \overset{H}{\leq} \{x\}(x \leq t \wedge A(x)). \end{aligned}$$

Now we define T^{MSP} and A^{MSP} as follows, $T^{MSP}(a_1, a_2) = a_1$ and $A^{MSP}(x, a_1, \alpha_1^{a_1}, a_2, \alpha_2^{a_2})$ is

$$\exists y \leq a_1(\{z\}(z \leq y) \wedge Bit(z, y) = 1) \overset{H}{=} \{z\}(z \leq a_2 \wedge \alpha_2^{a_2}(z)) \wedge y + x \leq a_1 \wedge \alpha_1^{a_1}(y+x)).$$

Then $(T^{MSP}(a_1, a_2), V^{MSP})$ represents MSP with respect to $(a_1, \alpha_1^{a_1})$ and $(a_2, \alpha_2^{a_2})$ and A^{MSP} is $\Delta_0^{1,b}$ with respect to $\overset{\circ}{U}{}^i_k(BD)$ and the defining formulas of MSP can be translated with respect to $T^{MSP}(a_1, a_2)$ and V^{MSP} and the result can be proved in $\overset{\circ}{U}{}^i_k(BD)$.

Now let $a_0, a_1, a_2, \ldots$ be all the free variables in R^i_{k+1} and $x_0, x_1, x_2, \ldots$ be all the bound variables in R^i_{k+1}. We assign $\alpha_i^{a_i}$ to a_i and $\varphi_i^{x_i}$ to x_i.

Let t be a term in R^i_{k+1}. As is in [12], we inductively define T^t and $A^t(x)$ so that if all free variables in t are $a_1, \ldots, a_n$, then all the free variables in T^t are $a_1, \ldots, a_n$, and all the free variables in $A^t(x)$ are $a_1, \ldots, a_n$ and $\alpha_1^{a_1}, \ldots, \alpha_n^{a_n}$.

(1) If t is 0, then T^t is 0 and $A^t(x)$ is $0 = 1$.

(2) If t is a_i, then T^t is a_i and $A^t(x)$ is $\alpha_i^{a_i}(x)$.

(3) We define V^t to be $\{x\}(x \leq T^t \wedge A^t(x))$.

(4) If t is $f(t_1, \ldots, t_n)$, then T^t is $T^f(T^{t_1}, \ldots, T^{t_n})$ and $A^t(x)$ is

$$A^f(x, T^{t_1}, V^{t_1}, \ldots, T^{t_n}, V^{t_n}).$$

Let A be a formula in R^i_{k+1}. We define A^H inductively as follows.

(1) If A is an atomic formula of the form $t_1 \leq t_2$ (or $t_1 = t_2$), then A^H is $\{x\}(x \leq T^{t_1} \wedge A^{t_1}(x)) \overset{H}{\leq} \{x\}(x \leq T^{t_2} \wedge A^{t_2}(x))$ (or $\{x\}(x \leq T^{t_1} \wedge A^{t_1}(x)) \overset{H}{=} \{x\}(x \leq T^{t_2} \wedge A^{t_2}(x))$).

(2) If A is $\neg B$, $B \cap C$, $B \cup C$, or $B \supset C$, then A^H is $\neg B^H$, $B^H \cap C^H$, $B^H \cup C^H$, or $B^H \supset C^H$ respectively.

(3) If A is $\forall x \leq tB(x)$ (or $\exists x \leq tB(x)$), then A^H is

$$\forall x \leq T^t \forall \varphi^x(\{y\}(y \leq x \wedge \varphi^x(y)) \overset{H}{\leq} \{y\}(y \leq T^t \wedge A^t(y)) \supset B^H(\varphi^x, x))$$

(or $\exists x \leq T^t \exists \varphi^x(\{y\}(y \leq x \wedge \varphi^x(y)) \overset{H}{\leq} \{y\}(y \leq T^t \wedge A^t(y)) \wedge B^H(\varphi^x, x))$, where $B^H(\alpha^a, a)$ is $(B(a))^H$.

(4) If A is $\forall x B(x)$ (or $\exists x B(x)$), then A^H is $\forall x \forall \varphi^x B^H(\varphi^x, x)$ (or $\exists x \exists \varphi^x B^H(\varphi^x, x)$) where $B^H(\alpha^a, a)$ is $(B(a))^H$.

Let $V_1 = \{x\}A(x)$ and $V_2 = \{x\}B(x)$. We say "V_1 and V_2 are equivalent in $\overset{\circ}{U}{}^i_k(BD)$" if $\forall x(A(x) \longleftrightarrow B(x))$ is provable in $\overset{\circ}{U}{}^i_k(BD)$.

In the same way as in [12], we have the following propositions.

Proposition 12. *Let t be a term in R^i_{k+1}. Then $A^t(a)$ is $\Delta_1^{1,b}$ with respect to $\overset{\circ}{U}{}^1_k(BD)$.*

Proposition 13. *The following sequent is provable in $\overset{\circ}{U}{}^i_k(BD)$.*

$$\{x\}(x \leq a \wedge \alpha^a(x)) \overset{H}{\leq} \{x\}(x \leq b \wedge \beta^b(x)) \rightarrow \{x\}(x \leq b \wedge x \leq a \wedge \alpha^a(x)) \overset{H}{=} \{x\}(x \leq a \wedge \alpha^a(x)).$$

Proposition 14. *Let $t_0(a)$ and s be terms in R^i_{k+1} and t be $t_0(s)$.*

(1) Let $T_0(a)$ be $T^{t_0(a)}$. Then

$$T^t = T_0(T^s).$$

(2) Let $\tilde{A}(x, a, \alpha^a)$ be $A^{t_0(a)}(x)$. Then $V^t = \{x\}(x \leq T^t \wedge A^t(x))$ is equivalent to $\{x\}(x \leq T^t \wedge \tilde{A}(x, T^s, V^s))$ in $\overset{\circ}{U}{}^1_k(BD)$.
(3) Let $F(a)$ be formula in R^i_{k+1} and $F^H(\alpha^a, a)$ be $(F(a))^H$. Then $(F(s))^H$ is equivalent to $F^H(V^s, T^s)$ in $\overset{\circ}{U}{}^1_k(BD)$.

Proposition 15. *(1) Let $t(a)$ be a term and $T(a)$ be $T^{t(a)}$ and $A(x, \alpha^a, a)$ be $A^{t(a)}(x)$. Then the following sequent is provable in $\overset{\circ}{U}{}^1_k(BD)$.*

$$\{x\}(x \leq a \wedge \alpha^a(x)) \overset{H}{=} \{x\}(x \leq b \wedge \beta^b(x))$$
$$\to \{x\}(x \leq T(a) \wedge A(x, \alpha^a, a)) \overset{H}{=} \{x\}(x \leq T(b) \wedge A(x, \beta^b, b)).$$

(2) Let $F(a)$ be a formula in R^i_{k+1} and $F^H(\alpha^a, a)$ be $(F(a))^H$. Then the following sequent is provable in $\overset{\circ}{U}{}^1_k(BD)$.

$$\{x\}(x \leq a \wedge \alpha^a(x)) \overset{H}{=} \{x\}(x \leq b \wedge \beta^b(x)), F^H(\alpha^a, a) \to F^H(\beta^b, b).$$

Let Γ be $F_1, \ldots, F_n$. Then Γ^H is defined to be $F_1^H, \ldots, F_n^H$. The following theorem is the main result of this section.

Theorem 16. *Let $\Gamma \to \Delta$ be provable in R^i_{k+1}. Then $\Gamma^H \to \Delta^H$ is provable in $\overset{\circ}{U}{}^i_k(BD)$.*

Proof. As in the proof of Theorem in §1 in [12], we prove this by induction on the number of inferences in R^i_{k+1}-proof of $\Gamma \to \Delta$. We treat only two cases which are not treated in [12].

Case 1. $\Gamma \to \Delta$ is an extensionality axiom of the form

$$|a_1| = |a_2|, \forall i < |a_1|(Bit(i, a_1) = Bit(i, a_2)) \to a_1 = a_2.$$

In this case, it is easily seen that in $\overset{\circ}{U}{}^1_k(BD)(|a_1| = |a_2|)^H$ is equivalent to

$$\exists x \leq a_1(x \leq a_2 \wedge \alpha_1^{a_1}(x) \wedge \alpha_2^{a_2}(x) \wedge \forall y \leq a_1(x < y \supset \neg\alpha_1^{a_1}(y))$$
$$\wedge \forall y \leq a_2(x < y \supset \neg\alpha_2^{a_2}(y)).$$

Under the hypothesis $(|a_1| = |a_2|)^H$, $\overset{\circ}{U}{}^1_k(BD)$ proves that $(\forall i < |a_1|(Bit(i, a_1) = Bit(i, a_2)))^H$ is equivalent to $\forall i \leq a_1(\alpha_1^{a_1}(i) \longleftrightarrow \alpha_2^{a_2}(i))$. Then what remains to show is

$$\forall i \leq \max(a_1, a_2)(\alpha_1^{a_1}(i) \longleftrightarrow \alpha_2^{a_2}(i))$$
$$\to \{x\}(x \leq a_1 \wedge \alpha_1^{a_1}(x)) \overset{H}{=} \{x\}(x \leq a_2 \wedge \alpha_2^{a_2}(x)).$$

This is easily shown in $\overset{\circ}{U}{}^1_k(BD)$.

Case 2. $\Gamma \to \Delta$ is obtained by the following inference

$$\frac{F([\frac{1}{2}a]),\ \Gamma_0 \to \Delta_0,\ F(a)}{F(0),\ \Gamma_0 \to \Delta_0,\ F(|t|)}$$

where $F(a)$ is Σ^b_i in R^i_3.

Let $(F(a))^H$ be $F^H(a, \alpha^a)$. Then $F^H(a, \alpha^a)$ is $\Sigma^{1,b}_i$ in $\overset{\circ}{U}{}^i_k(BD)$ and $(F([\frac{1}{2}a]))^H$ is equivalent to $F^H(a\dot{-}1, \{x\}(x \leq a\dot{-}1 \wedge \alpha^a(x+1)))$ in $\overset{\circ}{U}{}^i_k(BD)$. By induction hypothesis we have in $\overset{\circ}{U}{}^i_k(BD)$

$$F^H(a\dot{-}1, \{x\}(x \leq a\dot{-}1 \wedge \alpha^a(x+1))), \Gamma^H_0 \to \Delta^H_0, \quad F^H(a, \{x\}(x \leq a \wedge \alpha^a(x))).$$

Let $\tilde{A}(x, a)$ be

$$x \leq a \wedge \exists y \leq T^t(A^t(y) \wedge \forall z \leq T^t(y < z \supset \neg A^t(z)) \wedge Bit(|T^t|\dot{-}(a\dot{-}x), y+1) = 1.$$

Then $\tilde{A}(x+1,$ a) is

$$x+1 \leq a \wedge \exists y \leq T^t(A^t(y) \wedge \forall z \leq T^t(y < z \supset \neg A^t(z)) \wedge Bit(|T^t|\dot{-}(a\dot{-}(x+1)), y+1) = 1)$$

which is equivalent to

$$x \leq a\dot{-} \wedge \exists y \leq T^t(A^t(y) \wedge \forall z \leq T^t(y < z \supset \neg A^t(z)) \wedge Bit(|T^t|\dot{-}((a\dot{-}1)\dot{-}x), y+1) = 1)$$

i.e. $\tilde{A}(x, a\dot{-}1)$.

By substituting $\{x\}\tilde{A}(x, a)$ for α^a in the sequent obtained by induction hypothesis, we have $F^H(a\dot{-}1, \{x\}\tilde{A}(x, a\dot{-}1)), \Gamma_0 \to \Delta_0, F^H(a, \{x\}\tilde{A}(x, a))$. By $\Sigma^{1,b}_i$-LIND, we have $F^H(0, \{x\}\tilde{A}(x, 0)), \Gamma_0 \to \Delta_0, F^H(|T^t|, \{x\}\tilde{A}(x, |T^t|))$. Obviously $F^H(|T^t|, \{x\}\tilde{A}(x, |T^t|))$ is equivalent to $(F(|t|))^H$. Now $\tilde{A}(x, 0)$ is equivalent to $x = 0$ or $0 = 1$. Therefore $F^H(0, \{x\}\tilde{A}(x, 0))$ is equivalent to $(F(0))^H$ or $(F(1))^H$. Since we have $(F(0))^H, \Gamma^H_0 \to \Delta^H_0, (F(1))^H$ from $F(0), \Gamma_0 \to \Delta_0, F(1)$, we have in $\overset{\circ}{U}{}^i_2(BD)(F(0))^H, \Gamma^H_0 \to \Delta^H_0, (F(|t|))^H$. □

§4. An interpretation of $\overset{\circ}{U}{}^i_k(BD)$ in R^i_{k+1}.

In this section we assume that all free variables in $\overset{\circ}{U}{}^i_2(BD)$ are $a_0, a_1, \ldots, \beta^{t_0}_0,$ $\beta^{t_0}_1, \ldots$ and all bound variables in $\overset{\circ}{U}{}^i_2(BD)$ are $x_0, x_1, \ldots, \psi^{t_0}_0, \psi^{t_0}_1, \ldots$. We assign a free variable b_i to β_i and a bound variable y_i to ψ_i.

Let t be a semi-term in $\overset{\circ}{U}{}^i_k(BD)$. (See p. 197 in [10] for the definition of semi–term and semi–formula.) As in [12], we assign a semi-term, t^L to t inductively as follows.

(1) 0^L is 0 itself.

(2) a_i^L is $|a_i|$ and x_i^L is $|x_i|$.

(3) If t is of the form $f(t_1, \ldots, t_n)$, then t^L is $f(t_1^L, \ldots, t_n^L)$.

As in [11], we say "t is small" if $\exists x(t \leq |x|)$ or equivalently $\exists x(t = |x|)$. Then the meaning of the statement "$\#_{k+1}$ is a total function" is that $a\#_k b$ is small if a and b are small.

Let $t(a_1, \ldots, a_n)$ be a term in R_k^i in which all free variables are $a_1, \ldots, a_n$. Then $t(a_1, \ldots, a_n)$ is small in R_{k+1}^1 (but not in R_k^1) if all $a_1, \ldots, a_n$ are small.

In the following we use many Σ_1^b-definable functions in R_{k+1}^i. The thus extended system is a conservative extension of R_{k+1}^i which we identify with R_{k+1}^i.

Let $A(a)$ be Δ_1^b with respect to R_{k+1}^i. Then

$$\exists! x(|x| \leq |a| \land \forall i < |a|(Bit(i, x) = 1 \longleftrightarrow A(x)))$$

is provable in R_{k+1}^i, where $\exists! x$ is read "there exists a unique x". Therefore we can introduce a functional $K(|a|, \{x\}A(x))$ in R_{k+1}^i satisfying the condition

$$|K(|a|, \{x\}A(x))| \leq |a| \land \forall i < |a|(Bit(i, K(|a|, \{x\}A(x))) = 1 \longleftrightarrow A(i))).$$

Let all free variables in $A(a)$ be $a, a_1, \ldots, a_n$. *I.e.*, $A(a)$ is of the form $A(a, a_1, \ldots, a_n)$. Then $K(|a|, \{x\}A(x))$ is an abbreviation of $K_A(|a|, a_1, \ldots, a_n)$ where K_A is a new function symbol.

If t is small, then one can find a such that $|a| = t$. Therefore we often use $K(t, \{x\}A(x))$ in the place of $K(|a|, \{x\}A(x))$.

Let A be a semi-formula in $\overset{\circ}{U}{}_k^i(BD)$. We assign a semi-formula A^L to A inductively as follows.

(1) If A is of the form $t_1 \leq t_2$ of $t_1 = t_2$, then A^L is $t_1^L \leq t_2^L$ or $t_1^L = t_2^L$ respectively.

(2) If A is of the form $\beta^t(s)$ or $\psi^t(s)$, then A^L is $s^L \leq t^L \land Bit(s^L, b) = 1$ or $s^L \leq t^L \land Bit(s^L, y) = 1$ respectively.

(3) If A is of the form $\neg B$, $B \land C$, $B \lor C$, or $B \supset C$, then A^L is $\neg B^L$, $B^L \land C^L$, $B^L \lor C^L$, or $B^L \supset C^L$ respectively.

(4) Let $t(a)$ be a semi-term in $\overset{\circ}{U}{}_k^i(BD)$ and $(t(a))^L$ is $\tilde{t}(|a|)$. Then we define $t^L(a)$ to be $\tilde{t}(a)$. Let $F(a)$ be a semi-formula in $\overset{\circ}{U}{}_k^i(BD)$ and $(F(a))^L$ be $\tilde{F}(|a|)$. Then we define $F^L(a)$ to be $\tilde{F}(a)$.

(5) If A is of the form $\forall x \leq tB(x)$ or $\exists x \leq tB(x)$, then A^L is $\forall x \leq t^L B^L(x)$ or $\exists x \leq t^L B(x)$ respectively.

(6) If A is of the form $\forall x B(x)$ or $\exists x B(x)$, then A^L is $\forall x B^L(|x|)$ or $\exists x B^L(|x|)$ respectively.

(7) If A is of the form $\forall \psi^t B(\psi^t)$ or $\exists \psi^t B(\psi^t)$, then A^L is $\forall y < 2^{(t^L+1)}(B(\psi^t))^L$ or $\exists y < 2^{(t^L+1)}(B(\psi^t))^L$ respectively. Here we can use the notion $2^{(t^L+1)}$ since t^L is always small in R^i_{k+1}.

Obviously, if A is $\Sigma^{1,b}_i$ in $\overset{\circ}{U}{}^i_k(BD)$, then A^L is Σ^b_i in R^i_{k+1}. As in [12], the following lemma is easily proved.

Lemma 17. *(1)* $(t(s))^L$ *is* $t^L(s^L)$*. (2)* $(F(s))^L$ *is* $F^L(s^L)$*.*

Let Γ be $A_1, \ldots, A_n$. Then Γ^L is defined to be $A^L_1, \ldots, A^L_n$.

Theorem 18. *Let* $\Gamma \to \Delta$ *be provable in* $\overset{\circ}{U}{}^i_k(BD)$*. Then* $\Gamma^L \to \Delta^L$ *is provable in* R^i_{k+1}*.*

Proof. As in the proof of Theorem 1 in §2 in [12], we prove this by induction on the number of inferences in a proof of $\Gamma \to \Delta$ in $\overset{\circ}{U}{}^i_k(BD)$. We treat only the case which was not discussed in [12].

Case. The last inference to $\Gamma \to \Delta$ is

$$\frac{F(a),\ \Gamma_0 \to \Delta_0,\ F(a+1)}{F(0),\ \Gamma_0 \to \Delta_0,\ F(|t|)}.$$

By the induction hypothesis, we have in R^i_{k+1}

$$F^L(|a|), \Gamma^L_0 \to \Delta^L_0, F^L(|a|+1).$$

Hence we have in R^i_{k+1}

$$F^L(|\lfloor \tfrac{1}{2}a \rfloor|),\quad \Gamma^L_0 \to \Delta^L_0,\quad F^L(|a|)$$

and therefore

$$F^L(0), \Gamma^L_0 \to \Delta^L_0, F^L(||\tilde{t}||)$$

where $\tilde{t}$ satisfies that $|\tilde{t}| = t^L$ is provable in R^1_{k+1}.

¿From this follows in R^i_3

$$F^L(0),\quad \Gamma^L_0 \to \Delta^L_0,\quad F^L(|t^L|). \quad \square$$

§5. Relations between R^i_{k+1} and $\overset{\circ}{U}{}^i_k(BD)$..

We formalize the construction and proofs in Sects. 3 and 4 in S^1_2. First the operations H and L are obviously Σ^b_1-definable in S^1_2 and elementary constructions and elementary proofs of basic properties in H and L can also be carried out in S^1_2. Proofs of theorems in Sects. 3 and 4 are Σ^b_1-tree inductions on proof-trees and can be formalized in S^1_2. Obviously bounded proofs are transformed into bounded proofs. Therefore we have the following theorem.

Theorem 19.

$$S^1_2 \vdash Con(R^i_{k+1}) \longleftrightarrow Con(\overset{\circ}{U}{}^i_k(BD))$$

and

$$S^1_2 \vdash BDCon(R^i_{k+1}) \longleftrightarrow BDCon(\overset{\circ}{U}{}^i_k(BD)),$$

where by $Con(S)$ we mean "the consistency of S" and by $BDCon(S)$ we mean the statement "there are no bounded proof to a contradiction". (A proof P is bounded if no unbounded first order quantifiers occur in P.)

Now let A be a formula in R^i_{k+1}. We assume that the free variables in R^i_{k+1} are $a_0, a_1, a_2, \ldots$. By operation H, a_i becomes a_i and $\alpha^{a_i}_i$. In order to apply the operation L to A^H, one has to change α_i to β_i since we assume that all second order free variables in $\overset{\circ}{U}{}^i_k(BD)$ are of the form $\beta^t_0, \beta^t_1, \ldots$. By A^{HL} we mean the formula obtained from A^H by applying the operation L after renaming α to β. Now a_i in A becomes $|a_i|$, b_i in A^{HL}. By A^{HL*} we mean the formula obtained from A^{HL} by renaming b_i to a_i. We define $\tilde{K}(a)$ to be

$$K(|a|, \{x\}(x \leq |a| \wedge Bit(x, b) = 1))$$

and $\tilde{K}(a_i)$ is obtained from $\tilde{K}(a)$ by substituting a_i for a and b_i for b. □

In the same way as in §3 in [12], we have the following lemma and theorems.

Lemma 20. *Let $s(a_1, \ldots, a_n)$ be a term in R^1_{k+1}. Then R^1_{k+1} proves*

$$K(\tilde{T}, \{x\}(x \leq \tilde{T} \wedge \tilde{A}(x))) = s(\tilde{K}(a_1), \ldots, \tilde{K}(a_n))$$

where $\tilde{T}$ is $T^{s(a_1,\ldots,a_n)L}$ and $\tilde{A}(x)$ is $A^{s(a_1,\ldots,a_n)L}(x)$.

Theorem 21. *Let $A(a_1, \ldots, a_n)$ be a formula in R^1_{k+1}. Then R^1_{k+1} proves*

$$(A(a_1, \ldots, a_n))^{HL} \longleftrightarrow A(\tilde{K}(a_1), \ldots, \tilde{K}(a_n)).$$

Theorem 22. *Let A be a formula in R^1_{k+1}. Then $A^{HL*} \longleftrightarrow A$ is provable in R^1_{k+1}.*

A term is said to be simple if it is constructed from variables, SO, S, $+$, $\cdot$, and $\#_k$. For every term $t(a_1, \ldots, a_n)$ in $\overset{\circ}{U}{}^i_k(BD)$, there exists a simple term $s(a_1, \ldots, a_k)$ such that $t(a_1, \ldots, a_k) \leq s(a_1, \ldots, a_k)$ is provable without using logical inferences, induction, or any free variables other than $a_1, \ldots, a_k$. Since $a + b \leq (a+1) \cdot (b+1)$, $a \cdot b \leq (1\#a)\#(1\#b)$, and $1 + a \leq 1\#a$, we may even assume that $s(a_1, \ldots, a_k)$ is constructed form $1, a_1, \ldots, a_k$ and $\#_k$. It is easy to give a construction of such $s(a_1, \ldots, a_n)$ from $t(a_1, \ldots, a_k)$. Such $s(a_1, \ldots, a_k)$ is denoted by $t^*(a_1, \ldots, a_k)$ in this section.

Let A be a bounded formula in $\overset{\circ}{U}{}^i_k(BD)$. We assign to every bound variable x a simple term $s(\vec{a})$ in the following way where $\vec{a}$ is a sequence of all free variables in A.

If the bounded quantifier of x is of the form $Qx \leq s(\vec{a})$, then $s^*(\vec{a})$ is assigned to x. If the bounded quantifier of x is of the form

$$Qx \leq s(x_1, \ldots, x_i, \vec{a})$$

then $s^*(s_1(\vec{a}), \ldots, s_i(\vec{a}), \vec{a})$ is assigned to x, where $s^*_\ell(\vec{a})$ is a simple term assigned to x_ℓ. The assigned $s(\vec{a})$ is simple since $s_\ell(\vec{a})$ are assumed to be simple.

Let $\beta^{s(x_1,\ldots,x_i,\vec{a})}$ or $\psi^{s(x_1,\ldots,x_i,\vec{a})}$ occur in A. We call $s(x_1, \ldots, x_i, \vec{a})$ the bound of
$\beta^{s(x_1,\ldots,x_i,\vec{a})}$ or $\psi^{s(x_1,\ldots,x_i,\vec{a})}$. Since $\overset{\circ}{U}{}^i_k(BD)$ proves

$$s_1 \leq s_2 \rightarrow \forall x(\beta^{s_1}(x) \longleftrightarrow x \leq s_1 \wedge \beta^{s_2}(x)),$$

$\overset{\circ}{U}{}^i_k(BD)$ proves $A \longleftrightarrow \tilde{A}$, where $\tilde{A}$ is obtained from A by substituting

$$\{y\}(y \leq s(x_1, \ldots, x_i, \vec{a}) \wedge \beta^{s^*(s_1(\vec{a}),\ldots,s_i(\vec{a}),\vec{a})}(y))$$

for $\beta^{s(x_1,\ldots,x_i,\vec{a})}$ and

$$\{y\}(y \leq s(x_1, \ldots, x_i, \vec{a}) \wedge \psi^{s^*(s_1(\vec{a}),\ldots,s_i(\vec{a}),\vec{a})}(y))$$

for $\psi^{s(x_1,\ldots,x_i,\vec{a})}$. We can even replace $s^*(s_1(\vec{a}), \ldots, s_i(\vec{a}), \vec{a})$ by some other $\tilde{s}(\vec{a})$ if
$s^*(s_1(\vec{a}), \ldots, s_i(\vec{a}), \vec{a}) \leq \tilde{s}(\vec{a})$. By this method, we can obtain a bounded formula $\tilde{A}$ satisfying

(1) If A is $\sum_j^{1,b}$, then $\tilde{A}$ is $\sum_j^{1,b}$.

(2) There exists a unique, simple $s(\vec{a})$ that is the bound of all second order variables in A and does not have any bound variables. In this case, $s(\vec{a})$ is called the bound of the second order variables in A, and A is said to be in the second order variable normal form.

Let A be a bounded formula in $\overset{\circ}{U}{}^i_k(BD)$ and in the second order variable normal form and let $s(\vec{a})$ be the bound of the second order variables. We are going to define $A^{LH\Box}$. In order to define A^{LH}, we extend the definition of H by assigning $\beta_i^{b_i}$ to b_i and $\psi_i^{y_i}$ to y_i. The formula $A^{LH\Box}$ is obtained from A^{LH} by substituting $\{z\}(z \leq a \wedge z+1=a)$ for α^a and $s(\vec{a})$ for all b_i.

Proposition 23. *Let s and t be terms in $\overset{\circ}{U}{}^i_k(BD)$.*

Then $|s| \leq T^{s^L}$ is provable in $\overset{\circ}{U}{}^i_k(BD)$.

Proof. We prove this induction on s. If s is a_i, then s^L is $|a_i|$. Therefore T^{s^L} is $T^{|*|}(a_i)$, i.e. $|a_i|$. Now let s be $f(t_1,\ldots,t_n)$. We treat only the cases where f is $|*|$ or MSP. Let s be $|t|$. Then s^L is $|t^L|$ and T^{s^L} is $|T^{t^L}|$. Since $|t| \leq T^{t^L}$, by the induction hypotheses, we have $|s| \leq |T^{t^L}| = T^{s^L}$. Now let s be $MSP(t_1,t_2)$. Then T^{s^L} *is* $T^{t_1^L}$. We have $|t_1| \leq T^{t_1^L}$ by induction the hypothesis. Therefore we have $|s| \leq |t_1| \leq T^{t_1^L} \leq T^{s^L}$. □

Proposition 24. *Let s be a term in $\overset{\circ}{U}{}^i_k(BD)$. Then $\{z\}(z \leq T^{s^L} \wedge A^{s^L\Box}(z))$ and $\{z\}(z \leq |s| \wedge Bit(z,s)=1)$ are equivalent in $\overset{\circ}{U}{}^i_k(BD)$.*

Proof. We prove this by induction on s. Let s be a. Then T^{s^L} is $|a|$ and $A^{s^L}(z)$ is

$$z \leq |a| \wedge \exists y \leq a(\alpha^a(y) \wedge \forall x \leq a(y < x \supset \neg\alpha^a(x)) \wedge Bit(z,y+1)=1.$$

Therefore $A^{s^L\Box}(z)$ is

$$z \leq |a| \wedge \exists y \leq a(y+1=a \wedge \forall x \leq a(y<x \supset \neg x+1=a) \wedge Bit(z,y+1)=1)$$

which is equivalent to $z \leq |a| \wedge Bit(z,a)=1$. Now let s be $f(t_1,\ldots,t_n)$. We treat only the cases where f is $|*|$ or MSP.

Let s be $|t|$. Then s^L is $|t^L|$ and T^{s^L} is $|T^{t^L}|$. By induction hypothesis $(V^{t^L})^\Box$ is equivalent to $\{z\}(z \leq |t| \wedge Bit(z,t)=1)$ in $\overset{\circ}{U}{}^i_k(BD)$. Therefore

$\{z\}(z \leq T^{s^L} \wedge A^{s^L\Box}(z))$ is equivalent to

$$\{z\}(z \leq |T^{t^L}| \wedge \exists y \leq T^{t^L}((y \leq |t| \wedge Bit(y,t) = 1)$$
$$\wedge \forall x \leq T^{t^L}(y < x \supset \neg(x \leq |t| \wedge Bit(x,t) = 1)) \wedge Bit(z, y+1) = 1)$$

which is equivalent to $\{z\}(z \leq \|t\| \wedge Bit(z, |t|) = 1)$*i.e.* $\{z\}(z \leq |s| \wedge Bit(z,s) = 1)$. Now let s be $MSP(t_1, t_2)$. Then s^L is $MSP(t_1^L, t_2^L)$. T^{s^L} is $T^{t_1^L}$ and $\{x\}A^{s^L\Box}(x)$ is

$$\{x\}\exists y \leq T^{t_1^L}(\{z\}(z \leq |y| \wedge Bit(z,y) = 1) \overset{H}{=} \{z\}(z \leq T^{t_2^L} \wedge A^{t_2^L\Box}(z))$$
$$\wedge\, y + x \leq T^{t_1^L} \wedge A^{t_1^L}(y+x)).$$

By induction hypothesis, $\{z\}(z \leq T^{t_2^L} \wedge A^{t_2^L\Box}(z))$ is equivalent to

$$\{z\}(z \leq |t_2| \wedge Bit(z, t_2) = 1).$$

Therefore $\{x\}A^{s^L\Box}(x)$ is equivalent to

$$\{x\}\exists y \leq T^{t_1^L}(y = t_2 \wedge A^{t_1^L\Box}(y+x) \wedge y + x \leq T^{t_1^L}) \quad \text{i.e.}$$
$$\{x\}(t_2 + x \leq T^{t_1^L} \wedge A^{t_1^L\Box}(t_2 + x)).$$

By the induction hypothesis $\{x\}(x \leq T^{t_1^L} \wedge A^{t_1^L\Box}(x))$ is equivalent to $\{x\}(x \leq |t_1| \wedge Bit(x, t_1) = 1)$. Therefore $\{x\}A^{s^L\Box}(x)$ is equivalent to $\{x\}(t_2 + x \leq |t_1| \wedge Bit(t_2 + x, t_1) = 1)$. The latter is equivalent to $\{x\}(x \leq |MSP(t_1, t_2)| \wedge Bit(x, MSP(t_1, t_2)) = 1)$. $\Box$

Proposition 25. *Let s and t be terms in $\mathring{U}_k^i(BD)$. Then $\mathring{U}_k^i(BD)$ proves*
(1) $(s \leq t)^{LH\Box} \longleftrightarrow s \leq t$
(2) $(\beta^{t_0}(s))^{LH\Box} \longleftrightarrow \beta^{t_0}(s)$, where t_0 is the bound of the second order variables.

Proof. (1) By Proposition 24, $(s \leq t)^{LH\Box}$ is equivalent to

$$\{z\}(z \leq |s| \wedge Bit(z,s) = 1) \overset{H}{\leq} \{z\}(z \leq |t| \wedge Bit(z,t) = 1).$$

This is obviously equivalent to $s \leq t$.

(2) By definition $(\beta^{t_0}(s))^L$ is $s^L \leq t_0^L \wedge Bit(s^L, b) = 1$. Therefore $(\beta^{t_0}(s))^{LH\Box}$ is equivalent to $s \leq t_0 \wedge Bit(s^L, b) = 1)^{H\Box}$. Since $(Bit(a, b))^H$ is equivalent to

$$\exists y \leq b(\beta^b(y) \wedge |y| \leq a + 1 \wedge \forall z \leq a(Bit(z, y) = 1 \longleftrightarrow \alpha^a(z)),$$

by Proposition 24 $(Bit(s^L, b))^{H\Box}$ is equivalent to

$$\exists y \leq t_0(\beta^{t_0}(y) \wedge |y| \leq T^{s^L} + 1 \wedge \forall z \leq T^{s^L}(Bit(z, y) = 1 \longleftrightarrow z \leq |s| \wedge Bit(z, s) = 1)).$$

This is equivalent to $s \leq t_0 \wedge \beta^{t_0}(s)$. $\Box$

Theorem 26. *Let A be a bounded formula in a normal bound form in $\overset{\circ}{U}{}^i_k(BD)$. Then $A^{LH\Box}$ is equivalent to A in $\overset{\circ}{U}{}^i_k(BD)$.*

Proof. We prove this by induction on the number of logical symbols in A. If A is atomic, then the theorem follows from Proposition 25. If A is not atomic, we treat the case according to the outermost logical symbol of A. Since other cases are similar or trivial, we treat only the cases that A is of the form $\exists x \leq sB(x)$ or $\exists \psi^{t_0} B(\psi^{t_0})$, where t_0 is the normal bound of A.
(1) A is of the form $\exists x \leq sB(x)$.
In this case A^L is $\exists x \leq s^L B^L(x)$, where $(B(a))^L$ is $B^L(|a|)$. Then A^{LH} is

$$\exists x \leq T^{s^L} \exists \varphi^x(\{y\}(y \leq x \wedge \varphi^x(y)) \overset{H}{\leq} \{y\}(y \leq T^{s^L} \wedge A^{s^L}(y)) \wedge B^{LH}(\varphi^x, x)).$$

Therefore $A^{LH\Box}$ is

$$\exists x \leq T^{s^L} \exists \varphi^x(\{y\}(y \leq x \wedge \varphi^x(y)) \overset{H}{\leq} \{y\}(y \leq T^{s^L} \wedge A^{s^L\Box}(y)) \wedge B^{LH\Box}(\varphi^x, x)).$$

By Proposition 24 this is equivalent to

$$\exists x \leq T^{s^L} \exists \varphi^x(\{y\}(y \leq x \wedge \varphi^x(y)) \overset{H}{\leq} \{y\}(y \leq |s| \wedge Bit(z, s) = 1) \wedge B^{LH\Box}(\varphi^x, x)).$$

This is equivalent to $\exists x \leq sB^{LH\Box}(\{y\}(y \leq |x| \wedge Bit(y, x) = 1), x))$. Since $B^{LH\Box}(\{y\}(y \leq |a| \wedge Bit(y, a) = 1), a))$ is equivalent to $(B^{LH}(\alpha^a, a))^\Box$ and therefore to $(B(a))^{LH\Box}$*i.e.* to $B(a)$, $A^{LH\Box}$ is equivalent to $\exists x \leq sB(x)$.
(2) A is of the form $\exists \psi^{t_0} B(\psi^{t_0})$
A is equivalent to $\exists x \leq t_0 \exists \psi^x B(\psi^x)$. By (1), $(\exists x \leq t_0 \exists \psi^x B(\psi^x))^{LH\Box}$ is equivalent to $\exists a \leq t_0(\exists \psi^a B(\psi^a))^{LH\Box}$. Now $(\exists \psi^a B(\psi^a))^{LH\Box}$ is equivalent to

$$\exists y \leq T^{2^{|a|+1}} \exists \psi^y(\{z\}(z \leq y \wedge \psi^y(z)) \overset{H}{<} \{z\}(z \leq T^{2^{|a|+1}} \wedge A^{2^{|a|+1}}(z))^\Box \wedge (B(\psi^a))^{LH\Box}).$$

Now in the same way as Proposition 24, $\{z\}(z \leq T^{2^{|a|+1}} \wedge A^{2^{|a|+1}}(z))^{\Box}$ is equivalent to $\{z\}(z \leq a+2 \wedge Bit(z, 2^{a+1}) = 1$ i.e. $\{z\}(z = a+1)$. Since

$$\forall y \leq T^{2^{|a|+1}} \forall \psi^y(\{z\}(z \leq y \wedge \psi^y(z)) \overset{H}{<} \{z\}(z = a+1) \supset$$
$$\exists u \leq a \exists \zeta^u(\{z\}(z \leq y \wedge \psi^y(z)) \overset{H}{=} \{z\}\zeta^u(z))$$

and

$$\forall u \leq a \forall \zeta^u \exists y \leq T^{2|a|+1} \exists \psi^y(\{z\}(z \leq y \wedge \psi^y(z)) \overset{H}{<} \{z\}(z = a+1)$$
$$\wedge \{z\}(z \leq y \wedge \psi^y(z)) \overset{H}{=} \{z\}\zeta^u(z)),$$
$$(\exists \psi^a B(\psi^a))^{LH\Box}$$

is equivalent to $\exists y \leq a \exists \psi^y(B(\psi^a))^{LH\Box}$ and $\exists \psi^a(B(\psi^a))^{LH\Box}$. In order to see the form $(B(\psi^a))^{LH\Box}$, let us see $(\psi^a(s))^{LH\Box}$ where s is a term in $\overset{\circ}{U}{}^i_2(BD)$. Then $(\psi^a(s))^{LH\Box}$ is $(s^L \leq a^L \wedge Bit(s^L, y) = 1)^{H\Box}$. By using (1) of Proposition 25, this is equivalent to

$$(s \leq a \wedge Bit(s^L, y) = 1)^{H\Box}.$$

In the same way as in the proof of (2) of Proposition 25, $Bit(s^L, y) = 1)^{H\Box}$ is equivalent to

$$\exists z \leq y(\psi^y(z) \wedge |z| \leq s^L + 1 \wedge \forall u \leq s^L(Bit(u, z) = 1 \longleftrightarrow u \leq |s| \wedge Bit(u, s) = 1)$$

which is equivalent to $\psi^y(s)$. Therefore $(\psi^a(s))^{LH\Box}$ is equivalent to $s \leq a \wedge \psi^y(s)$. Under the hypothesis $y \leq a$, this is equivalent to $\psi^a(s)$. Therefore $(B(\psi^a))^{LH\Box}$ is equivalent to $B^{LH\Box}(\psi^a)$ and to $B(\psi^a)$ by the induction hypothesis. In general s might have bound variables. However, by the proof of (1), $(\exists x \leq sB(x))^{LH\Box}$ is equivalent to $\exists a \leq s(B(a))^{LH\Box}$. Therefore we may assume that s does not have any bound variables. $\Box$

Theorem 27. *Let A be a formula in R^i_{k+1}. A is provable in R^i_{k+1} iff A^H is provable in $\overset{\circ}{U}{}^i_k(BD)$.*

Proof. If A is provable in R^i_{k+1}, then A^H is provable in $\overset{\circ}{U}{}^i_k(BD)$ by Theorem 16. If A^H is provable in $\overset{\circ}{U}{}^i_k(BD)$, then A^{HL} and A^{HL^*} is provable in R^i_{k+1}. Since $A \longleftrightarrow A^{HL^*}$ is provable in R^i_{k+1}, A is provable in R^i_{k+1}. $\Box$

Theorem 28. *Let A be a bounded formula in $\overset{\circ}{U}{}^i_k(BD)$. A is provable in $\overset{\circ}{U}{}^i_k(BD)$ iff A^L is provable in R^i_{k+1}.*

Proof. Without loss of generality, we assume that A has a normal bound t_0. If A is provable in $\overset{\circ}{U}{}^i_k(BD)$, then A^L is provable in R^i_{k+1} by Theorem 18. If A^L is provable in R^i_{k+1}, then A^{LH} and $A^{LH\Box}$ is provable in $\overset{\circ}{U}{}^i_k(BD)$. Since $A^{LH\Box} \longleftrightarrow A$ is provable in $\overset{\circ}{U}{}^i_k(BD)$, A is probable in $\overset{\circ}{U}{}^i_k(BD)$. □

In the same way, we have the following theorem.

Theorem 29. *(1) Let A be a formula in S^i_{k+1}. A is provable in S^i_{k+1} iff A^H is provable in $\overset{\circ}{V}{}^i_k(BD)$.*

(2) Let A be a bounded formula in $\overset{\circ}{V}{}^i_k(BD)$. A is provable in $\overset{\circ}{V}{}^i_k(BD)$ iff A^L is probable in S^i_{k+1}.

J. Krajíček proved the following theorem in [6].

Theorem 30. *Let $\mathcal{M} = (M, X)$ be a model of V_2 not satisfying EXP, i.e. $\mathcal{M} \models V_2 + \neg EXP$. Then $\mathcal{M} \models B\Sigma^o_1$*

The following related theorem is proved by the method we have developed though it is not its immediate corollary.

Theorem 31. *Let $A(a, b)$ be a Σ^b_i formula. Then V^{i+1}_2 proves*

$$\forall x \leq a \exists y (A(x, y) \wedge |y| \leq b) \rightarrow \exists t \forall x \leq a \exists y \leq t A(x, y).$$

Proof. In this proof we use (a, α) to represent $\Sigma_{i<a} \epsilon_i 2^i$ where $\epsilon_i = 1$ if $\alpha(i)$ and $\epsilon_i = 0$ otherwise. We assign $\alpha, \beta, \ldots, \varphi, \psi, \ldots$ to $a, b, \ldots, x, y, \ldots$ respectively as before and make a correspondence between a and $(|a|, \alpha)$ with $F(a, \alpha)$ where $F(a, \alpha)$ is $\forall x \leq |a| (\alpha(x) \longleftrightarrow Bit(x, a) = 1)$. Then V^1_2 proves

$$F(a, \alpha), F(b, \beta) \rightarrow a \leq b \longleftrightarrow (|a|, \alpha) \overset{H}{\leq} (|b|, \beta).$$

Therefore we can replace $\forall x \leq t$ by $\forall x \leq |t| \forall \varphi$ provided that the occurrences of x and $\leq$ will be replaced by (x, φ) and $\overset{H}{\leq}$ respectively. In this way we can find a second order version $\tilde{A}(a, \alpha, b, \beta)$ of $A(a, b)$ such that V^1_2 proves

$$F(a, \alpha), F(b, \beta) \rightarrow A(a, b) \longleftrightarrow \tilde{A}(|a|, \alpha, |b|, \beta),$$

and $\tilde{A}(|a|, \alpha, |b|, \beta)$ is a $\Sigma_i^{1,b}$ formula. Then in V_2^1, $\forall x \leq a \exists y(A(x, y) \wedge |y| \leq b)$ implies $\forall x \leq |a| \forall \varphi((x, \varphi) \leq a \supset \exists \psi \tilde{A}(x, \varphi, b, \psi))$ where $(x, \varphi) \leq a$ is an abbreviation of

$$(x, \varphi) \overset{H}{\leq} (|a|, \{y\}(Bit(y, a) = 1)).$$

Let $B(x, \varphi, c, \xi)$ be the statement saying "(c, ξ) is the minimum satisfying $\tilde{A}(x, \varphi, c, \xi)$", i.e. let $B(x, \varphi, c, \xi)$ be

$$\tilde{A}(x, \varphi, c, \xi) \wedge \forall x \leq c \forall \eta((x, \eta) \overset{H}{<} (c, \xi) \supset \neg \tilde{A}(x, \varphi, c, \xi)).$$

$B(x, \varphi, c, \xi)$ is a $\Sigma_i^{1,b} \wedge \Pi_i^{1,b}$ formula. Therefore V_2^{i+1} proves

$$\forall x \leq |a| \forall \varphi((x, \varphi) \leq a \supset \exists c \leq b \exists \xi(B(x, \varphi, c, \xi) \wedge \exists y(y = (c, \xi))),$$

where $y = (c, \xi)$ is an abbreviation of

$$(c, \xi) \overset{H}{=} (|y|, \{z\}(Bit(z, y) = 1)).$$

Now let $\tilde{c} \leq b$ be the maximum among $\tilde{c}$ satisfying $\exists x \leq a \exists \varphi((x, \varphi) \leq a \wedge \exists \tilde{\xi} B(x, \varphi, \tilde{c}, \tilde{\xi}))$. Since $B(x, \varphi, c, \xi) \wedge \exists y(y = (c, \xi))$ and $(\tilde{c}, \tilde{\xi})$ is the minimum satisfying $\tilde{A}(x, \varphi, \tilde{c}, \tilde{\xi})$, there exists $\tilde{y} = (\tilde{c}, \tilde{\xi})$. Let $t = 2\tilde{y}$. Then $\forall x \leq a \exists y \leq t A(x, y)$ and the theorem is proved. □

If R_{k+1}^i and S_{k+1}^i are equivalent with respect to the bounded formulas, then R_{k+1}^i and S_{k+1}^i are equivalent. It is easily seen that PIND on the bounded formulas can be expressed by

$$A(0) \wedge \forall x \leq a(A(\lfloor \frac{1}{2} a \rfloor) \supset A(x)) \supset A(a)$$

where $A(a)$ is a bounded formula. In the same way, if $U_k^i(BD)$ and $V_k^i(BD)$ are equivalent with respect to the bounded formulas, then they are equivalent.

Corollary 32. *R_{k+1}^i and S_{k+1}^i are equivalent iff $U_k^i(BD)$ and $V_k^i(BD)$ are equivalent.*

Corollary 33. *R_{k+1}^{i+1} and S_k^i are equivalent iff $U_k^{i+1}(BD)$ and $V_k^i(BD)$ are equivalent.*

Corollary 34. *(1)* $V_k^i \preceq_{\Sigma_\ell^{1,b}(\#_k)} V_k^j$ *iff* $S_{k+1}^i \preceq_{\Sigma_\ell^b(\#_{k+1})} S_{k+1}^j$.
(2) $U_k^i \preceq_{\Sigma_\ell^{1,b}(\#_k)} U_k^j$ *iff* $R_{k+1}^i \preceq_{\Sigma_\ell^b(\#_{k+1})} R_{k+1}^j$.
(3) $U_k^i \preceq_{\Sigma_\ell^{1,b}(\#_k)} V_k^j$ *iff* $R_{k+1}^i \preceq_{\Sigma_\ell^b(\#_{k+1})} S_{k+1}^j$.
(4) $V_k^i \preceq_{\Sigma_\ell^{1,b}(\#_k)} U_k^j$ *iff* $S_{k+1}^i \preceq_{\Sigma_\ell^b(\#_{k+1})} R_k^j$.

Corollary 35. *If* $\Sigma_{i+2}^b(\#_{k+1}) \neq \Pi_{i+2}^b(\#_{k+1})$, *then* $V_k^{i+1}(BD)$ *and* $V_k^i(BD)$ *are not equivalent with respect to* $\Sigma_{i+2}^{1,b}(\#_k)$.

Proof. By Theorem B in Krajíček–Pudlák–Takeuti [7], $\Sigma_{i+2}^b(\#_{k+1}) \neq \Pi_{i+2}^b(\#_{k+1})$ implies T_{k+1}^i and S_{k+1}^{i+1} are not equivalent with respect to $\Sigma_{i+2}^b(\#_{k+1})$ and therefore so are not S_{k+1}^i and S_{k+1}^{i+1}. From this follows under the same assumption that $V_k^{i+1}(BD)$ and $V_k^i(BD)$ are not equivalent with respect to $\Sigma_{i+2}^{1,b}(\#_k)$. □

Corollary 36. *TNC is equivalent to* S_2^1 *iff* $U_1^1(BD)$ *is equivalent to* $V_1^1(BD)$.

Corollary 37. *If* $V_1^1(BD)$ *and* $U_1^1(BD)$ *are equivalent with respect to boolean combination of* $\Sigma_1^{1,b}$, *then* $NC = P$.

Proof. If $V_1^1(BD)$ and $U_1^1(BD)$ are equivalent with respect to $\Sigma_1^{1,b} \cup \Pi_1^{1,b}$, then S_2^1 and TNC are equivalent with respect to $\Sigma_1^b \cup \Pi_1^b$. A predicate in P is Δ_1^b with respect to S_2^1 and hence Δ_1^b with respect to TNC and therefore belongs to NC. □

Open problem. Is T_{k+1}^i a subsystem of R_{k+1}^{i+1}? The affirmative answer of this question implies that S_{k+1}^{i+1} is equivalent to R_{k+1}^{i+1} with respect to Σ_{i+1}^b by a result in Buss [3] and therefore that $V_k^{i+1}(BD)$ is equivalent to $U_k^{i+1}(BD)$ with respect to $\Sigma_{i+1}^{1,b}$. We conjecture that this is not the case.

References

[1]. B. Allen, *Arithmetizing Uniform NC*, Ann. Pure Appl. Logic **53** (1991), 1–50.
[2]. S. Buss, *Bounded Arithmetic*, Bibliopolis, Napoli, 1986.
[3]. S. Buss, *Axiomatizations and Conservation Results for Fragments of Bounded Arithmetic*, Logic and Computation ed. W. Sieg, Contemporary Mathematics 106, pp. 57–84, 1990, AMS.
[4]. P. Clote, *A first order theory for the parallel complexity class NC*, Tech. Rep BCCS-8901, Boston College, Jan. 1989.
[5]. P. Clote and G. Takeuti, *Bounded Arithmetic for NC, AlogTIME, L and NL*, to appear in Annals of Pure and Applied Logic.
[6]. J. Krajíček, *Exponentiation and second order bounded arithmetic*, Annals of Pure and Applied Logic **48** (1990), 261–276.

[7]. J. Krajíček, P. Pudlák, and G. Takeuti, *Bounded Arithmetic and the Polynomial Hierarchy*, Annals of Pure and Applied Logic **52** (1991), 143–153.

[8]. J. Krajíček and G. Takeuti, *On Bounded* $\sum_1^1$Polynomial Induction, Feasible Mathematics, ed. S. Buss and P. Scott, pp. 259–280, 1990, Birkhäuser.

[9]. A. A. Razborov, *An Equivalence Between Second Order Bounded Domain Bounded Arithmetic and First Order Bounded Arithmetic*, this volume.

[10]. G. Takeuti, *Proof Theory Second edition*, North–Holland, 1987.

[11]. G. Takeuti, *Bounded Arithmetic and Truth Definition*, Annals of Pure and Applied Logic **39** (1988), 75–104.

[12]. G. Takeuti, S_3^i *and* $\overset{\circ}{V}{}_2^i(BD)$, Archive for Mathematical Logic **29** (1990), 149–169.

Gaisi Takeuti
Department of Mathematics
University of Illinois
1409 West Green Street
Urbana, IL 61801
email: takeuti@symcom.math.uiuc.ed

Feasible interpretability

Rineke Verbrugge*

Abstract

In PA, or even in $I\Delta_0 + EXP$, we can define the concept of feasible interpretability. Informally stated, U feasibly interprets V iff:

> for some interpretation, U proves the interpretations of all axioms of V by proofs with Gödel numbers of length polynomial in the length of the Gödel numbers of those axioms.

Here both U and V are Σ_1^b-axiomatized theories.

Many interpretations encountered in everyday mathematics (e.g. the interpretation of $ZF + \mathbf{V} = \mathbf{L}$ into ZF) are feasible. However, by fixed point constructions we can find theories that are interpretable in PA in the usual sense but not by a feasible interpretation. By making polynomial analogs of the usual proofs, we show that the bimodal interpretability logic ILM is sound for feasible interpretability over the base theory PA. Here, $A \rhd B$ is translated as $PA + A^* \rhd_f PA + B^*$, where * is the translation. Moreover, we can prove in PA a polynomial version of Orey's theorem for feasible interpretability. This paves the way for a polynomial adaptation of Berarducci's proof of arithmetical completeness of ILM with respect to PA. Thus, we show that ILM is arithmetically sound and complete with respect to feasible interpretability over PA.

1 Introduction

In this paper, we investigate a new concept of interpretability – we call it feasible interpretability – in which the complexity of the interpretation is

*Research supported by the Netherlands Organization for Scientific Research (NWO).

bounded in a certain way. The concept was invented by Albert Visser, who called it effective interpretability in his paper [Vi].

In order to define this concept, we first review the usual definition of interpretability. Let U, V be two Σ_1^b-axiomatized theories, where V is axiomatized by the Σ_1^b-formula α_V. An interpretation K of V into U is given as usual by a formula $\delta(x)$ of L_U defining the universe, and a function from the relation and function symbols of L_V to formulas of L_U, respecting the original arities. In the sequel we take the image of $=$ to be $=$, though this is not essential for the results. We can extend K in the usual way to map all formulas φ of L_V into formulas φ^K of L_U; in fact we can, in an intensionally correct way, Δ_1^b-define in $I\Delta_0 + \Omega_1$ a function K corresponding to this mapping. For ease of reading, we will write a^K even if a is a Gödel number. Thus $U \rhd V$ can be defined in $I\Delta_0 + \Omega_1$ as follows:

$$\begin{aligned} I\Delta_0 + \Omega_1 \vdash U \rhd V \quad &\leftrightarrow \exists K(\text{“}K \textit{is an interpretation}\text{”} \\ &\wedge \forall a(\alpha_V(a) \to \exists p Prf_U(p, a^K))). \end{aligned}$$

Similarly, we would like to define a concept of *feasible interpretability*, given half-formally as

$$\begin{aligned} U \rhd_f V \leftrightarrow \exists K \exists P(&\text{“}K \textit{is an interpretation and } P \textit{is a polynomial}\text{”} \\ &\wedge \forall a(\alpha_V(a) \to \exists p(\text{“}|p| \le P(|a|)\text{”} \wedge Prf_U(p, a^K)))) \end{aligned} \tag{1}$$

If we want to formalize this concept, we need an evaluation function for coded polynomials and we need to be able to prove that the *exp* of this function is total. We remind the reader that *exp*(the values of polynomials in $|x|$) corresponds to the values of #-terms in x, where $x\#y = exp(|x| \cdot |y|)$ as defined in Buss[86]. Thus, since there is an evaluation function for formalized terms containing # that is provably total in $I\Delta_0 + EXP$, we see that the formalization of feasible interpretability can be carried out in $I\Delta_0 + EXP$. We will not carry out the details, and for ease of reading we will keep using the half-formal definition(1).

However, it is clear that the formula $U \rhd_f V$ is Σ_2^0. As we know that, for reasonable theories U extending PA, $\{A \,|\, U \rhd U + A\}$ is a Π_2^0-complete predicate, it would be interesting to find out whether $\{A \,|\, U \rhd_f U + A\}$ is Σ_2^0-complete. We haven't yet found the answer to this question.

In [Vi], Visser gave proof sketches to show that ILM is arithmetically sound with respect to feasible interpretability over PA. Moreover, he gave an Orey-Hájek like characterization for feasible interpretability over PA^*, where PA^* is defined as follows:

> C is an axiom of PA^* iff C is the conjunction of the first n axioms of PA for some n.

He then surmised that, using this characterization, Berarducci's arguments from [Be 90] could be adapted to show that ILM is the modal interpretability logic for feasible interpretability over PA^*.

In this paper, we show that ILM is indeed arithmetically sound and complete with respect to feasible interpretability over PA itself.

The rest of the paper is organized as follows. In section 2, we show that some well-known interpretations from the contexts of set theory and bounded arithmetic are feasible. For the subsequent sections, the horizon is narrowed down to Peano Arithmetic. Thus we prove in section 3 and section 5 that ILM is exactly the modal interpretability logic for feasible interpretability over PA. Section 4, meanwhile, gives two counterexamples to show that, for reasonable theories U extending PA, feasible interpretability over U is a definitely stricter concept than normal interpretability.

2 Feasible interpretations in various settings

For an intuitive introduction to feasible interpretability, it is useful to define feasible interpretability also for settings other than arithmetic. The informal definition is as follows.

> $U \rhd_f V$ if and only if there is an interpretation K of V into U which is feasible, i.e. for which there is a polynomial P such that for all axioms φ of V, there is a proof of length $\leq P(|\varphi|)$ in U of φ^K

Here $|\varphi|$ denotes the length of φ. In this section, we look at some well-known interpretations from different settings and show that they are feasible. As a first remark, it is clear that every interpretation of a finitely axiomatized theory into some other theory is feasible: a constant polynomial, namely the maximum of the lengths of the proofs of the interpreted axioms, suffices. We first prove an easy lemma which can be used to show that many well-known interpretations are feasible.

Remark 2.1 Of course the definitions of $|\varphi|$ and of the lengths of proofs depend on the setting. For example, it is not always convenient to define $|\varphi|$ as "the length of the binary expression for the Gödel number of φ".

However, we have to keep in mind that a few conditions on the definition of the lengths of formulas and proofs are necessary to make lemma 2.2 applicable.

The length of formulas should be defined in such a way that the following conditions hold:

1. $|\neg\psi| \geq |\psi| + 1$,
2. $|\psi \circ \chi| \geq |\psi| + |\chi| + 1$ for $\circ \in \{\wedge, \vee, \rightarrow, \leftrightarrow\}$,
3. $|Qx\psi| \geq |\psi| + 1$ for $Q \in \{\forall, \exists\}$, and
4. for all formulas φ, $|\varphi| \geq 2$.

The last of these conditions is not necessary, but it just simplifies the computations by allowing us to work with polynomials $P(n)$ of the form n^d only.

Moreover, we suppose that the proof system and the corresponding length of a proof is defined in such a way that applications of $\wedge$-rules and Modus Ponens do not make the proofs explode to an inordinate length; e.g. we suppose that we do not use a tableau system or a sequent calculus. A sufficient condition is the following.

There is a constant c such that the following conditions hold:

1. if $|A| \leq |B|$, and we have a proof of length l_A of the formula A, and a proof of length $l_{A\rightarrow B}$ of $A \rightarrow B$, then there is a proof of length $\leq l_A + l_{A\rightarrow B} + c \cdot |B|$ of the formula B; and
2. if $|A|, |B| \leq |C|$, and we have a proof of length l_A of A, a proof of length l_B of B and a proof of length $l_{A\wedge B\rightarrow C}$ of $A \wedge B \rightarrow C$, then we have a proof of length $\leq l_A + l_B + l_{A\wedge B\rightarrow C} + c \cdot |C|$ of the formula C.

Lemma 2.2 *Let L be a language and U a theory satisfying the conditions in Remark 2.1. Let F be a function from L into L_U such that*

there is a polynomial P such that for all $\varphi \in L$, $|F(\varphi)| \leq P(|\varphi|)$.

Moreover, suppose that U proves the following by proofs of length $\leq P(|\varphi|)$, resp. $\leq P(|\neg\psi|)$, resp. $\leq P(|\psi \circ \chi|)$, resp. $\leq P(|Qx\psi|)$:

1. *$F(\varphi)$ for all atomic $\varphi \in L$;*

2. *$F(\psi) \to F(\neg\psi)$ for all $\psi \in L$;*

3. *$F(\psi) \wedge F(\chi) \to F(\psi \circ \chi)$ for all $\psi, \chi \in L$ and $\circ \in \{\wedge, \vee, \to, \leftrightarrow\}$;*

4. *$F(\psi) \to F(Qx\psi)$ for all $\psi \in L$ and $Q \in \{\forall, \exists\}$.*

Then there is a polynomial R such that for all $\varphi \in L$, $U \vdash F(\varphi)$ by a proof of length $\leq R(|\varphi|)$.

Proof. We do not need to find the smallest possible polynomial bound R, which makes the proof quite simple. Take a constant $d \geq 2$ such that

1. for all $n \geq 2$, $P(n) \leq n^d$ and

2. for all $\varphi \in L$, $c \cdot |F(\varphi)| \leq |\varphi|^d$, where c is as in Remark 2.1 in the condition on the length of proofs.

Define the polynomial $R(n) := n^{d+2}$. We will prove by induction on the construction of φ that for all $\varphi \in L$, $U \vdash F(\varphi)$ by a proof of length $\leq R(|\varphi|)$.

Basic step By the assumption we know that for atomic formulas φ, $U \vdash F(\varphi)$ by a proof of length $\leq P(|\varphi|)$. But by definition of d, $P(|\varphi|) \leq |\varphi|^d \leq |\varphi|^{d+2}$.

$\neg$-step Suppose as induction hypothesis that $U \vdash F(\psi)$ by a proof of length $\leq |\psi|^{d+2}$. By assumption, $U \vdash F(\psi) \to F(\neg\psi)$ by a proof of length $\leq P(|\neg\psi|) \leq |\neg\psi|^d$ (where the last inequality holds because of clause 1 of the definition of d). Therefore by the first clause in the condition on the length of proofs in Remark 2.1, we have $U \vdash F(\neg\psi)$ by a proof of length $\leq |\psi|^{d+2} + |\neg\psi|^d + c \cdot |F(\neg\psi)| \leq |\psi|^{d+2} + |\neg\psi|^d + |\neg\psi|^d$ (where the last inequality holds by clause 2 of the definition of d). Since we assume that $|\neg\psi| \geq |\psi| + 1$, we have $|\psi|^{d+2} + |\neg\psi|^d + |\neg\psi|^d \leq |\neg\psi|^{d+2}$ by an easy computation using the binomial theorem and the fact that $d \geq 2$.

Connective step Let $\circ \in \{\wedge, \vee, \to, \leftrightarrow\}$. Suppose as induction hypothesis that $U \vdash F(\psi)$ by a proof of length $\leq |\psi|^{d+2}$, and $U \vdash F(\chi)$ by a proof of length $\leq |\chi|^{d+2}$. By assumption, $U \vdash F(\psi) \wedge F(\chi) \to F(\psi \circ \chi)$ by a proof of length $\leq P(|\psi \circ \chi|) \leq |\psi \circ \chi|^d$.

The second clause in the condition on the length of proofs in Remark 2.1 now implies that $U \vdash F(\psi \circ \chi)$ by a proof of length $\leq |\psi|^{d+2} + |\chi|^{d+2} + |\psi \circ \chi|^d + c \cdot |F(\psi \circ \chi)| \leq |\psi|^{d+2} + |\chi|^{d+2} + |\psi \circ \chi|^d + |\psi \circ \chi|^d$ (where the last inequality holds by clause 2 in the definition of d).

Since we assume that $|\psi \circ \chi| \geq |\psi| + |\chi| + 1$, we can again use the binomial theorem to show that $|\psi|^{d+2} + |\chi|^{d+2} + |\psi \circ \chi|^d + |\psi \circ \chi|^d \leq |\psi \circ \chi|^{d+2}$, as desired.

Quantifier step The quantifier steps are analogous to the $\neg$-step, so we leave them to the reader.

QED

Remark 2.3 For applications of lemma 2.2, we usually take $F(\psi)$ to be a schema in ψ and ψ^K, where K is an interpretation. When we want to prove that some interpretation K of V into U is feasible, we can often use lemma 2.2 in the following way. Suppose all axioms of V have the form $\Phi(\psi)$, where Φ is a formula scheme. The feature we need in order to apply lemma 2.2 is the fact that both $|\Phi(\psi)|$ and $|\psi^K|$ are polynomial in $|\psi|$.

An adapted version of lemma 2.2 works in case attention is restricted to Δ_1^0-formulas. An example of a function F for which the Δ_1^0-version of lemma 2.2 could be applied is the scheme $F(\psi) := \psi \leftrightarrow \psi^K$ for $\psi \in \Delta_1^0$, where K is a fixed interpretation such that $|\psi^K|$ is polynomial in $|\psi|$.

As a first example, in which we do not yet need lemma 2.2, we will show that the usual interpretation of $\mathrm{I}\Delta_0 + \Omega_1$ into $\mathrm{I}\Delta_0$ by a cut is feasible.

Theorem 2.4 *$I\Delta_0 \rhd_f I\Delta_0 + \Omega_1$ by a cut.*

Proof. Let J be a cut constructed by Solovay's methods such that $\mathrm{I}\Delta_0$ proves that J is a cut closed under $+, \cdot$, and ω_1 (see [HP, Ch. V.5]). Define φ^J to be the formula φ with all quantifiers restricted to J. It is well-known that J is an interpretation of $\mathrm{I}\Delta_0 + \Omega_1$ into $\mathrm{I}\Delta_0$; so to show that it is a feasible interpretation, it suffices to find a polynomial P such that for all Δ_1^0-formulas φ, the following holds by proofs of length $\leq P(|\ulcorner\varphi\urcorner|)$:

$$\mathrm{I}\Delta_0 \vdash [\varphi(0) \wedge \forall x(\varphi(x) \rightarrow \varphi(Sx)) \rightarrow \forall x \varphi(x)]^J$$

First, it is easy to see that there is a polynomial P_1 such that for all Δ_1^0-formulas φ, $\mathrm{I}\Delta_0 \vdash J(a) \rightarrow (\varphi(a) \leftrightarrow \varphi(a)^J)$ and $\mathrm{I}\Delta_0 \vdash \forall x\varphi \rightarrow (\forall x\varphi)^J$

by proofs of length $\leq P_1(|\ulcorner\varphi\urcorner|)$. Second, there is a polynomial P_2 such that for all Δ_1^0-formulas φ, the following holds by proofs of length $\leq P_2(|\ulcorner\varphi\urcorner|)$:

$$\mathrm{I}\Delta_0 \vdash \forall a\,[\varphi(0) \wedge \forall x \leq a\,(\varphi(x) \rightarrow \varphi(Sx)) \rightarrow \forall x \leq a\,\varphi(x)]$$

In fact one uses only the induction axiom for $\forall x \leq a\,\varphi(x)$, the fact that $\forall a \forall x(Sx \leq a \rightarrow x \leq a)$, and some predicate logic. Combining P_1 with P_2, we then find a polynomial P_3 such that for all Δ_1^0-formulas φ, the following holds by proofs of length $\leq P_3(|\ulcorner\varphi\urcorner|)$:

$$\mathrm{I}\Delta_0 \vdash (\forall a[\varphi(0) \wedge \forall x \leq a(\varphi(x) \rightarrow \varphi(Sx)) \rightarrow \forall x \leq a\,\varphi(x)])^J$$

Now it is easy to find a polynomial P from P_3 such that for all Δ_1^0-formulas φ, the following holds by proofs of length $\leq P(|\ulcorner\varphi\urcorner|)$:

$$\mathrm{I}\Delta_0 \vdash [\varphi(0) \wedge \forall x(\varphi(x) \rightarrow \varphi(Sx)) \rightarrow \forall x \varphi(x)]^J$$

We use only the fact that $\forall a(a \leq a)$ and some predicate logic. Thus, J is a feasible interpretation of $\mathrm{I}\Delta_0 + \Omega_1$ into $\mathrm{I}\Delta_0$. QED

Next, we will prove that the usual interpretation of $ZF + \mathbf{V} = \mathbf{L}$ into ZF is feasible. Because ZF consists of a finite list of axioms plus the schemata of separation and replacement, we can restrict our attention to feasibly proving these schemata relativized to the universe $\mathbf{L}$ of constructible sets. We will first prove that the schema of separation relativized to $\mathbf{L}$ follows feasibly from the reflection theorem for $\mathbf{L}$, and then give a feasible proof of the reflection theorem itself. We will try to follow the elegant proof in terms of closed unbounded collections, which unfortunately becomes much less elegant when forced into the strait-jacket of the calculation of lengths. We will not stray far from the straightforward presentation given in [Ku 80], where all details about the constructible universe that we omit here can be found. The length $|\varphi|$ of a formula φ of ZF is defined as the number of appearances of symbols in φ; without loss of generality, we can take the length of all variables to be 1. Likewise, we define the length of a proof in ZF to be the total number of symbols appearing in the proof. In the following lemmas, quantifiers in greek letters range over the ordinals, while those in roman letters range over all sets.

The next lemma corresponds to lemma IV.2.5 of [Ku 80].

Lemma 2.5 *ZF proves the following by proofs of length polynomial in* $|\varphi|$*:*

$$\begin{aligned}\forall z, \overline{v} \in \mathbf{L} \quad & \{x \in z \mid \varphi^{\mathbf{L}}(x, z, \overline{v})\} \in \mathbf{L} \rightarrow \\ & \forall z, \overline{v} \in \mathbf{L}\, \exists y \in \mathbf{L}\, [x \in y \leftrightarrow x \in z \wedge \varphi^{\mathbf{L}}(x, z, \overline{v})]\end{aligned}$$

Proof. Straightforward; we do not even need the fact that $\mathbf{L}$ is transitive. Note that by absoluteness of atomic formulas for $\mathbf{L}, \mathbf{V}$, the succedent is feasibly equivalent to the comprehension schema for φ, relativized to $\mathbf{L}$. QED

The following lemma corresponds to a part of lemma VI.2.1 of [Ku 80].

Lemma 2.6 *ZF proves the following by proofs of length polynomial in* $|\varphi|$*:*

$$\begin{aligned}\forall \alpha\, \exists \beta\, \forall z, x, \overline{v} \in \mathbf{L}_\beta \quad & [\varphi^{\mathbf{L}}(x, z, \overline{v}) \leftrightarrow \varphi^{\mathbf{L}_\beta}(x, z, \overline{v})] \rightarrow \\ & \forall z, \overline{v} \in \mathbf{L}\, \{x \in z \mid \varphi^{\mathbf{L}}(x, z, \overline{v})\} \in \mathbf{L}\end{aligned}$$

Proof. It is easy to see that the usual proof in ZF is feasible: suppose

1. $\forall \alpha\, \exists \beta\, \forall z, x, \overline{v} \in \mathbf{L}_\beta\, [\varphi^{\mathbf{L}}(x, z, \overline{v}) \leftrightarrow \varphi^{\mathbf{L}_\beta}(x, z, \overline{v})]$ and

2. $z, \overline{v} \in \mathbf{L}$

From 2 it follows that there is an α such that $z, \overline{v} \in \mathbf{L}_\alpha$. Now let $\beta > \alpha$ be such that $\forall x \in \mathbf{L}_\beta\, [\varphi^{\mathbf{L}}(x, z, \overline{v}) \leftrightarrow \varphi^{\mathbf{L}_\beta}(x, z, \overline{v})]$. Then, using the fact that $\mathbf{L}$ is transitive and that $x \in z$ is absolute for $\mathbf{L}_\beta, \mathbf{L}$, we find that

$$\begin{aligned}\{x \in z \mid \varphi^{\mathbf{L}}(x, z, \overline{v})\} \quad & = \{x \in \mathbf{L}_\beta \mid (x \in z \wedge \varphi(x, z, \overline{v}))^{\mathbf{L}_\beta}\} \\ & \in \mathit{Def}(\mathbf{L}_\beta) = \mathbf{L}_{\beta+1},\end{aligned}$$

so $\{x \in z \mid \varphi^{\mathbf{L}}(x, z, \overline{v})\} \in \mathbf{L}$. QED

From lemma 2.5 and lemma 2.6, we conclude that in order to feasibly prove the comprehension schema, we only need polynomial length proofs of

$$\forall \alpha\, \exists \beta\, \forall z, x, \overline{v} \in \mathbf{L}_\beta\, [\varphi^{\mathbf{L}}(x, z, \overline{v}) \leftrightarrow \varphi^{\mathbf{L}_\beta}(x, z, \overline{v})].$$

For a proof of this reflection theorem, we need a few more definitions.

Definition 2.7 A collection $\mathcal{C}$ of ordinals is

- *unbounded* iff $\forall \alpha\, \exists \beta > \alpha\, (\beta \in \mathcal{C})$;
- *closed* iff $\forall a\, (a \neq \emptyset \wedge a \subseteq \mathcal{C} \rightarrow \sup\; a \in \mathcal{C})$;
- *closed unbounded* (c.u.b.) iff $\mathcal{C}$ is both closed and unbounded.

Lemma 2.8 *$ZF \vdash$ "If $\mathcal{C}$ and $\mathcal{D}$ are c.u.b., then $\mathcal{C} \cap \mathcal{D}$ is c.u.b. as well"*

Proof. An easy application of lemma II.6.8 of [Ku 80]. QED

Definition 2.9 A collection $\mathcal{C}$ of ordinals is *closed unbounded for* φ iff

1. $\mathcal{C}$ is closed unbounded, and
2. $\mathbf{L}_\alpha$ reflects φ for all ordinals $\alpha \in \mathcal{C}$, i.e.
   $\forall \alpha\, (\alpha \in \mathcal{C} \rightarrow \forall \overline{v} \in \mathbf{L}_\alpha\, [\varphi^{\mathbf{L}}(\overline{v}) \leftrightarrow \varphi^{\mathbf{L}_\alpha}(\overline{v})])$

Suppose φ is a formula and $\mathcal{D}$ is a first-order definable collection of ordinals. Using definition 2.7, we are able to construct new first-order formulas $CUB_{\mathcal{D}}$, $CUB_{\mathcal{D},\varphi}$ and REF_φ with the following intended meanings:

1. $CUB_{\mathcal{D}} :=$ "$\mathcal{D}$ is closed unbounded"
2. $CUB_{\mathcal{D},\varphi} :=$"$\mathcal{D}$ is closed unbounded for φ"

3. $REF_\varphi :=$ " there is some collection of ordinals that is closed unbounded for φ"

The next lemma roughly corresponds to theorem IV.7.5 of [Ku 80].

Lemma 2.10 (Reflection theorem) *ZF proves the following by proofs of length polynomial in* $|\varphi|$*:*

$$\forall\alpha\,\exists\beta\,\forall z, x, \overline{v} \in \mathbf{L}_\beta\,[\varphi^{\mathbf{L}}(x, z, \overline{v}) \leftrightarrow \varphi^{\mathbf{L}_\beta}(x, z, \overline{v})]$$

Proof. First we note that ZF proves $(\alpha < \beta \to \mathbf{L}_\alpha \subseteq \mathbf{L}_\beta)$, "if γ is a limit ordinal, then $\mathbf{L}_\gamma = \bigcup_{\alpha<\gamma}\mathbf{L}_\alpha$" and $\mathbf{L} = \bigcup_{\alpha\in OR}\mathbf{L}_\alpha$.

We will prove the reflection theorem by induction on the construction of φ. A straightforward application of lemma 2.2 implies that for the reflection theorem to have a proof of length polynomial in $|\varphi|$, it is sufficient to find a polynomial bounding the lengths of the induction steps. Thus, we need to find a polynomial P such that by proofs of length $\leq P(|\varphi|)$, resp. $\leq P(|\neg\psi|)$, resp. $\leq P(|\psi\circ\chi|)$, resp. $\leq P(|Qx\psi|)$, ZF proves the following:

1. for atomic φ:

$$\forall\alpha\,\exists\beta>\alpha\,\forall z, x \in \mathbf{L}_\beta\,[\varphi^{\mathbf{L}}(x, z) \leftrightarrow \varphi^{\mathbf{L}_\beta}(x, z)] \wedge CUB_{OR,\varphi}$$

2. the $\neg$-step:

$$\forall\alpha\,\exists\beta>\alpha\,\forall\overline{v}\in\mathbf{L}_\beta\,[\psi^{\mathbf{L}}(\overline{v}) \leftrightarrow \psi^{\mathbf{L}_\beta}(\overline{v})] \wedge REF_\psi \quad\to$$
$$\forall\alpha\,\exists\beta>\alpha\,\forall\overline{v}\in\mathbf{L}_\beta\,[\neg\psi^{\mathbf{L}}(\overline{v}) \leftrightarrow \neg\psi^{\mathbf{L}_\beta}(\overline{v})] \wedge REF_{\neg\psi}$$

3. the connective step, where $\circ \in \{\wedge, \vee, \to, \leftrightarrow\}$:

$$\forall\alpha\,\exists\beta>\alpha\,\forall\overline{v}\in\mathbf{L}_\beta\,[\psi^{\mathbf{L}}(\overline{v}) \leftrightarrow \psi^{\mathbf{L}_\beta}(\overline{v})] \wedge REF_\psi \wedge$$
$$\forall\alpha\,\exists\beta>\alpha\,\forall\overline{w}\in\mathbf{L}_\beta\,[\chi^{\mathbf{L}}(\overline{w}) \leftrightarrow \chi^{\mathbf{L}_\beta}(\overline{w})] \wedge REF_\chi \to$$
$$\forall\alpha\,\exists\beta>\alpha\,\forall\overline{v}\in\mathbf{L}_\beta\,[\psi^{\mathbf{L}}\circ\chi^{\mathbf{L}}(\overline{v},\overline{w}) \leftrightarrow \psi^{\mathbf{L}_\beta}\circ\chi^{\mathbf{L}_\beta}(\overline{v},\overline{w})] \wedge REF_{\psi\circ\chi}$$

4. the quantifier step, where $Q \in \{\exists, \forall\}$:

$$\forall\alpha\,\exists\beta>\alpha\,\forall z, \overline{v}\in\mathbf{L}_\beta\,[\psi^{\mathbf{L}}(z,\overline{v}) \leftrightarrow \psi^{\mathbf{L}_\beta}(z,\overline{v})] \wedge REF_\psi \to$$
$$\forall\alpha\,\exists\beta>\alpha\,\forall\overline{v}\in\mathbf{L}_\beta[Qz\in\mathbf{L}\,\psi^{\mathbf{L}}(z,\overline{v}) \leftrightarrow Qz\in\mathbf{L}_\beta\,\psi^{\mathbf{L}_\beta}(z,\overline{v})] \wedge REF_{Qz\psi}$$

Finding polynomials bounding the lengths of the proofs of 1, 2 and 3 is very easy: we can use the feasibly provable fact that atomic formulas are absolute for any $\mathbf{L}_\alpha, \mathbf{L}$, some propositional reasoning independent on the specific ψ, χ, and an application of lemma 2.8 for step 3. We will show how the proofs of the $\exists$-case in step 4 can be bounded by a polynomial; a bound for the $\forall$-step then follows by some uses of the bounds for the $\neg$-step and the $\exists$-step.

Define

$$\mathcal{D} := \{\beta \mid \forall \overline{v} \in \mathbf{L}_\beta\, [\exists z \in \mathbf{L}\, \psi^{\mathbf{L}}(z, \overline{v}) \to \exists z \in \mathbf{L}_\beta\, \psi^{\mathbf{L}}(z, \overline{v})]\}.$$

It is easy to see that ZF proves the following by proofs of length polynomial in $|\exists z \psi|$:

$$\begin{aligned}&\forall \alpha\, \exists \beta > \alpha\, \forall z, \overline{v} \in \mathbf{L}_\beta\, [\psi^{\mathbf{L}}(z, \overline{v}) \leftrightarrow \psi^{\mathbf{L}_\beta}(z, \overline{v})] \wedge CUB_{\mathcal{C}_\psi,\, \psi} \wedge CUB_{\mathcal{D}} \to \\ &\forall \alpha\, \exists \beta > \alpha\, \forall \overline{v} \in \mathbf{L}_\beta\, [\exists z \in \mathbf{L}\, \psi^{\mathbf{L}}(z, \overline{v}) \leftrightarrow \exists z \in \mathbf{L}_\beta\, \psi^{\mathbf{L}}(z, \overline{v})] \\ &\wedge CUB_{\mathcal{C}_\psi \cap \mathcal{D},\, \exists z \psi}\end{aligned}$$

In fact, we only use lemma 2.8 and the fact that $\forall \beta\, (\mathbf{L}_\beta \subseteq \mathbf{L})$. Thus we need to find a polynomial P such that $ZF \vdash CUB_{\mathcal{D}}$ by a proof of length $\leq P(|\exists z \psi|)$. Immediately from the definition, it is clear that $ZF \vdash$ "$\mathcal{D}$ is closed" by a proof of length polynomial in $|\exists z \psi|$. Thus, it suffices to show by a proof of length polynomial in $|\exists z \psi|$ that ZF proves that $\mathcal{D}$ is unbounded, that is:

$$\forall \alpha\, \exists \beta > \alpha\, \forall \overline{v} \in \mathbf{L}_\beta\, [\exists z \in \mathbf{L}\, \psi^{\mathbf{L}}(z, \overline{v}) \to \exists z \in \mathbf{L}_\beta\, \psi^{\mathbf{L}}(z, \overline{v})],$$

i.e.

$$\forall \alpha\, \exists \beta > \alpha\, \forall \overline{v} \in \mathbf{L}_\beta\, \exists z \in \mathbf{L}_\beta\, [\exists z \in \mathbf{L}\, \psi^{\mathbf{L}}(z, \overline{v}) \to \psi^{\mathbf{L}}(z, \overline{v})].$$

We will reason in ZF, taking care that all steps are applications of general ZF-theorems that do not depend on the specific formula ψ. Take any ordinal α. We know using only predicate logic that

$$\forall \overline{v} \in \mathbf{L}_\alpha \, \exists z \in \mathbf{L} \, [\exists z \in \mathbf{L} \, \psi^{\mathbf{L}}(z, \overline{v}) \rightarrow \psi^{\mathbf{L}}(z, \overline{v})];$$

therefore,

$$\forall \overline{v} \in \mathbf{L}_\alpha \, \exists ! \alpha_{\overline{v}} \, (\alpha_{\overline{v}} = \bigcap \{\beta > \alpha \mid \exists z \in \mathbf{L}_\beta \, [\exists z \in \mathbf{L} \, \psi^{\mathbf{L}}(z, \overline{v}) \rightarrow \psi^{\mathbf{L}}(z, \overline{v})]\})$$

By the unrelativized replacement and union axioms, there is a β_1 such that $\beta_1 = \sup\{\alpha_{\overline{v}} | \overline{v} \in \mathbf{L}_\alpha\}$. Continuing in this way, we can define by recursion a sequence β_p for $p \in \omega$, where for all $p \in \omega$,

$$\forall \overline{v} \in \mathbf{L}_{\beta_p} \, \exists z \in \mathbf{L}_{\beta_{p+1}} \, [\exists z \in \mathbf{L} \, \psi^{\mathbf{L}}(z, \overline{v}) \rightarrow \psi^{\mathbf{L}}(z, \overline{v})] \tag{2}$$

Define $\beta := \sup\{\beta_p \, | p \in \omega\}$. Because $\alpha = \beta_0 < \beta_1 < \beta_2 < \ldots$, we infer that β is a limit ordinal $> \alpha$. Now using(2) and the fact that $\mathbf{L}_\beta = \bigcup_{\gamma < \beta} \mathbf{L}_\gamma$, we find that

$$\forall \overline{v} \in \mathbf{L}_\beta \, \exists z \in \mathbf{L}_\beta \, [\exists z \in \mathbf{L} \, \psi^{\mathbf{L}}(z, \overline{v}) \rightarrow \psi^{\mathbf{L}}(z, \overline{v})],$$

as desired. QED

Lemma 2.11 *For all φ, ZF feasibly proves the comprehension schema for φ, relativized to* **L**; *i.e. by proofs of length polynomial in $|\varphi|$, ZF proves the following:*

$$\forall z, \overline{v} \in \mathbf{L} \, \exists y \in \mathbf{L} \, \forall x \in \mathbf{L} \, [x \in y \leftrightarrow x \in z \wedge \varphi^{\mathbf{L}}(x, z, \overline{v})]$$

Proof. Combine lemmas 2.5, 2.6 and 2.10. QED

Lemma 2.12 *For all φ, ZF feasibly proves the replacement schema for φ, relativized to* $\mathbf{L}$*; i.e. by proofs of length polynomial in $|\varphi|$, ZF proves the following:*

$$\begin{aligned}\forall a, \overline{v} \in \mathbf{L} \quad & [\forall x \in a\, \exists! y \in \mathbf{L}\, \varphi^{\mathbf{L}}(x, y, \overline{v}) \rightarrow \\ & \exists c \in \mathbf{L}\, \forall y \in \mathbf{L}\, (y \in c \leftrightarrow \exists x \in a\, \varphi^{\mathbf{L}}(x, y, \overline{v}))]\end{aligned}$$

Proof. We already have feasible proofs of the relativized comprehension schema for the formula $y \in b \wedge \exists x \in a\, \varphi(x, y, \overline{v})$. So we can (feasibly) prove that it suffices to show the following by proofs of length polynomial in $|\varphi|$:

$$\begin{aligned}ZF \vdash \forall a, \overline{v} \in \mathbf{L}\, [\quad & \forall x \in a\, \exists! y \in \mathbf{L}\, \varphi^{\mathbf{L}}(x, y, \overline{v}) \rightarrow \\ & \exists b \in \mathbf{L}\, (\forall x \in a\, \exists y \in b\, \varphi^{\mathbf{L}}(x, y, \overline{v}))]\end{aligned}$$

The last proof works, as in lemma 2.10, by general theorems of ZF that do not depend on the specific φ. Work in ZF and suppose $a, \overline{v} \in \mathbf{L}$ and $\forall x \in a\, \exists! y \in \mathbf{L}\, \varphi^{\mathbf{L}}(x, y, \overline{v})$. Now

$$\forall x \in a\, \exists! \beta_x\, (\beta_x = \bigcap \{\alpha \mid \exists y \in \mathbf{L}_\alpha\, \varphi^{\mathbf{L}}(x, y, \overline{v})\});$$

then by replacement and the union axiom we find β such that $\beta = \bigcup \{\beta_x | x \in a\}$, and we let b be $\mathbf{L}_\beta$. Then

$$\forall x \in a\, \exists y \in b\, \varphi^{\mathbf{L}}(x, y, \overline{v}).$$

QED

Theorem 2.13 $ZF \rhd_f ZF + \mathbf{V} = \mathbf{L}$

Proof. We take the usual interpretation $^{\mathbf{L}}$ of $ZF + \mathbf{V} = \mathbf{L}$ into ZF. Because ZF is axiomatized by a finite list of axioms plus the schemata of comprehension and replacement, the lemmas 2.11 and 2.12 immediately imply that $^{\mathbf{L}}$ is a feasible interpretation. QED

Remark 2.14 Looking carefully at the proofs of the lemmas leading up to theorem 2.13 and using an analogue of lemma 2.2 for polynomial time instead of polynomial length, one can observe that theorem 2.13 can be strengthened: the proofs in ZF of the interpreted $ZF + \mathbf{V} = \mathbf{L}$–axioms $\varphi^{\mathbf{L}}$ are not only of length polynomial in $|\varphi|$. There is even a deterministic polynomial time Turing machine M such that if the input of M is the code of an axiom φ of $ZF + \mathbf{V} = \mathbf{L}$, then M outputs the code of a ZF-proof of $\varphi^{\mathbf{L}}$.

It is a known result that $PRA \vdash Con(ZF) \rightarrow Con(ZF + \mathbf{V} = \mathbf{L})$ (see [Sm 77, Corollary 5.2.4]). One of the referees suggested that by theorem 2.13 this could perhaps be strengthened to $\mathrm{I}\Delta_0 + \exp \vdash Con(ZF) \rightarrow Con(ZF + \mathbf{V} = \mathbf{L})$. The observation about the polynomial time computability of the proofs of the interpreted axioms, however, even leads to the conjecture that $\mathrm{I}\Delta_0 + \Omega_1 \vdash \forall a(\alpha_{ZF+\mathbf{V}=\mathbf{L}}(a) \rightarrow \exists p Prf_{ZF}(p, a^{\mathbf{L}}))$, where $\alpha_{ZF+\mathbf{V}=\mathbf{L}}$ is a Δ_1^b-formula axiomatizing $ZF + \mathbf{V} = \mathbf{L}$ (cf. [Bu 86, Theorem 5.6]). Then, by a standard argument, we would have $\mathrm{I}\Delta_0 + \Omega_1 \vdash Con(ZF) \rightarrow Con(ZF + \mathbf{V} = \mathbf{L})$.

Contrary to our expectations, the usual interpretation of $ZF + \mathbf{V} \neq \mathbf{L}$ into $ZF(M)$ (by forcing with generic extensions), although much more complex, is still feasible. We checked this following the lines of the proof in [Ku 80]. Our proof relies so heavily on the many details of Kunen's proof, that it would be incomprehensible to readers not conversant with that book. Therefore, we do not give it here.

In the literature there are also proofs of $ZF \rhd ZF + \mathbf{V} \neq \mathbf{L}$ and $ZF + AC \rhd ZF + AC + \neg CH$ which entirely avoid the use of the transitive countable collection M. A sketch of such a proof can be found in [Co 66, Section IV.11], and a completely different full proof in [VH 72, Ch. V, VI]. It appears that these proofs can also be analyzed to show that the interpretations in question are feasible.

Other well-known interpretations, such as the one of PA into ZF, and those of $\mathrm{I}\Delta_0 + \Omega_n$ into Q [HP, Section V.5] are also feasible, as the reader may check for her/himself. All in all it seems that the only examples of theories U and V such that $U \rhd V$ but not $U \rhd_f V$ are contrived theories obtained by fixed-point constructions like the ones in section 4. It would be nice to find a more natural counterexample.

It would also be interesting to investigate severely restricted kinds of interpretability which do distinguish between interpretations used in everyday mathematics. For example, one could restrict the complexity of formulas allowed to occur in the proofs of the interpreted axioms.

Sam Buss suggested the following restricted definition of feasible interpretability to us:

$$\begin{aligned} U \rhd_{fm} V \leftrightarrow \exists K \exists M (&\text{“}K \textit{ is an interpretation and } M \textit{ is a} \\ &\textit{deterministic polynomial time Turing Machine”} \\ &\wedge \forall a (\alpha_V(a) \rightarrow Prf_U(M(a), a^K))). \end{aligned} \tag{3}$$

This definition is more in line with the conventional use of the word "feasible" in the context of polynomial time computability. The clause $Prf_U(M(a), a^K)$ in (3) is a P-like formula, while the clause $\exists p(\text{“}|p| \leq P(|a|)\text{”} \wedge Prf_U(p, a^K))$ in the definition of feasible interpretability used in this paper is an NP-like formula. However, all interpretations considered in this section can also be shown to be feasible in Buss' sense: as in remark 2.14, we only need an easy analogue of lemma 2.2.

3 ILM is sound for feasible interpretability over PA

In this section, we restrict our attention to feasible interpretability over PA. We show that the modal interpretability logic ILM is PA-sound even if the intended meaning of $A \rhd B$ is "$PA + A$ feasibly interprets $PA + B$".

Definition 3.1 The modal interpretability logic ILM contains, besides all formulas having the form of a propositional tautology, the usual axioms for the provability logic L and the rules modus ponens and necessitation, the following axioms:

J1 $\Box(A \rightarrow B) \rightarrow (A \rhd B)$

J2 $(A \rhd B) \wedge (B \rhd C) \rightarrow (A \rhd C)$

J3 $(A \rhd C) \wedge (B \rhd C) \rightarrow (A \vee B \rhd C)$

J4 $(A \rhd B) \rightarrow (\Diamond A \rightarrow \Diamond B)$

J5 $\Diamond A \rhd A$

M $(A \rhd B) \rightarrow (A \wedge \Box C \rhd B \wedge \Box C)$

Definition 3.2 A *feasibility interpretation* is a map * which assigns to every propositional variable p a sentence p^* of the language of PA, and which is extended to all modal formulas as follows:

1. $(A \rhd B)^* = PA + A^* \rhd_f PA + B^*$
2. $(\Box A)^* = Prov_{PA}(A^*)$
3. * commutes with the propositional connectives.

Here $\rhd_f$ abbreviates the formalization of feasible interpretability

We will prove that ILM is arithmetically sound for feasible interpretability, i.e. that for all modal formulas A, if $ILM \vdash A$, then for all feasibility interpretions *, $PA \vdash A^*$. Thus, we have to check that the axioms J1 to J5 are valid in PA when $A \rhd B$ is read as $PA + A \rhd_f PA + B$. Whenever possible, we will prove generalizations of these axioms to theories $U, V \supseteq PA$. Also we prove a generalization of the property M, where an infinite set of Σ_1^0-sentences can be added on both sides instead of one $\Box$-sentence only.

Lemma 3.3 *PA proves all feasibility translations of J1 to J5.*

Proof. The proofs for J1 through J4 can be found almost verbatim in [Vi]. We reason in PA.

J1 Suppose for some theory V and some p that $Prf_V(p, \ulcorner A \urcorner)$. Then by the identity interpretation and the polynomial bound $P(n) = n + 3 \cdot |p|$, $V \rhd_f V + A$. So in particular, if $\Box_{PA}(A \rightarrow B)$, then $PA + A \rhd_f PA + A + B$, and surely $PA + A \rhd_f PA + B$.

J2 Suppose

- $U \rhd_f V$ by interpretation K_1 and polynomial P_1, and
- $V \rhd_f W$ by the interpretation K_2 and polynomial P_2.

As in the usual case, $U \rhd W$ by the interpretation $K_2 \circ K_1$. We need to show that there is a polynomial bound for the proofs of the translated axioms. So let b code an axiom of W, and p a proof in V of b^{K_2} with $|p| \leq P_2(|b|)$.

If we take the K_1- translations of all formulas appearing in the proof coded by p, and add some intermediate steps, we can construct a U-proof of $(b^{K_2})^{K_1}$ from K_1-translations of axioms of V as assumptions; the number of steps in this proof will be $\leq k \cdot |p|$, where k is a constant depending on the translation K_1. Now we only have to add proofs of the translated V-axioms; the axioms themselves have codes of length $\leq |p|$, so their K_1-translations have proofs with codes of length $\leq P_1(|p|) \leq P_1(P_2(|b|))$.

All in all, even in the worst case where the U-proof of $(b^{K_2})^{K_1}$ consists wholly of assumptions, there is a q with $|q| \leq k \cdot P_2(|b|) \cdot P_1(P_2(|b|))$ such that $Prf_U(q, (b^{K_2})^{K_1})$. In particular, if $PA + A \rhd_f PA + B$ and $PA + B \rhd_f PA + C$, then $PA + A \rhd_f PA + C$.

J3 Suppose

- $U + A \rhd_f V$ by interpretation K_1 and polynomial P_1, and
- $U + B \rhd_f V$ by interpretation K_2 and polynomial P_2.

As in the usual case, we have $U + A \vee B \rhd V$ by the disjunctive interpretation M which equals K_1 in case A holds and equals K_2 in case $\neg A$ holds. To find a polynomial bound, we observe that for all C, $\vdash A \rightarrow (C^M \leftrightarrow C^{K_1})$ and $\vdash \neg A \rightarrow (C^M \leftrightarrow C^{K_1})$ by proofs of length $\leq P(|C|)$, where the polynomial P depends on K_1 and K_2. Now suppose that c codes an axiom of V, that p_1 codes a $U + A$-proof of c^{K_1} with $|p_1| \leq P_1(|c|)$, and that p_2 codes a $U + B$-proof of c^{K_2} with $|p_2| \leq P_2(|c|)$. But then there is a constant k such that

- we can find p_1' such that $Prf_U(p_1', \ulcorner A \rightarrow \urcorner c^M)$ with $|p_1'| \leq k \cdot (P(|c|) + P_1(|c|))$; and
- we can find p_2' such that $Prf_U(p_2', \ulcorner \neg A \wedge B \rightarrow \urcorner c^M)$ with $|p_2'| \leq k \cdot (P(|c|) + P_2(|c|))$.

Combining p_1' and p_2' and their respective polynomial bounds, we find p and P' such that $Prf_U(p, \ulcorner A \vee B \rightarrow \urcorner c^M)$ with $|p| \leq P'(|c|)$. In particular, we have: if $PA + A \rhd_f PA + C$ and $PA + B \rhd_f PA + C$, then $PA + A \vee B \rhd_f PA + C$.

J4 Because $(PA + A \rhd_f PA + B) \rightarrow (PA + A \rhd PA + B)$, we have by the soundness of J4 for normal interpretability immediately $(PA + A \rhd_f PA + B) \rightarrow (\Diamond A \rightarrow \Diamond B)$.

J5 In an easier variation of lemma 5.11, we use a claim proved in [Vi 89], which is stated in our paper as lemma 5.10. Suppose β is a Σ_1^b-formula axiomatizing a subset U of a Σ_1^b-language L. We will prove that $Q + \Diamond_\beta \top \rhd_f U$ i.e. $Q + \Diamond_U \top \rhd_f U$.

By lemma 5.10, we have

$$\begin{aligned} PA \vdash \; & \Box_{Q+Con(\beta)} Con(\beta) \rightarrow \\ & \exists K \forall a \in Sent(L) Polprov_{Q+Con(\beta),|a|}(\ulcorner \Box_\beta a \rightarrow \urcorner a^K). \end{aligned}$$

Of course we also know that $PA \vdash \Box_{Q+Con(\beta)} Con(\beta)$, so

$$PA \vdash \exists K \forall a \in Sent(L) Polprov_{Q+Con(\beta),|a|}(\ulcorner \Box_\beta a \rightarrow \urcorner a^K).$$

On the other hand, we have by provable Σ_1^b-completeness

$$PA \vdash \forall a(\beta(a) \rightarrow Polprov_{Q+Con(\beta),|a|}(\ulcorner \Box_\beta a \urcorner).$$

Combining the last two results, we have

$$PA \vdash \exists K \forall a(\beta(a) \rightarrow Polprov_{Q+Con(\beta),|a|}(a^K),$$

so $PA \vdash (Q + \Diamond_\beta \top) \rhd_f U$. In particular, we have for any sentence A:

$$PA \vdash (Q + \Diamond_{PA} A) \rhd_f PA + A,$$

so especially

$$PA \vdash PA + \Diamond_{PA} A \rhd_f PA + A.$$

QED

We want to prove that Montagna's property M holds for feasible interpretability over PA in its general version, where we can add an infinite set of Σ_1^0-sentences on both sides. In order to ensure that the usual arguments can indeed be polynomialized, we do not formulate the proof in the usual model-theoretic way, and we give many details that are not given in most proofs of Montagna's property for normal interpretability over

PA. The example we give in theorem 4.1 of a set $\mathcal{S}$ of formulas such that $PA \vdash PA \rhd PA + \mathcal{S}$ but $\omega \not\models PA \rhd_f PA + \mathcal{S}$ also relies heavily on these details.

Suppose $U \supseteq PA$, $V \supseteq PA$. Now suppose $U \rhd_f V$ by the interpretation K (preserving $=$) with domain δ, and polynomial P. We want to find a polynomial Q such that for every Σ_1^0-sentence σ there is a $U + \sigma$-proof p of σ^K with $|\ulcorner p \urcorner| \leq Q(|\ulcorner \sigma \urcorner|)$. First, we need some definitions and lemmas. Fix U, V, K, P as given above.

Definition 3.4 Define *pism(s)* for "s is a partial isomorphism" and the function $G(j, y)$ as follows:

$$\begin{array}{lll} pism(s) & := & seq(s) \wedge (s)_0 = 0^K \wedge \forall i < lh(s) - 1((s)_{i+1} = S^K(s)_i) \\ G(j, y) & := & \exists s(pism(s) \wedge lh(s) = j + 1 \wedge (s)_j = y) \end{array}$$

Lemma 3.5 *$U \vdash \forall j \exists! s(\mathrm{pism}(s) \wedge \mathrm{lh}(s) = j + 1)$ and thus $U \vdash \forall j \exists! y G(j, y)$. Therefore, there is a function g corresponding to G.*

Proof. By induction. QED

Lemma 3.6 *U proves that g is injective, and $U \vdash \forall j \forall y(G(j, y) \rightarrow \delta(y))$.*

Proof. By induction. QED

Lemma 3.7 *U proves that g preserves $0, S, +, \cdot,$ and $\leq$.*

Proof. We will give some of the preservation proofs. It follows immediately from the definition of *pism(s)* that $U \vdash g(0) = 0^K$ and $U \vdash \forall x(g(Sx) = S^K(g(x)))$.

We now prove by induction that g preserves $+$; the other proofs are analogous. We have $U \vdash g(x + 0) = g(x) = g(x) +^K 0^K = g(x) +^K g(0)$ and

$U \vdash g(x+y) = g(x) +^K g(y) \rightarrow g(x+Sy) = g(S(x+y)) = S^K(g(x+y)) = S^K(g(x) +^K g(y)) = g(x) +^K S^K(g(y)) = g(x) +^K g(Sy)$, So by induction (with x as parameter) $U \vdash \forall x \forall y(g(x+y) = g(x) +^K g(y))$. QED

Lemma 3.8 *The range of g is 'closed downwards', i.e.* $U \vdash \forall x \forall u(\delta(u) \wedge u <^K g(x) \rightarrow \exists y < x(u = g(y)))$.

Proof. Before we start the proof proper, we note a useful fact. V includes PA and K is an interpretation of V into U. Thus, as

1. $PA \vdash \forall x \forall u(u < x+1 \rightarrow u < x \vee u = x)$ and
2. $U \vdash \forall x(g(x) +^K 1^K = g(x+1))$, we also have
3. $U \vdash \forall x \forall u(\delta(u) \wedge u <^K g(x+1) \rightarrow u <^K g(x) \vee u = g(x))$.

Now we can start with the proof by induction on x of $U \vdash \forall x \forall u(\delta(u) \wedge u <^K g(x) \rightarrow \exists y < x(u = g(y)))$.

x = 0 We have $U \vdash \neg \exists u(\delta(u) \wedge u <^K g(0))$, so $U \vdash \forall u(\delta(u) \wedge u <^K g(0) \rightarrow \exists y < 0(u = g(y)))$.

Induction step Work in U and suppose $\forall u(\delta(u) \wedge u <^K g(x) \rightarrow \exists y < x(u = g(y)))$ (induction hypothesis). Moreover, suppose $\delta(u) \wedge u <^K g(x+1)$. Then, by 3, $u <^K g(x) \vee u = g(x)$. So by the induction hypothesis $\exists y < x(u = g(y)) \vee u = g(x)$, i.e. $\exists y < x+1(u = g(y))$.

QED

Remark 3.9 Let $I(x)$ be the formula $\exists y(x = g(y))$. Note that if $U \nvdash \forall x \delta(x)$, then U does not prove that I defines a cut. For, suppose that $U \vdash I(0) \wedge \forall x(I(x) \rightarrow I(x+1))$. Then induction gives $U \vdash \forall x \exists y(x = g(y))$, thus, by lemma 3.6, $U \vdash \forall x \delta(x)$, contradicting our assumption.

On the other hand, by the previous lemma we do have $U \vdash \forall x \forall u(\delta(u) \wedge I(x) \wedge u <^K x \rightarrow I(u))$.

Lemma 3.10 *For all formulas $\varphi \in \Delta_1^0$, U proves the following by proofs of length polynomial in $|\ulcorner\varphi(a_1, \ldots, a_n)\urcorner|$:*

$$\varphi(a_1, \ldots, a_n) \leftrightarrow (\varphi^K)(g(a_1), \ldots, g(a_n)).$$

Proof. By induction on the construction of φ. We will see below that the proofs for the atomic formulae ψ are obviously of length linear in $|\ulcorner\psi\urcorner|$, and that all induction steps follow a given proof scheme in which the particular formulas at hand can be plugged in. So, because every φ has at most $|\ulcorner\varphi\urcorner|$ subformulas, there is a polynomial R such that for all φ, the U- proof of $\varphi(a_1, \ldots, a_n) \leftrightarrow (\varphi^K)(g(a_1), \ldots, g(a_n))$ is of length $\leq R(|\ulcorner\varphi\urcorner|)$. We will do the atomic step and the $\forall x \leq t$-step of the proof, and leave the others to the reader.

Atomic step By lemma 3.7, we have for all terms t by proofs of length polynomial in $|\ulcorner t\urcorner|$:

$$U \vdash g(t(a_1, \ldots, a_n)) = (t^K)(g(a_1), \ldots, g(a_n)).$$

So suppose φ is the formula $t_1(a_1, \ldots, a_n) = t_2(a_1, \ldots, a_n)$ where $a_1, \ldots, a_n$ include all variables appearing in t_1 and t_2. Then, because U proves that g is an injective function,

$$\begin{array}{lrcl} U \vdash & t_1(a_1, \ldots, a_n) & = & t_2(a_1, \ldots, a_n) \\ & \leftrightarrow g(t_1(a_1, \ldots, a_n)) & = & g(t_2(a_1, \ldots, a_n)) \\ & \leftrightarrow (t_1^K)(g(a_1), \ldots, g(a_n)) & = & (t_2^K)(g(a_1), \ldots, g(a_n)) \\ & \leftrightarrow ((t_1 & = & t_2)^K)(g(a_1), \ldots, g(a_n)) \end{array}$$

$\forall$x $\leq$ t-step Suppose that

$$\varphi(a_1, \ldots, a_n) = \forall x \leq t(a_1, \ldots, a_n)\psi(x, a_1, \ldots, a_n),$$

and that

$$U \vdash \psi(x, a_1, \ldots, a_n) \leftrightarrow (\psi^K)(g(x), g(a_1), \ldots, g(a_n))$$

(induction hypothesis). We will use the fact that, because of lemmas 3.6, 3.7 and 3.8,

$$\begin{array}{ll} U \vdash \forall u, a_1, \ldots, a_n & (\exists x\, [x \leq t(a_1, \ldots, a_n) \wedge u = g(x)] \\ & \leftrightarrow \delta(u) \wedge u \leq^K t^K(g(a_1), \ldots, g(a_n))). \end{array}$$

Thus, we have the following equivalences:

$$\begin{aligned}
&U \vdash \varphi(a_1, \ldots, a_n) \\
&\leftrightarrow \forall x \le t(a_1, \ldots, a_n)\psi(x, a_1, \ldots, a_n) \\
&\leftrightarrow \forall x \le t(a_1, \ldots, a_n)(\psi^K)(g(x), g(a_1), \ldots, g(a_n)) \quad \text{(by ind. hyp.)} \\
&\leftrightarrow \forall u(\delta(u) \wedge u \le^K t^K(g(a_1), \ldots, g(a_n)) \rightarrow \psi^K(u, g(a_1), \ldots, g(a_n))) \\
&\leftrightarrow (\forall x \le t\, \psi)^K(g(a_1), \ldots, g(a_n)) \qquad \text{(by def. of } K) \\
&\leftrightarrow (\varphi)^K(g(a_1), \ldots, g(a_n))
\end{aligned}$$

QED

Now we can finish the proof of the uniform version of Montagna's property M for feasible interpretability.

Theorem 3.11 *Suppose*

- *U satisfies full induction,*
- *V extends PA and*
- *$U \rhd_f V$ by interpretation K (preserving $=$) and polynomial P.*

Then there is a polynomial Q such that for every Σ_1^0-sentence σ there is a $U + \sigma$-proof p of σ^K with $|\ulcorner p \urcorner| \le Q(|\ulcorner \sigma \urcorner|)$.

Thus, $U + \mathcal{S} \rhd_f V + \mathcal{S}$ where $\mathcal{S}$ is a finite or infinite set of Σ_1^0-sentences.

Proof. Suppose $\sigma \in \mathcal{S}$ is the Σ_1^0-sentence $\exists x \varphi(x)$, where $\varphi \in \Delta_1^0$. By lemma 3.10, there is a polynomial R such that we can prove the following by a proof of length $\le R(|\ulcorner \sigma \urcorner|)$:

$$\begin{aligned}
U \vdash \exists x \varphi(x) \quad &\rightarrow \exists x \varphi^K(g(x)) \\
&\rightarrow \exists y(\delta(y) \wedge \varphi^K(y)) \\
&\rightarrow (\exists x \varphi(x))^K.
\end{aligned}$$

Now we have $U + \mathcal{S} \rhd_f V + \mathcal{S}$ by the interpretation K and polynomial $Q := P + R$. QED

All results of this section also hold if we add the function symbol *exp* to the language of U and V, which we need in theorem 4.1. Let g be as defined in lemma 3.5. We will only give the result which needs some adaptation. The following preservation lemma corresponds to lemma 3.7:

Lemma 3.12 *Suppose* $exp \in L_U$. *Then* U *proves that* g *preserves* 0, S, $+$, $\cdot$, $\leq$, *and* exp.

Proof. We already have a preservation proof for $\cdot$ by lemma 3.7. Preservation of exp then follows in the same way as preservation of $+$ was proved from preservation of S in lemma 3.7. QED

4 Interpretability does not imply feasible interpretability

Theorem 4.1 *There is a set* $\mathcal{S}$ *of* $\Delta_1^0(exp)$*-sentences such that* $\omega \models PA \rhd PA + \mathcal{S}$, *but* $\omega \not\models PA \rhd_f PA + \mathcal{S}$.

Proof. Define by Gödel's diagonalization theorem (or rather by the free variable version as formulated by Montague) a $\Delta_1^0(exp)$-formula $\varphi(y)$ such that

$$PA \vdash \varphi(y) \leftrightarrow \forall x \leq exp(y)\, \neg Prf(x, \ulcorner\varphi(\dot{y})\urcorner).$$

It is easy to see that if we diagonalize directly, there is a polynomial O such that for each n, $|n| < |\ulcorner\varphi(\dot{n})\urcorner| \leq O(|n|)$. Moreover, if $\varphi(\dot{n})$ were false, then by definition we would have a proof of the $\Delta_1^0(exp)$-sentence $\varphi(\dot{n})$; so $\varphi(\dot{n})$ must be true. But then, since $\varphi(\dot{n})$ is $\Delta_1^0(exp)$, we have the following:

1. PA proves $\varphi(\dot{n})$, though
2. because $\varphi(\dot{n})$ is true, PA does not prove $\varphi(\dot{n})$ by any proof whose Gödel number is of length $\leq n$.

Define $\mathcal{S} := \{\varphi(\dot{n}) \mid n \in \omega\}$. Then, by the identity interpretation, $\omega \models PA \rhd PA + \mathcal{S}$. Actually, as in [JM 88, section 6], we even have $PA \vdash \forall y Prov(\ulcorner\varphi(\dot{y})\urcorner)$, so $PA \vdash PA \rhd PA + \mathcal{S}$.

Now suppose, in order to derive a contradiction, that $\omega \models PA \rhd_f PA + \mathcal{S}$ by interpretation K and polynomial P. Thus, for all n,

$$PA \vdash \varphi(\dot{n})^K \text{by a proof of length} \leq P(|\ulcorner\varphi(\dot{n})\urcorner|).$$

We also know by lemma 3.10 (with $U = V = PA$) that there is a polynomial R such that for every n,

$$PA \vdash \varphi(\dot{n}) \leftrightarrow \varphi(\dot{n})^K \text{by a proof of length} \leq R(|\ulcorner \varphi(\dot{n}) \urcorner|).$$

Now can construct from R and P a polynomial Q such that for all n,

$$PA \vdash \varphi(\dot{n}) \text{by a proof of length} \leq Q(|\ulcorner \varphi(\dot{n}) \urcorner|).$$

However, there will be n such that $n > Q(O(|n|)) \geq Q(|\ulcorner \varphi(\dot{n}) \urcorner|)$, and we have a contradiction with 2. QED

A salient feature of the counterexample above is the trivial identity interpretation by which PA interprets $PA+\mathcal{S}$. To prove that interpretability does not imply feasible interpretability, it is not essential that the set of formulas added to PA be infinite like $\mathcal{S}$ above. In theorem 4.2 we show a counterexample where one sentence can be normally but not feasibly interpreted over PA. Of course in this case the normal interpretation cannot be the identity. The counterexample also shows that in general we cannot feasibly merge two compatible feasible interpretations; i.e. it is not true that if $U \rhd_f V$, $U \rhd_f B$ and $U \rhd V + B$, then $U \rhd_f V + B$ (take $U = V = PA$, $B = A(\dot{n})$ or $B = E^*$ as below in the proof of theorem 4.2).

Theorem 4.2 *There is a sentence $A(\dot{n})$ such that $\omega \models PA \rhd PA + A(\dot{n})$, but $\omega \not\models PA \rhd_f PA + A(\dot{n})$.*

In order to prove theorem 4.2, we need a well-known result and its proof, given in theorem 4.4 below. Solovay proved that the set $\{A | PA \rhd PA+A\}$ is Π^0_2-complete [So]. This result inspired Hájek to prove that, for every n, the set $\{A \mid A \text{ is } \Pi^0_{n+1}\text{-conservative over } PA\}$ is also Π^0_2-complete [Há 79].

We have adapted the proof of theorem 4.4 from Visser's unpublished rendition [Vi 90] of an alternative proof by Lindström of Hájek's general result. First we need a definition.

Definition 4.3 Define $\Box_{U,x}B$ for "there is a proof of the formula B which only uses those axioms of U with Gödel number $\leq x$." Then define $\Box^*_U B := \exists x \Box_{U,x} B$.

Theorem 4.4 *Suppose U and V are theories extending PA, such that for all B, $PA \vdash \forall x \Box_U(\Box_{U,x}B \to B)$ (reflection for U). Then for every Π_2-predicate P, there is a formula A such that*

$$PA \vdash \Diamond_U \top \to (U \rhd U + A \leftrightarrow P)$$

Proof. Reflection for U implies in particular that $PA \vdash \forall x \Box_U \Diamond_{U,x} \top$. Thus by the Orey-Hájek theorem, $PA \vdash U \rhd V \leftrightarrow \forall x \Box_U \Diamond_{V,x} \top$. The rest of the proof is taken almost verbatim from [Vi 90].

Let P be any Π_2-predicate, say $P = \forall x S(x)$, with $S \in \Sigma^0_1$. Pick R by diagonalization such that $PA \vdash R \leftrightarrow S \preceq \Box_U R$. Let $Q := \Box_U R \preceq S$. (We suppress free variables when convenient). Now we can prove the following:

$$PA \vdash \Box_U R \leftrightarrow S \vee \Box_U \bot. \tag{4}$$

For, work inside PA and suppose $\Box_U R$. Then either R or Q holds. In case that R holds we have S by definition. In case that Q holds we have $\Box_U Q$ by Σ^0_1-completeness, and hence by definition both $\Box_U R$ and $\Box_U \neg R$, thus $\Box_U \bot$.

For the other direction, suppose S. Again we have either R or Q. From R we find $\Box_U R$ by Σ^0_1-completeness. From Q we immediately derive $\Box_U R$. Finally $\Box_U \bot$ gives $\Box_U R$ as well. This finishes the proof of (4).

Define A by diagonalization such that $PA \vdash A \leftrightarrow \Box^*_U \neg A \preceq \exists y \neg R(y)$. Note that by (4) we have $PA \vdash \Diamond_U \top \to [\forall x \Box_U R(x) \to P]$ and $PA \vdash P \to \forall x \Box_U R(x)$. For this A, we can prove

$$PA \vdash \Diamond_U \top \to (U \rhd U + A \leftrightarrow P) \tag{5}$$

For, work in PA and suppose $\Diamond_U \top$.

→-side Suppose $U \rhd U{+}A$. Then by the Orey-Hájek theorem $\forall x \Box_U \Diamond_{U,x} A$. We will prove $\forall x \Box_U R(x)$. Pick any x. We have $\Box_U [Q(x) \to \neg R(x)]$; therefore by definition of A,

$$\Box_U [Q(x) \to \neg A \vee \Box_{U,x} \neg A]$$

and hence by reflection

$$\Box_U [Q(x) \to \neg A].$$

But then there is a y such that

$$\Box_{U,y} [Q(x) \to \neg A],$$

so by Σ^0_1- completeness

$$\Box_U \Box_{U,y} [Q(x) \to \neg A].$$

Also by Σ^0_1-completeness, there is a z such that

$$\Box_U [Q(x) \to \Box_{U,z} Q(x)].$$

Combining the previous two facts, we find a u such that

$$\Box_U [Q(x) \to \Box_{U,u} \neg A,]$$

and thus, by the assumption, $\Box_U \neg Q(x)$. It follows that

$$\Box_U [\Box_U R(x) \to R(x)],$$

hence by Löb's theorem $\Box_U R(x)$. We may conclude $\forall x \Box_U R(x)$, thus, because we have $\Diamond_U \top$, we conclude P.

←-side Suppose P. Then $\forall x \Box_U R(x)$ and thus $\forall x \Box_U (\forall y < x\, R(y))$. It follows by definition of A that

$$\forall x \Box_U (\Box_{U,x} \neg A \to A).$$

On the other hand, we have

$$\forall x \Box_U (\Box_{U,x} \neg A \to \neg A)$$

by reflection, hence $\forall x \Box_U (\Diamond_{U,x} A)$. But then by the Orey-Hájek theorem $U \rhd U + A$.

This finishes the proof of (5), and thus of the theorem. QED

Proof of theorem 4.2. Let $P(x)$ be some Π^0_2-complete formula, say $P(x) = \forall y S(x,y)$, with $S \in \Sigma^0_1$. Define the formulas R and A by diagonalization such that

$$PA \vdash R(x,y) \leftrightarrow S(x,y) \preceq \Box_{PA} R(x,y)$$

and

$$PA \vdash A(x) \leftrightarrow \Box^*_{PA} \neg A(x) \preceq \exists y \neg R(x,y),$$

where $\Box^*$ is as defined in definition 4.3.

Carrying out the proof of theorem 4.4 in True Arithmetic, and taking the theory U mentioned there to be PA, we find the following result: if PA is consistent (as we believe it to be), then

$$\omega \models \forall x (PA \rhd PA + A(x) \leftrightarrow P(x)).$$

Now suppose, to derive a contradiction, that

$$\omega \models \forall x [(PA \rhd PA + A(x)) \leftrightarrow (PA \rhd_f PA + A(x))].$$

Then

$$\omega \models \forall x [(PA \rhd_f PA + A(x)) \leftrightarrow P(x)].$$

However, it is easy to see that $PA \rhd_f PA + A(x)$ is a Σ^0_2-predicate, contradicting the Π^0_2-completeness of P. Therefore, there is an $n \in \omega$ such that

- $\omega \models PA \rhd PA + A(\dot{n})$ but

- $\omega \not\models PA \rhd_f PA + A(\dot{n})$.

By this method we do not immediately find the value of a particular n that works, however.

A. Visser pointed out that we can make a specific counterexample in a more direct way using the Lindström method. Because $\{e \mid e$ is the Gödel number of a sentence E such that $\neg(PA \rhd_f PA + E)\} \in \Pi^0_2$, we can construct a formula A as in theorem 4.4 for which the following holds: for all sentences E,

$$\omega \models \neg(PA \rhd_f PA + E) \leftrightarrow PA \rhd PA + A(\ulcorner E \urcorner).$$

Now let E^* be the sentence constructed by the fixed point theorem such that

$$PA \vdash E^* \leftrightarrow A(\ulcorner E^* \urcorner).$$

Then

$$\omega \models \neg(PA \rhd_f PA + E^*) \leftrightarrow PA \rhd PA + E^*.$$

Therefore,

$$\omega \models PA \rhd PA + E^* \text{ and } \omega \not\models PA \rhd_f PA + E^*.$$

5 ILM is the interpretability logic of feasible interpretability over PA

In this section, we will show that Berarducci's proof of the arithmetic completeness of ILM with respect to interpretability over PA can be adapted to prove that ILM is also arithmetically complete with respect to *feasible* interpretability over PA.

We have already proved in chapter 3 that for all modal formulas in the language of ILM we have:

$$\text{if } ILM \vdash \varphi, \text{then for all feasibility interpretations } ^*, PA \vdash \varphi^*.$$

Therefore, we will only need to show the converse:

$$\text{if } ILM \not\vdash \varphi, \text{then there is a feasibility interpretation } ^* \text{ with } PA \not\vdash \varphi^*.$$

We suppose that the reader has a copy of [Be 90] at hand in order to follow the original proofs. For the lemmas 5.5 up to 5.7, knowledge of [Pu 86] or [Pu 87] or [Ve 89] will be helpful to the reader. As in [Pu 87], we take the logical complexity of a formula to be its quantifier depth. We can then adapt the results obtained in [Pu 87] to find for every standard n a formula Sat_n, a satisfaction predicate for formulas of logical complexity $\leq n$, such that Sat_n is of length linear in n. Subsequently, we can find proofs of length quadratic in n of the Tarski conditions and of the truth lemma for these satisfaction predicates Sat_n. Moreover, all these results can be formalized in PA. In the formalized case, we read Sat_n and $True_n$ as Gödel numbers found as function value in n. We will not go into the details here but refer the reader to the papers by Pudlák and Verbrugge.

First, we define some of the concepts that we use in the subsequent lemmas.

Definition 5.1 Formally, we define the following concepts:

- *Sent*(a) for "a is the Gödel number of a sentence";
- *Fmla*(a) for "a is the Gödel number of a formula";
- $Fmla_n(a)$ for "a is the Gödel number of a formula of logical complexity $\leq n$";
- *Cl*(a) for "the Gödel number of the universal closure of the formula with Gödel number a"; note that *Cl* denotes a function;
- $Indax_n(b)$ for "b is the Gödel number of an induction axiom of logical complexity $\leq n$", i.e.

$$\begin{aligned} Indax_n(b) \Longleftrightarrow \quad & Fmla_n(b) \wedge \exists y[Fmla(y) \wedge \\ & b = Sub(y, \ulcorner v_1 \urcorner, \ulcorner 0 \urcorner) \\ & \ulcorner \wedge \forall v_1(\urcorner y \ulcorner \rightarrow \urcorner Sub(y, \ulcorner v_1 \urcorner, \ulcorner Sv_1 \urcorner) \ulcorner) \rightarrow \forall v_1 \urcorner y. \end{aligned}$$

We need to discriminate between a few different kinds of restricted provability, as defined below. In this section, provability means provability in PA, unless we explicitly state otherwise.

Definition 5.2 We formally define the following:

- $BPrf_n(x, y)$ for "x codes a proof of the formula coded by y, where only formulas of logical complexity $\leq n$ appear in the proof";
- $P\text{-}Polprov_n(x)$ for "x codes a formula that is provable by a proof of length $\leq P(n)$" where P is a polynomial;
- $Polprov_n(x)$ for "there is a polynomial P such that" $\forall n \exists p(|p| \leq P(n) \wedge Prf(p, x))$;
- $Polprov_{W,n}(x)$ for "there is a polynomial P such that" $\forall n \exists p(|p| \leq P(n) \wedge Prf_W(p, x))$;
- $Prov_n(x)$ for "x codes a formula that is provable by a proof which only uses those axioms of PA with Gödel number $\leq n$"; abbreviation $\Box_n \varphi$ for $Prov_n(\ulcorner \varphi \urcorner)$;
- $Prov_{W,v}(x)$ for "x codes a formula that is provable by a proof which only uses those axioms of W with Gödel number $\leq n$"; abbreviation $\Box_{W,n}\varphi$ for $Prov_{W,n}(\ulcorner \varphi \urcorner)$.

In the context of satisfaction predicates $Sat_n(x, w)$, we need a few more concepts.

Definition 5.3 We formally define the following:

- $Evalseq(w, x)$ for "w encodes an evaluation sequence for the formula or term with Gödel number x; i.e. the length of the sequence w exceeds any i for which a variable v_i occurs in the formula or term coded by x";

- $s^*(i, x, w)$ for "the sequence which is identical to w, except that x appears in the i-th place"; note that s^* denotes a function;
- $True_n(x)$ for $\forall w(Evalseq(w, x) \rightarrow Sat_n(x, w))$;

Remark 5.4 When we prove formalized results, we read $True_n$ as a Gödel number just as Sat_n. So in that case the appropriate definition is as follows:

$$True_n(x) \text{ for } \ulcorner\forall w(Evalseq(w, x) \rightarrow\urcorner Sat_n(x, w)\ulcorner)\urcorner.$$

Lemma 5.5 (feasible subformula property) *There is a polynomial P such that*

$$\begin{aligned} PA \vdash \forall k \forall a \quad & (Fmla(a) \rightarrow \\ & P\text{-}Polprov_{|k|+|a|}[Prov_k(a) \rightarrow \exists q(BPrf_{|k|+|a|}(q, a))]) \end{aligned}$$

Proof. In [Ta 75], Takeuti gives a proof of the free cut-elimination theorem for PA, where PA is formulated as a Gentzen system. Free cut-elimination works in such a way that all principal formulas of induction inferences in the new free cut-free proof are substitution instances of principal formulas of induction inferences in the old proof. From this result Takeuti derives a proof of the corresponding subformula property for PA.

The proof of the subformula property can be adapted to the natural deduction formulation of PA, and can subsequently be formalized in PA. Thus, we can substitute any k bounding the Gödel numbers of axioms used, and any Gödel number a of a formula into the proof of the subformula property. Therefore, there is a polynomial P such that PA proves the following by proofs of length $\leq P(|k| + |a|)$:

$$PA \vdash Prov_k(a) \rightarrow \exists q(BPrf_{|k|+|a|}(q, a)).$$

Now this statement can again be formalized, so that we find

$$\begin{aligned} PA \vdash \forall k \forall a \quad & (Fmla(a) \rightarrow \\ & P\text{-}Polprov_{|k|+|a|}[Prov_k(a) \rightarrow \exists q(BPrf_{|k|+|a|}(q, a))]), \end{aligned}$$

as desired. QED

Lemma 5.6 *There is a polynomial P such that*

$$PA \vdash \forall k \forall a \quad (Fmla(a) \rightarrow \\ P\text{-}Polprov_{|k|+|a|}[\exists q(BPrf_{|k|+|a|}(q,a) \rightarrow True_{|k|+|a|}(a)])$$

Proof. First, we work informally by induction on the construction of q. We work in PA, and we take any k and an a such that a is the Gödel number of a formula. We have to prove by polynomial length proofs (where the polynomial is fixed from outside) that $True_{|k|+|a|}$ preserves the axioms and rules as applied to formulas of logical complexity $\leq |k| + |a|$.

As an example, we show how this works for the induction schema. We take v_1 as the induction variable in all our instances of the induction axioms. So suppose b codes an induction axiom of logical complexity $\leq |k| + |a|$, e.g. $b = (Sub(y, \ulcorner v_1 \urcorner, \ulcorner 0 \urcorner)^\ulcorner \wedge \forall v_1(\urcorner y^\ulcorner \rightarrow \urcorner Sub(y, \ulcorner v_1 \urcorner, \ulcorner Sv_1 \urcorner)^\ulcorner) \rightarrow \forall v_1 \urcorner y)$. We have to prove the following by a proof of length polynomial in $n := |k| + |a|$:

$$True_n(Sub(y, \ulcorner v_1 \urcorner, \ulcorner 0 \urcorner) \\ \ulcorner \wedge \forall v_1(\urcorner y^\ulcorner \rightarrow \urcorner Sub(y, \ulcorner v_1 \urcorner, \ulcorner Sv_1 \urcorner)^\ulcorner) \rightarrow \forall v_1 \urcorner y). \tag{6}$$

By a proof of length quadratic in n of the Tarski properties for Sat_n and a proof of length quadratic in n of a call by name / call by value lemma for Sat_n (cf. the proofs of lemmas 3.12 and 3.16 in [Ve 89],(6) is equivalent to the following:

$$\forall w[Sat_n(y, s^*(1,0,w)) \\ \wedge \forall x(Sat_n(y, s^*(1,x,w)) \rightarrow Sat_n(y, s^*(1,Sx,w))) \\ \rightarrow \forall x(Sat_n(y, s^*(1,x,w))]. \tag{7}$$

The formulas (7) are themselves instances of induction of length linear in n, so they are provable by proofs of length linear in n. A polynomial of the form $P(n) = K \cdot n^3$ should now suffice to carry out the proofs of (6).

Again, we can formalize the argument to derive the following:

$$PA \vdash \forall k \forall a \quad (\mathit{Fmla}(a) \rightarrow \\ P\text{-}\mathit{Polprov}_{|k|+|a|}[\forall b(\mathit{Indax}_{|k|+|a|}(b) \rightarrow \mathit{True}_{|k|+|a|}(b))]).$$

Similarly, we can show by polynomially short proofs that the other axioms of logical complexity $\leq |k|+|a|$ are true, and that the rules preserve truth. We leave these proofs and their formalizations to the reader. QED

Lemma 5.7 *There is a polynomial P such that*

$$PA \vdash \forall k \forall a(\mathit{Fmla}(a) \rightarrow P\text{-}\mathit{Polprov}_{|k|+|a|}(\mathit{True}_{|k|+|a|}(a)^{\ulcorner} \rightarrow {}^{\urcorner}\mathit{Cl}(a))$$

Proof. By a formalized Tarski's snowing lemma; see for example lemma 3.10 of [Ve 89]. QED

The following theorem corresponds to the reflection theorem; see theorem 1.6 in [Be 90].

Theorem 5.8 (feasible reflection theorem) *There is a polynomial P such that*

$$PA \vdash \forall k \forall a(\mathit{Sent}(a) \rightarrow P\text{-}\mathit{Polprov}_{|k|+|a|}(^{\ulcorner}\mathit{Prov}_k(a) \rightarrow {}^{\urcorner}a))$$

Proof. Combine lemmas 5.5, 5.6 and 5.7. QED

In the following lemmas and theorems, $\exists K$ abbreviates $\exists K$("K codes an interpretation" $\wedge \ldots$).

The next lemma was proved by Albert Visser [Vi 89, Chapter 6, Claim 3] in the course of a formalized Henkin construction in $I\Delta_0 + \Omega_1$.

Lemma 5.9 *Suppose β axiomatizes some subset of a Σ_1^b-language L. Then there is an r such that*

$$\begin{aligned} I\Delta_0 + \Omega_1 \vdash \ & \Box_U Con(\beta) \rightarrow \\ & \exists K \forall a \in Sent(L) \exists p < \omega_1^r(a) Prf_U(p, \ulcorner \Box_\beta a \rightarrow \urcorner a^K). \end{aligned}$$

Proof. See [Vi 89]. QED

Because of the correspondence between the values of ω_1-terms in a and *exp*(the values of polynomials in $|a|$), lemma 5.9 implies the following lemma:

Lemma 5.10 *Suppose β axiomatizes some subset of a Σ_1^b-language L. Then there is a polynomial P such that*

$$\begin{aligned} I\Delta_0 + \Omega_1 \vdash \ & \Box_U Con(\beta) \rightarrow \\ & \exists K \forall a \in Sent(L) P\text{-}Polprov_{U,|a|}(\ulcorner \Box_\beta a \rightarrow \urcorner a^K). \end{aligned}$$

The following theorem corresponds to Orey's theorem; see for example [Be 90, Theorem 2.9]

Theorem 5.11 (feasible Orey's theorem) *Suppose that $U \supseteq PA$ and W is axiomatized by α, where α is a Σ_1^b-formula. Then*

$$PA \vdash \forall x Polprov_{U,|x|}(\ulcorner \Diamond_{\alpha,x} \top \urcorner) \rightarrow U \rhd_f W.$$

Proof. Work in PA and suppose

$$\forall x Polprov_{U,|x|}(\ulcorner \Diamond_{\alpha,x} \top \urcorner).$$

In U, we will do a Henkin construction for the Feferman proof predicate for W. First define:

$$\beta(x) := \alpha(x) \wedge \Diamond_{\alpha,x+1}\top.$$

As in Feferman's original proof, we can prove that

$$\Box_U Con(\beta).$$

(For, reason in U and suppose $Prf_\beta(x, \perp)$, then for the axiom of β coded by the biggest Gödel number y to appear in x we have $\alpha(y) \wedge \neg\Diamond_{\alpha,y+1}\top$, thus $\neg\beta(y)$: a contradiction.)

On the other hand, by provable Σ_1^b-completeness for $\alpha(a)$ and by the assumption $\forall x Polprov_{U,|x|}(\ulcorner\Diamond_{\alpha,x}\top\urcorner)$, we have:

$$\forall a(\alpha(a) \rightarrow Polprov_{U,|a|}(\ulcorner\alpha(a) \wedge \Diamond_{\alpha,x+1}\top\urcorner)).$$

So, by definition of β, we have the following:

$$\forall a(\alpha(a) \rightarrow Polprov_{U,|a|}(\ulcorner\beta(a)\urcorner)). \tag{8}$$

But, using $\Box_U Con(\beta)$ we can apply lemma 5.10 to first derive

$$\exists K \forall a \in Sent(L) Polprov_{U,|a|}(\ulcorner\Box_\beta a \rightarrow\urcorner a^K),$$

and thus

$$\exists K \forall a \in Sent(L) Polprov_{U,|a|}(\ulcorner\beta(a) \rightarrow\urcorner a^K). \tag{9}$$

Finally we can combine 8 and 9 to get the desired conclusion

$$\exists K \forall a(\alpha(a) \to Polprov_{U,|a|}(a^K)),$$

i.e. $U \rhd_f W$. QED

Now we can start the proof of the arithmetical completeness of ILM with respect to feasible interpretations (cf. definition 3.2) over PA.

Theorem 5.12 *If $ILM \nvdash B$, then there is a feasibility interpretation * such that $PA \nvdash B^*$.*

The proof will in most places be identical to the one in [Be 90]. First we will sketch the outline of the proof, then we will prove the propositions that we need in the feasible case but differ essentially from those used in [Be 90].

Proof sketch. Suppose $ILM \nvdash B$, and take, by modal completeness of ILM with respect to simplified models, a provably primitive recursive ILM-Kripke model $V =< V, R, S, b, \Vdash>$, with $b = 1$ and $1 \nVdash B$. Extend V with a new root 0 with $0Rx$ for all $x \in V$, as in definition 5.1 of [Be 90]. Adapting definition 5.2 of [Be 90], we define a *feasibility interpretation* * such that for all propositional variables p,

$$p^* := \text{“}\exists x \in V \cup \{0\} : L = x \wedge x \Vdash p\text{”},$$

where L is defined as the limit of the Solovay function F, which is in turn defined in definition 5.7 of [Be 90]. We want to prove the following:

$$\text{whenever } 1 \nVdash A, \text{then } PA \nvdash A^*, \tag{10}$$

Then we will be done, as we have chosen V such that $1 \nVdash B$. To prove (10), we need to prove in PA a few properties of F and its limit L. Subsequently we need to prove by induction on the construction of the formula that for all formulas A, the feasibility interpretation * is *faithful* on A, i.e.

$$PA \vdash \forall x \in V(x \Vdash A \wedge L = x \to A^*) \text{ and}$$

$$PA \vdash \forall x \in V(x \Vdash \neg A \wedge L = x \to \neg A^*).$$

It is clear from the definition of F that * is faithful on atomic formulas. Moreover, the induction steps for the propositional connectives and $\Box$ immediately follow from the proofs in [Be 90]. Even the "negative" induction step for $\rhd$ has a straightforward proof:

> Work in PA and suppose $x \in V$, $x \Vdash \neg(A \rhd B)$, and $L = x$; then by part 2 in the proof of lemma 5.6 of [Be 90] and by the induction hypothesis, $\neg(A^* \rhd B^*)$. But then surely $\neg(A^* \rhd_f B^*)$, thus, as * is a feasibility interpretation, $\neg(A \rhd B)^*$.

For the "positive" direction, we need two extra lemmas. First we will prove in PA that F satisfies a feasible adaptation of Berarducci's property S , which we then use to finish the induction step for $\rhd$.

For $x \in V$, let $\mathit{rank}(x, n)$, the rank of x at stage n, be defined as in definition 5.7 of [Be 90]. The following proposition is an analogue of proposition 5.14 in [Be 90].

Proposition 5.13 (F has feasible property S) *PA proves the following:*

$$\begin{aligned} PA \vdash \quad & \forall x \in V \cup \{0\}[L = x \to \\ & \mathit{Polprov}_{|k|}(\ulcorner \forall y, z \in V \cup \{0\} \\ & (L = y \wedge xRz \wedge ySz \to \Diamond_{\underline{k}} L = z)\urcorner)] \end{aligned}$$

Proof. We will prove the proposition by combining a few facts that are easy to check. For brevity's sake, we will leave out "$\in V \cup \{0\}$" after quantifiers $\forall x, \forall y, \forall z$. Likewise, capital P, with or without subscript, refers to formalized polynomials.

Fact 1 $PA \vdash Polprov_{|k|}(\ulcorner \forall y(L = y \rightarrow \Diamond_{\dot{k}} L = y) \urcorner)$

Proof. Immediately from the reflection theorem 5.8. The formula $L = y$ has a fixed length, so the polynomial found in the proof of the reflection theorem in this case depends only on $|k|$. QED

Fact 2 $PA \vdash Polprov_{|k|}(\ulcorner \forall y(L = y \rightarrow \forall n(\dot{k} < rank(y, n))) \urcorner)$

Proof. Immediately from fact 1 and the definition of rank. The appearance of k as an efficient numeral keeps the length of the proof polynomial in $|k|$. (This is also the case in the other facts below) QED

Fact 3 $PA \vdash Polprov_{|k|}(\ulcorner \forall z(\Box_{\dot{k}} L \neq z \rightarrow \exists m \forall n \geq m(rank(z, n) \leq \dot{k})) \urcorner)$

Proof. Immediately from the definition of rank. QED

Fact 4 $PA \vdash Polprov_{|k|}(\ulcorner \forall y \forall z(L = y \wedge \Box_{\dot{k}} L \neq z \rightarrow \exists n(F(n) = y$
$\wedge\, n \text{ codes } y \wedge rank(z, n) \leq \dot{k} \wedge rank(z, n) < rank(y, n))) \urcorner))$

Proof. From the definition of limit and fact 3: just take n big enough. We can take care that n codes y because we have an infinitely repetitive primitive recursive coding of the elements of $V \cup \{0\}$. Finally, to prove $rank(z, n) < rank(y, n)$, we use fact 2. QED

Fact 5 $PA \vdash \forall x(L = x \rightarrow Polprov_{|k|}(\ulcorner \exists j(j \geq \dot{k} \wedge F(j) = x) \urcorner))$

Proof. Immediate from the definition of the limit L of F. QED

Fact 6 We have the following:

$$PA \vdash \forall x(L = x \rightarrow Polprov_{|k|}(\ulcorner \forall y \forall z(L = y \wedge \Box_{\dot{k}} L = z \wedge xRz \wedge ySz$$
$$\rightarrow \exists n \exists j(F(n) = y \wedge n \text{ codes } y \wedge rank(z, n) < rank(y, n)$$
$$\wedge rank(z, n) \leq k \leq j \wedge F(j) = x$$
$$\wedge F(rank(z, n))SxRz \wedge F(rank(z, n))Rz)) \urcorner))$$

Proof. For the part up to $F(j) = x$, we combine facts 5 and 4. For the last two conjuncts, we use the monotonicity of F and the property corresponding to M of Veltman $ILM-$frames. QED

Fact 7 We have the following:

$$PA \vdash \forall x(L = x \rightarrow Polprov_{|k|}(\ulcorner \forall y \forall z(L = y \wedge \Box_{\dot{k}} L = z \wedge xRz \wedge ySz \rightarrow \exists n(F(n) = y \wedge F(n+1) = z))\urcorner)$$

Proof. Immediate from fact 6 and the definition of the function F, clause 2. QED

Now we can wrap up the proof: we see that $\exists n(F(n) = y \wedge F(n+1) = z)$ is inconsistent with $L = y$, so in fact we have what we were looking for:

$$PA \vdash \forall x[L = x \rightarrow Polprov_{|k|}(\ulcorner \forall y \forall z(L = y \wedge xRz \wedge ySz \rightarrow \Diamond_{\dot{k}} L = z)\urcorner)]$$

QED

The following proposition corresponds to part 1 of Lemma 5.6 of [Be 90].

Proposition 5.14 (positive induction step for $\rhd$) *Let * be the feasibility interpretation defined in the proof sketch of theorem 5.12. Suppose as induction hypothesis that*

$$PA \vdash \forall y(L = y \rightarrow (y \Vdash A \leftrightarrow A^*)) \text{ and}$$

$$PA \vdash \forall z(L = z \rightarrow (z \Vdash B \leftrightarrow B^*)).$$

Then

$$PA \vdash \forall x(L = x \wedge x \Vdash A \rhd B \rightarrow (A \rhd B)^*).$$

Proof. Let b be such that

$$PA \vdash \forall y(L = y \rightarrow (y \Vdash A \leftrightarrow A^*)) \text{ and}$$

$$PA \vdash \forall z(L = z \rightarrow (z \Vdash B \leftrightarrow B^*)),$$

both by proofs that use axioms of Gödel number up to b. Moreover suppose c is such that

$$PA \vdash \forall z(z \Vdash B \rightarrow \Box_c(z \Vdash B));$$

for this, any $c \geq$ the Gödel number of the biggest axiom of Robinson's arithmetic Q will do. Define $d := max(b, c)$. By theorem 5.11, the feasible version of Orey's theorem, it is sufficient to prove the following:

$$PA \vdash \forall x(L = x \wedge x \Vdash A \rhd B \rightarrow \forall k \geq d \; Polprov_{|k|}(\ulcorner A^* \rightarrow \Diamond_{\underline{k}} B^* \urcorner).$$

Again, we will state a list of easily provable facts from which the result immediately follows.

Fact 1 $PA \vdash \forall x(L = x \wedge x \Vdash A \rhd B \rightarrow \Box[A^* \rightarrow \exists y(L = y \wedge xRy \wedge y \Vdash A \wedge x \Vdash A \rhd B)])$

Proof. $L = x \rightarrow \Box\exists y(L = y \wedge xRy)$ by property $(\neg R)$, $\Box(A^* \wedge L = y \rightarrow y \Vdash A)$ by the induction hypothesis, and $\Box(x \Vdash A \rhd B)$ by provable Σ^0_1-completeness. QED

Fact 2 $PA \vdash \forall x(L = x \wedge x \Vdash A \rhd B \rightarrow \Box[A^* \rightarrow \exists y \exists z(L = y \wedge xRy \wedge y \Vdash A \wedge x \Vdash A \rhd B \wedge xRz \wedge ySz \wedge z \Vdash B)])$

Proof. From fact 1 and the definition of $x \Vdash A \rhd B$. QED

Fact 3 $PA \vdash \forall z \forall k \geq d \; Polprov_{|k|}(z \Vdash B \rightarrow \Box_{\underline{k}} z \Vdash B)$

Proof. From the assumption, and the fact that k appears only as efficient numeral. QED

Fact 4 $PA \vdash \forall x(L = x \wedge x \Vdash A \rhd B \rightarrow \forall k \geq d \; Polprov_{|k|}(\ulcorner A^* \rightarrow \exists y \exists z(L = y \wedge xRy \wedge xRz \wedge ySz \wedge \Diamond_{\underline{k}} L = z \wedge \Box_{\underline{k}} z \Vdash B)\urcorner))$

Proof. From fact 2 for $A^* \rightarrow \exists y \exists z(L = y \wedge xRy \wedge xRz \wedge ySz \wedge z \Vdash B)$; fact 3 for a proof of length polynomial in k of $z \Vdash B \rightarrow \Box_{\underline{k}} z \Vdash B$, and proposition 5.13 for a proof of length polynomial in $|k|$ of $L = y \wedge xRy \wedge xRz \wedge ySz \rightarrow \Diamond_{k} L = z$. QED

Fact 5 $PA \vdash \forall x(L = x \wedge x \Vdash A \rhd B \rightarrow \forall k \geq d\ Polprov_{|k|}(\ulcorner A^* \rightarrow \exists z \Diamond_{k}(L = z \wedge z \Vdash B)\urcorner))$

Proof. If k is big enough (and $k \geq d$ will do), then by an easily formalized property of modus ponens, we have the following by proofs of length polynomial in $|k|$: $PA \vdash \forall z([\Box_k(z \Vdash B \rightarrow L \neq z) \wedge \Box_k z \Vdash B] \rightarrow \Box_k L \neq z)$, and thus $PA \vdash \forall z(\Diamond_k L = z \wedge \Box_k z \Vdash B \rightarrow \Diamond_k(L = z \wedge z \Vdash B))$. This argument can be formalized and combined with fact 4 to derive fact 5. QED

Fact 6 $PA \vdash \forall x(L = x \wedge x \Vdash A \rhd B \rightarrow \forall k \geq d\ Polprov_{|k|}(\ulcorner A^* \rightarrow \Diamond_{k} B^* \urcorner)$

Proof. From fact 5 and the induction hypothesis; the fact that $k \geq d$ is used at this place. QED

From fact 5 and the feasible version of Orey's theorem, we may indeed derive

$$PA \vdash \forall x(x \Vdash A \rhd B \wedge L = x \rightarrow (A \rhd B)^*),$$

as desired.

QED

Proof sketch of theorem 5.12, continued. Concluding by induction that * is faithful on all formulas A, we have proved that $ILM \nvdash B^*$. Therefore, ILM is arithmetically complete with respect to feasible interpretability over PA.

QED.

Acknowledgements

I would like to thank Peter Hájek, Marc Jumelet, Edith Spaan, and Albert Visser for stimulating discussions and helpful comments. I would also like to thank the two anonymous referees for their suggestions.

References

[Be 90] A. Berarducci, The interpretability logic of Peano Arithmetic, *The Journal of Symbolic Logic*, vol. 55 (1990), pp. 1059-1089.

[Bu 86] S. Buss, *Bounded Arithmetic*, Bibliopolis, edizioni di filosofia e scienze, Napoli, 1986.

[Co 66] P.J. Cohen, *Set Theory and the Continuum Hypothesis*, Benjamin, New York, 1966.

[Fe 60] S. Feferman, Arithmetization of metamathematics in a general setting, *Fundamenta Mathematicae*, vol. 49 (1960), pp. 33-92.

[Há 79] P. Hájek, On partially conservative extensions of arithmetic, *Logic Colloquium '78* (M. Boffa et al., editors), North Holland, Amsterdam, 1979, pp. 225-234.

[HP] P. Hájek and P. Pudlák, *Metamathematics of first-order logic*, to appear.

[JM 88] D. de Jongh and F. Montagna, Provable fixed points, *Zeitschrift für mathematische Logik und Grundlagen der Mathematik*, vol. 34 (1988), pp. 229-250.

[JV 90] D. de Jongh and F. Veltman, Provability logics for relative interpretability, *Mathematical Logic* (*Proceedings, Chaika, Bulgaria, 1988*; P.P. Petkov, editor), Plenum Press, New York, 1990, pp. 31-42.

[Ku 80] K. Kunen, *Set Theory: An Introduction to Independence Proofs*, North-Holland, Amsterdam, 1980.

[Pu 85] P. Pudlák, Cuts, consistency statements and interpretability, *The Journal of Symbolic Logic*, vol. 50 (1985), pp. 423-441.

[Pu 86] P. Pudlák, On the length of proofs of finitistic consistency statements in first order theories, *Logic Colloquium '84* (J.B. Paris et al., editors), North Holland, Amsterdam, 1986, pp 165-196.

[Pu 87] P. Pudlák, Improved bounds on the length of finitistic consistency statements, *Logic and Combinatorics* (S.G. Simpson, editor), Contemporary Mathematics 35, American Mathematical Society, Providence, 1987, pp. 309-332.